The behaviour of magnetic impurities in metals has posed problems to challenge the condensed matter theorist over the past thirty years. This book deals with the concepts and techniques which have been developed to meet this challenge, and with their application to the interpretation of experiments.

After an introduction to the basic theoretical models, Kondo's explanation of the resistance minimum is described, which was the first of the major puzzles to be solved. As Kondo's perturbational calculations break down at low temperatures a non-perturbational approach is needed to predict the low temperature behaviour of the models, the so-called Kondo problem. The author surveys in some detail the many-body techniques, scaling, renormalization group, Fermi liquid and Bethe ansatz, which lead to a solution of this problem for most of the theoretical models. The book also deals with special techniques for N-fold degenerate models for rare earth impurities (including mean field and $1/N$ expansions). The theoretical framework having been established, a comparison is made between the theoretical predictions and the experimental results on particular systems in the penultimate chapter.

With the success of the many-body techniques developed to deal with impurity problems the new challenge is the extension of these strong correlation techniques to models with periodicity in order to understand the behaviour of heavy fermion and high T_c superconducting compounds. The work which has provided insights into heavy fermion behaviour is reviewed in the last chapter, together with the questions that need to be answered in future work.

This book will be of interest to condensed matter physicists, particularly those interested in strong correlation problems. The detailed discussions of advanced many-body techniques should make it of interest and useful to theoretical physicists in general.

The Kondo Problem to Heavy Fermions

CAMBRIDGE STUDIES IN MAGNETISM

Edited by

DAVID EDWARDS
Department of Mathematics, Imperial College of Science and Technology

and

DAVID MELVILLE
Vice Chancellor, Middlesex University

Titles in this series

K. H. Fischer & J. A. Hertz: *Spin Glasses*
A. C. Hewson: *The Kondo Problem to Heavy Fermions*

The Kondo Problem
to Heavy Fermions

A. C. Hewson

Reader, Department of Mathematics,
Imperial College

PUBLISHED BY THE PRESS SYNDICATE OF THE UNIVERSITY OF CAMBRIDGE
The Pitt Building, Trumpington Street, Cambridge CB2 1RP, United Kingdom

CAMBRIDGE UNIVERSITY PRESS
The Edinburgh Building, Cambridge CB2 2RU, United Kingdom
40 West 20th Street, New York, NY 10011-4211, USA
10 Stamford Road, Oakleigh, Melbourne 3166, Australia

Printed in the United Kingdom at the University Press, Cambridge

A catalogue record for this book is available from the British Library

Library of Congress Cataloguing in Publication data

Hewson, A. C. (Alexander Cyril)
The Kondo problem to heavy fermions / A. C. Hewson
p.   cm. – (Cambridge studies in magnetism : 2)
Includes bibliographical references and index.
ISBN 0-521-36382-9
1. Kondo effect.   2. Anderson model.   3. Fermi liquid theory.
4. Fermions.   I. Title.   II. Series.
QO176.8.E35H49   1992
530.4'12–dc20   92-8101   CIP

ISBN 0 521 36382 9 hardback
ISBN 0 521 59947 4 paperback

CONTENTS

Preface

This book charts the progress in the theory of magnetic impurities since the late 50s, from the early developments leading to the Kondo impurity problem and its solution to the challenging problems posed by the recent work on heavy fermions and the high temperature superconductors. The first eight chapters cover, largely in chronological order, the techniques which have been developed to deal with single impurity problems. Some of these techniques, such as Green's functions, Feynman diagrams and perturbation theory, are covered in standard many-body texts (for example, Fetter and Walecka, 1971: Abrikosov, Gorkov and Dzyaloshinski, 1975: Mahan, 1990). Others may be less familiar so for these techniques I have included general introductory sections in the relevant chapters, and some appendices with further details, in order to make the text as self-contained as possible. The aim has been to make the book readable at two levels. At the higher level I have tried to present the general development of ideas, the emphasis being on the results of the theory and the general physical picture that emerges. The equations at this level are included to make it clear how these results are obtained. I have tried to make it readable also at a second more detailed level by including enough information in the text and appendices so that one can work from one equation to the next. To do so might require quite an effort on the reader's part, and some further hints may be necessary from some of the references cited. If the reader is prepared to make the effort I think he/she will gain more of a working knowledge of the subject. I have in mind second and third year postgraduate students who might wish to extend their range of many-body techniques. What is to be learnt from the Kondo problem is that no one technique has all the answers. Different formulations and different techniques have clarified different aspects of the problem, a composite picture has emerged which cannot be encompassed by one approach alone.

Chapter 1 deals with the basic models on which most of the work in this field has been based. It also covers some of the early theoretical concepts, such as the idea of a virtual bound state, and includes a derivation of the fundamental theorem of Friedel, known as the Friedel sum rule. Chapter 2 deals with the calculation of the resistivity due to impurities and gives in outline the perturbational calculation of Kondo that led to an understanding of the resistance minimum. Chapter 3 covers the period immediately after Kondo's discovery and the search for an acceptable theory of the low temperature behaviour of the s-d and Anderson models. The ideas of scaling, which provided the framework for the eventual solution of the problem, are described in this chapter and the 'poor man's' method of Anderson is considered in detail. This leads to the renormalization group approach of Wilson, which is introduced in chapter 4. This approach, which gave definitive results for the $S = \frac{1}{2}$ s-d model, is considered in detail based mainly on the seminal 1975 *Reviews of Modern Physics* paper. Insights into the nature of this solution are obtained via Fermi liquid theory, with a remarkably simple derivation of some of the basic results. This is described in chapter 5 where we give a derivation of the phenomenological theory of Nozières, and then outline the steps in the microscopic derivation by Yamada and Yosida. We later found a synthesis of these two approaches which we term 'renormalized perturbation theory', and this is described in appendix L.

The Bethe ansatz approach, which has been used to generate exact solutions for so many of the important magnetic impurity models, is introduced in chapter 6 with a survey of the results for the $S = \frac{1}{2}$ s-d model. The results for the non-degenerate Anderson model in the physically relevant parameter regimes are also described. Techniques for the N-fold degenerate model appropriate for systems with the rare earth impurities Ce and Yb are described in chapters 7 and 8. There is such a variety of techniques that can be applied to this model from the Bethe ansatz, Fermi liquid and various realizations of the large N or $1/N$ expansion methods that I decided to divide the material into two chapters. Chapter 7 gives an introduction to the $1/N$ expansion and the slave boson approaches. Chapter 8 deals with further developments, the self-consistent summation of diagrams (the non-crossing approximation, NCA), the effects of Gaussian fluctuations about the mean field theory, and the extensive calculations of Gunnarsson and Schönhammer on photoemission spectra. This completes the section devoted mainly to the models and theoretical techniques. For those interested mainly in the results there is a brief summary of the main conclusions from chapters 3 to 8 in appendix K.

There is a different emphasis in the remaining chapters which are much less self-contained. Chapter 9 is concerned with the relation between theory and experiment. The theory for magnetic impurities has been successful in accounting for the broad features of the experimental results, the resistance minimum, the power law behaviour at low temperatures, the specific heat peak, and local moment behaviour at higher temperatures. There have been some detailed comparisons between theory and experiment for some specific systems where the agreement is very satisfactory. However, there is scope for more detailed work, particularly for the 3d transition elements. The aim of chapter 9 has been to see what is required for a closer dialogue between theory and experiment. To consider the likely modifications to the results of the simpler models on introducing crystal fields, orbital degeneracy, multiplet splittings, and to identify gaps where the present theory is incomplete. The experiments have been classified on various energy scales, but to keep the chapter within reasonable bounds (it is still rather long) only a selection of experiments has been considered on each energy scale. There are regrettably some obvious omissions such as electron spin resonance (ESR), and the effects of magnetic impurities on superconductors, and some types of experiments have been considered only very briefly.

The final chapter introduces a new generation of problems associated with systems where transition metal or rare earth ions interact with conduction electrons at many sites either in an alloy or a periodic lattice. Lattice models have been formulated to describe these systems in order to explain the behaviour of anomalous rare earth compounds and alloys. These anomalies are the most marked in compounds and alloys known as heavy fermion systems. Similar models have also been proposed to explain the behaviour of the high T_c superconductors. Having spent most of the time on problems which have now been largely solved I thought the balance should be restored by considering the problems which are largely unsolved: to end with more questions than answers. Techniques developed for the single impurity problem have provided some partial answers but much remains to be done. My hope is that the reader might be inspired by the imagination, ingenuity and physical insight of those who took on the challenge of the earlier problems, leading to our present understanding outlined in chapters 1–8, to take on the challenge of these new problems.

Acknowledgments

I am indebted to present and former colleagues, Dennis Newns, David Edwards, Jawaid Rasul, Nick Read, Ule Desgranges, Lysandros Lysandrou, Theo Costi and Sarah Evans, and many others who over the years have been a source of stimulation and encouragement, and have increased my depth of understanding of this subject. I am also grateful to Natan Andrei, David Edwards and Veljko Zlatić for their encouragement and careful reading of early drafts of the manuscript. I also wish to thank Theo Costi for use of his renormalization group calculations prior to publication which have been used for several of the diagrams, and Nick Read for unpublished notes used in appendix D.

A few words of explanation on references

In covering subjects like the Kondo problem and heavy fermions for which there is such a vast literature the problem of which references to cite is an especially difficult one. I have found no good solution to this problem, but I have adopted a few guiding principles. In the course of describing a particular approach or work I have cited one or two papers (to keep these to a minimum) where the readers can if they wish find further details. The citation of a reference does not indicate priority for a particular work unless this is explicitly stated. The papers have been selected as ones most closely related to the approach I have used, or the ones most familiar to me. The earliest paper of an author on a particular topic is not always the one cited; a later paper may be referenced if I feel that is likely to be more helpful to the reader. Inevitably in trying to keep the number of references to within reasonable bounds much interesting related work will appear to have been overlooked (I apologize to any contributors to the field who feel that this applies to their work). The reader, however, is encouraged to use the references given as 'seed' references for use with the citation index. Used in this way, together with the reference list of the review articles quoted, they should be able to build up a comprehensive reference list to the whole field or any part of it which is of particular interest to them.

Imperial College, April 1992.   A.C. Hewson

Preface to paperback edition

I welcome the opportunity that the new edition of this book gives for me to correct a number of minor slips and omissions that escaped my notice in the original text. I am grateful to the help given me by Dr Jan von Delft in compiling this list of corrections.

I have also used this occasion to add a short section in the form of an Addendum to cover some recent developments. The subject of strongly correlated electron systems continues to be a very active field of research so within the short space at my disposal I could only briefly mention some results that are particularly related to the topics covered in the original edition. However, I have also included references to some more recent review articles where a fuller discussion of these topics can be found, as well as references to other related work.

Imperial College, October 1995.   A.C. Hewson

Brief History

There have been significant developments in the theory of magnetic impurities in metals in a period extending over more than 30 years. By magnetic impurities we mean those impurities that contribute a Curie–Weiss term to the susceptibility χ,

$$\chi = \frac{C}{T + \theta},$$

where T is the temperature and θ is a constant with a value in the thermal energy range ($0 < \theta < 300$ K). This term is in addition to the largely temperature independent Pauli susceptibility χ_P of the host metal. An isolated 'local moment', such as that due to a localized, but otherwise free, spin S, gives a Curie law ($\theta = 0$, $C = 4\mu_B^2 S(S+1)/3k_B$). Impurities which show Curie–Weiss behaviour are from the 3d transition element series or from the 4f rare earth series in the periodic table. Well studied examples are Fe in Cu, and Ce in $LaAl_2$ and LaB_6. The basic questions are: How does such a moment survive in the metallic environment ? How does it affect the conduction electrons of the host metal? Experimentally it is observed that such impurities give anomalous contributions to many metallic properties, particularly to the transport properties such as resistivity and thermopower, but also to the thermodynamic behaviour.

One manifestation of the effect magnetic impurities has been known since the early 30s. This is the observation of a resistance minimum in some metals (in most metals the resistivity monotonically decreases with decrease of temperature because it is dominated by phonon scattering which decreases rapidly at low temperatures). It was only recognized later that this minimum was an impurity effect associated with

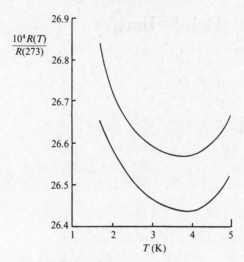

Figure 1 The minimum in the electrical resistivity of Au (de Haas, de Boer and van den Berg, 1934).

3d transition metal impurities such as Fe, dependent on the impurity concentration.

The resistance minimum as observed in Au is shown in figure 1, reproduced from the 1953 edition of *The Theory of Metals* by A.H. Wilson, one of the standard texts of this period. The reason for the minimum was not known at that time and Wilson comments, 'the cause of the minimum is entirely obscure and constitutes a most striking departure from Mathiessens's rule, according to which the ideal and residual resistances are additive — some new physical principle seems to be involved'. A very significant advance in the theory of magnetic impurities was an explanation of this effect by J. Kondo in 1964.

Early theoretical work on impurities in metals in the late 50s by J. Friedel and associates concentrated on explaining the trends in the behaviour as the impurity elements are varied across the transition element series. The most important concept to emerge from this work was that of 'virtual bound states' ; states which are almost localized due to resonant scattering at the impurity site. A different formulation of this idea was put forward by P.W. Anderson (1961), in a version now known as the 'Anderson model'; this model has played a very important role in the later developments of the theory. The model contains, in addition to a narrow resonance associated with the impurity states, a short range interaction U between the localized electrons. This interaction is needed to explain the observation of localized magnetic moments. Kondo's calculation of the resistivity, which was to explain this minimum, was based on a model where it is assumed that there is already a local magnetic moment associated with a spin S which is coupled via an exchange interaction J with the conduction electrons. This is known as the s-d model; it can be deduced from the Anderson model in the appropriate parameter regime. Kondo (1964) showed, using third order perturbation theory in the coupling J, that this interaction leads to singular scattering of the conduction electrons near the Fermi level and a $\ln T$ contribution to the resistivity. The $\ln T$ term increases at low temperatures for an antiferromagnetic coupling and when this term is included with the phonon contribution to the resistivity it is sufficient to explain the observed resistance minimum. Hence the solution of a longstanding puzzle.

There were difficulties with the theory, however, as $\ln T$ terms diverge as $T \to 0$ so that it was clear that Kondo's perturbation calculations could not be valid at low temperatures. A more comprehensive theory was needed to explain the low temperature behaviour of systems giving resistance minima. The search for such a theory became known as the 'Kondo problem' and attracted a lot of theoretical interest in the late 60s and early 70s. Extension of the perturbation approach to the summation of the leading order logarithmically divergent terms, using many-body techniques developed in the 50s and 60s, proved to be inadequate. In the antiferromagnetic case this summation leads to a divergence at a finite temperature T_K, known as the Kondo temperature. The perturbation theory provided a good description of the magnetic impurity systems for $T \gg T_K$, but could not be extended to the region $T \ll T_K$. The challenge was, therefore, to find a non-perturbational technique to predict the behaviour for $T < T_K$. The basic models, the s-d model and Anderson model, having only a local two-body interaction, seemed simple enough to hold out the possibility of finding exact solutions. Hence the wide theoretical appeal of the problem.

Experimental work during this period provided some clues for the theory. It was not a straightforward task to determine the behaviour associated with a single impurity, due to interimpurity interactions. These interactions become more of a problem at low temperatures, particularly for transition metals. Careful elimination of these interimpurity effects revealed that the impurity contributions to both thermodynamic and transport properties give power laws in T; the impurity resistivity and the susceptibility, for example, deviating from their $T = 0$ values by T^2 terms.

The theoretical framework for understanding these results was introduced by Anderson in the late 60s. The key idea was that of scaling. Anderson and coworkers showed that if the higher order excitations were eliminated perturbatively to give an effective model valid on a lower energy scale, the effective coupling between the local moment and the conduction electrons increased. Their approach being perturbational broke down when the coupling became large so that it could not be carried out down to the lowest energy scales, those which determine the behaviour in the regime $T \ll T_K$. Nevertheless, close analogies with other systems indicated that this scaling could be continued, and that the coupling would increase indefinitely as the energy scale is reduced. Such a scaling behaviour would imply a ground state with an infinite coupling in which the impurity is bound to a conduction electron in a singlet state. The weak residual interactions, due to virtual excitations to the triplet state, which would come into play on a low energy scale, could then account for the observed power law behaviour. The behaviour at low temperatures would be similar to that of a non-magnetic impurity (the impurity spin having been compensated) but with enhanced coefficients in the power laws.

The important contribution by K.G. Wilson (1974, 1975), which was recognized in the award of a Nobel prize in 1982, was to devise a non-perturbative way, the 'numerical renormalization group', of confirming this hypothesis. Wilson took renormalization group ideas from field theory and scaling ideas from condensed matter, and constructed a powerful tool which he applied initially to problems of phase transitions, particularly to the calculation of critical exponents, and then later to the Kondo problem. Wilson obtained definitive results for the ground state and low temperature behaviour for the spin $S = \frac{1}{2}$ s-d model. One particularly simple result was for the χ/γ ratio of the impurity (γ is the low temperature specific heat coefficient) which he found was enhanced over that for non-interacting electrons by a factor of two. P. Nozières (1974, 1975) gave an interpretation of Wilson's low temperature results in terms of a form of Landau Fermi liquid theory, and also gave an appealingly simple derivation of the χ/γ result, as well as a exact calculation for the T^2 coefficient for the resistivity. A microscopic derivation of this Fermi liquid theory was derived by K. Yamada (1975) based on the Anderson model.

This period marked the end of a phase. The Kondo problem having been solved it appeared that all that was required was a tidying up operation, generalizing Wilson's results to deal with the complexities of more

realistic models for magnetic impurities, to include such terms as the orbital degeneracy of the 3d electrons and the effects of crystal fields. The
method was applied to the non-degenerate Anderson model but generalization to models with more degrees of freedom was not an easy task,
requiring greatly increased computing resources making it difficult to
get satisfactory results. Both theoreticians and experimentalists looked
for more exciting tasks.

Two developments in the late 70s and early 80s attracted interest back
to the field, one theoretical, the other experimental. The theoretical development was the discovery of exact solutions to the s-d model $S = \frac{1}{2}$ by
N. Andrei (1980) and P.B. Wiegmann (1980) using the Bethe ansatz, a
hypothesis first used by H.A. Bethe in 1931 to solve the one dimensional
Heisenberg model. This approach gave analytic results for the high and
low temperature behaviour, and a set of integral equations from which
the thermodynamic behaviour of the model could be calculated over the
full temperature range. These results confirmed Wilson's calculations.
More importantly the method proved to be generalizable to models of
greater physical interest.

The experimental development was the study of dilute and concentrated alloys with rare earth elements such as cerium and ytterbium.
These systems gave Kondo-like anomalies, similar to single impurity
type behaviour, over a wide concentration range. The theoretical developments were timely because a generalization of the Anderson and
s-d models appropriate to Ce and Yb systems, the Coqblin–Schrieffer
model, could be diagonalized via the Bethe ansatz and exact results
generated for comparison with experiment. There is a degeneracy factor N associated with the ground state multiplet of the impurity in the
Coqblin–Schrieffer model and exact results revealed qualitatively different behaviour as a function of N, the susceptibility as a function of T,
for example, develops a maximum for $N > 3$ (Ce, $N = 6$; Yb, $N = 8$).

For comparison with many of the experimental results, such as photoemission and neutron scattering, dynamic response functions are required. These functions cannot be calculated via the Bethe ansatz.
Many approximate techniques, however, were developed in the 80s based
on treating $1/N$ as a small parameter. Asymptotically exact results were
generated in the limit $N \to \infty$, and good approximations obtained for
finite N by a variety of methods; variational, diagram summation, and
slave boson mean field calculations with Gaussian corrections. These
methods were used to calculate the one electron density of states and
the dynamic susceptibility. What they clearly showed was the build up

of a very narrow many-body resonance in the density of states at the Fermi level in the Kondo regime, known as the Kondo resonance. This very sharp resonance accounts for the low temperature anomalies caused by the magnetic impurities. As this resonance is the key insight into understanding the effects of magnetic impurities in metals it has been chosen as the cover illustration of this volume. The many-body calculations, giving exact results for the models or approximate ones within controlled approximations, have extended the range of theoretical predictions and enabled some quantitative comparison between theory and experiment to be made.

The similarities in the behaviour of a class of cerium intermetallic compounds and concentrated alloys to that of impurities and dilute alloys has led to these systems being known loosely as 'Kondo lattices'. Some of these compounds have such dramatically marked low temperature anomalies that they have become known as 'heavy fermions'. This is due to the very large specific heat coefficients which correspond to a large effective mass m^*, of the order 1000 times that of a free electron. Several actinide compounds, mainly of U, show similar behaviour and are included in this class. They show diverse forms of low temperature behaviour; some appear to be unconventional superconductors, some order magnetically, some do both and others seem to remain paramagnetic down to the lowest temperatures measured. There is evidence in certain compounds of a very weak form of antiferromagnetic order with tiny magnetic moments of the order of 10^{-2} Bohr magnetons. De Haas–van Alphen measurements in the paramagnetic phase show that these systems are Fermi liquids. There is as yet no generally accepted explanation for this range of behaviour. Self-consistent band calculations appear to account for the Fermi surface measurements in some cases but not for the very large effective masses observed. Many-body calculations, based on techniques developed for the impurity problem, predict 'renormalized bands' at the Fermi level, similar to the many-body Kondo resonance in the impurity density of states, but composed of coherent states due to the translational invariance of the lattice. These results may explain the enhanced masses; the superconducting and magnetic behaviour are still fully to be understood. It is a very active field in current research.

The anomalies in the magnetic impurity, Kondo lattice, and heavy fermion systems, are basically due to the strong correlations induced by the short range Coulomb interaction U at the transition metal and rare earth sites. There are similarities in the models used to describe these systems and those for the new high temperature superconductors, such as $La_{2-x}Sr_xCuO_4$ and $YBa_2Cu_3O_{7-x}$. The very high T_c materials have CuO_2 planes in common and the effects of the Coulomb interaction U at the copper sites is believed by many to be the key to understanding their anomalous behaviour. They are likely to be technologically very important and are of great interest at the moment. Their behaviour so far is not well understood. Hence, the story of strongly correlated systems is an on-going one, throwing up fresh challenges. There is likely to be as much intellectual excitement in work in this field in the future as there has been over the past three decades.

1

Models of Magnetic Impurities

1.1 First Principles Model

An impurity in a metallic host can be described by a very general Hamiltonian specifying all the electrons and their interactions. Such a Hamiltonian, which looks deceptively simple, has the form,

$$H = \sum_{i=1}^{N_0} \left(\frac{\mathbf{p}_i^2}{2m} + U(\mathbf{r}_i) + V_{\mathrm{imp}}(\mathbf{r}_i) \right) + \frac{1}{2} \sum_{i \neq j}^{N_0} \frac{e^2}{|\mathbf{r}_i - \mathbf{r}_j|} + \sum_{i=1}^{N_0} \lambda(\mathbf{r}_i)\, \mathbf{l}_i \cdot \boldsymbol{\sigma}_i ,$$

$$(1.1)$$

for N_0 electrons. The first term represents the kinetic energy of the electrons, the second term the periodic potential of the host metal due to the nuclei (before the impurity is introduced). The third term corresponds to the *additional* potential due to the nucleus of the impurity. The fourth term is the Coulomb interaction term between the electrons, and the last term is the spin-orbit interaction, which is a relativistic correction. A first principles calculation based on (1.1) is very difficult due to the strong Coulomb interaction which cannot be treated perturbatively. The only feasible method would appear to be some form of self-consistent field approach in which the problem is reduced to a single electron moving in some averaged potential of all the other electrons, which has to be determined self-consistently. Remarkably it was shown by Hohenberg & Kohn (1964) and Kohn & Sham (1965) that the *ground state properties* can be calculated exactly in principle by such an approach. The theorem proved by Hohenberg & Kohn, which is quite general, asserts that for any system of interacting electrons the ground state energy is a universal functional of the electron density $n(\mathbf{r})$. They also proved that the density corresponds to a minimum of this functional with respect to $n(\mathbf{r})$. Using this stationary property Kohn and Sham showed that the density can be calculated exactly from single particle equations for the electrons moving within an appropriate effective potential. The functional, and hence the potential, is non-local and is not known exactly. Calculations in practice are based on a local approximation, and good results for the ground state properties have been obtained for many systems using this approach. It is not an approach that we shall use here. The local approximation for the functional does not work well for systems which are strongly correlated due to a strong local Coulomb interaction, such as systems with incomplete f-shells. It is this type of strong correlation system which we shall be concerned with for transition metal and rare earth impurities, so we shall explore alternative approaches. Also we shall be more interested in the behaviour of the system as a function of temperature rather than ground state properties and in the response to dynamic probes. These involve probing the excitation spectrum of the system and there is no guarantee that this is given correctly in the local density functional approach. Instead of a first principles treatment using the full Hamiltonian (1.1) we will attempt to derive simpler model Hamiltonians, which describe the low energy excitations associated with the impurity and ignore features that are not directly relevant to the calculation of impurity effects. We shall do this in stages.

As mentioned in the introduction we will be concerned almost exclusively with simple metals which have broad conduction bands, such as those derived from s and p states. Conduction electrons in wide bands can be assumed to behave approximately as independent particles moving within a periodic potential. The long range Coulomb interaction between the electrons is screened (contributing to plasma excitations, which are of too high an energy to concern us) so that the electrons are essentially quasi-particles, electrons together with their screening cloud. The short range interactions between the quasi-particles should be small for a wide conduction band as the states are more delocalized than in narrow bands. We know also from Landau Fermi liquid theory (which we shall discuss more fully later), that the lifetimes of single particle excitations near the Fermi level are very long. It is a reasonable approximation to neglect the quasi-particle interactions, and to describe the host metal conduction electrons by a one-electron Hamiltonian,

$$H = \sum_{\mathbf{k},\sigma} \epsilon_{\mathbf{k}} c_{\mathbf{k},\sigma}^{\dagger} c_{\mathbf{k},\sigma}, \qquad (1.2)$$

where $c_{\mathbf{k},\sigma}^{\dagger}$ and $c_{\mathbf{k},\sigma}$ are creation and annihilation operators for Bloch states $\phi_{\mathbf{k},\sigma}(\mathbf{r})$ of wavevector $\mathbf{k}$ and spin component σ corresponding to an energy eigenvalue $\epsilon_{\mathbf{k}}$. The creation and annihilation operators obey the standard fermion anticommutation rules

$$[c_{\mathbf{k},\sigma}, c_{\mathbf{k}',\sigma'}^{\dagger}]_{+} = \delta_{\mathbf{k},\mathbf{k}'}\delta_{\sigma,\sigma'} \qquad [c_{\mathbf{k},\sigma}, c_{\mathbf{k}',\sigma'}]_{+} = 0. \qquad (1.3)$$

We shall not be concerned with the band calculations leading to energy states $\epsilon_{\mathbf{k}}$ in (1.2). We shall assume these conduction states to be characterized by a density of states $\rho_0(\epsilon)$ at energy ϵ given by

$$\rho_0(\epsilon) = \sum_{\mathbf{k}} \delta(\epsilon - \epsilon_{\mathbf{k}}), \qquad (1.4)$$

where $\delta(x)$ is the Dirac delta function. In the ground state these levels will be filled to the Fermi level ϵ_{F}.

As a consequence of the independent particle picture of the conduction electrons their thermodynamic properties can be easily calculated from the partition function Z_0,

$$Z_0 = \mathrm{Tr}\, e^{-\beta(H-\mu N)} = \prod_{\mathbf{k}} \left(1 + e^{-\beta(\epsilon_{\mathbf{k}}-\mu)}\right)^2, \qquad (1.5)$$

where $\beta = 1/k_{\mathrm{B}}T$, k_{B} is the Boltzmann constant and μ the chemical potential. The exponent 2 in (1.5) is due to the two spin components. Using the standard thermodynamic relations the specific heat $C(T)$ can be calculated and, with use of the Sommerfeld expansion,

$$\int_{-\infty}^{\infty} f(\epsilon)g(\epsilon)\, d\epsilon = \int_{-\infty}^{\mu} g(\epsilon)\, d\epsilon + \frac{\pi^2}{6}(k_{\mathrm{B}}T)^2 g'(\mu) + \mathrm{O}(T^4), \qquad (1.6)$$

where $f(\epsilon)$ is the Fermi–Dirac distribution function $1/\left(1 + e^{\beta(\epsilon-\mu)}\right)$, one finds

$$C(T) = \gamma T \left\{1 - \frac{\pi^2}{30}(k_{\mathrm{B}}T)^2 \left(\left(\frac{\rho_0'}{\rho_0}\right)^2 - \left(\frac{\rho_0''}{\rho_0}\right)\right)\right\}_{\epsilon=\epsilon_{\mathrm{F}}} + \mathrm{O}(T^5), \qquad (1.7)$$

where the linear coefficient of specific heat γ is given by

$$\gamma = \frac{2\pi^2 k_{\mathrm{B}}^2}{3}\rho_0(\epsilon_{\mathrm{F}}). \qquad (1.8)$$

On coupling to a magnetic field H the energy ϵ becomes spin dependent $\epsilon_{\pm} = \epsilon \mp \mu_{\mathrm{B}}H$, where μ_{B} is the Bohr magneton. Calculation of the zero field paramagnetic susceptibility $\chi(T)$ gives

$$\chi(T) = \chi(0) \left\{1 - \frac{\pi^2}{6}(k_{\mathrm{B}}T)^2 \left(\left(\frac{\rho_0'}{\rho_0}\right)^2 - \left(\frac{\rho_0''}{\rho_0}\right)\right)\right\}_{\epsilon=\epsilon_{\mathrm{F}}} + O(T^4), \qquad (1.9)$$

where

$$\chi(0) = 2\mu_{\mathrm{B}}^2 \rho_0(\epsilon_{\mathrm{F}}). \qquad (1.10)$$

1.2 Potential Scattering Model and the Friedel sum rule

In considering the effects of the impurity we could attempt to work within the same independent particle picture used for the conduction electrons and introduce an effective potential $V_{imp}^{eff}(\mathbf{r})$ due to the impurity. If the impurity has a charge in excess of that of the host metal atoms then this will be a Coulomb potential. However, from classical field theory we know that no macroscopic electric field can exist within a metal, so the charge must be screened by the conduction electrons on an atomic scale. The screening effects can be estimated within the Thomas–Fermi approximation. This gives a very localized effective interaction and in a typical metal the interaction is screened within a distance of the order of 1 Å. A somewhat better approximation would be to use the random phase approximation (RPA) for the dielectric function in estimating the screened interaction. This would have the advantage of including the long range oscillations of the screening cloud of the conduction electrons which are not included in the Thomas–Fermi calculation (these are the Friedel oscillations due to the discontinuity in electron occupation at Fermi surface; we will have more to say about these later). This approach will not prove to be sufficient for describing most transition metal and rare earth impurities. However it will be useful, at least initially, to explore the consequences of this approach and see why it does not give an adequate description for the case of 'magnetic' impurities later. The combined Hamiltonian of the host conduction electrons plus the one-body impurity potential in terms of Bloch states is

$$H = \sum_{\mathbf{k},\sigma} \epsilon_{\mathbf{k}} c_{\mathbf{k},\sigma}^{\dagger} c_{\mathbf{k},\sigma} + \sum_{\mathbf{k},\mathbf{k}',\sigma} V_{\mathbf{k},\mathbf{k}'} c_{\mathbf{k},\sigma}^{\dagger} c_{\mathbf{k}',\sigma} , \qquad (1.11)$$

where $V_{\mathbf{k},\mathbf{k}'} = \langle \mathbf{k} | V_{imp}^{eff} | \mathbf{k}' \rangle$. This Hamiltonian can in principle be diagonalized, and written in the form

$$H = \sum_{\alpha,\sigma} \epsilon_{\alpha} c_{\alpha,\sigma}^{\dagger} c_{\alpha,\sigma} , \qquad (1.12)$$

where $c_{\alpha,\sigma}^{\dagger}$ and $c_{\alpha,\sigma}$ create and annihilate eigenstates of (1.12) of energy ϵ_{α} and spin σ.

If we denote the density of states of (1.12) by $\rho(\epsilon)$, which is given by (1.4) but with a sum over α rather than $\mathbf{k}$, then the specific heat and magnetic susceptibility of host plus impurity will be again given by (1.7) and (1.9) with $\rho(\epsilon)$ replacing $\rho_0(\epsilon)$. It is clear from these equations that the addition of an impurity will only change the specific heat and the paramagnetic susceptibility if it significantly modifies the density of states in the region of the Fermi level ϵ_F. To investigate this we need to be able to calculate the change in density of states in the presence of the impurity. This we can do most conveniently using the resolvent Green's function $G(t - t')$ which is defined by the equation,

$$\left(i\hbar\frac{\partial}{\partial t} - H\right) G(t - t') = \delta(t - t'), \tag{1.13}$$

subject to appropriate boundary conditions. Equation (1.13) is an operator equation and can be expressed as a standard Green's function equation for $G(\mathbf{r}, \mathbf{r}' : t - t')$ by taking matrix elements with respect to bra and ket states $\langle\mathbf{r}|$ and $|\mathbf{r}'\rangle$ in the position representation and using $\langle\mathbf{r}|\mathbf{r}'\rangle = \delta(\mathbf{r} - \mathbf{r}')$. The functions of primary physical interest, due to causality, are the retarded functions $G^+(t - t')$ which vanish if $t < t'$. Advanced Green's functions $G^-(t - t')$ vanish for $t > t'$. Their Fourier transforms are defined by

$$G^\pm(\epsilon) = \int_{-\infty}^{\infty} G^\pm(t - t')e^{i\epsilon^\pm(t-t')/\hbar} \, d(t - t') \tag{1.14}$$

where $\epsilon^\pm = \epsilon \pm is$ and $s \to 0$. The equations for $G^\pm(\epsilon)$ are

$$(\epsilon \pm is - H)G^\pm(\epsilon) = I. \tag{1.15}$$

Equations (1.13)–(1.15) are operator equations. We can work with a particular representation for G by taking matrix elements with respect to the basis vectors. Using the eigenstates of (1.12) as a basis and taking matrix elements of (1.15) with respect to the states $\langle\alpha|$ and $|\alpha'\rangle$ gives

$$G^\pm_{\alpha\alpha'}(\epsilon) = \langle\alpha|\, G^\pm(\epsilon)|\alpha'\rangle = \frac{\delta_{\alpha\alpha'}}{(\epsilon \pm is - \epsilon_\alpha)}. \tag{1.16}$$

Using this result and the representation for the delta function,

$$\lim_{s\to 0^+} \frac{s}{(\epsilon - \epsilon_\alpha)^2 + s^2} = \pi\delta(\epsilon - \epsilon_\alpha), \tag{1.17}$$

it can be shown straightforwardly that the density of states $\rho(\epsilon)$ is given by

$$\rho(\epsilon) = \mp\frac{1}{\pi}\text{Im} \,\text{Tr}\, G^\pm(\epsilon), \tag{1.18}$$

which is independent of the particular representation used in the derivation.

To calculate the density of states perturbatively from (1.18) we introduce $G^\pm_0(\epsilon)$, the resolvent Green's function for the conduction electrons, which satisfies (1.15) with H given by the conduction electron Hamiltonian (1.2). We will denote this by H_0 and then write the Hamiltonian (1.11) as $H = H_0 + V$, where V is the additional interaction due to the impurity which will be treated perturbatively. Rewriting (1.15) as

$$(\epsilon \pm is - H_0 - V)G^\pm = I \tag{1.19}$$

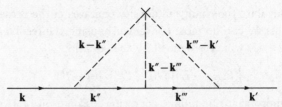

Figure 1.1 The perturbation series for $G^+_{\mathbf{kk'}}(\epsilon) = \langle \mathbf{k}|G^+(\epsilon)|\mathbf{k'}\rangle$ in powers of the impurity interaction $V_{\mathbf{kk'}} = \langle \mathbf{k}|V|\mathbf{k'}\rangle$. The cross x indicates the impurity, the solid line, the propagator $\langle \mathbf{k}|G^+_0(\epsilon)|\mathbf{k'}\rangle$, and the dashed line, the interaction $V_{\mathbf{kk'}}$.

and multiplying by $G^{\pm}_0$ gives an equation for G in the form,

$$G^{\pm} = G^{\pm}_0 + G^{\pm}_0 V G^{\pm}. \qquad (1.20)$$

From this point onwards we shall restrict the discussion to the case of the retarded function G^+ only; equations for the advanced function follow simply by changing is into $-is$. Equation (1.20) can be solved iteratively to obtain the solution as a power series in V, which can be expressed in the form,

$$G^+ = G^+_0 + G^+_0 T(\epsilon^+) G^+_0, \qquad (1.21)$$

where $\epsilon^+ = \epsilon + is$, and T is the T matrix given by

$$T(\epsilon^+) = V + V G^+_0 V + V G^+_0 V G^+_0 V + \ldots = V(I - G^+_0 V)^{-1}. \qquad (1.22)$$

The perturbation series is indicated diagrammatically in figure 1.1.

The change in density of states $\rho_{\text{imp}}(\epsilon) = \rho(\epsilon) - \rho_0(\epsilon)$ due to the addition of the impurity can be deduced from (1.18),

$$\Delta\rho(\epsilon) = -\frac{1}{\pi}\text{Im}\,\text{Tr}(G^+(\epsilon) - G^+_0(\epsilon)). \qquad (1.23)$$

This change in the density of states due to the impurity can be expressed in terms of the phase shift of the scattering due to the impurity potential. Using the identity,

$$\text{Tr}\,G^+(\epsilon) = -\frac{\partial}{\partial\epsilon}\ln\det G^+(\epsilon), \qquad (1.24)$$

which follows from the matrix relation $\text{Tr}\ln\mathbf{A} = \ln\det\mathbf{A}$ and equation (1.16), equation (1.23) can be re-expressed in the form,

$$\Delta\rho(\epsilon) = \frac{1}{\pi}\,\text{Im}\,\frac{\partial}{\partial\epsilon}\ln\left(\det(G^+(\epsilon)\,(G^+_0(\epsilon))^{-1})\right). \qquad (1.25)$$

The solution of (1.20) is $G^+ = (I - G^+_0 V)^{-1}G^+_0$ and on substituting this into (1.25) we find

$$\rho_{\text{imp}}(\epsilon) = \frac{1}{\pi}\,\text{Im}\,\frac{\partial}{\partial\epsilon}\ln\left(\det(I - G^+_0 V)^{-1}\right). \qquad (1.26)$$

We can replace $(I - G_0^+ V)^{-1}$ by $T(\epsilon^+)$ in the argument of (1.26) as V is independent of ϵ and hence,

$$\Delta\rho(\epsilon) = \frac{1}{\pi} \mathrm{Im}\, \frac{\partial}{\partial\epsilon} \ln\left(\det T(\epsilon^+)\right). \tag{1.27}$$

If the phase shift $\eta(\epsilon)$ is defined by

$$\eta(\epsilon) = \arg \det T(\epsilon^+), \tag{1.28}$$

(we refer readers more familiar with the definition of the phase shift in terms of the asymptotic form of the scattered wavefunction to appendix A) then (1.27) becomes

$$\Delta\rho(\epsilon) = \frac{1}{\pi} \frac{\partial\eta(\epsilon)}{\partial\epsilon}. \tag{1.29}$$

The change in the number of electrons n_{imp} due to the introduction of the impurity is obtained by integrating the density of states difference (1.29) to the Fermi level ϵ_{F}. At $T = 0$,

$$n_{\mathrm{imp}} = \int_{-\infty}^{\epsilon_{\mathrm{F}}} \Delta\rho(\epsilon)\, d\epsilon = \frac{\eta(\epsilon_{\mathrm{F}})}{\pi}, \tag{1.30}$$

taking $\eta(-\infty) = 0$.

The extra states that are induced below the Fermi level by the impurity must be such as to accommodate the charge required to screen the impurity. If the impurity nucleus plus the core electrons has an excess charge $\Delta Z e$ (relative to the host metal) then $n_{\mathrm{imp}} = \Delta Z$. This relation with n_{imp} given by (1.30) is known as the *Friedel Sum Rule* (Friedel, 1956). If the full long range Coulomb interactions between the impurity and conduction electrons, and between the conduction electrons themselves, are taken into account then screening of the impurity must occur. If, in this simplified picture, these interactions are described by a screened one-body interaction V, then the Friedel sum rule must be *imposed* as a self-consistency condition on V, so that the local screening cloud contains the right number of electrons.

In the case of a spherically symmetric impurity potential, and plane waves $\phi_{\mathbf{k}}(\mathbf{r})$, $T(\epsilon^+)$ will be block diagonal, i.e. $\langle k, l, m | T | k, l', m' \rangle = t_l(k)\delta_{l,l'}\delta_{m,m'}$ for partial wave states $|k, l, m\rangle$ corresponding to angular momentum quantum number l and z-component m. If η_l is the phase shift of the lth partial wave, $\eta_l = \arg t_l(k)$, then

$$\eta(k) = 2\sum_l (2l+1)\eta_l(k), \tag{1.31}$$

where the factor of 2 is due to the spin degeneracy and $(2l + 1)$ due to the $2l + 1$ values of m. Similar factorizations will occur for other space group symmetries such as cubic symmetry.

Though the impurity is essentially screened on an atomic scale, there are oscillations or ripples in the screening cloud that fall off relatively slowly with distance. These were referred to earlier and are known as *Friedel oscillations* (Friedel, 1952). To demonstrate this, we need to calculate the charge density at a point $\mathbf{r}$ from the impurity. This can be calculated from the Green's function G or, more conveniently, from the wavefunction $\psi_{\mathbf{k}}(\mathbf{r})$. In terms of the wavefunction the change in charge density due to the introduction of the impurity is given by

$$\Delta\rho(\mathbf{r}) = 2 \int_0^{k_{\mathrm{F}}} (|\psi_{\mathbf{k}}(\mathbf{r})|^2 - |\phi_{\mathbf{k}}(\mathbf{r})|^2) \frac{d^3k}{(2\pi)^3}. \quad (1.32)$$

Using the asymptotic forms for $\psi_{\mathbf{k}}(\mathbf{r})$ and $\phi_{\mathbf{k}}(\mathbf{r})$ (taking this to be a plane wave) given in appendix A, after performing the angular integration using (A.24) this becomes for large r,

$$\Delta\rho(r) \sim$$

$$\int_0^{k_{\mathrm{F}}} \sum_l \frac{(2l+1)}{\pi^2 r^2} \left\{ \sin^2\left(kr + \eta_l(k) - \frac{l\pi}{2} \right) - \sin^2\left(kr - \frac{l\pi}{2} \right) \right\} dk.$$

$$(1.33)$$

The leading asymptotic contribution to the integral in (1.33) can be found by an integration by parts, and does not require the dependence of $\eta_l(k)$ upon k to be known, and gives

$$\Delta\rho(r) \sim -\frac{1}{2\pi^2 r^3} \sum_l (2l+1)(-1)^l \cos(2k_{\mathrm{F}} r + \eta_l(\epsilon_{\mathrm{F}})) \sin\eta_l(\epsilon_{\mathrm{F}}). \quad (1.34)$$

The wavelength of the oscillation $1/2k_{\mathrm{F}}$ is due to the sharp Fermi surface at $T = 0$.

1.3 Virtual Bound States

A local impurity potential V_{imp} may be sufficiently attractive to induce a bound state below the conduction band of the host metal. An electron in this state will be localized in the vicinity of the impurity and its wavefunction will fall off exponentially with the distance r from the impurity as $r \to \infty$. A potential which is not sufficiently attractive to produce a bound state below the conduction band may nevertheless tend to localize the conduction electrons for a time in the vicinity of the impurity. This is due to resonant scattering at the impurity site which induces a narrow peak in the conduction band density of states known as a *virtual bound state resonance*. In such a state the conduction electrons spend a relatively larger proportion of the time in the region of the impurity, but it is not a bound state because the wave functions become Bloch states far from the impurity. If such a resonance occurs in the region of the Fermi level, it is clear from (1.7) and (1.9) that it will enhance the impurity contribution to the specific heat and paramagnetic susceptibility. Early interpretations of transition metal impurity effects in metal were based on the virtual bound state concept (see Friedel, 1958: Blandin & Friedel, 1959).

We can illustrate this concept and make explicit calculations of the Green's function and phase shift in the case of a very localized potential,

$$V_{\mathbf{k},\mathbf{k}'} = V/N_s \qquad \epsilon_{\mathbf{k}}, \epsilon_{\mathbf{k}'} < D\,, \qquad (1.35)$$

where D is the band width of the conduction density of states and N_s is the total number of sites.

Taking matrix elements of (1.15) with respect to the Bloch states gives

$$(\epsilon + is - \epsilon_{\mathbf{k}})G^{+}_{\mathbf{k},\mathbf{k}'}(\epsilon) = \delta_{\mathbf{k},\mathbf{k}'} + \frac{V}{N_s}\sum_{\mathbf{k}''}G^{+}_{\mathbf{k}'',\mathbf{k}'}(\epsilon). \qquad (1.36)$$

Solving (1.36) for $\sum_{\mathbf{k}''}G^{+}_{\mathbf{k}'',\mathbf{k}'}$ gives the result for $G^{+}_{\mathbf{k},\mathbf{k}'}(\epsilon)$,

$$G^{+}_{\mathbf{k},\mathbf{k}'}(\epsilon) = \frac{\delta_{\mathbf{k},\mathbf{k}'}}{(\epsilon + is - \epsilon_{\mathbf{k}})}$$
$$+ \frac{V/N_s}{(\epsilon + is - \epsilon_{\mathbf{k}})\left(1 - V/N_s\sum_{\mathbf{k}''}1/(\epsilon + is - \epsilon_{\mathbf{k}''})\right)(\epsilon + is - \epsilon_{\mathbf{k}'})}\,,$$
$$(1.37)$$

which corresponds to (1.21) with

$$\langle\mathbf{k}|T(\epsilon^{+})|\mathbf{k}'\rangle = \frac{V_{\mathbf{k},\mathbf{k}'}}{\left(1 - V/N_s\sum_{\mathbf{k}''}1/(\epsilon^{+} - \epsilon_{\mathbf{k}''})\right)}. \qquad (1.38)$$

The corresponding phase shift $\eta(\epsilon)$ is given by

$$\eta(\epsilon) = \tan^{-1}\left(\frac{\pi\rho_0(\epsilon)V}{V\Lambda(\epsilon) - 1}\right)\,, \qquad (1.39)$$

on using the standard result,

$$\sum_{\mathbf{k}}\frac{1}{(\epsilon - \epsilon_{\mathbf{k}} + is)} = P\sum_{\mathbf{k}}\frac{1}{(\epsilon - \epsilon_{\mathbf{k}})} - i\pi\sum_{\mathbf{k}}\delta(\epsilon - \epsilon_{\mathbf{k}})$$
$$= N_s(\Lambda(\epsilon) - i\pi\rho_0(\epsilon)), \qquad (1.40)$$

where $N_s\Lambda(\epsilon) = P\sum_{\mathbf{k}}1/(\epsilon - \epsilon_{\mathbf{k}})$ and P denotes a principal part integration. Figure 1.2(i) gives the form of $\Lambda(\epsilon)$ for the conduction density of states shown in figure 1.2(ii).

The change in density of states $\Delta\rho(\epsilon)$ due to the impurity potential can be calculated from (1.37) by taking diagonal elements, summing over

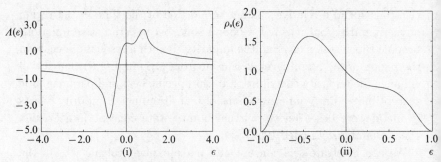

Figure 1.2 $\Lambda(\epsilon)$ plotted against ϵ in (i) for the conduction electron density of states $\rho_0(\epsilon)$ shown in (ii).

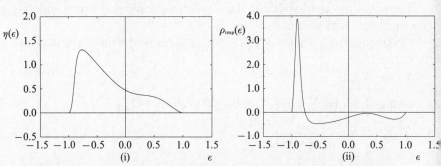

Figure 1.3 (i) The phase shift $\eta(\epsilon)$ against ϵ for the conduction density of states $\rho_0(\epsilon)$ shown in figure 1.2 (ii) with $V = -0.22$, (ii) the corresponding impurity density of states $\rho_{\mathrm{imp}}(\epsilon)$. A virtual bound state resonance is induced at at the bottom of the band. For $V < -0.25$ this becomes a bound state below the bottom of the band.

k, and then substituting into (1.18), or by using (1.39) in the Friedel sum rule (1.29). The result is

$$\Delta\rho(\epsilon) = -\frac{V}{N_s} \frac{\rho_0'(\epsilon)(1 - V\Lambda(\epsilon)) + V\rho_0(\epsilon)\Lambda'(\epsilon)}{(1 - V\Lambda(\epsilon))^2 + (\pi\rho_0(\epsilon)V)^2}, \qquad (1.41)$$

where the prime indicates a derivative with respect to ϵ. This expression is valid provided $1 - V\Lambda(\epsilon) = 0$ has no roots which lie outside the conduction band. If there is such a root, say at $\epsilon = \epsilon_B$, then there is an additional term $\delta(\epsilon - \epsilon_B)$ which corresponds to a bound state outside the band. If the potential is sufficiently attractive (or sufficiently repulsive) there will be a bound state below (above) the band. There may be solutions of $1 - V\Lambda(\epsilon) = 0$, when V is not so large, that lie within the band. If such a solution occurs at $\epsilon = \epsilon_0$ and we expand $\Lambda(\epsilon)$ in a Taylor's series about ϵ_0 in (1.41) then, in the region near $\epsilon = \epsilon_0$, $\Delta\rho(\epsilon)$

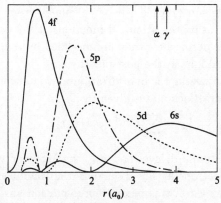

Figure 1.4 Wavefunctions of Ce in local density functional calculations in the configuration $4f^1 5d^1 6s^2$ (from Gunnarsson & Schönhammer, 1987). The 4f states are seen to be localized largely within the 5s, 5p core and well within the Wigner–Seitz radius for α and γ-Ce (indicated by arrows).

will have the Lorentzian form,

$$\Delta\rho(\epsilon) = \frac{\Delta/\pi}{((\epsilon - \epsilon_0)^2 + \Delta^2)}, \qquad (1.42)$$

where $\Delta = \pi\rho_0/\Lambda'(\epsilon_0)$. If the resonance width Δ is small compared with the band width D then the density of states has a large peak ($\Lambda'(\epsilon_0) > 0$). This is an example of a virtual bound state resonance. An example is shown in figure 1.3(ii) calculated from (1.41) for the density of states $\rho_0(\epsilon)$ of figure 1.2(ii). The integrated induced density of states is zero because no states are added to the system on introducing the impurity scattering term V.

1.4 The Non-Interacting Anderson Model

Virtual bound state resonances occur for transition metal or rare earth impurities when the d or f levels lie within the conduction band of the host metal. In the radial Schrödinger equation for a state of angular momentum l there is an effective potential due to the angular momentum equal to $l(l + 1)/r^2$. This additional 'potential' term has the effect of localizing the higher angular momentum states for $l = 2, 3$ within the potential barrier (see the 4f states of Ce in figure 1.4) so that these states in an impurity ion resemble atomic d or f states. There will be some probability for the electrons to tunnel through this barrier so that these states will be virtual bound states. The resonances in these cases, as shown by Anderson (1961), can be calculated from a rather different starting point using the atomic d functions of the isolated impurity ion, and then considering how they are modified by the presence of the neighbouring metal ions in the host metal.

There will be an overlap or hybridization matrix element $V_{\mathbf{k}}$ with the conduction electron Bloch states given by

$$V_{\mathbf{k}} = \sum_{\delta} e^{i\mathbf{k}\cdot\mathbf{d}_{\delta}} \langle \phi_{\mathrm{d}}|H|\psi_{\mathbf{d}_{\delta}}\rangle\,, \qquad (1.43)$$

where ϕ_{d} is the atomic d level and $\psi_{\mathbf{d}_{\delta}}$ the Wannier wavefunction of the conduction electrons at site $\mathbf{d}_{\delta}$ and H the Hamiltonian (1.1). The Hamiltonian expressed in this mixed representation has the form,

$$H = \sum_{\sigma} \epsilon_{\mathrm{d}} c_{\mathrm{d},\sigma}^{\dagger} c_{\mathrm{d},\sigma} + \sum_{\mathbf{k},\sigma} \epsilon_{\mathbf{k}} c_{\mathbf{k},\sigma}^{\dagger} c_{\mathbf{k},\sigma} + \sum_{\mathbf{k},\sigma} (V_{\mathbf{k}} c_{\mathrm{d},\sigma}^{\dagger} c_{\mathbf{k},\sigma} + V_{\mathbf{k}}^{*} c_{\mathbf{k},\sigma}^{\dagger} c_{\mathrm{d},\sigma})\,,$$
$$(1.44)$$

where ϵ_{d} is the energy of the d level of the impurity ion and $c_{\mathrm{d},\sigma}^{\dagger}$ and $c_{\mathrm{d},\sigma}$ are creation and annihilation operators for an electron in this state. For the moment we will keep things as simple as possible and ignore the orbital degeneracy of this state, and treat it as a state with spin degeneracy only.

The localized state and the conduction electron states are not orthogonal. However, for a localized d or f state in a conduction band constructed mainly from states of s or p symmetry, it is reasonable to neglect this and to assume $\langle d|\mathbf{k}\rangle = 0$ (this is a simplification and lack of orthogonality can be taken into account if necessary).

The Hamiltonian (1.44) corresponds to the *Anderson model* without any explicit interelectron interaction terms (the full Anderson model will be described in the next section). A similar model was introduced by Fano (1961) in a different context. We can set up equations that determine the Green's functions for this model by taking appropriate matrix elements of (1.15). This leads to the coupled equations,

$$(\epsilon - \epsilon_{\mathrm{d}}) G_{\mathrm{d},\mathrm{d}}^{+}(\epsilon) = 1 + \sum_{\mathbf{k}} V_{\mathbf{k}}^{*} G_{\mathbf{k},\mathrm{d}}^{+}(\epsilon)\,, \qquad (1.45)$$

$$(\epsilon - \epsilon_{\mathbf{k}}) G_{\mathbf{k},\mathrm{d}}^{+}(\epsilon) = V_{\mathbf{k}}^{*} G_{\mathrm{d},\mathrm{d}}^{+}(\epsilon)\,, \qquad (1.46)$$

where $G_{p,q}^{+} = \langle p|G^{+}(\epsilon)|q\rangle$, $p, q = \mathrm{d}, \mathbf{k}$. As the spin does not play any explicit role at this point we have dropped the spin index σ. Solving (1.45) and (1.46) for the impurity level Green's function $G_{\mathrm{d},\mathrm{d}}^{+}(\epsilon)$,

$$G_{\mathrm{d},\mathrm{d}}^{+}(\epsilon) = \frac{1}{\left(\epsilon + is - \epsilon_{\mathrm{d}} - \sum_{\mathbf{k}} |V_{\mathbf{k}}|^{2}/(\epsilon + is - \epsilon_{\mathbf{k}})\right)}. \qquad (1.47)$$

Taking further matrix elements of (1.15) gives the equations,

$$(\epsilon - \epsilon_{\mathbf{k}})G^+_{\mathbf{k},\mathbf{k}'}(\epsilon) = \delta_{\mathbf{k},\mathbf{k}'} + V^*_{\mathbf{k}}G^+_{\mathrm{d},\mathbf{k}'}(\epsilon), \qquad (1.48)$$

$$(\epsilon - \epsilon_{\mathrm{d}})G^+_{\mathrm{d},\mathbf{k}'}(\epsilon) = \sum_{\mathbf{k}''} V_{\mathbf{k}''}G^+_{\mathbf{k}'',\mathbf{k}'}(\epsilon), \qquad (1.49)$$

and leads to the solution for the conduction electron Green's function $G^+_{\mathbf{k},\mathbf{k}'}(\epsilon)$,

$$G^+_{\mathbf{k},\mathbf{k}'}(\epsilon) = \frac{\delta_{\mathbf{k},\mathbf{k}'}}{(\epsilon^+ - \epsilon_{\mathbf{k}})} + \frac{V^*_{\mathbf{k}}}{(\epsilon^+ - \epsilon_{\mathbf{k}})}G^+_{\mathrm{d},\mathrm{d}}(\epsilon)\frac{V_{\mathbf{k}'}}{(\epsilon^+ - \epsilon_{\mathbf{k}'})}. \qquad (1.50)$$

Comparing this with (1.21) we find for the matrix element of the T matrix,

$$\langle \mathbf{k}|T(\epsilon)|\mathbf{k}'\rangle = V^*_{\mathbf{k}}\, G^+_{\mathrm{d},\mathrm{d}}(\epsilon)\, V_{\mathbf{k}'}. \qquad (1.51)$$

We can calculate the change in the density of states $\Delta\rho(\epsilon)$ using (1.51) and (1.27), or from (1.18) using a direct evaluation of the trace of G^+,

$$\begin{aligned}
\mathrm{Tr}\, G^+(\epsilon) &= \sum_{\mathbf{k}} G^+_{\mathbf{k},\mathbf{k}}(\epsilon) + G^+_{\mathrm{d},\mathrm{d}}(\epsilon) \\
&= \sum_{\mathbf{k}} \frac{1}{(\epsilon^+ - \epsilon_{\mathbf{k}})} + \frac{\partial}{\partial\epsilon}\ln\left(\epsilon^+ - \epsilon_{\mathrm{d}} - \sum_{\mathbf{k}} \frac{|V_{\mathbf{k}}|^2}{(\epsilon^+ - \epsilon_{\mathbf{k}})}\right).
\end{aligned} \qquad (1.52)$$

Using (1.52) the Friedel sum rule can be verified explicitly, the change in density of states $\rho_{\mathrm{imp}}(\epsilon)$ is given by equation (1.29) with the phase shift $\eta(\epsilon)$ given by

$$\eta(\epsilon) = \frac{\pi}{2} - \tan^{-1}\left(\frac{\epsilon_{\mathrm{d}} + \Lambda(\epsilon) - \epsilon}{\Delta(\epsilon)}\right), \qquad (1.53)$$

where in this case $\Delta(\epsilon)$ and $\Lambda(\epsilon)$ are defined by

$$\Lambda(\epsilon) = \mathrm{P}\sum_{\mathbf{k}} \frac{|V_{\mathbf{k}}|^2}{(\epsilon - \epsilon_{\mathbf{k}})} \qquad \Delta(\epsilon) = \pi\sum_{\mathbf{k}} |V_{\mathbf{k}}|^2\delta(\epsilon - \epsilon_{\mathbf{k}}). \qquad (1.54)$$

If we evaluate $\Lambda(\epsilon)$ and $\Delta(\epsilon)$ for a flat conduction band with a density of states,

$$\rho_0(\epsilon) = \rho_0 \qquad \text{for} \quad -D < \epsilon < D, \qquad (1.55)$$

and neglect the $\mathbf{k}$ dependence of $V_{\mathbf{k}}$ we find $\Delta(\epsilon) = \pi\rho_0|V|^2 = \Delta$ for $-D < \epsilon < D$ and zero elsewhere, and

$$\Lambda(\epsilon) = \rho_0|V|^2\ln\left|\frac{D + \epsilon}{D - \epsilon}\right|. \qquad (1.56)$$

If ϵ_{d} lies well within the conduction band then the phase shift near the impurity level is approximately given by

$$\eta(\epsilon) = \frac{\pi}{2} - \tan^{-1}\frac{(\tilde{\epsilon}_{\mathrm{d}} - \epsilon)}{\Delta}, \qquad (1.57)$$

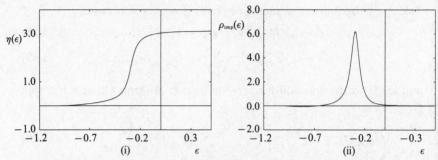

Figure 1.5 (i) The phase shift $\eta(\epsilon)$ against ϵ and (ii) the impurity density of states $\Delta\rho(\epsilon)$ deduced from (1.53) for $\epsilon_d = -0.3$ and $\Delta(\epsilon) = 0.01\pi\rho_0(\epsilon)$, for $\rho_0(\epsilon)$ shown in figure 1.2(ii). The small region where $\rho_{imp}(\epsilon) < 0$ in (ii) disappears if we use the flat band density of states (1.55) for $\rho_0(\epsilon)$.

where $\tilde{\epsilon}_d$ is the solution of $\epsilon - \epsilon_d - \Lambda(\epsilon) = 0$. The local density of states $\Delta\rho(\epsilon)$ is given by

$$\Delta\rho(\epsilon) = \frac{1}{\pi}\frac{\partial\eta}{\partial\epsilon} = \frac{\Delta/\pi}{(\epsilon - \tilde{\epsilon}_d)^2 + \Delta^2}, \qquad (1.58)$$

and corresponds to a virtual bound state resonance. With the flat band assumption it is easy to show that this density of states arises entirely from $G^+_{d,d}(\epsilon)$ for large D, when substituted into (1.18). Hence, the change in number of electrons due to the introduction of the impurity in this case is n_d, the expectation value for the d level occupation number, which at $T = 0$, is given by

$$n_d = 2\int_{-\infty}^{\epsilon_F} \Delta\rho(\epsilon)\,d\epsilon = 1 - \frac{2}{\pi}\tan^{-1}\left(\frac{\tilde{\epsilon}_d - \epsilon_F}{\Delta}\right). \qquad (1.59)$$

This corresponds to the Friedel sum rule (1.30) with the phase shift (1.57) (the factor of 2 is due to the spin degeneracy). There are no long range Coulomb interactions included in the Anderson model so that the parameters of the model must be such as to localize the required number of electrons to screen the impurity charge.

In figure 1.5 we show $\eta(\epsilon)$ and $\Delta\rho(\epsilon)$ as deduced from (1.57) for the density of states $\rho_0(\epsilon)$ shown in figure 1.2(ii). This is well approximated by the Lorentzian form (1.58). The integrated induced density of states is unity in this case (compare with figure 1.3(ii)) because the d level of the impurity is added to the system.

So far we have only considered a single impurity. This will contribute terms of the order of $1/N_s$ where N_s is the number of sites. In taking the thermodynamic limit $N_s \to \infty$ there must be a finite concentration of impurities $c_{imp} = N_{imp}/N_s$, where N_{imp} is the total number of impurities, to obtain an impurity contribution. If this concentration is very small, so that all impurity interactions can be neglected, then the leading contributions to the specific heat C_{imp} and paramagnetic susceptibility χ_{imp} are given by

$$C_{imp} = \frac{2\pi^2}{3} k_B^2 T c_{imp} \Delta\rho(\epsilon_F) \quad \text{and} \quad \chi_{imp} = 2\mu_B^2 c_{imp} \Delta\rho(\epsilon_F). \quad (1.60)$$

The ratio R of these quantities, where R is defined by

$$R = \frac{\pi^2 k_B^2}{s(s+1)(g\mu_B)^2} \frac{\chi_{imp}}{\gamma_{imp}}, \quad (1.61)$$

with $g = 2$ and $s = \frac{1}{2}$, is equal to unity, independent of the impurity concentration and the density of states.

We noted in the introduction that impurities in metals that are classified as 'magnetic' give a Curie–Weiss term to the susceptibility at higher temperatures,

$$\chi_{imp} \approx \frac{C}{T+\theta} \quad \text{for} \quad T > \theta. \quad (1.62)$$

A Curie-like contribution is difficult to explain within the one-body approach we have developed so far. The leading term in the paramagnetic susceptibility (1.60) is independent of T, and the corrections to this, which can be obtained from (1.9), are of the order T^2. These corrections for a free electron gas are very small, of the order of $(k_B T/\epsilon_F)^2$. If there is some sharp feature in the density of states in the region of the Fermi level, which could be induced by the presence of an impurity, then these corrections could be significant. If the impurity induced a narrow resonance at the Fermi level, a Curie contribution to the susceptibility can result. However, this would require the resonance to be extremely narrow, its width Δ must be such that $\Delta \sim k_B\theta$. The Curie term emerges in the regime $k_B T > \Delta$, the expansion to order T^2 (1.9) is only appropriate in the region $k_B T \ll \Delta$. Within the one-body formalism a very narrow resonance does not seem likely. Estimates of the hybridization parameter $V_\mathbf{k}$ for the Anderson model indicate Δ for transition element impurities to be of the order of 0.1–0.7 eV ($\sim$2000–8000 K), which would be far too high to explain the observation of a magnetic moment in many systems at very much lower temperatures. Another factor that suggests the one-body approach is not sufficient is that the approach cannot explain the resistance minimum observed in systems with magnetic impurities. We will consider this fully in the next chapter.

1.5 The s-d Exchange Model

An alternative approach to the question of magnetic effects due to impurities is to pursue the analogy of magnetic insulators. Magnetic insulators are usually described by some form of Heisenberg model, local magnetic moments interacting via some form of exchange interaction. The moments are free at high temperatures giving a Curie law,

$$\chi = \frac{(g_j \mu_B)^2 j(j+1)}{3k_B T}, \qquad (1.63)$$

for a magnetic moment associated with the angular momentum j in the ground state configuration of the magnetic ions with an associated g-factor g_j. In transition metal ions the orbital contribution to the angular momentum is often quenched, due to the effects of the local environment (crystal fields), in which case the moment is due to the spin only, $j = s$ with $g_s = 2$. For rare earth ions the crystal field effects are less strong, and the spin-orbit interaction is more important, so that j is the angular momentum quantum number associated with the total angular momentum, spin plus orbital, and g_j is the Landé g-factor. For rare earth ions at lower temperatures if there are significant crystal field terms, the moment may be associated with the lowest or lower crystal field split configurations.

For a transition metal ion, one that displays evidence of local moment behaviour as an impurity in a metal, it might be reasonable to describe it as in an insulator, and then consider the effects of the ion magnetic moment on the conduction electrons. A model of this form was proposed by Zener (1951) and is known as the s-d model. For a single impurity in a metal the interaction takes the form,

$$H_{sd} = \sum_{\mathbf{k},\mathbf{k'}} J_{\mathbf{k},\mathbf{k'}} (S^+ c^\dagger_{\mathbf{k},\downarrow} c_{\mathbf{k'},\uparrow} + S^- c^\dagger_{\mathbf{k},\uparrow} c_{\mathbf{k'},\downarrow} + S_z(c^\dagger_{\mathbf{k},\uparrow} c_{\mathbf{k'},\uparrow} - c^\dagger_{\mathbf{k},\downarrow} c_{\mathbf{k'},\downarrow})),$$

$$(1.64)$$

where S_z and $S^\pm$ $(= S_x \pm iS_y)$ are the spin operators for a state of spin S. It represents a Heisenberg exchange interaction between a local moment and the conduction electrons with a coupling constant $J_{\mathbf{k},\mathbf{k'}}$. There are terms in this interaction in which the spin of the conduction electron is flipped on scattering with the impurity (see figure 1.6 for $S = \frac{1}{2}$). There is in addition the potential scattering term (1.11) so the complete Hamiltonian is (1.64) plus (1.11). The potential scattering term can be formally eliminated by a transformation to eigenstates $|\alpha\rangle$ of (1.12). The creation and annihilation operators for the conduction electrons, $c^\dagger_{\mathbf{k},\sigma}$ and $c_{\mathbf{k'},\sigma}$, get replaced by $c^\dagger_{\alpha,\sigma}$ and $c_{\alpha',\sigma}$ with a modified

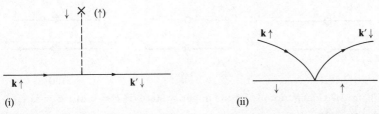

(i)   (ii)

Figure 1.6 (i) Spin flip scattering of a localized impurity moment by a conduction electron. (ii) The same scattering process where the base line indicates the spin state of the impurity. The representation (ii) is more convenient for describing repeated scatterings when the impurity has an internal degree of freedom (spin in this case).

coupling $J_{\alpha,\alpha'}$. The ground state configuration of the impurity ion is assumed to be fully quenched so that the state is characterized by a total spin quantum number S. We will give a derivation of (1.64) in section 1.7 for the case $S = \frac{1}{2}$. A more appropriate form of the model for rare earth impurities will be given in section 1.10.

When the coupling constant $J_{\mathbf{k},\mathbf{k}'}$ is put equal to zero, we have a free ion moment which will give the Curie law susceptibility (1.63). We should expect, therefore, to get back this Curie law at high enough temperatures for the interacting system. One might speculate that a perturbation calculation at high temperatures should give an impurity susceptibility consistent with the Curie–Weiss form (1.62) with θ proportional to the coupling constant J. This turns out not to be the case when we look at the perturbation theory for the susceptibility in chapter 3. In the next chapter when we look at Kondo's perturbation calculation of the electrical conductivity for this model we find that anomalous terms arise. We turn for the moment to the problem of understanding how a local moment can survive in a metallic environment.

1.6 The Anderson Model ($U \neq 0$)

To describe a local moment in the framework of the Anderson model (Anderson, 1961), which we introduced in section 1.4, we need to include the Coulomb interaction U between the electrons in the impurity ion d states,

$$U = \int \phi_{\mathrm{d}}^*(\mathbf{r})\phi_{\mathrm{d}}^*(\mathbf{r}')\frac{e^2}{|\mathbf{r} - \mathbf{r}'|}\phi_{\mathrm{d}}(\mathbf{r})\phi_{\mathrm{d}}(\mathbf{r}')\,d\mathbf{r}\,d\mathbf{r}'. \tag{1.65}$$

In atomic d shells this Coulomb interaction can be very large, of the order of 30 eV. For 3d electrons in an ion in a metallic environment it might

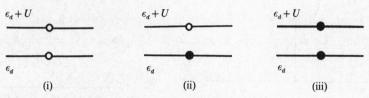

Figure 1.7 Configurations for the impurity state (i) $E = 0$, (ii) $E = \epsilon_{d,\uparrow(\downarrow)}$, (iii) $E = 2\epsilon_d + U$.

be expected to be reduced somewhat, partly due to the delocalization of the orbital, and partly due to screening by the other electrons. Theoretical estimates (Herbst, Watson & Wilkins, 1978) and experimental results based on photoemission and absorption spectra, which we shall discuss fully in chapter 9, give values in the range 1–7 eV. The same considerations apply to rare earth ions though, because the 4f levels are more compact and closer to the nucleus (see figure 1.4), values at the higher end of the range are to be expected in these cases.

The Hamiltonian for the interacting Anderson model consists of (1.44), which describes the d levels of the impurity hybridized with the conduction electrons, plus the interaction term $U n_{d,\uparrow} n_{d,\downarrow}$,

$$H = \sum_\sigma \epsilon_d n_{d,\sigma} + U n_{d,\uparrow} n_{d,\downarrow} + \sum_{\mathbf{k},\sigma} \epsilon_{\mathbf{k}} c^\dagger_{\mathbf{k},\sigma} c_{\mathbf{k},\sigma}$$

$$+ \sum_{\mathbf{k},\sigma} (V_{\mathbf{k}} c^\dagger_{d,\sigma} c_{\mathbf{k},\sigma} + V^*_{\mathbf{k}} c^\dagger_{\mathbf{k},\sigma} c_{d,\sigma}),$$

$$(1.66)$$

where we have taken the simplest case of a non-degenerate 'd' orbital, which has at the most double occupancy with a spin $\uparrow$ and a spin $\downarrow$ electron. To get some insight into this model let us consider first of all the case $V_{\mathbf{k}} = 0$, which is a case in which the model can be trivially solved, because the localized d states are uncoupled from the conduction electrons. There are three total energy configurations for the d states: (i) zero occupation with a total energy $E_0 = 0$; (ii) single occupation by a spin σ with a total energy $E_{1,\sigma} = \epsilon_d$ where $\sigma = \uparrow, \downarrow$; (iii) double occupation with a spin $\uparrow$ and a spin $\downarrow$ electron with a total energy $E_2 = 2\epsilon_d + U$ (see figure 1.7). If the ground state corresponds to single occupation then the state has two-fold degeneracy corresponding to spin $\frac{1}{2}$. It will have an associated magnetic moment which will give a Curie law contribution to the susceptibility. The other two configurations are non-degenerate and consequently have no magnetic moment. In this 'atomic' limit the condition for a 'local moment' to exist is that the singly occupied configuration lies lowest, which requires $\epsilon_d < \epsilon_F$, so that it is favourable to add one electron, and $\epsilon_d + U > \epsilon_F$ so that it is unfavourable to add a second. The question arises as to what happens to this local moment when the local d states and the conduction states are mixed for $V_k \neq 0$. We show in the next section that in this regime, if V_k is sufficiently small, then the Anderson model is equivalent to the s-d model.

1.7 Relation between the Anderson and s-d Models

If the ground state configuration of the Anderson model for $V_k = 0$ is the singly occupied one, then the other configurations are higher excited states. We can attempt to derive an effective Hamiltonian by taking into account virtual excitations to these excited states within lowest order perturbation theory. If we write the total wavefunction ψ as the sum of terms ψ_0, ψ_1 and ψ_2, where ψ_n is the component in which the 'd' occupation is n, then the Schrödinger equation $H\psi = E\psi$ can be expressed in the form

$$\begin{bmatrix} H_{00} & H_{01} & H_{02} \\ H_{10} & H_{11} & H_{12} \\ H_{20} & H_{21} & H_{22} \end{bmatrix} \begin{bmatrix} \psi_0 \\ \psi_1 \\ \psi_2 \end{bmatrix} = E \begin{bmatrix} \psi_0 \\ \psi_1 \\ \psi_2 \end{bmatrix}, \qquad (1.67)$$

where $H_{nn'} = P_n H P_{n'}$ and P_n is a projection operator on to the subspace with d occupation n. As there is no term in the Hamiltonian in which two d electrons are simultaneously removed or added then $H_{02} = H_{20} = 0$. In terms of the occupation numbers of the d states the projection operators are

$$P_0 = (1 - n_{d,\uparrow})(1 - n_{d,\downarrow}), \qquad P_1 = n_{d,\uparrow} + n_{d,\downarrow} - 2n_{d,\uparrow}n_{d,\downarrow},$$

$$P_2 = n_{d,\uparrow}n_{d,\downarrow}. \qquad (1.68)$$

The off diagonal matrix elements arise from the hybridization term, and are given by

$$H_{10} = \sum_{k,\sigma} V_k c_{d,\sigma}^\dagger (1 - n_{d,-\sigma}) c_{k,\sigma}, \qquad H_{12} = \sum_{k,\sigma} V_k c_{d,\sigma}^\dagger n_{d,-\sigma} c_{k,\sigma}, \qquad (1.69)$$

and the matrix is Hermitian so $H_{12} = H_{21}^\dagger$ and $H_{01} = H_{10}^\dagger$.

We are interested in an effective Hamiltonian in the local moment limit in which the d state is singly occupied so we eliminate ψ_0 and ψ_2 from (1.67) to obtain

$$[H_{11} + H_{12}(E - H_{22})^{-1}H_{21} + H_{10}(E - H_{00})^{-1}H_{01}]\psi_1 = E\psi_1. \qquad (1.70)$$

So far no approximation has been used and diagonalization of (1.70) will give all the energy states of the full Hamiltonian. Using (1.69) the second term on the left hand side of (1.70) can be written in the form,

$$H_{12}(E - H_{22})^{-1}H_{21} =$$

$$\sum_{\mathbf{k},\mathbf{k}'\sigma\sigma'} \frac{-V_{\mathbf{k}}^*V_{\mathbf{k}'}}{(U + \epsilon_{\mathrm{d}} - \epsilon_{\mathbf{k}'})} \left(1 - \frac{(E - \epsilon_{\mathrm{d}} - H_0)}{(U + \epsilon_{\mathrm{d}} - \epsilon_{\mathbf{k}'})}\right)^{-1} c_{\mathbf{k},\sigma}^\dagger c_{\mathbf{k}',\sigma'} c_{\mathrm{d},\sigma} c_{\mathrm{d},\sigma'}^\dagger n_{\mathrm{d},-\sigma'},$$

$$(1.71)$$

where H_0 is the conduction electron Hamiltonian (1.2). The effects of virtual excitations from the $n_{\mathrm{d}} = 1$ ground state to the $n_{\mathrm{d}} = 2$ subspace can be taken into account to lowest order in $V_{\mathbf{k}}$ by neglecting the second term in brackets in (1.71). Then looking at the term $\sigma =\downarrow$, $\sigma' =\uparrow$ and using the fact that in the $n_{\mathrm{d}} = 1$ subspace $c_{\mathrm{d},\uparrow}^\dagger c_{\mathrm{d},\downarrow}$ can be replaced by the spin operator S^+ for spin $\frac{1}{2}$, we find that (1.71) corresponds to the first term in (1.64). Similarly the term $\sigma =\uparrow$, $\sigma' =\downarrow$ gives the second term of (1.64). The terms $\sigma =\uparrow$, $\sigma' =\uparrow$, and $\sigma =\downarrow$, $\sigma' =\downarrow$ can be shown to correspond to the third term in (1.64) together with a pure potential scattering term by using the relations,

$$S_z = \frac{1}{2}(n_{\mathrm{d},\uparrow} - n_{\mathrm{d},\downarrow}), \qquad n_{\mathrm{d},\uparrow} + n_{\mathrm{d},\downarrow} = 1, \qquad n_{\mathrm{d},\uparrow}n_{\mathrm{d},\downarrow} = 0, \quad (1.72)$$

which hold in the $n_{\mathrm{d}} = 1$ subspace.

Treating the third term in (1.70) in a similar way (this will involve virtual excitations to the $n_{\mathrm{d}} = 0$ subspace) and collecting all the terms together it can be seen that to lowest order in $V_{\mathbf{k}}$ (1.70) is equivalent to the s-d model with an effective exchange coupling $J_{\mathbf{k},\mathbf{k}'}$ given by

$$J_{\mathbf{k},\mathbf{k}'} = V_{\mathbf{k}}^* V_{\mathbf{k}'} \left\{ \frac{1}{(U + \epsilon_{\mathrm{d}} - \epsilon_{\mathbf{k}'})} + \frac{1}{(\epsilon_{\mathbf{k}} - \epsilon_{\mathrm{d}})} \right\}, \qquad (1.73)$$

together with a potential scattering term,

$$\sum_{\mathbf{k},\mathbf{k}',\sigma,\sigma'} K_{\mathbf{k},\mathbf{k}'} c_{\mathbf{k},\sigma}^\dagger c_{\mathbf{k}',\sigma}, \qquad (1.74)$$

where

$$K_{\mathbf{k},\mathbf{k}'} = \frac{V_{\mathbf{k}}^* V_{\mathbf{k}'}}{2} \left\{ \frac{1}{(\epsilon_{\mathbf{k}} - \epsilon_{\mathrm{d}})} - \frac{1}{(U + \epsilon_{\mathrm{d}} - \epsilon_{\mathbf{k}'})} \right\}. \qquad (1.75)$$

These are valid for $|\epsilon_{\mathbf{k}}| \ll |\epsilon_{\mathrm{d}} - \epsilon_{\mathrm{F}}|$ and $|\epsilon_{\mathbf{k}}| \ll |U + \epsilon_{\mathrm{d}} - \epsilon_{\mathrm{F}}|$. (We have changed notation at this point from (1.11) for the potential scattering term so as to avoid possible confusion of $V_{\mathbf{k},\mathbf{k}'}$ with $V_{\mathbf{k}}$.) Hence we see that in the local moment regime $U + \epsilon_{\mathrm{d}} > \epsilon_{\mathrm{F}}$, $\epsilon_{\mathrm{d}} < \epsilon_{\mathrm{F}}$ the effective exchange coupling between the localized spin and conduction electrons is antiferromagnetic for scattering of conduction electrons in the region of the Fermi level $\epsilon_{\mathbf{k}} \sim \epsilon_{\mathrm{F}}$, $\epsilon_{\mathbf{k}'} \sim \epsilon_{\mathrm{F}}$. A contribution to the exchange coupling $J_{\mathbf{k},\mathbf{k}'}$ can also arise from the exchange terms in the Coulomb interaction between the impurity d electron and the conduction electrons, which we have not considered explicitly. This is normally ferromagnetic in sign. The antiferromagnetic contribution will be expected to dominate when there is a tendency for the ion to display valence instabilities, because in these cases the levels lie closer to the Fermi level. The results (1.73)–(1.75) were first obtained by Schrieffer & Wolff (1966), who derived them using a canonical transformation.

1.8 Parameter Regimes of the Anderson Model

We have established that there is a parameter regime $\epsilon_{\mathrm{d}} + U \gg \epsilon_{\mathrm{F}}$, $\epsilon_{\mathrm{d}} \ll \epsilon_{\mathrm{F}}$ with $|\epsilon_{\mathrm{d}} + U - \epsilon_{\mathrm{F}}|$, $|\epsilon_{\mathrm{F}} - \epsilon_{\mathrm{d}}| \gg \Delta$ in which the Anderson model has a *local moment* and an antiferromagnetic exchange interaction with the conduction electrons. Anderson in his original paper on this model set out to obtain a precise criterion for a local moment by treating the interaction term in the Hartree–Fock approximation. In this approximation the interaction is replaced by an effective one-body term,

$$U n_{\mathrm{d},\uparrow} n_{\mathrm{d},\downarrow} \longrightarrow U n_{\mathrm{d},\uparrow} \langle n_{\mathrm{d},\downarrow} \rangle + U \langle n_{\mathrm{d},\uparrow} \rangle n_{\mathrm{d},\downarrow} - U \langle n_{\mathrm{d},\uparrow} \rangle \langle n_{\mathrm{d},\downarrow} \rangle, \quad (1.76)$$

where $\langle n_{\mathrm{d},\uparrow} \rangle$, $\langle n_{\mathrm{d},\downarrow} \rangle$ are expectation values of the operators $n_{\mathrm{d},\downarrow}$, $n_{\mathrm{d},\downarrow}$ which have to be calculated self-consistently. With this approximation the model is formally equivalent to the non-interacting model (1.44) with an effective d level, $\epsilon_{\mathrm{d},\sigma} = \epsilon_{\mathrm{d}} + U \langle n_{\mathrm{d},-\sigma} \rangle$. Self-consistent solutions for $\langle n_{\mathrm{d},\uparrow} \rangle$ and $\langle n_{\mathrm{d},\downarrow} \rangle$ were calculated using (1.59) for each spin component,

$$\langle n_{\mathrm{d},\sigma} \rangle = \frac{1}{2} - \frac{1}{\pi} \tan^{-1} \left(\frac{\tilde{\epsilon}_{\mathrm{d},\sigma} - \epsilon_{\mathrm{F}}}{\Delta} \right), \quad (1.77)$$

where $\tilde{\epsilon}_{\mathrm{d},\sigma}$ satisfies $\epsilon - \epsilon_{\mathrm{d}} + U \langle n_{\mathrm{d},-\sigma} \rangle - \Lambda(\epsilon) = 0$. The onset of solutions which have $\langle n_{\mathrm{d},\uparrow} \rangle \neq \langle n_{\mathrm{d},\downarrow} \rangle$, which leads to a Curie term in the impurity susceptibility, was taken to define the local moment regime. These Hartree–Fock solutions are, however, not satisfactory because they correspond to breaking a local symmetry of the Hamiltonian, in conflict with rigorously established theorems. In the exact solution $\langle n_{\mathrm{d},\uparrow} \rangle = \langle n_{\mathrm{d},\downarrow} \rangle$, so the criterion used by Anderson is not one that can be used generally, though the Hartree–Fock results do indicate approximately the region in which the model is equivalent to an s-d model. The experimental criterion for a local moment is that a Curie–Weiss term (1.62) is to be found in the impurity susceptibility, and this can depend on the temperature range being considered. Early experimental observations of *Fe* impurities in various *Mo–Nb* and *Mo–Re* alloys are shown in figure 1.8.

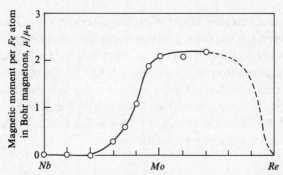

Figure 1.8.   The magnetic moment in μ_B of Fe in various Mo–Nb and Mo–Re alloys as a function of alloy composition (Clogston et al, 1962).

The concepts leading to a fuller understanding of the impurity contribution to the susceptibility will be developed in chapters 3–5, and we shall compare theory and experiment in chapter 9. Theoretically we define the local moment regime as the regime in which the s-d model is a valid approximation to the Anderson model.

Starting from the local moment regime, and then letting either ϵ_d or $\epsilon_d + U$ approach the Fermi level so that $\epsilon_d - \epsilon_F$ or $\epsilon_d + U - \epsilon_F$ become comparable with the resonance width Δ (see figure 1.9 (ii) and (iv)), the charge fluctuations of the impurity become important, and they can no longer be treated as virtual processes so the Schrieffer–Wolff transformation to the s-d model breaks down. This regime is known as the *intermediate valence* regime and is of interest for certain rare earth compounds, for example SmB_6, where the f levels lie near the Fermi level. The Sm ions do not have an integral number of f electrons and are in an ionic state which is a dynamic admixture of Sm^{2+} and Sm^{3+}, and hence they are described as being in a state of intermediate valency. This parameter regime is more of interest for the lattice models rather than the impurity models, and we shall consider it later in chapter 10.

Finally there are two non-magnetic regimes, one in which the impurity level is predominately in the state with no electrons, $\epsilon_d - \epsilon_F \gg \Delta$, sometimes known as the *empty orbital regime*; the other in the state of double occupancy $\epsilon_d + U - \epsilon_F \ll \Delta$. These regimes are probably of least interest because the levels are not close enough to the Fermi level for charge fluctuations to be important, and the ground state is non-degenerate for $V_k = 0$ so there are no real problems in applying perturbation theory in the hybridization parameter V_k.

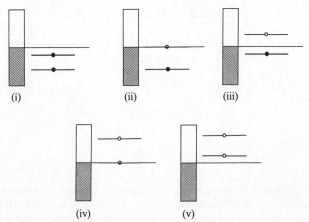

Figure 1.9. Various possible ground states for the Anderson model for $V_{\mathbf{k}} = 0$: (i) and (v) non-magnetic, (ii) and (iv) mixed valent, and (iii) magnetic.

1.9 The Ionic Model

The s-d model (1.64) and the non-degenerate Anderson model (1.66) are the simplest models for the investigation of the effects of magnetic impurities in simple metals, and most theoretical work has been based on them. We know, however, for impurities with unfilled d or f shells that we should take into account the orbital degeneracy of these states, the Hund's rule couplings, spin orbit interactions and local environment effects. At high temperatures the Curie term which is observed in the susceptibility for magnetic impurities is very close to that for the ground and low lying configurations of the d and f shells in the isolated impurity ion. This suggests a way of taking these interactions into account would be to start with the low lying ionic configurations of the isolated ion, and then consider how they are modified when they are mixed with the conduction electrons of the host metal, in the spirit of the Anderson model. If we denote the multiplets of the configurations of the incomplete shells of the impurity ion by $|n, \mathbf{m}_n\rangle$, where n indicates the number of electrons in the shell and $\mathbf{m}_n$ the set of quantum numbers characterizing the multiplet, and the corresponding energy by $E_{n,\mathbf{m}_n}$, then a zero order model of the impurity plus host metal is

$$H_0 = \sum_{n,\mathbf{m}_n} E_{n,\mathbf{m}_n} |n, \mathbf{m}_n\rangle\langle n, \mathbf{m}_n| + \sum_{\mathbf{k},\sigma} \epsilon_{\mathbf{k}} c_{\mathbf{k},\sigma}^{\dagger} c_{\mathbf{k},\sigma}, \qquad (1.78)$$

where in the Dirac bra-ket notation $|n, \mathbf{m}_n\rangle\langle n, \mathbf{m}_n|$ plays the role of

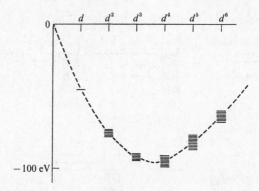

Figure 1.10 Sketch of the atomic level system for a transition metal d shell. The dashed curve is the parabola $E_n = (\epsilon_d - \mu)n + Un(n-1)/2$ (Hubbard, 1964).

a projection operator for the state $|n, \mathbf{m}_n\rangle$. If we describe the multiplets within the Russell–Saunders coupling scheme then $\mathbf{m}_n$ will denote the total orbital angular momentum quantum number l, the total spin quantum number s, and the total spin plus orbital angular momentum quantum number j. Alternatively if the local crystal field environment is taken into account then $\mathbf{m}_n$ will correspond to the classification of the states in terms of the irreducible representations of the crystal field symmetry group.

The one-electron hybridization term (1.43) between the d or f electrons and the conduction electrons will have matrix elements connecting the states $|n, \mathbf{m}_n\rangle$ and $|n', \mathbf{m}'_{n'}\rangle$ for $n' = n \pm 1$. This mixing term can be written in the form,

$$H_{\mathrm{mixing}} = \sum_{\mathbf{k}\sigma n} \sum_{\mathbf{m}_{n+1}\mathbf{m}'_n} V^{\mathbf{k}\sigma}_{\mathbf{m}_{n+1}\mathbf{m}'_n} |n+1, \mathbf{m}_{n+1}\rangle\langle n, \mathbf{m}'_n| c_{\mathbf{k},\sigma}$$
$$+ V^{*\mathbf{k}\sigma}_{\mathbf{m}_{n+1}\mathbf{m}'_n} c^{\dagger}_{\mathbf{k},\sigma} |n, \mathbf{m}'_n\rangle\langle n+1, \mathbf{m}_{n+1}|.$$

(1.79)

The Hamiltonian H_0 (1.78) plus the mixing interaction H_{mixing} (1.79) is known as the *ionic model* (Hirst, 1978), and is essentially a generalization of the Anderson model (1.44). The model looks rather complicated as it requires a number of parameters to specify it. If we neglect to first order the Hund's rules and other splittings within a configuration the configuration energy E_n for n electrons is given by

$$E_n = (\epsilon_d - \mu)n + \frac{n(n-1)}{2}U.$$

(1.80)

A plot of E_n against n is shown in figure 1.10 and we see that these energies are, in general, widely spaced with interconfigurational splittings of the order of U and it will be sufficient to take the three lowest configurations into account, the ground state configuration and the configurations with one more and one less electron. For $\mu - \epsilon_d = Un$, the configurations with energies E_n and E_{n+1} are degenerate and this corresponds to a mixed valence or intermediate valence situation with an excitation from E_n to E_{n+1} in the region of the Fermi level. The splittings between the configurations are such that in calculating the thermodynamic behaviour it is usually sufficient to consider only the lowest multiplet of each of these configurations. The higher states within a configuration are seen in the final states of photoemission spectra and so have to be taken into account when calculating these spectra. This will be discussed more fully in chapter 9.

It is possible to express the non-degenerate Anderson model in the form (1.78)–(1.79). We noted earlier that for $V_\mathbf{k} = 0$ that the localized d states could be classified in terms of the many electron configurations, $|0,0\rangle$, $|1,\uparrow\rangle$, $|1,\downarrow\rangle$ and $|2,\uparrow\downarrow\rangle$ with energies E_0, $E_{1,\uparrow}$, $E_{1,\downarrow}$, E_2 respectively, where $E_{1,\sigma} - E_0 = \epsilon_{d,\sigma}$ and $E_2 - E_0 = 2\epsilon_d + U$. Matrix elements of the creation operators between these states are

$$\langle 1,\uparrow |c_{d,\uparrow}^\dagger|0,0\rangle = 1, \qquad \langle 2,\uparrow\downarrow |c_{d,\uparrow}^\dagger|1,\downarrow\rangle = 1,$$
$$\langle 1,\downarrow |c_{d,\downarrow}^\dagger|0,0\rangle = 1, \qquad \langle 2,\uparrow\downarrow |c_{d,\downarrow}^\dagger|1,\uparrow\rangle = -1, \tag{1.81}$$

where the -1 in the last term arises from the sign in the definition of fermion creation and annihilation operators. Hence, in terms of the many body states, the creation operators can be expressed as

$$c_{d,\uparrow}^\dagger = |1,\uparrow\rangle\langle 0,0| + |2,\uparrow\downarrow\rangle\langle 1,\downarrow|, \quad c_{d,\downarrow}^\dagger = |1,\downarrow\rangle\langle 0,0| - |2,\uparrow\downarrow\rangle\langle 1,\uparrow|,$$
$$\tag{1.82}$$

with similar expressions for the annihilation operators, on taking the Hermitian conjugate. We can substitute these into the Anderson model (1.44) to express it in the form (1.78)–(1.79). In the case where the Coulomb interaction U is very large, the doubly occupied configuration is a very high energy state, and it might be a reasonable approximation to eliminate this configuration and take the limit $U \to \infty$, retaining only the two configurations $|0,0\rangle$ and $|1,\sigma\rangle$. In this limit the Anderson model can be written in the form,

$$H = E_0|0,0\rangle\langle 0,0| + \sum_\sigma E_{1,\sigma}|1,\sigma\rangle\langle 1,\sigma|$$
$$+ \sum_{\mathbf{k},\sigma} \epsilon_\mathbf{k} c_{\mathbf{k},\sigma}^\dagger c_{\mathbf{k},\sigma} + \sum_{\mathbf{k},\sigma}(V_\mathbf{k}|1,\sigma\rangle\langle 0,0|c_{\mathbf{k},\sigma} + V_\mathbf{k}^* c_{\mathbf{k},\sigma}^\dagger|0,0\rangle\langle 1,\sigma|).$$
$$\tag{1.83}$$

There is a generalization of this two configuration model which is applicable to many rare earth impurities. Let us consider the case of 4f electrons in the configurations $4f^n$ and $4f^{n+1}$ which differ by one electron, and take the lowest multiplet of each. To be more specific consider the case of a Ce ion with configurations $n = 0$ (Ce^{4+}) and $n = 1$ (Ce^{3+}). The lowest lying level of the $4f^1$ configuration has orbital and spin angular momentum quantum numbers $l = \frac{5}{2}$, $s = \frac{1}{2}$, and this is split by the spin-orbit interaction such that the state with total angular momentum quantum number $j = l - s = \frac{3}{2}$ lies lowest. This multiplet has a degeneracy $2j + 1$ corresponding the different eigenvalues m of the azimuthal component of angular momentum j_z. We will denote a particular state by $|1, m\rangle$ and its energy by $E_{1,m}$. The $4f^0$ corresponds to an empty 4f shell which we shall denote by $|0, 0\rangle$ and is non-degenerate. In using these states it will be convenient to change the basis of the conduction electron states. Instead of classifying these states by $\mathbf{k}$ and σ we expand the Bloch states in terms of partial waves about the impurity site. Each component of angular momentum l can be combined with the spin component so that the states are eigenstates of total angular momentum operators $\mathbf{j}$ and j_z about the impurity site. The one-body hybridization term (1.43), on the assumption of spherical symmetry about the impurity site, will mix only with conduction states of the same j and m. The conduction states with other values of j will only be weakly coupled due to deviations from spherical symmetry. If these terms are neglected then the conduction states with other values of j play only a passive role, and need not be included explicitly. The Hamiltonian can then be written in the form,

$$H = E_0|0, 0\rangle\langle 0, 0| + \sum_m E_{1,m}|1, m\rangle\langle 1, m|$$
$$+ \sum_{k,m} \epsilon_k c_{k,m}^\dagger c_{k,m} + \sum_{k,m}(V_k|1, m\rangle\langle 0, 0|c_{k,m} + V_k^* c_{k,m}^\dagger|0, 0\rangle\langle 1, m|),$$

$$(1.84)$$

where the summation is over the $2j + 1$ values of m. We will denote $2j + 1$ by N, where N is the degeneracy factor. The $U = \infty$ Anderson model (1.83) is formally equivalent to (1.84) with $N = 2$. The same model can be used for the Yb ion in terms of holes in the f shell rather than electrons with $j = \frac{7}{2}$ or $N = 8$. The fact that $1/N$ is significantly less than unity means that it is a suitable expansion parameter and special techniques have been devised to exploit this which are described fully in chapters 7 and 8. With slight modifications this model can be used for one of the other ions Sm^{2+}/Sm^{3+} which displays significant interconfigurational mixing in compounds such as SmB_6 (V_k becomes dependent on m, $n = 6$). One exception is the case of Tm^{2+}/Tm^{3+}, $n = 12$. The lowest configurations for $4f^{12}$ and $4f^{13}$ are both degenerate and consequently the Hamiltonian has a more complicated form than (1.84). Models for this situation will be discussed in chapter 9.

In working with the ionic model it is sometimes useful to adopt the Hubbard X-operator notation, $X_{p;q}$ to denote $|p\rangle\langle q|$, where $|p\rangle$ and $|q\rangle$ are states in a complete set of many particle states (Hubbard, 1965). The diagonal elements $X_{p;p}$ are projection operators for the state $|p\rangle$ and the completeness relation is $\sum_p X_{p;p} = I$. Products of the operators can be contracted according to

$$X_{p;q}X_{p';q'} = \delta_{p',q}X_{p;q'}. \tag{1.85}$$

We can illustrate this in the case of the $U = \infty$ Anderson model (1.83) with the states $|0,0\rangle$ and $|1,\sigma\rangle$. In this notation (1.83) becomes

$$H = E_0 X_{0,0;0,0} + \sum_\sigma E_{1,\sigma} X_{1,\sigma;1,\sigma}$$

$$+ \sum_{\mathbf{k},\sigma} \epsilon_{\mathbf{k}} c_{\mathbf{k},\sigma}^\dagger c_{\mathbf{k},\sigma} + \sum_{\mathbf{k},\sigma} (V_{\mathbf{k}} X_{1,\sigma;0,0} c_{\mathbf{k},\sigma} + V_{\mathbf{k}}^* c_{\mathbf{k},\sigma}^\dagger X_{0,0;1,\sigma}).$$

$$\tag{1.86}$$

Operators which involve the creation or annihilation of an odd number of electrons anticommute with creation and annihilation operators for the conduction electrons. In terms of the creation and annihilation operators for the localized d states the operators $X_{1,\sigma;0,0}$ and $X_{0,0;1,\sigma}$ are given by

$$X_{1,\sigma;0,0} = c_{\mathrm{d},\sigma}^\dagger (1 - n_{\mathrm{d},-\sigma}) \quad \text{and} \quad X_{0,0;1,\sigma} = c_{\mathrm{d},\sigma}(1 - n_{\mathrm{d},-\sigma}). \tag{1.87}$$

In terms of the X operators the spin operators for spin $\frac{1}{2}$ are

$$S^+ = X_{\uparrow\downarrow}, \quad S^- = X_{\downarrow\uparrow}, \quad S_z = \frac{1}{2}(X_{\uparrow\uparrow} - X_{\downarrow\downarrow}), \tag{1.88}$$

where we have simplified the notation because all the states have the same number of electrons.

1.10 The Coqblin–Schrieffer Model

We can derive the equivalent of the s-d model for these more general models using the approach described in section 1.7. We will illustrate this in the case of the model (1.84) for the rare earth ions Ce and Yb. The degenerate state will be taken to be the ground state and virtual fluctuations to the non-degenerate state will be taken into account to lowest order in the hybridization matrix element V_k. We use equation (1.67) with $H_{12} = 0$, as we have no doubly occupied state to take into account. We have $P_0 = X_{0,0;0,0}$ and $P_1 = 1 - X_{0,0;0,0}$ and the off diagonal terms H_{10}, H_{01} arise from the mixing term (1.79). Using the algebra of the Hubbard operators (1.85), and the same form of approximation as in (1.71), we obtain an effective Hamiltonian in the form,

$$H = \sum_{k,m} \epsilon_k c_{k,m}^\dagger c_{k,m} + \sum_{k,k',m,m'} J_{k,k'} X_{m,m'} c_{k',m'}^\dagger c_{k,m}, \qquad (1.89)$$

where

$$J_{k,k'} = \frac{V_k^* V_{k'}}{(\epsilon_k - E_1 + E_0)}. \qquad (1.90)$$

This model which was first derived by Coqblin & Schrieffer (1969) is a natural generalization of the s-d model (1.64). It is invariant under special unitary transformations $SU(N)$ in the 'spin' space, the s-d model corresponds to the $SU(2)$ case. To be equivalent exactly to the s-d model for $N = 2$ we must add a potential scattering term so that the model becomes

$$H = \sum_{k,m} \epsilon_k c_{k,m}^\dagger c_{k,m}$$
$$+ \sum_{k,k',m,m'} J_{k,k'} X_{m,m'} c_{k',m'}^\dagger c_{k,m} - \frac{1}{N} \sum_{k,k',m} J_{k,k'} c_{k',m}^\dagger c_{k,m}.$$

$$(1.91)$$

The main difference between the s-d exchange model for spin S and the Coqblin–Schrieffer model for a 'spin' $j = S \, (> \frac{1}{2})$ is that in the scattering the change of the z-component of angular momentum of the impurity is restricted to $\Delta m = 0, \pm 1$ in the s-d model, but is unrestricted in the Coqblin–Schrieffer model. The behaviour of this model will be fully discussed in chapters 7 and 8.

2

Resistivity Calculations and the Resistance Minimum

2.1 Multiple Impurity Scattering

One of the most important ways in which impurities affect the behaviour of a metal is in their contribution to the electrical resistivity at low temperatures. The conductivity is infinite for electrons in a perfect lattice as there is no scattering, and consequently no current dissipation. At finite temperatures the scattering with phonons usually provides the most important mechanism for the dissipation of the electron current. At low temperatures there are few phonons, so the scattering by impurities and defects becomes more important, and becomes the dominant dissipative mechanism as $T \to 0$, leading to a finite conductivity in this limit. In calculating the electrical conductivity due to impurity scattering we shall need to consider a finite concentration of impurities, and take into account the scattering of electrons by different impurities. To consider this, we must generalize our models to include a finite concentration of impurities. If we generalize the potential scattering model (1.11) for N_{imp} impurities at sites $\mathbf{R}_i$, by replacing $V_{\mathrm{imp}}^{\mathrm{eff}}(\mathbf{r})$ by $\sum_i V_{\mathrm{imp}}^{\mathrm{eff}}(\mathbf{r} - \mathbf{R}_i)$, then we obtain a multi-impurity Hamiltonian in terms of Bloch states in the form,

$$H = \sum_{\mathbf{k},\sigma} \epsilon_{\mathbf{k}} c_{\mathbf{k},\sigma}^{\dagger} c_{\mathbf{k},\sigma} + \sum_{\mathbf{k},\mathbf{q},\sigma} \sum_{\mathbf{R}_i} e^{i\mathbf{q}\cdot\mathbf{R}_i} V_{\mathbf{q}} c_{\mathbf{k}+\mathbf{q},\sigma}^{\dagger} c_{\mathbf{k},\sigma} . \qquad (2.1)$$

We can again consider a perturbational approach to calculate G in powers of V from (1.20) for a given distribution of impurities. As a distribution in any real situation is likely to be random there is no point in attaching any significance to a particular distribution. The results for any physical quantity will be averaged over an ensemble of random

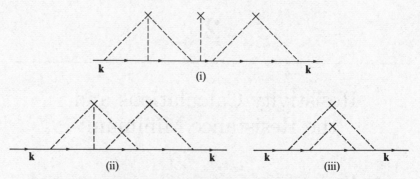

Figure 2.1 Diagrams for the configurationally averaged Green's function $\langle G_{\mathbf{kk}}^+(\epsilon)\rangle$ with multiple impurity scattering. The separate xs in each diagram indicate distinct impurities. The wave vector $\mathbf{k}$ is conserved in the overall scattering at each impurity site.

distributions of impurity sites. Terms of first order in V, when configurationally averaged, give

$$V(\mathbf{q})\langle e^{i\mathbf{q}\cdot\mathbf{R}_i}\rangle_{\mathrm{cfig}} = N_{\mathrm{imp}}V(0)\delta_{\mathbf{q},0} \tag{2.2}$$

and of second order,

$$V(\mathbf{q})V(\mathbf{q}')\langle e^{i\mathbf{q}\cdot\mathbf{R}_i}e^{i\mathbf{q}'\cdot\mathbf{R}_j}\rangle_{\mathrm{cfig}} = N_{\mathrm{imp}}V(\mathbf{q})V(\mathbf{q}')\delta_{\mathbf{q}+\mathbf{q}',0}$$
$$+N_{\mathrm{imp}}(N_{\mathrm{imp}}-1)V(0)^2\delta_{\mathbf{q},0}\delta_{\mathbf{q}',0}. \tag{2.3}$$

In (2.3) the first term on the right hand side arises from $\mathbf{R}_j = \mathbf{R}_i$, where the electron is scattered by the same impurity. For the second term $\mathbf{R}_j \neq \mathbf{R}_i$ and the scattering involves two different impurities. For a finite concentration c_{imp} of impurities $N_{\mathrm{imp}} \gg 1$ and the factor $N_{\mathrm{imp}}(N_{\mathrm{imp}}-1)$ can be replaced by N_{imp}^2. The scattering processes, which involve only a single impurity, can be represented by the diagrams of figure 1.1 with an overall conservation of the $\mathbf{k}$ vector, $\mathbf{k} = \mathbf{k}'$, due to the configurational average. The potential for zero momentum transfer $V(0)$ is uniform and can be absorbed into $\epsilon_{\mathbf{k}}$, and hence put equal to zero in the diagrammatic analysis. Generalizing the configurational average to higher order terms leads to diagrams of the type shown in figure 2.1.

The scattering processes of the type shown in figure 2.1(i), in which the scattering events at any two impurities are disconnected, can be summed to give $\langle G_{\mathbf{k},\mathbf{k}}^+(\epsilon)\rangle$, the configurational average of the Green's function $G_{\mathbf{k},\mathbf{k}}^+(\epsilon)$. These diagrams give a geometric series which on summing gives

$$\langle G_{\mathbf{k},\mathbf{k}}^+(\epsilon)\rangle = \frac{1}{(\epsilon + is - \epsilon_{\mathbf{k}} - \Sigma(\mathbf{k},\epsilon^+))}, \tag{2.4}$$

where the 'self-energy' $\Sigma(\mathbf{k}, \epsilon^+) = c_{imp}\langle\mathbf{k}|T(\epsilon^+)|\mathbf{k}\rangle$. The configurational average only leads to a limited momentum conservation. The Bloch state of a given $\mathbf{k}$ acquires a finite lifetime as can be seen by examining the imaginary part of the self-energy $\Sigma(\mathbf{k}, \epsilon_{\mathbf{k}}^+)$. From equation (A.21) in appendix A, $\langle\mathbf{k}|T(\epsilon^+)|\mathbf{k}\rangle$ for free particles is related to the total scattering cross-section $\sigma(k)$,

$$\text{Im}\,\Sigma(\mathbf{k}, \epsilon_{\mathbf{k}}^+) = c_{imp}\,\text{Im}\,\langle\mathbf{k}|T(\epsilon_{\mathbf{k}}^+)|\mathbf{k}\rangle = -\frac{c_{imp}\hbar^2 k\sigma(k)}{2m}. \qquad (2.5)$$

The term $c_{imp}v_{\mathbf{k}}$, where $v_{\mathbf{k}} = \hbar k/m$ can be interpreted as the rate at which the electrons encounter the impurities so $\tau(k)$, where $\tau(k)^{-1} = c_{imp}v_k\sigma(k)$, has the dimensions of time and can be interpreted as the lifetime in a particular $\mathbf{k}$ state, and $l_k = v_k\tau(k)$, which has the dimensions of length, as the corresponding mean free path. Hence,

$$\text{Im}\,\Sigma(\mathbf{k}, \epsilon_{\mathbf{k}}^+) = -\frac{\hbar}{2\tau(k)}, \qquad (2.6)$$

and corresponds to a finite lifetime. The real part of the self-energy is not so significant and can be absorbed into $\epsilon_{\mathbf{k}}$.

Diagrams of the type shown in figure 2.1(ii), which involve interference in the scattering between two impurity sites, have not been included in the partial summation leading to (2.6). It seems reasonable to expect that these will not be important at low concentrations, and careful examination of the contributions from such diagrams confirms this. They are of the order $1/k_F l_{k_F}$ compared with the diagrams taken into account. As $1/l_{k_F} = c_{imp}\sigma(k_F)$, one can justify the neglect of these terms at low concentrations such that $c_{imp}\sigma(k_F) \ll k_F$. Diagrams of the type shown in figure 2.1(iii) can be included by replacing G_0 by G in the T matrix (1.22), and then calculating G self-consistently. This will also give higher order correction terms, in powers of c_{imp}, to the self-energy (2.5) and, at low concentrations of impurities, these can also be neglected.

One might be tempted on the basis of the simple Drude theory of electron conduction to identify $\tau(k_F)$, defined by equation (2.6), as the transport lifetime τ_{tr} of the conduction electrons and substitute in the expression for the electrical conductivity,

$$\sigma = \frac{ne^2\tau_{tr}}{m}, \qquad (2.7)$$

where n is the number of conduction electrons per unit volume. In general this is not correct as we shall see in the next section.

2.2 Conductivity and the Boltzmann Equation

Here we shall briefly review a derivation of the impurity conductivity based on the Boltzmann equation. This derivation has the advantage of being intuitive and direct. An alternative derivation based on linear response theory and Green's functions, where the approximations used are more explicit, will be given in the section following.

In the Boltzmann theory the electrons are described by a semi-classical distribution function $f_E(\mathbf{k})$ in the presence of a static electric field $\mathbf{E}$. The rate of change of $f_E(\mathbf{k})$ is governed by the equation of dynamic equilibrium,

$$\frac{df_E(\mathbf{k})}{dt} = \left(\frac{\partial \mathbf{k}}{\partial t}\right) \cdot \nabla_k f_E(\mathbf{k}) = \left(\frac{df_E(\mathbf{k})}{dt}\right)_{\text{col}}. \qquad (2.8)$$

where the right hand side of the equation is the rate of change of $f_E(\mathbf{k})$ due to collisions.

The rate of change of the momentum $\hbar\mathbf{k}$ with time in the semi-classical approximation is the acceleration due to the electric field given by

$$\hbar \frac{\partial \mathbf{k}}{\partial t} = -e\mathbf{E}. \qquad (2.9)$$

If it is assumed that the rate of change due to collisions is proportional to the difference between $f_E(\mathbf{k})$ and the equilibrium distribution function in the absence of the field, the Fermi–Dirac distribution function $f(\mathbf{k})$ (defined following (1.6)), then

$$\left(\frac{df_E(\mathbf{k})}{dt}\right)_{\text{col}} = -\frac{(f_E(\mathbf{k}) - f(\mathbf{k}))}{\hbar \tau_1(\mathbf{k})}, \cdot \qquad (2.10)$$

where $\tau_1(\mathbf{k})$ is a relaxation time, to be calculated later.

From equations (2.8), (2.9) and (2.10),

$$f_E(\mathbf{k}) = f(\mathbf{k}) + e\tau_1(\mathbf{k})\,\mathbf{E} \cdot \nabla_k f(\mathbf{k}), \qquad (2.11)$$

where $f_E(\mathbf{k})$ has been replaced by $f(\mathbf{k})$ in the second term on the right hand side of (2.11) as only terms linear in $\mathbf{E}$ are required. The current $\mathbf{j}$ is related to the average velocity $\mathbf{v}_k$ by

$$\mathbf{j} = -e\langle \mathbf{v}_k \rangle = -2e \int f_E(\mathbf{k}) \frac{\hbar\mathbf{k}}{m} \frac{d\mathbf{k}}{(2\pi)^3}, \qquad (2.12)$$

where the factor of 2 is due to the two spin components. As the current in the absence of the field is zero then,

$$\mathbf{j} = -2 \int \tau_1(k) \mathbf{v_k}(\mathbf{E} \cdot \mathbf{v_k}) \frac{\partial f}{\partial \epsilon_\mathbf{k}} \frac{d\mathbf{k}}{(2\pi)^3}, \qquad (2.13)$$

assuming the system to be isotropic and using,

$$\nabla_k f(\mathbf{k}) = \frac{\mathbf{k}}{m} \frac{\partial f}{\partial \epsilon_k}. \qquad (2.14)$$

For an isotropic system $\mathbf{j}$ is parallel to the field so $\mathbf{j} = \sigma(T)\mathbf{E}$, and on averaging the angular integrations in (2.14) we obtain an expression for $\sigma(T)$,

$$\sigma(T) = -\frac{2e^2}{3} \int v_k^2 \tau_1(k) \frac{\partial f}{\partial \epsilon_k} \frac{d\mathbf{k}}{(2\pi)^3}. \tag{2.15}$$

At $T = 0$, $\partial f / \partial \epsilon_k = -\delta(\epsilon_k - \epsilon_F)$ as the Fermi–Dirac function becomes a step function in this limit, then for free electrons ($\epsilon_k = \hbar^2 k^2 / 2m$) we find,

$$\sigma = \frac{ne^2 \tau_1(k_F)}{m}, \tag{2.16}$$

which is the same as (2.7) with $\tau_1(k_F) = \tau_{\mathrm{tr}}$ and $n = k_F^3 / 3\pi^2$, the electron density of the host metal.

To calculate $\tau_1(k)$ the rate of change of $f_E(\mathbf{k})$ due to collisions must be estimated. As the impurities are rigid with no internal modes the scattering is elastic. The scattering rate $W_{\mathbf{k}\mathbf{k}'}$ from the state $\mathbf{k}$ to $\mathbf{k}'$ is derived in terms of the T matrix in appendix A, equation (A.17). This must be multiplied by $f_E(\mathbf{k})$, the average occupation of the state $\mathbf{k}$, and $(1 - f_E(\mathbf{k}))$, the probability that the state $\mathbf{k}'$ is unoccupied, and then summed over the available states $\mathbf{k}'$. A similar term with $\mathbf{k}$ and $\mathbf{k}'$ interchanged gives the rate for the reverse process. Hence the net scattering rate is

$$\left(\frac{df_E(\mathbf{k})}{dt}\right)_{\mathrm{coll}} = -\frac{2\pi c_{\mathrm{imp}}}{\hbar} \int \delta(\epsilon_k - \epsilon_{k'}) \{|T_{\mathbf{k}\mathbf{k}'}|^2 f_E(\mathbf{k})(1 - f_E(\mathbf{k}'))$$

$$-|T_{\mathbf{k}'\mathbf{k}}|^2 f_E(\mathbf{k}')(1 - f_E(\mathbf{k}))\} \frac{d\mathbf{k}'}{(2\pi)^3}. \tag{2.17}$$

As $T_{\mathbf{k}\mathbf{k}'}$ is symmetric, the right hand side of (2.17) can be simplified, and on equating to (2.8) gives

$$\frac{2\pi c_{\mathrm{imp}}}{\hbar} \int \delta(\epsilon_k - \epsilon_{k'})|T_{\mathbf{k}\mathbf{k}'}|^2 (f_E(\mathbf{k}) - f_E(\mathbf{k}')) \frac{d\mathbf{k}'}{(2\pi)^3} = -e\mathbf{E} \cdot \nabla_k f_E(\mathbf{k}), \tag{2.18}$$

which is an integro-differential equation for $f_E(\mathbf{k})$.

The relaxation hypothesis (2.10) provides the solution of this equation to first order in $\mathbf{E}$. Substituting (2.11) into (2.18) and using (2.14), we find that (2.11) is the solution if $\tau_1(k)$ is given by

$$\frac{1}{\tau_1(k)} = 2\pi c_{\mathrm{imp}} \int \delta(\epsilon_k - \epsilon_{k'})|T_{\mathbf{k}\mathbf{k}'}|^2 (1 - \cos\theta') \frac{d\mathbf{k}'}{(2\pi)^3}, \tag{2.19}$$

where it has been assumed that the T matrix depends only on the angle θ' between $\mathbf{k}$ and $\mathbf{k}'$.

For a spherically symmetric scattering potential, and free electrons, it is possible to write the T matrix in terms of the phase shifts for the partial waves corresponding to angular momentum l. Using equations (A.22) and (A.24) in appendix A for $T_{\mathbf{kk'}}$ in terms of the phase shifts, substituting into (2.19) and performing the angular integrations, gives

$$\frac{1}{\tau_1(k)} = \frac{4\pi c_{\text{imp}}}{mk} \sum_{l=1}^{\infty} l \sin^2(\eta_l(k) - \eta_{l-1}(k)). \qquad (2.20)$$

This can be substituted into (2.16) to give an expression for the conductivity in terms of phase shifts, evaluated at the Fermi momentum k_{F}. In general the transport relaxation lifetime $\tau_{\text{tr}}(k)$ differs from the lifetime in a particular $\mathbf{k}$ state, $\tau(k)$, due to the extra factor $1 - \cos\theta'$ in (2.19). The higher angle scattering is more effective in dissipating the momentum of the electrons and, hence, contributes more significantly to the resistance than the smaller angle scattering. For pure s wave scattering, $\eta_0 \neq 0$, $\eta_l = 0$ for $l \neq 0$, and the two relaxation times are the same. This is because the $(1 - \cos\theta')$ factor gives no contribution on performing the angular integration. If there is scattering in only a single channel so that $\eta_l = 0$ except for $l = l_0$, then τ_1 is proportional to τ in this case. (This follows from using (A.22) and (A.26) in (2.5) and comparing with (2.20).)

We can use the results of section 1.3 to estimate the resistance for s wave scattering potential (1.35), substituting (1.39) as the phase shift $\eta_0(\epsilon)$ into (2.20). The inverse lifetime $1/\tau(k_{\text{F}})$ is a maximum, and consequently so is the resistivity, when there is a virtual bound state resonance near the Fermi level so $\eta(\epsilon_{\text{F}}) \sim \pi/2$. To lowest order in V, which corresponds to estimating the scattering in the first Born approximation, we find

$$\frac{1}{\tau(k_{\text{F}})} = \frac{3\pi c_{\text{imp}} n V^2}{2\hbar\epsilon_{\text{F}}}, \qquad (2.21)$$

and for the impurity resistivity R_{imp},

$$R_{\text{imp}} = \frac{3\pi c_{\text{imp}} m V^2}{2e^2 \hbar \epsilon_{\text{F}}}. \qquad (2.22)$$

2.3 Conductivity and Linear Response Theory

The conductivity can also be calculated using diagrammatic perturbation theory. The starting point is the Kubo formula (Kubo, 1957) for the

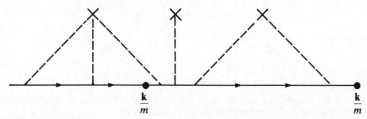

Figure 2.2 A typical diagram contributing to equation (2.26) for the electrical conductivity σ after configurational averaging.

frequency dependent conductivity $\sigma(\omega, T)$ for an isotropic conductor,

$$\sigma(\omega, T) = \frac{1}{3} \int_0^\infty \int_0^\beta \langle \mathbf{j}(0) \cdot \mathbf{j}(t + i\lambda\hbar) \rangle e^{-(i\omega+s)t} \, dt \, d\lambda, \qquad (2.23)$$

in the limit $s \to +0$, where $\langle \hat{O} \rangle$ indicates a thermal average of the operator $\hat{O}$,

$$\langle \hat{O} \rangle = \frac{\text{Tr}\left(e^{-\beta(H-\mu N_0)} \hat{O}\right)}{\text{Tr}\, e^{-\beta(H-\mu N_0)}}, \qquad (2.24)$$

and $\mathbf{j}(t)$ is the current operator in the Heisenberg representation,

$$\mathbf{j}(t) = e^{iHt/\hbar}\left(\frac{-e\mathbf{p}}{m}\right)e^{-iHt/\hbar}, \qquad (2.25)$$

where $\mathbf{p}$ is the linear momentum.

The Kubo formula is based on linear response theory, the current being required to first order in the electric field $\mathbf{E}$. A brief discussion of linear response theory and a derivation of (2.23) is given in appendix B. Within the one-body formalism we can re-express (2.23) in terms of the resolvent Green's functions $G^+(\epsilon)$, defined in the previous chapter, equation (1.15). For the static response ($\omega = 0$) we find

$$\sigma(T) = \frac{e^2\hbar}{12\pi m^2} \sum_{\nu=x,y,z} \text{Tr} \int \sum_{\alpha,\beta=+,-} \alpha\beta \, G^\alpha(\epsilon) p_\nu G^\beta(\epsilon) p_\nu \frac{\partial f}{\partial \epsilon} \, d\epsilon,$$

$$(2.26)$$

where the trace is taken with respect to a complete set of one-electron states (we shall use the conduction states $|\mathbf{k}\rangle$). Details of the derivation of (2.26) from (2.23) are given in appendix B. The result has to be configurationally averaged over a random set of impurity sites.

After configurational averaging the contributions to (2.26) in perturbation theory can be represented in diagrams of the form shown in figure 2.2. These are similar to the diagrams shown in figure 2.1 except that there are two lines to represent the two Green's functions in (2.26). In performing the configurational average these two lines can be linked as momentum can be transmitted from one electron line to the other when

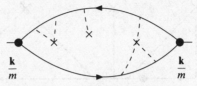

Figure 2.3 The same diagram as in figure 2.2 but drawn as a closed bubble by bending back the second Green's function line.

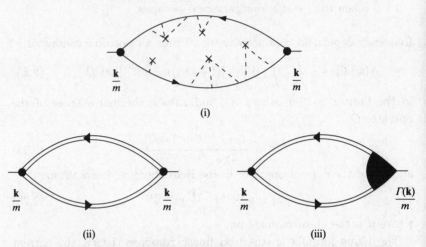

Figure 2.4 (i) A diagram with no cross-links, (ii) the sum of all diagrams without cross terms, where the double line indicates the exact Green's function $\langle G^{+}_{\mathbf{kk}}(\epsilon)\rangle$, and (iii) the sum of all diagrams contributing to (2.26), where $\Gamma^{\alpha\beta}(\mathbf{k})$ represents the vertex part.

the scattering is from the same impurity. In the overall scattering at a single impurity site the momentum $\mathbf{k}$ is still conserved. The filled circles in the diagram represent the two p_ν factors in (2.26). The matrix elements of these with respect to the Bloch states are diagonal, $\langle\mathbf{k}|p_\nu|\mathbf{k}'\rangle = \hbar k_\nu \delta_{\mathbf{k},\mathbf{k}'}$. It is convenient to redraw the diagrams as shown in figure 2.3 by bending back the second Green's function to form a closed bubble.

Diagrams of the form shown in figure 2.4(i), which have no cross-links, correspond to two separately configurationally averaged Green's functions $\langle G_{\mathbf{kk}}(\epsilon)\rangle\langle G_{\mathbf{kk}}(\epsilon)\rangle$. These can be estimated using the results (2.4) and (2.5), derived earlier, which are valid in the low concentration limit. The combination $\langle G^{+}_{\mathbf{kk}}(\epsilon)\rangle\langle G^{-}_{\mathbf{kk}}(\epsilon)\rangle$ is given by

$$\langle G^{+}_{\mathbf{kk}}(\epsilon)\rangle\langle G^{-}_{\mathbf{kk}}(\epsilon)\rangle = \frac{1}{(\epsilon-\epsilon_{\mathbf{k}})^2 + \hbar^2/4\tau^2(k)}. \quad (2.27)$$

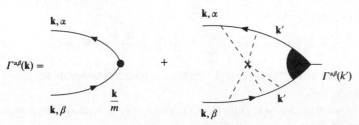

Figure 2.5 The integral equation for the vertex part $\Gamma^{\alpha\beta}(\mathbf{k})$ as a sum of ladder diagrams.

In the low concentration limit $c_{\mathrm{imp}} \to 0$, this term gives a contribution,

$$\langle G_{\mathbf{kk}}^+(\epsilon)\rangle\langle G_{\mathbf{kk}}^-(\epsilon)\rangle \to \frac{\pi\tau(k)}{\hbar}\delta(\epsilon - \epsilon_{\mathbf{k}}). \tag{2.28}$$

This term is proportional to c_{imp}^{-1} as conductivity is infinite in the zero concentration limit. The terms $\langle G_{\mathbf{kk}}^+(\epsilon)\rangle\langle G_{\mathbf{kk}}^+(\epsilon)\rangle$ and $\langle G_{\mathbf{kk}}^-(\epsilon)\rangle\langle G_{\mathbf{kk}}^-(\epsilon)\rangle$ give contributions which are not singular in c_{imp} and, hence, negligible in comparison with the contribution from (2.27) for low concentration. If we evaluate (2.26) using these terms only, then we obtain the result (2.16) at $T = 0$ with $\tau(k_{\mathrm{F}})$ instead of $\tau_1(k_{\mathrm{F}})$, and so differs from the Boltzmann result. This is due to the neglect of cross-linked diagrams which also give terms singular in c_{imp}. The cross-linked diagrams can be absorbed into a vertex part $\Gamma^{\alpha\beta}(\mathbf{k}, \epsilon)$ $(\alpha, \beta = +, -)$, and the set of such diagrams is indicated in figure 2.4(iii). In terms of this vertex part (2.26) can be written in the form,

$$\sigma(T) = \frac{e^2\hbar^3}{12\pi m^2}\sum_{\mathbf{k}}\int\sum_{\alpha,\beta=+,-}\alpha\beta\,\langle G_{\mathbf{kk}}^\alpha(\epsilon)\rangle\langle G_{\mathbf{kk}}^\beta(\epsilon)\rangle\mathbf{k}\cdot\Gamma^{\alpha\beta}(\mathbf{k},\epsilon)\frac{\partial f}{\partial\epsilon}\,d\epsilon. \tag{2.29}$$

In the absence of cross-links $\Gamma^{\alpha\beta}(\mathbf{k}, \epsilon) = \mathbf{k}$.

An integral equation can be set up for $\Gamma^{\alpha\beta}(\mathbf{k}, \epsilon)$ which sums a set of ladder diagrams corresponding to a sequence of multiple scatterings from a single impurity. The equation for $\Gamma^{+-}(\mathbf{k}, \epsilon)$, which is illustrated in figure 2.5, is

$$\Gamma^{+-}(\mathbf{k},\epsilon) = \mathbf{k} + c_{\mathrm{imp}}\sum_{\mathbf{k}'}\langle G_{\mathbf{k}'\mathbf{k}'}^+(\epsilon)\rangle\langle G_{\mathbf{k}'\mathbf{k}'}^-(\epsilon)\rangle|\langle\mathbf{k}|T(\epsilon)|\mathbf{k}'\rangle|^2\Gamma^{+-}(\mathbf{k}',\epsilon). \tag{2.30}$$

If $\mathbf{k}\cdot\Gamma^{+-}(\mathbf{k}, \epsilon) = \gamma(k, \epsilon)k^2$, then taking the scalar product of (2.30) with $\mathbf{k}$, gives an integral equation for $\gamma(k, \epsilon)$,

$$\gamma(k,\epsilon) = 1 + c_{\mathrm{imp}}\sum_{\mathbf{k}'}\langle G_{\mathbf{k}'\mathbf{k}'}^+(\epsilon)\rangle\langle G_{\mathbf{k}'\mathbf{k}'}^-(\epsilon)\rangle|\langle\mathbf{k}|T(\epsilon)|\mathbf{k}'\rangle|^2\frac{\mathbf{k}\cdot\mathbf{k}'}{k^2}\gamma(\epsilon,k'). \tag{2.31}$$

Using (2.28) in the limit $c_{\text{imp}} \to 0$ leads to

$$\gamma(k, \epsilon_{\mathbf{k}}) = 1 + 2\pi c_{\text{imp}} \sum_{\mathbf{k}'} \tau(k) |T_{\mathbf{k}\mathbf{k}'}|^2 \frac{\mathbf{k} \cdot \mathbf{k}'}{k^2} \gamma(k', \epsilon_{\mathbf{k}}) . \qquad (2.32)$$

The terms in $\mathbf{\Gamma}^{++}(\mathbf{k}, \epsilon)$ and $\mathbf{\Gamma}^{--}(\mathbf{k}, \epsilon)$ are higher order in c_{imp} and will be neglected.

The solution for $\gamma(k, \epsilon)$ for $k = k_{\text{F}}$ and $\epsilon = \epsilon_{\text{F}}$ is required in calculating the conductivity at $T = 0$. This is given by

$$\gamma(k_{\text{F}}, \epsilon_{\text{F}}) = 1 + 2\pi c_{\text{imp}} \tau(k_{\text{F}}) \sum_{\mathbf{k}'} |T_{\mathbf{k}\mathbf{k}'}|^2 \frac{\mathbf{k} \cdot \mathbf{k}'}{k^2} \gamma(k_{\text{F}}, \epsilon_{\text{F}}) . \qquad (2.33)$$

As $\mathbf{k} \cdot \mathbf{k}' = k^2 \cos \theta'$ using (2.19) equation (2.33) can be written as

$$\gamma_1 = 1 + \tau(k_{\text{F}}) \left(\frac{1}{\tau_1(k_{\text{F}})} - \frac{1}{\tau(k_{\text{F}})} \right) \gamma_1 , \qquad (2.34)$$

where $\gamma_1 = \gamma(k_{\text{F}}, \epsilon_{\text{F}})$. Hence the solution $\gamma_1 = \tau(k_{\text{F}})/\tau_1(k_{\text{F}})$ when substituted into (2.29) gives the Boltzmann result (2.16) and (2.19).

The discussion in this section has been restricted to the case of purely potential scattering. Equation (2.26) is not valid when the Hamiltonian cannot be written in the single particle form (1.12). A more general derivation of the Boltzmann equation result starting from the Kubo formula (2.23) can be found in standard many-body texts books such as that of Mahan (1990). The resulting calculations are very similar to those described in this section but in terms of the single and two particle Green's functions which we introduce in chapter 5, rather than the resolvent Green's function (1.13). Mahan also considers in detail phonon contributions, electron–electron Umklapp scattering and inelastic scattering by impurities.

2.4 Kondo's Explanation of the Resistance Minimum

Potential scattering, which we have considered so far, gives a finite low temperature conductivity which is essentially temperature independent. Temperature dependent corrections can be calculated using the Sommerfeld expansion (1.6) but the corrections, which are of the order of $(T/T_{\text{D}})^2$, are not significant for conduction electrons with a high degeneracy temperature T_{D} ($= \epsilon_{\text{F}}/k_{\text{B}}$). The contributions from the phonon scattering, which increase rapidly with temperature, dominate the temperature dependence. The resistivity due to phonons, on the assumption of a Debye spectrum, increases with temperature as T^5. The total resistivity due to the impurities and phonons increases monotonically with T so the theory based on purely potential scattering by the

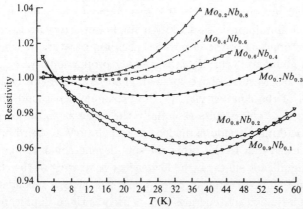

Figure 2.6 Resistance minima for Fe in a series of $Mo-Nb$ alloys (from Sarachik et al, 1964). Compare the depths of the minima with the corresponding moments in figure 1.8.

impurities cannot explain the resistance minimum, observed in many dilute alloys. As discussed in the introduction, the resistance minimum was a long standing theoretical puzzle since its first experimental observation in Au by de Haas et al (1934). The later observation, that the minimum depended on the impurity concentration, indicated it as being an impurity phenomenon.

A possible basis for an explanation of the minimum emerged with the experimental observation of a correlation between the existence of a Curie–Weiss term in the impurity susceptibility (a local moment), and the occurrence of the resistance minimum. This was the motivation for Kondo's calculation in 1964 of higher terms in the conductivity in the s-d model, the model that describes scattering of conduction electrons by a local moment. Kondo in his paper cited the experimental work of Sarachik, Corenzwit & Longinotti (1964) on dilute alloys of Fe in a series of $Nb-Mo$ alloys as host metals as particularly convincing evidence for this correlation. This correlation can be seen by comparing the depths of the minima of the resistivity as shown in figure 2.6 for Fe in $Mo-Nb$ alloys with the corresponding moments in figure 1.8.

This evidence suggests that the resistance minimum is a consequence of the interaction between the spins of the localized and conduction electrons, rather than other features of the impurity, such as the impurity charge or crystal field effects. Kondo argued further that, as the depth of the minimum, measured relative to the value at $T = 0$ K, was found to be roughly proportional to the impurity concentration c_{imp}, and its ratio with the resistivity at $T = 0$ K was found to be independent of c_{imp}, that the minimum is essentially a single impurity effect, and not due to the interaction of the impurities. The temperature at which the minimum is found to occur, T_{min}, is roughly proportional to $c_{imp}^{1/5}$, and consequently not very sensitive to small changes in the concentration. Calculations of the conductivity for the s-d model to second order in the coupling $J_{kk'}$ give results essentially the same as those for purely potential scattering. Kondo extended his calculations to third order in $J_{kk'}$ and revealed that to this order the spin degeneracy of the impurity has a very significant effect on the low energy scattering of the conduction electrons. This scattering gives a marked energy dependence to the relaxation time at low energies and, consequently, a significant temperature dependent correction to the resistivity. We now look at this calculation in some detail.

To calculate the transport lifetime $\tau_1(\epsilon, \mathbf{k})$ from (2.19) we need the matrix elements of the T matrix, $\langle \mathbf{k}', \sigma' | T(\epsilon^+) | \mathbf{k}, \sigma \rangle$, between states of Slater determinants in which a conduction electron is scattered from a state $\mathbf{k}, \sigma$ to a state $\mathbf{k}', \sigma'$. The calculation differs from that for the purely potential scattering model as the spin degrees of freedom have to be considered explicitly. Scattering processes can occur in which the spins of the conduction and impurity are flipped. If we neglect the interaction between the impurities we can use the single impurity s-d model,

$$H_{sd} = \sum_{\mathbf{k}, \mathbf{k}'} J_{\mathbf{k}, \mathbf{k}'}(S^+ c_{\mathbf{k}, \downarrow}^\dagger c_{\mathbf{k}', \uparrow} + S^- c_{\mathbf{k}, \uparrow}^\dagger c_{\mathbf{k}', \downarrow} + S_z(c_{\mathbf{k}, \uparrow}^\dagger c_{\mathbf{k}', \uparrow} - c_{\mathbf{k}, \downarrow}^\dagger c_{\mathbf{k}', \downarrow})),$$

$$(2.35)$$

which we introduced in section 1.5, equation (1.64).

The matrix element of $\langle \mathbf{k}', \uparrow | T(\epsilon^+) | \mathbf{k}, \uparrow \rangle$ to lowest order in $J_{kk'}$ is

$$\langle \mathbf{k}', \uparrow | T(\epsilon^+) | \mathbf{k}, \uparrow \rangle_{(1)} = J_{\mathbf{k}\mathbf{k}'} S_z, \qquad (2.36)$$

as $T = H_{sd}$ to lowest order in $J_{kk'}$ from (1.22), and similarly for the spin flip term, $\langle \mathbf{k}', \uparrow | T(\epsilon^+) | \mathbf{k}, \downarrow \rangle$,

$$\langle \mathbf{k}', \uparrow | T(\epsilon^+) | \mathbf{k}, \downarrow \rangle_{(1)} = J_{\mathbf{k}\mathbf{k}'} S^-. \qquad (2.37)$$

If we consider a localized $\mathbf{k}$ independent interaction,

$$J_{\mathbf{k}, \mathbf{k}'} = J/N_s \qquad \epsilon_{\mathbf{k}}, \epsilon_{\mathbf{k}'} < D \qquad (2.38)$$

and calculate $\tau_1(k)$ from (2.19), we find, on taking into account all the scattering processes to order J^2,

$$\frac{1}{\tau(k_F)} = \frac{3nc_{imp}J^2 S(S+1)}{2e^2 \hbar \epsilon_F}, \qquad (2.39)$$

for free electron dispersion $\epsilon(\mathbf{k}) = \hbar^2 \mathbf{k}^2/2m$, using the relation $2S_z^2 + S^+ S^- + S^- S^+ = S(S+1)$. The contribution to the resistivity is

$$R_{\text{imp}} = \frac{3\pi m J^2 S(S+1)}{2e^2 \hbar \epsilon_F}, \qquad (2.40)$$

which is essentially the same as that calculated for the potential scattering model (1.11) and (1.35). The calculation to this order corresponds to calculating the scattering by the impurity potential in the first Born approximation.

To calculate the resistivity to third order in J we will need the T matrix to second order in J,

$$\langle \mathbf{k}', \sigma' | T(\epsilon^+) | \mathbf{k}, \sigma \rangle_{(2)} = \langle \mathbf{k}', \sigma' | H_{\text{sd}} G_0^+(\epsilon) H_{\text{sd}} | \mathbf{k}, \sigma \rangle, \qquad (2.41)$$

from (1.22). There are many contributions to this matrix element, the most important terms are the ones in which the spins of the conduction and localized electron are flipped. At this point we have to go beyond the single particle formalism as we are dealing with a real many-body problem. The s-d Hamiltonian cannot be reduced to the single particle form (1.12) as it always can, at least in principle, for pure potential scattering. The modification required in the calculation of the T matrix in the many-body case can be stated quite simply; we have to replace the occupation numbers for electrons (or holes) in the intermediate $\mathbf{k}$ states by the Fermi factor $f(\epsilon_{\mathbf{k}})$ $((1-f(\epsilon_{\mathbf{k}}))$ for holes). This form of many-body perturbation theory will be discussed fully in chapter 7. Let us look at terms which contribute to $\langle \mathbf{k}', \uparrow | T(\epsilon^+) | \mathbf{k}, \uparrow \rangle_{(2)}$. One contribution is

$$\frac{J^2}{N_s^2} \sum_{\mathbf{k}_1 \mathbf{k}'_1 \mathbf{k}_2 \mathbf{k}'_2} \langle \mathbf{k}', \uparrow | S^- c_{\mathbf{k}_1, \uparrow}^\dagger c_{\mathbf{k}'_1, \downarrow} (\epsilon + is - H_0)^{-1} S^+ c_{\mathbf{k}_2, \downarrow}^\dagger c_{\mathbf{k}'_2, \uparrow} | \mathbf{k}, \uparrow \rangle.$$
$$(2.42)$$

This is non-vanishing if $\mathbf{k}_1 = \mathbf{k}'$, $\mathbf{k}'_1 = \mathbf{k}_2$, $\mathbf{k}'_2 = \mathbf{k}$ and gives

$$\frac{J^2}{N_s^2} \sum_{\mathbf{k}_2} S^- S^+ \frac{(1 - f(\epsilon_{\mathbf{k}_2}))}{(\epsilon + is - \epsilon(\mathbf{k}_2))}. \qquad (2.43)$$

This scattering process is indicated diagramatically in figure 2.7(i) for $S = \frac{1}{2}$, with the state $|\mathbf{k}, \uparrow\rangle$ on the left, then the two scattering terms, and the final state $|\mathbf{k}', \uparrow\rangle$ on the right. The diagrams differ from the purely potential scattering terms in figure 1.1, because we have to keep track of the spin state of the impurity which is indicated in the base line which runs through the diagram (see figure 1.6). The $\mathbf{k} \uparrow$ conduction electron scatters with spin flip into an unoccupied hole state $\mathbf{k}_2 \downarrow$. The $\mathbf{k}_2 \downarrow$ electron then scatters into the final $\mathbf{k}' \uparrow$ state. Intermediate $\mathbf{k}$ lines running from left to right carry a hole factor $1 - f(\epsilon_{\mathbf{k}})$, the probability

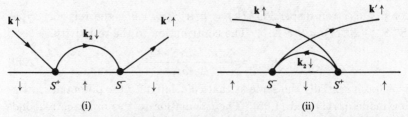

Figure 2.7 (i) A second order spin flip contribution to the matrix element $\langle \mathbf{k}', \uparrow |T(\epsilon^+)|\mathbf{k}, \uparrow\rangle$ corresponding to equation (2.43), (ii) a second spin flip contribution to the same matrix element corresponding to equation (2.45).

that they are initially unoccupied, and intermediate lines running from right to left carry a factor $f(\epsilon_\mathbf{k})$.

The second possible contribution to $\langle \mathbf{k}', \uparrow |T(\epsilon^+)|\mathbf{k}, \uparrow\rangle_{(2)}$ is

$$\frac{J^2}{N_\mathrm{s}^2} \sum_{\mathbf{k}_1 \mathbf{k}'_1 \mathbf{k}_2 \mathbf{k}'_2} \langle \mathbf{k}', \uparrow |S^+ c_{\mathbf{k}_2,\downarrow}^\dagger c_{\mathbf{k}'_2,\uparrow}(\epsilon + is - H_0)^{-1} S^- c_{\mathbf{k}_1,\uparrow}^\dagger c_{\mathbf{k}'_1,\downarrow}|\mathbf{k}, \uparrow\rangle,$$

$$(2.44)$$

which contributes if $\mathbf{k}'_2 = \mathbf{k}'$, $\mathbf{k}_2 = \mathbf{k}'_1$, $\mathbf{k}_2 = \mathbf{k}$ giving

$$\frac{J^2}{N_\mathrm{s}^2} \sum_{\mathbf{k}_2} S^+ S^- \frac{f(\epsilon_{\mathbf{k}_2})}{(\epsilon + is - \epsilon(\mathbf{k}_2))}. \qquad (2.45)$$

This second term is represented by figure 2.7(ii). In this case an electron in an occupied state $\mathbf{k}_2 \downarrow$ is scattered with a spin flip into the state $\mathbf{k}' \uparrow$, and the remaining hole $\mathbf{k}_2 \downarrow$ is annihilated by the initial $\mathbf{k} \uparrow$ with another spin flip, leaving the final state $|\mathbf{k}' \uparrow\rangle$.

The sum of these two terms can be written in the form,

$$(S^2 - S_z)\frac{J^2}{N_\mathrm{s}^2} \sum_{\mathbf{k}_2} \frac{1}{(\epsilon + is - \epsilon(\mathbf{k}_2))} + 2S_z \frac{J^2}{N_\mathrm{s}^2} \sum_{\mathbf{k}_2} \frac{f(\epsilon_{\mathbf{k}_2})}{(\epsilon + is - \epsilon(\mathbf{k}_2))}, \quad (2.46)$$

using

$$S^+ S^- = S^- S^+ + 2S_z, \qquad S^- S^+ = S^2 - S_z^2 - S_z. \qquad (2.47)$$

If we consider the two corresponding processes with the longitudinal rather than the transverse terms in (2.41), these are similar to the terms shown in figure 2.7(i)(ii) but with an S_z factor and no spin flip in the intermediate states, we get contributions of the form (2.43) and (2.45) except that each term has a local spin factor S_z^2. The factor $f(\epsilon_{\mathbf{k}_2})$ associated with the intermediate state cancels out when we sum these to give

$$S_z^2 \frac{J^2}{N_\mathrm{s}^2} \sum_{\mathbf{k}_2} \frac{1}{(\epsilon + is - \epsilon(\mathbf{k}_2))}. \qquad (2.48)$$

In repeated longitudinal scattering to any order in J, as in potential scattering, the factor $f(\epsilon_\mathbf{k})$ cancels out. Such terms, including the first term in (2.46), when substituted into (2.19) contribute correction factors to the inverse lifetime $1/\tau(\mathbf{k})$ of order J/ϵ_F or higher. As the physical situations correspond to $J \ll \epsilon_\mathrm{F}$ then these corrections are negligible.

The factor $f(\epsilon_\mathbf{k})$ in the second term of (2.46), which did not cancel out due to the non-commutivity of S^+ and S^-, leads to a logarithmically singular contribution to the low energy scattering $\epsilon \sim \epsilon_\mathrm{F}$ at low temperatures. Collecting the first order, and the second order terms which we have retained, together,

$$\langle \mathbf{k}', \uparrow \,|T(\epsilon^+)|\mathbf{k}, \uparrow \rangle = S_z \frac{J}{N_\mathrm{s}}(1 - 2Jg(\epsilon))\,, \qquad (2.49)$$

where

$$g(\epsilon) = \frac{1}{N_\mathrm{s}} \sum_\mathbf{k} \frac{f(\epsilon_\mathbf{k})}{\epsilon_\mathbf{k} - \epsilon - is}. \qquad (2.50)$$

Similar terms arise in calculating $\langle \mathbf{k}', \downarrow \,|T(\epsilon^+)|\mathbf{k}, \downarrow \rangle$, $\langle \mathbf{k}', \uparrow \,|T(\epsilon^+)|\mathbf{k}, \downarrow \rangle$ and $\langle \mathbf{k}', \downarrow \,|T(\epsilon^+)|\mathbf{k}, \uparrow \rangle$. Collecting all these terms together and substituting into (2.19) gives

$$\frac{1}{\tau(\mathbf{k})} = \frac{3nc_\mathrm{imp}J^2 S(S+1)}{2e^2\hbar\epsilon_\mathrm{F}}(1 - 2J(g(\epsilon_\mathbf{k}) + g^*(\epsilon_\mathbf{k})))\,, \qquad (2.51)$$

to order J^3, using $S^+S^- + S^-S^+ + 2S_z^2 = S(S+1)$.

As the integral $g(\epsilon)$ is singular as $T \to 0$ we need to evaluate the contribution from this term rather carefully. The real part of $g(\epsilon)$, which we denote by $\overline{g}(\epsilon)$, is given by

$$\overline{g}(\epsilon) = P \int \frac{f(\epsilon_\mathbf{k})}{\epsilon_\mathbf{k} - \epsilon} \frac{d\mathbf{k}}{(2\pi)^3}\,, \qquad (2.52)$$

which we evaluate for free electron dispersion $\epsilon(\mathbf{k}) = \hbar^2 k^2/2m$. This becomes

$$\overline{g}(\epsilon) = P \int_0^D \frac{C\sqrt{\epsilon'}f(\epsilon')}{\epsilon' - \epsilon}\,d\epsilon'\,, \qquad (2.53)$$

where the conduction density of states is $\rho_0 = C\sqrt{\epsilon}$, $C = m^{3/2}/\sqrt{2}\pi^2\hbar^3$.

On integrating by parts,

$$\overline{g}(\epsilon) = -C \int_0^D \frac{\partial f(\epsilon')}{\partial \epsilon'}\left(2\sqrt{\epsilon'} + \sqrt{\epsilon}\ln\left(\frac{\sqrt{\epsilon'} - \sqrt{\epsilon}}{\sqrt{\epsilon'} + \sqrt{\epsilon}}\right)\right)\,d\epsilon'\,, \qquad (2.54)$$

assuming $D \gg \epsilon_\mathrm{F}$. For free electrons,

$$\sigma(T) = -\frac{4e^2}{3m}\int \rho_0(\epsilon)\epsilon\tau(\epsilon)\frac{\partial f(\epsilon)}{\partial \epsilon}\,d\epsilon. \qquad (2.55)$$

When we substitute (2.51) into (2.55) the logarithmic terms from (2.54) lead to a term proportional to the integral,

$$\int_0^D \int_0^D \frac{\partial f(\epsilon)}{\partial \epsilon} \frac{\partial f(\epsilon')}{\partial \epsilon'} \ln \left| \frac{\epsilon' - \epsilon}{4\epsilon_F} \right| d\epsilon \, d\epsilon' = \ln \left(\frac{k_B T}{D} \right) + \text{constant}, \quad (2.56)$$

where the non-singular part of the integrand has been evaluated at $\epsilon = \epsilon' = \epsilon_F$ due to the sharp delta function form of $\partial f / \partial \epsilon$ as $T \to 0$. The $\ln T$ dependence $(k_B T \ll D)$ can be extracted by making the substitution $x = (\epsilon - \epsilon_F)/k_B T$, $x' = (\epsilon' - \epsilon_F)/k_B T$. The final result for the impurity resistivity, retaining only the $\ln T$ in (2.56), is

$$R_{\text{imp}}^{\text{spin}} = \frac{3\pi m J^2 S(S+1)}{2e^2 \hbar \epsilon_F} \left(1 - 4 J \rho_0(\epsilon_F) \ln \left(\frac{k_B T}{D} \right) \right). \quad (2.57)$$

The extra third order term in $\ln T$ increases as T decreases for an antiferromagnetic coupling of the local moment with the conduction electrons. If this is combined with the other contributions to the resistivity it provides a satisfactory explanation for the resistance minimum. The total resistivity may be assumed to be of the form,

$$R(T) = aT^5 + c_{\text{imp}} R_0 - c_{\text{imp}} R_1 \ln (k_B T/D), \quad (2.58)$$

where the first term is the phonon contribution, the second term is the temperature independent contribution from the impurity scattering, and the final term the third order contribution from the spin scattering with the local moment. This expression has a minimum at a temperature T_{min} given by

$$T_{\text{min}} = \left(\frac{R_1}{5a} \right)^{1/5} c_{\text{imp}}^{1/5}, \quad (2.59)$$

in general agreement with the experimental observations.

A detailed comparison of the form (2.58) with the resistivity for Fe impurities in Au (MacDonald, Pearson & Templeton, 1962) at three different concentrations was given in Kondo's original paper and is reproduced in figure 2.8. A logarithmic term in the temperature clearly gives a good fit to the experimental data.

The calculation of the exchange coupling J based on the Anderson model, section 1.6, provides an explanation for the occurrence of an antiferromagnetic coupling, rather than the ferromagnetic coupling expected from direct Coulomb exchange. We leave further discussion of the comparison of theoretical predictions with experimental results to chapter 9. Clearly Kondo's calculation provided a convincing explanation of the resistance minimum in dilute magnetic alloys. The $\ln T$ term, however, diverges as $T \to 0$ so the perturbation result is not applicable

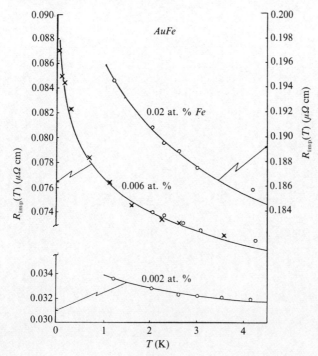

Figure 2.8 The comparison of experimental results for the resistivity of dilute Fe in Au at very low temperatures with logarithmic form (2.58), reproduced from the paper of Kondo in 1964.

down to $T = 0$. The problem of finding a solution valid in the low temperature regime $T \to 0$ is the 'Kondo problem'. It is considered in the next chapter.

3

The Kondo Problem

3.1 Perturbation Theory

We found in the last chapter that perturbation theory for the s-d model to finite order in J breaks down due to the appearance of $\ln T$ terms which diverge in the limit $T \to 0$. These terms occurred in calculating the resistivity but they also occur in calculating other physical quantities. For example, if we calculate in a magnetic field H the magnetization due to the presence of the impurity from

$$M_{\text{imp}}(H) = g\mu_{\text{B}}(\langle S_z + S_z^{\text{c}} \rangle - \langle S_z^{\text{c}} \rangle_0), \qquad (3.1)$$

where $\langle \, \rangle$ indicates a thermal average with respect to the full Hamiltonian, and $\langle \, \rangle_0$ with respect to the conduction electron Hamiltonian (1.2), and S_z^{c} is the z-component of the total spin $\mathbf{S}^{\text{c}}$ for the conduction electrons, then we find to second order in $J\rho_0$ and linear in H,

$$M_{\text{imp}}(H) = \frac{(g\mu_{\text{B}})^2 S(S+1) H}{3k_{\text{B}}T} \{1 - 2J\rho_0$$
$$+ (2J\rho_0)^2 \ln(k_{\text{B}}T/D) + c_2(2J\rho_0)^2\},$$

$$(3.2)$$

for a conduction band of width $2D$, where c_2 is a coefficient which depends on the form used for the conduction density of states. We have assumed the same g-factor for the impurity and conduction electrons. The zero field susceptibility to this order is therefore,

$$\chi_{\text{imp}} = \frac{(g\mu_{\text{B}})^2 S(S+1)}{3k_{\text{B}}T} \{1 - 2J\rho_0$$
$$+ (2J\rho_0)^2 \ln(k_{\text{B}}T/D) + c_2(2J\rho_0)^2\},$$

$$(3.3)$$

At high temperatures the logarithmic term gives a reduction of the effective magnetic moment compared with that for a free spin. As $T \to 0$ this effective moment diverges.

Perturbation calculations give fourth order logarithmic terms to the entropy $S_{\mathrm{imp}}(T)$,

$$S_{\mathrm{imp}}(T) = k_B \ln(2S + 1) - \frac{\pi^2}{3} k_B S(S + 1)(2J\rho_0)^3 \{$$
$$1 - 3(2J\rho_0)\ln(k_B T/D) + 6(2J\rho_0)^2 \ln^2(k_B T/D) + ...\}.$$

(3.4)

The impurity specific heat deduced from this is

$$C_{\mathrm{imp}}(T) = T \frac{dS_{\mathrm{imp}}}{dT} = k_B \pi^2 S(S + 1)(2J\rho_0)^4 \{1 - 8J\rho_0 \ln(k_B T/D) + ...\},$$

(3.5)

where the leading order logarithmic term here is fifth order in $J\rho_0$.

Perturbation calculations can be made at $T = 0$ in an applied magnetic field H giving logarithmic terms of the form $\ln(\mu_B H/D)$. The leading terms in the impurity magnetization deduced using (3.1) are

$$M_{\mathrm{imp}}(H) = g\mu_B S\{1 - 2J\rho_0 + (2J\rho_0)^2 \ln(\mu_B H/D) + c_2'(2J\rho_0)^2\}. \quad (3.6)$$

One of the first questions to explore in looking for a satisfactory theory in the low temperature regime is whether the divergences arising from the logarithmic terms can be removed by summing the higher order logarithmic terms in the perturbation expansion. A general nth order perturbation term gives $\ln T$ contributions of the form $(J\rho_0)^n (\ln T)^m$ for $n > n_0$, $m \leq n - n_0$, where n_0 depends on the property being calculated. The leading order divergent terms for a given n correspond to $m = n - n_0$. Abrikosov (1965) carried out a summation of these leading order terms for the resistivity. Though infinite order perturbation methods were developed in the 50s and 60s for quantum statistical many-body problems, there are complications in applying these techniques to the s-d model. This is because most of these expansions are about an unperturbed system of non-interacting electrons. The expansion is then in powers of the interparticle interaction, which is expressed in terms of standard fermion creation and annihilation operators, $a_l^\dagger$, a_l. These have a c-number commutator, $[a_l^\dagger, a_{l'}]_+ = \delta_{l,l'}$, so that Wick's theorem can be applied. The individual contributions can then be represented by standard Feynman diagrams and a linked cluster theorem holds. For the s-d model, however, the interaction is in terms of spin operators which satisfy the commutation rule, $[S^+, S^-]_- = 2S_z$, and the usual Wick's theorem cannot be used. To overcome this problem Abrikosov introduced a fermion representation for the spin operators, $S = a_\alpha^\dagger S_{\alpha,\beta} a_\beta$, where $S_{\alpha,\beta}$ are the spin matrices,

$$(S_z)_{\alpha,\beta} = \alpha \delta_{\alpha,\beta}, \quad (S^\pm)_{\alpha,\beta} = \delta_{\alpha,\beta\pm1}[(S \mp \beta)(S \pm \beta + 1)]^{1/2}, \quad (3.7)$$

where α, β take $2S + 1$ values corresponding to the eigenvalues of S_z. This is a valid representation of the spin operators provided the total fermion occupation number is unity. A projection technique has to be introduced to eliminate the unphysical states which occur when the total occupation of the fermion states deviates from one. Abrikosov (1965) introduced an effective level λ for the fermions, and to eliminate the contributions from the non-physical states in an average over the states he multiplied each term by $e^{\lambda/k_BT}/(2S+1)$ and took the limit $\lambda \to \infty$. Standard perturbation theory can then be applied. With the projection procedure, however, unlinked diagrams do not factorize and the linked cluster theorem does not hold. However, diagrams which give the leading order logarithmic terms can be identified and unlinked diagrams contribute only to lower order. The leading order diagrams are known as *parquet* diagrams and are characterized by having only one particle or hole in an intermediate state. If these are summed, with the contribution from each term being evaluated to leading order logarithmic accuracy only, then the result for the impurity susceptibility is

$$\chi_{\text{imp}}(T) = \frac{(g\mu_B)^2 S(S+1)}{3k_BT} \left\{ 1 - \frac{2J\rho_0}{1 + 2J\rho_0 \ln(k_BT/D)} + c_2(2J\rho_0)^2 \right\}. \quad (3.8)$$

The $\ln T$ terms are now in the denominator due to the summation of a geometric series. As $T \to 0$ in the ferromagnetic case, $J < 0$, the contribution from the interaction terms goes asymptotically to zero leaving a free spin Curie law. For the antiferromagnetic case, $J > 0$, the difficulties arising from the $\ln T$ terms become more severe as there is now a divergence at a finite temperature T_K,

$$k_BT_K \sim De^{-1/2J\rho_0}, \quad (3.9)$$

known as the *Kondo temperature*.

Other physical properties can be calculated within the same approximation of summing leading order logarithmic terms. The result for the entropy generalizes (3.4) to

$$S_{\text{imp}}(T) = k_B\ln(2S+1) - \frac{\pi^2}{3} \frac{k_B S(S+1)(2J\rho_0)^3}{[1 + 2J\rho_0\ln(k_BT/D)]^3} + \dots \,. \quad (3.10)$$

For $J < 0$ this is positive, decreases as T decreases, and is well behaved at all temperatures. For $J > 0$, however, it increases as T decreases and diverges at $T = T_K$. It is clearly unphysical as T approaches T_K. The corresponding contribution to the impurity specific heat is

$$C_{\text{imp}}(T) = \frac{\pi^2 k_{\text{B}} S(S+1)(2J\rho_0)^4}{[1 + 2J\rho_0 \ln(k_{\text{B}}T/D)]^4} + ..., \tag{3.11}$$

which also diverges strongly at $T = T_{\text{K}}$ in the antiferromagnetic case. We will not give details of the derivations of these results. We refer the review of Kondo (1969) for a fuller survey and the details. We will derive (3.8) later by an alternative approach.

Clearly the leading order logarithmic sum fails to provide a satisfactory theory for the case of antiferromagnetic coupling. The challenge to find a satisfactory solution to this problem stimulated an enormous amount of activity in the late 60s and early 70s. The aim theoretically was to find non-perturbative techniques for calculating such quantities as χ_{imp}, C_{imp} and transport properties as $T \to 0$ for the s-d and Anderson models. Experimentally the aim was to clarify the behaviour of these quantities as a function of T for the range $T \ll T_{\text{K}}$ in the dilute alloys which have a resistance minimum.

3.2 Beyond Perturbation Theory

As the leading order logarithmic sum of the parquet diagrams gave unphysical results for $J > 0$, the next question to be investigated is whether a complete summation of the parquet diagrams can provide a more satisfactory approximation in the low temperature regime $T \leq T_{\text{K}}$. Coupled integral equations that effectively sum this class of diagrams were first formulated by Suhl (1965). Essentially equivalent equations were derived by a decoupling of Green's function equations of motion by Nagaoka (1965). Solutions to these equations look promising in that they are non-singular at $T = T_{\text{K}}$ and can be continued down to $T = 0$ (Suhl & Wong, 1967; Hamann, 1967; Bloomfield & Hamann, 1967; Zittartz & Müller-Hartmann, 1968). Hamann obtained an expression for the impurity resistivity, $R_{\text{imp}}(T)$ for $J > 0$ in the form,

$$R_{\text{imp}}(T) = \frac{R_0}{2} \left\{ 1 - \frac{\ln(T/T_{\text{K}})}{[(\ln(T/T_{\text{K}}))^2 + \pi^2 S(S+1)]^{1/2}} \right\}, \tag{3.12}$$

with T_{K} defined by (3.9). At very high temperatures this has the expansion,

$$R_{\text{imp}}(T) = \frac{R_0 \pi^2 S(S+1)}{4(\ln(T/T_{\text{K}}))^2} \left\{ 1 - \frac{3\pi^2 S(S+1)}{4(\ln(T/T_{\text{K}}))^2} ... \right\}. \tag{3.13}$$

The leading order term in this result was first derived by Abrikosov. At low temperatures such that $|\ln(T/T_{\text{K}})| \gg 1$, (3.12) can be expanded to give

$$R_{\text{imp}}(T) = R_0 \left\{ 1 - \frac{\pi^2 S(S+1)}{4(\ln(T/T_{\text{K}}))^2} + \frac{3(\pi^2 S(S+1))^2}{16(\ln(T/T_{\text{K}}))^4} \cdots \right\}. \tag{3.14}$$

The limiting value as $T \to 0$, R_0, corresponds to the maximum value possible for s wave scattering (the unitarity limit, $\eta = \pi/2$, see appendix A, equation (A.26)).

The result for the specific heat at very high and very low temperatures, such that $|\ln(T/T_{\text{K}})| \gg 1$, is

$$C_{\text{imp}}(T) \sim \frac{\pi^2 k_{\text{B}} S(S+1)}{[\ln(T/T_{\text{K}})]^4}, \tag{3.15}$$

for $J > 0$, in agreement with (3.11) in the high temperature regime. With increase of T from $T = 0$ the specific heat increases rapidly with a maximum for $T \sim T_{\text{K}}/3$ and over a wide temperature range about T_{K} was found to behave by Bloomfield and Hamann (1967) as $C_{\text{imp}}(T) \sim (T/T_{\text{K}})^{0.57}$. We note that all these expressions are universal functions of T/T_{K}. This is a general feature which will be established later.

Results for the susceptibility at high temperatures correspond to (3.8). At very low temperatures they depend sensitively on the approach used and results for the $S = \frac{1}{2}$ model giving a range of behaviour from less than complete screening of the impurity moment, to complete, and over complete screening, can be obtained (Brenig & Zittartz, 1973). Mattis (1967) proved a theorem that the ground state for $J > 0$ for a system at finite volume corresponds to a spin value $S - \frac{1}{2}$. Hence for $S = \frac{1}{2}$ the ground state should be a singlet with complete spin compensation, as was originally conjectured by Nagaoka (1967). This result would imply a total entropy change down to $T = 0$,

$$\Delta S_{\text{imp}} = k_{\text{B}} \ln \left(\frac{2S+1}{2S} \right), \tag{3.16}$$

which disagrees with an exact result for ΔS_{imp} of Zittartz & Müller-Hartmann (1968) based on Nagaoka's equations. This indicates that the Nagaoka equations do not constitute a good enough approximation to describe the low temperature behaviour of the model. The discrepancy with (3.16) is not conclusive, however, as Mattis's proof is for a finite volume whereas Nagaoka's equations are for the thermodynamic limit in which the infinite volume limit is taken before the limit $T \to 0$.

Experimental evidence, however, does not support the strong temperature dependence found at low temperatures in the solutions of the Nagaoka–Suhl equations. The predictions are in reasonably good agreement with experiment for $T > T_{\text{K}}$ but this agreement breaks down

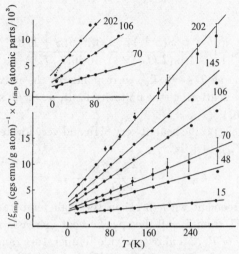

Figure 3.1 $1/\chi_{\mathrm{imp}}(T)$ versus T for Fe impurities in Cu at the concentrations indicated in parts per million. The results are all fitted with the same Curie–Weiss law (Hurd, 1967).

progressively with decreasing temperature. The impurity resistivity at low temperatures is best fitted to a simple power law,

$$R_{\mathrm{imp}}(T) = R_0 \left\{ 1 - \left(\frac{T}{\theta_R}\right)^2 + \mathrm{O}\left(\frac{T}{\theta_R}\right)^4 \right\}, \qquad (3.17)$$

rather than the form (3.14) involving $\ln T$ terms. The impurity specific heat behaves linearly with T at low temperatures and has a maximum at higher temperatures, $T \sim T_{\mathrm{K}}$. The impurity susceptibility can usually be fitted well by a Curie–Weiss type law at higher temperatures.

We give an example in figure 3.1 for dilute concentrations of Fe impurities in Cu which can all be fitted with the same Curie–Weiss form with $\theta = 32 \pm 2K$, and a Curie constant $C = 4.5\mu_{\mathrm{B}}^2$. A Curie–Weiss term is consistent with the high temperature result (3.8), as this log form can be approximated by

$$\chi_{\mathrm{imp}}(T) = \frac{(g\mu_{\mathrm{B}})^2 S(S+1)/1.22}{3k_{\mathrm{B}}(T + 4.5T_{\mathrm{K}})}, \qquad (3.18)$$

to better than $\frac{1}{2}\%$ accuracy for $7 < T/T_{\mathrm{K}} < 100$ (Heeger, 1969). At very low temperatures, however, a power law behaviour is observed experimentally; the measurements of Triplett & Phillipps (1971) for example for Fe in Cu fit the form,

$$\chi_{\mathrm{imp}}(T) = \chi_{\mathrm{imp}}(0) \left\{ 1 - \left(\frac{T}{\theta_\chi}\right)^2 + \mathrm{O}\left(\frac{T}{\theta_\chi}\right)^4 \right\}, \qquad (3.19)$$

with $\theta_\chi = 7.5$ K.

A finite impurity susceptibility at $T = 0$ implies a fully screened singlet ground state. To show this we differentiate (3.1) with respect to H and then put $H = 0$, we find

$$\chi_{\text{imp}}(T) = \frac{(g\mu_B)^2(\langle(\mathbf{S} + \mathbf{S}^c)^2\rangle - \langle(\mathbf{S}^c)^2\rangle_0)}{3k_B T}, \qquad (3.20)$$

Hence, if $T\chi_{\text{imp}}(T) \to 0$ as $T \to 0$, then $\langle(\mathbf{S} + \mathbf{S}^c)^2\rangle = 0$, as $\langle(\mathbf{S}^c)^2\rangle_0 = 0$ at $T = 0$, in the ground state which implies a singlet. The energy gain through hybridization leading to the formation of a singlet ground state can be demonstrated by exact diagonalization for the Anderson model in the zero band width limit where the conduction states are represented by a single level at the Fermi level. This calculation, though very simple, brings out some general features and so we give it in detail in appendix C.

Variational methods provide a general way of investigating the nature of the ground state and can generate non-perturbative solutions. The simplest form of variational wavefunction for the $S = \frac{1}{2}$ model, which was considered by Yosida (1966), is

$$\Psi_{\text{s,t}} = \sum_{k > k_F} \alpha_{\mathbf{k}}(c_{\mathbf{k},\downarrow}^{\dagger}\chi_{\uparrow} \mp c_{\mathbf{k},\uparrow}^{\dagger}\chi_{\downarrow})|\Omega\rangle, \qquad (3.21)$$

where $|\Omega\rangle$ is the Fermi sphere of conduction electrons, the subscripts s,t correspond to singlet and triplet states, $\chi_{\uparrow}$, $\chi_{\downarrow}$, are the local spin states and $\alpha_{\mathbf{k}}$ is the variational parameter. Inserting this into the Schrödinger equation for a state of energy E, in the singlet case, gives the equation,

$$\alpha_{\mathbf{k}}(\epsilon_{\mathbf{k}} - E) - \frac{3J}{2N_s} \sum_{k > k_F} \alpha_{\mathbf{k}} = 0, \qquad (3.22)$$

which has a non-zero solution if

$$\frac{1}{N_s} \sum_{k > k_F} \frac{1}{(\epsilon_{\mathbf{k}} - E)} = \frac{2}{3J}. \qquad (3.23)$$

For a flat band density of states, $\rho_0(\epsilon) = \rho_0$, $-D < \epsilon < D$, the lowest solution is

$$E_s^{(0)} = -De^{-2/3J\rho_0}. \qquad (3.24)$$

Similar equations for the triplet case give $E_t^{(0)} = -De^{2/J\rho_0}$. Hence for the antiferromagnetic case a singlet ground state is predicted with this ansatz. The ansatz can be generalized to include states with particle–hole pairs in the Fermi sphere $|\Omega\rangle$. The inclusion of one and two particle–hole pairs gives results of the form (3.23) but with different numerical factors in the exponent. Yosida & Yoshimori (1973) succeeded in including an infinite number of particle–hole pairs to leading order. They obtained a factor $\frac{1}{2}$ in the exponent instead of the factor $\frac{2}{3}$ so that $E_{\mathrm{s}}^{(0)} = -k_{\mathrm{B}}T_{\mathrm{K}}$, with T_{K} given by (3.9). Ishii & Yosida (1967) included a magnetic field and deduced an expression for the ground state susceptibility for $S = \frac{1}{2}$,

$$\chi_{\mathrm{imp}} = \frac{(g\mu_{\mathrm{B}})^2}{4k_{\mathrm{B}}T_{\mathrm{K}}}, \tag{3.25}$$

with T_{K} given by (3.9).

Another approach, giving similar results, by Takano & Ogawa (1966) used the Abrikosov representation of the spin operators, which for $S = \frac{1}{2}$ simplifies to

$$S^+ = a_\uparrow^\dagger a_\downarrow, \quad S^- = a_\downarrow^\dagger a_\uparrow, \quad S_z = \frac{1}{2}(a_\uparrow^\dagger a_\uparrow - a_\downarrow^\dagger a_\downarrow). \tag{3.26}$$

The terms in the s-d interaction were then decoupled in a generalized Hartree–Fock scheme according to

$$a_\uparrow^\dagger a_\downarrow c_{\mathbf{k},\downarrow}^\dagger c_{\mathbf{k}',\uparrow} \to \langle a_\uparrow^\dagger c_{\mathbf{k}',\uparrow}\rangle a_\downarrow c_{\mathbf{k},\downarrow}^\dagger + a_\uparrow^\dagger c_{\mathbf{k}',\uparrow}\langle a_\downarrow c_{\mathbf{k},\downarrow}^\dagger\rangle. \tag{3.27}$$

This is analogous to the Gorkov treatment of superconductivity in that anomalous expectation values, such as $\langle a_\uparrow^\dagger c_{\mathbf{k}',\uparrow}\rangle$, are retained (which should be zero as the fermion number, $a_\uparrow^\dagger a_\uparrow + a_\downarrow^\dagger a_\downarrow$, commutes with the Hamiltonian) and calculated self-consistently. In this approximation the s-d model becomes equivalent to the *non-interacting* Anderson model if the constraint, $a_\uparrow^\dagger a_\uparrow + a_\downarrow^\dagger a_\downarrow = 1$, is required to be satisfied only on the average by putting the fermion level λ at the Fermi level. It is then straightforward to diagonalize the Hamiltonian and to derive self-consistent equations for the anomalous averages. There is a non-trivial solution only in the case $J > 0$. The susceptibility at $T = 0$ is given by

$$\chi_{\mathrm{imp}} = \frac{(g\mu_{\mathrm{B}})^2}{2\pi\Delta_0}, \tag{3.28}$$

where Δ_0 is the width of the effective resonance at the Fermi level. This is determined self-consistently, and at $T = 0$ is given by $\pi\Delta_0 = De^{-2/3J\rho_0}$ for a flat band density of states.

Some of these methods will be explored further in chapters 7 and 8 for the N-fold degenerate models.

Another line of attack on the Kondo problem was developed in a series of papers by Anderson and coworkers (Anderson, 1967; Anderson & Yuval, 1969; Anderson, Yuval & Hamann, 1970; Yuval & Anderson, 1970) and leads logically to the approach used later by Wilson (1975), which we consider in the next chapter. The calculations are based on the anisotropic s-d model, in which the transverse couplings $J_\pm$ can be

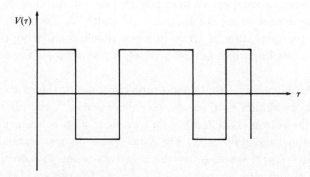

Figure 3.2   The spin dependent potential acting on an up spin electron as a function of 'time' τ.

different from the longitudinal coupling J_z. The interaction term in this model is then

$$H = \sum_{\mathbf{k},\mathbf{k}'} J_+ S^+ c^\dagger_{\mathbf{k},\downarrow} c_{\mathbf{k}',\uparrow} + J_- S^- c^\dagger_{\mathbf{k},\uparrow} c_{\mathbf{k}',\downarrow} + J_z S_z (c^\dagger_{\mathbf{k},\uparrow} c_{\mathbf{k}',\uparrow} - c^\dagger_{\mathbf{k},\downarrow} c_{\mathbf{k}',\downarrow}).$$

$$(3.29)$$

For $J_\pm = 0$ there is no spin flip scattering and there are two possible ground states in the absence of a field, with the local spin either up or down in the case $S = \frac{1}{2}$. The model can be solved in these two distinct cases using the potential scattering methods of chapter 1. If one then applies finite temperature perturbation theory in powers of the transverse coupling $J_\pm$ spin flips are generated at a sequence of imaginary times τ_1, τ_2... from one spin state to the other reversing the sign of the potential, due to the longitudinal term in (3.29), for electrons of given spin, as illustrated in figure 3.2. There is purely potential scattering by this longitudinal term between spin flips.

The partition function Z is calculated by summing the contributions for all possible numbers and positions of the spin flips. This corresponds to all possible paths connecting the spin up and spin down states and leads to a representation of the partition function as a Feynman path integral. The partition function can be shown to be equivalent to that for a classical one dimensional system of charged particles with a long range interaction that is logarithmic. At finite temperatures the system is finite and corresponds to a ring of length $1/k_B T$. The spin flips correspond to the charged particles, those flipping one way to positive charges, those

flipping the other way to negative charges. There is a minimum distance
of approach corresponding to a hard core at a cut-off 'distance' τ_0 of the
order of the inverse of the conduction band width D. There is another
equivalent representation in terms of a one dimensional array of Ising
spins with a coupling which falls off inversely as the square of the distance
as $T \to 0$.

In this formulation of the Kondo problem some of the steps in the
calculation are similar to those in the calculation of X-ray absorption
in which an electron is excited from a core shell of an atom or ion in a
metal into the conduction band. The X-ray spectrum near threshold has
the form $(\omega - \omega_T)^\alpha$, where α may be negative leading to a divergence,
or positive giving zero, at the threshold $\omega = \omega_T$ (Nozières & Dominicis,
1969; Mahan, 1974). This power law behaviour arises from the long
time response due to the change in local potential caused by the sudden
creation of the hole as the core electron is excited, modifying all the
conduction electron states. It was pointed out by Anderson (1967) that
the ground states of the conduction electrons with and without the core
hole potential are orthogonal in the limit of an infinite system. This
result is known as the *orthogonality catastrophe*. This effect leads to a
positive contribution to α of the form, $2\sum_{l'}(2l'+1)(\eta_{l'}(\epsilon_F)/\pi)^2$, where
$\eta_{l'}(\epsilon_F)$ is the phase shift of the partial wave state of angular momentum l'
at the Fermi level. The other term in α is a negative term, $2\eta_l(\epsilon_F)/\pi$, due
to the extra electron in a state of angular momentum l in the conduction
band. The phase shift is that due to the *additional* potential due to the
core hole. In the Kondo problem there is a change in the local scattering
potential, arising from the longitudinal term in (3.29), every time the
impurity spin is flipped by a conduction electron. The calculation of the
electron scattering following a spin flip is similar to that following the
sudden appearance of a core hole. The conduction electron propagator
in the long time limit between spin flips similarly depends on the phase
shift due to the change in local potential. Consequently the effective
interaction between the spin flips (the charged particles or Ising spins)
depends on this phase shift (for further details we refer the reader to
the original papers). It is the presence of spin flip terms which make the
Kondo problem a more difficult problem to tackle theoretically than the
X-ray one.

What has been achieved by converting one problem into another ap-
parently just as difficult problem? One direct result was a numerical
calculation of $\chi_{imp}(T)$ by Schotte & Schotte (1971). As the equivalent problem is a classical one a Monte Carlo simulation is possible and

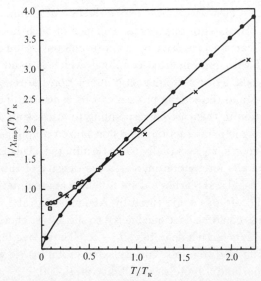

Figure 3.3 Results for $1/\chi_{imp}(T)T_K$ vs T/T_K in different approximations. The Monte Carlo results of Schotte & Schotte (1971) based on the equivalent charge particle system are shown for three different values of T_K (open circle, cross and square). The full circles are results of Ting (1970) based on the Suhl equations (from Anderson & Yuval, 1973).

Schotte and Schotte used this approach to calculate the thermodynamics of the interacting charge problem. Their calculations were restricted to relatively small systems with up to a maximum of 50 charged particles. The susceptibility of the impurity model $\chi_{imp}(T)$ was then deduced from the correlation functions of the charged particles over a range of temperatures extending below T_K. The length of the ring increases as $T \to 0$ so the restriction to a maximum of 50 charges limits the accuracy at lower temperatures.

The results, which are shown in figure 3.3, are markedly different from those generated from the Suhl–Nagaoka equations and extrapolate to a finite value at $T = 0$.

More generally the equivalent charged particle system provided the framework for the introduction of scaling and renormalization ideas. For a low density of spin flips or charged particles, which corresponds to the weak coupling limit, the number of closely spaced particles is expected to be small. By changing the energy scale and extending the cut-off from τ_0 to $\tau_0 + \delta\tau_0$, those that occur in the range $\delta\tau_0$ can be eliminated or 'renormalized' away provided that the interaction between the particles is adjusted accordingly by changing the couplings, $J_z \to J_z + \delta J_z$ and

$J_\pm \rightarrow J_\pm + \delta J_\pm$. This was shown explicitly by Anderson, Yuval & Hamann (1970). By integrating over a sequence of such infinitesimal steps the cut-off can be increased by a finite amount resulting in an equivalent model with renormalized couplings which is valid over the reduced energy scale. For a ferromagnetic interaction the renormalized coupling decreases and the scaling process can be continued indefinitely leading to a solution in this case corresponding to a free spin at $T = 0$. For an antiferromagnetic interaction the renormalized couplings increase and the scaling process has eventually to be terminated. This is because as the strength of the interaction increases the density of the spin flip particles also increases and reaches a point at which approximations used are no longer valid. It was conjectured by Anderson Yuval & Hamann that, if this process could be continued down to the lowest energy scales, the coupling would increase indefinitely, $J \rightarrow \infty$. The strong interaction then induced would lead to a fully compensated ground state with non-magnetic behaviour and a finite susceptibility at $T \rightarrow 0$.

Subsequently Anderson (1970) found a more direct way of carrying out the scaling. He progressively reduced the band width D of the conduction electrons and calculated perturbatively the renormalized interactions due to the elimination of virtual excitations to the band edges. The method is similar to the one we adopted to eliminate the higher order excitations from the Anderson model to derive the s-d model in section 1.7. Anderson termed this a 'poor man's scaling approach'. We shall describe this method in some detail in the next section.

3.3 Poor Man's Scaling

The terms associated with the breakdown of perturbation theory are the log terms, $\ln(k_B T/D)$ and $\ln(\epsilon/D)$, which depend on the conduction band width via $\ln D$. If the cut-off dependent terms in the perturbation theory depended only on $1/D$ or $1/D^2$, then it is possible that the limit $D \rightarrow \infty$ might be taken with impunity. This would be the case if the very high energy excitations can be ignored in calculating the lower lying energy levels. With $\ln D$ terms this is clearly not possible implying that the higher energy excitations are important and have to be taken into account. The essence of the scaling approach is that these higher energy excitations can be absorbed as a renormalization of the couplings. The $\ln D$ terms enter only in the parameters of the effective Hamiltonian for the calculation of the low lying levels.

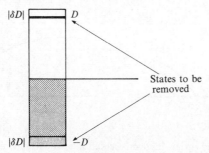

Figure 3.4 The particle and hole states which are removed from the conduction band on reducing the band width by $|\delta D|$.

To carry out this scaling progressively for the s-d model we divide the conduction band into states, $0 < |\epsilon_{\mathbf{k}}| < D - |\delta D|$, which are retained, and states within $|\delta D|$ of the band edge which are to be eliminated (see figure 3.4).

To eliminate excitations that fall within the band edges we follow the procedure given in section 1.7. The wavefunction ψ_1 here describes the component of the wavefunction ψ in which there are no conduction electrons in the upper band edge, $D - |\delta D| < \epsilon_{\mathbf{k}} < D$, or holes in the lower band edge, $-D < \epsilon_{\mathbf{k}} < -D + |\delta D|$; the component ψ_0 a state which has at least one hole in the lower band edge, and ψ_2 the state with at least one conduction electron excited into the upper band edge. The component of the Hamiltonian H_{12} which scatters a conduction electron into a state $\mathbf{q}$ in the unoccupied upper band edge is

$$H_{12} = \sum_{\mathbf{q},\mathbf{k}'} J_+ S^+ c_{\mathbf{q},\downarrow}^\dagger c_{\mathbf{k}',\uparrow} + J_- S^- c_{\mathbf{q},\uparrow}^\dagger c_{\mathbf{k}',\downarrow} + J_z S_z (c_{\mathbf{q},\uparrow}^\dagger c_{\mathbf{k}',\uparrow} - c_{\mathbf{q},\downarrow}^\dagger c_{\mathbf{k}',\downarrow}).$$

(3.30)

The Hamiltonian H_{01} is similar with $\mathbf{k}$ and $\mathbf{q}$ reversed and $\mathbf{q}$ is then a state in the lower band edge. There are matrix elements connecting states in ψ_0 and ψ_2, H_{02}, but these can be ignored for large D, as they do not contribute to leading order in $1/D$. The lowest order corrections to H due to virtual scattering of conduction electrons into the band edge are of order J^2. We can represent these terms by essentially the same diagrams used earlier in section 2.4 for calculating the second order contributions to the matrix elements of $\langle \mathbf{k}\sigma | T(\epsilon) | \mathbf{k}'\sigma' \rangle$ of the T matrix. Consider first of all the scattering of a conduction electrons from the state $\mathbf{k}$ with spin $\uparrow$ into an intermediate state $\mathbf{q} \downarrow$ in the upper band edge and then to a final state $\mathbf{k}' \uparrow$. This is shown in figure 3.5(i) (corresponding to figure 2.7) with a dashed line to represent the electron in the intermediate state in the band edge region which is to be eliminated.

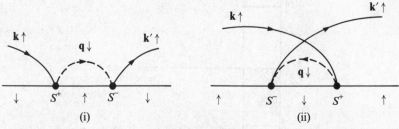

Figure 3.5   Second order diagrams which involve either a particle, diagram (i), or a hole, diagram (ii), in an intermediate state **q** at a band edge (dashed line).

The corresponding term from (1.70) is

$$J_+ J_- \sum_q S^- c_{\mathbf{k}',\uparrow}^\dagger c_{\mathbf{q},\downarrow} (E - H_{22})^{-1} \sum_q S^+ c_{\mathbf{q}',\downarrow}^\dagger c_{\mathbf{k},\uparrow}. \qquad (3.31)$$

As the band edge state is unoccupied in the initial and final states $c_{\mathbf{q},\downarrow} c_{\mathbf{q},\downarrow}^\dagger = \delta_{\mathbf{q},\mathbf{q}'}$, and to lowest order $H_{22} = H_0$, where H_0 is the conduction electron Hamiltonian, and so

$$J_+ J_- \rho_0 |\delta D| S^- S^+ c_{\mathbf{k}',\uparrow}^\dagger c_{\mathbf{k},\uparrow} (E - D + \epsilon_{\mathbf{k}} - H_0)^{-1}, \qquad (3.32)$$

where we have used $\epsilon_{\mathbf{q}} = D$. If the energy E is measured relative to the ground state of the conduction electron gas then H_0 can effectively be set equal to zero. Summing over the states in the lower band edge and using the relation, $S^- S^+ = 1/2 - S_z$, for operators corresponding to spin $S = \frac{1}{2}$, then (3.32) becomes approximately,

$$J_+ J_- \rho_0 |\delta D| (1/2 - S_z) c_{\mathbf{k},\uparrow}^\dagger c_{\mathbf{k}',\uparrow} (E - D + \epsilon_{\mathbf{k}})^{-1}, \qquad (3.33)$$

With similar approximations the contribution from figure 3.5(ii), corresponding to the scattering process in which there is a hole in the lower band edge at $-D$ in the intermediate state, gives

$$J_+ J_- \rho_0 |\delta D| (1/2 + S_z) c_{\mathbf{k}',\uparrow} c_{\mathbf{k},\uparrow}^\dagger (E - D - \epsilon_{\mathbf{k}'})^{-1}, \qquad (3.34)$$

on using $S^+ S^- = 1/2 + S_z$.

Processes in which an $\uparrow$ electron is scattered without spin flip arise from the longitudinal part of the interaction. These give contributions,

$$\frac{-J_z^2 \rho_0 |\delta D| c_{\mathbf{k}',\uparrow}^\dagger c_{\mathbf{k},\uparrow}}{4(E - D + \epsilon_{\mathbf{k}})} \quad \text{and} \quad \frac{-J_z^2 \rho_0 |\delta D| c_{\mathbf{k},\uparrow} c_{\mathbf{k}',\uparrow}^\dagger}{4(E - D - \epsilon_{\mathbf{k}'})} \qquad (3.35)$$

corresponding to an electron and hole in the intermediate state respectively.

Finally there are scattering processes involving both the longitudinal and transverse terms in which the conduction electron is scattered from an initial state $\mathbf{k} \uparrow$ to a final state $\mathbf{k}' \downarrow$ as shown in figure 3.6(i) and (ii).

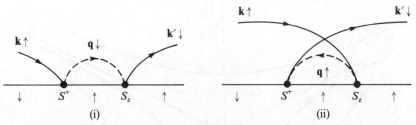

Figure 3.6 Second order diagrams with spin flip scattering and a particle (i), or a hole (ii), in an intermediate state at a band edge (dashed line).

The contributions from these diagrams are

$$-\frac{J_+J_z\rho_0|\delta D|S^+c^\dagger_{\mathbf{k}',\downarrow}c_{\mathbf{k},\uparrow}}{2(E-D+\epsilon_{\mathbf{k}})} \quad \text{and} \quad \frac{J_+J_z\rho_0|\delta D|S^+c_{\mathbf{k},\uparrow}c^\dagger_{\mathbf{k}',\downarrow}}{2(E-D-\epsilon_{\mathbf{k}'})} \qquad (3.36)$$

on using $S_zS^+ = S^+/2$. There are two similar contributions arising from the terms in which the longitudinal and transverse scatterings occur in the reverse order for which we use $S^+S_z = -S^+/2$.

Combining these results with similar terms arising from the scattering of a $\downarrow$ electron, and collecting all the terms together, we see that the elimination of the virtual scattering to the band edges in lowest order results in a Hamiltonian of the same form but with modified couplings, $J_\pm \to J_\pm + \delta J_\pm$ and $J_z \to J_z + \delta J_z$, where

$$\delta J_\pm = -J_\pm J_z\rho_0|\delta D|\left\{\frac{1}{(E-D+\epsilon_{\mathbf{k}})}+\frac{1}{(E-D-\epsilon_{\mathbf{k}'})}\right\}, \qquad (3.37)$$

$$\delta J_z = -J_+J_-\rho_0|\delta D|\left\{\frac{1}{(E-D+\epsilon_{\mathbf{k}})}+\frac{1}{(E-D-\epsilon_{\mathbf{k}'})}\right\}. \qquad (3.38)$$

The E dependence in these equations reflects the fact that the effective interaction is retarded. However, for low energy excitations relative to D the E dependence can be neglected. Similarly for scattering of conduction electrons near the Fermi surface, $\epsilon_{\mathbf{k}}$ and $\epsilon_{\mathbf{k}'}$, can also be neglected relative to D leading to the scaling equations,

$$\frac{dJ_\pm}{d\ln D} = -2\rho_0 J_z J_\pm \quad \text{and} \quad \frac{dJ_z}{d\ln D} = -2\rho_0 J_\pm^2. \qquad (3.39)$$

Note δD is negative here. Dividing these equations and integrating leads to the scaling trajectory,

$$J_z^2 - J_\pm^2 = \text{const.}, \qquad (3.40)$$

corresponding to hyperbolic curves sketched in figure 3.7.

On reducing the band width cut-off D, J_z always increases. As a consequence ferromagnetic models with $J_z < 0$ and $|J_z| \geq J_\pm$ scale to $J_\pm = 0$. The antiferromagnetic models and the ferromagnetic models

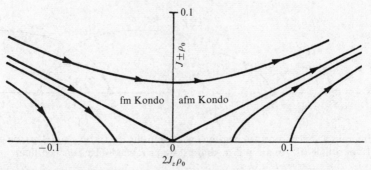

Figure 3.7   Scaling trajectories for the anisotropic s-d model, calculated in
lowest order perturbation theory (Anderson, 1970).

with $J_\pm > |J_z|$ scale away eventually from the axis $J_\pm = 0$ to stronger
and stronger couplings, beyond the regime of validity of the perturbation
method used in the derivation.

For the antiferromagnetic isotropic model, $J_\pm = J_z = J$,

$$\frac{dJ}{d\ln D} = -2\rho_0 J^2. \tag{3.41}$$

On integrating this from an initial cut-off D and coupling J to a new
cut-off $\tilde{D}$ and renormalized coupling $\tilde{J}$, assuming a constant density of
states ρ_0, we find

$$De^{-1/2J\rho_0} = \tilde{D}e^{-1/2\tilde{J}\rho_0} \sim k_{\mathrm{B}}T_{\mathrm{K}}. \tag{3.42}$$

The trajectories are characterized by the single parameter $T_{\mathrm{K}}(J, D)$,
which plays the role of a *scaling invariant*, provided we are in the weak
coupling regime $\rho_0 J \ll 1$. Systems with different parameters D, J but
which lie on the same trajectory are equivalent and have the same low
energy behaviour. Consequently the thermodynamic functions depend
only on the single energy scale T_{K}. Hence, for example, the impurity
susceptibility can be expressed in the form,

$$\chi_{\mathrm{imp}} = \frac{1}{T} F\left(\frac{T}{T_{\mathrm{K}}}\right). \tag{3.43}$$

where $F(x)$ is a universal function.

We have calculated the scaling equations only to order J^2. Terms
of order J^3 can also be calculated in the poor man's approach. The
calculation, which is rather lengthy, is outlined in appendix D, where it
has been generalized to the N-fold degenerate model (1.91). We quote
the results for the $S = \frac{1}{2}$ s-d model,

$$\frac{dJ\rho_0}{d\ln D} = -2(J\rho_0)^2 + 2(J\rho_0)^3 + \mathrm{O}((J\rho_0)^4), \tag{3.44}$$

or on inverting,

$$\int \left(-\frac{1}{2(J\rho_0)^2} - \frac{1}{2(J\rho_0)} + O(J\rho_0) \right) d(J\rho_0) = \int d\ln D, \qquad (3.45)$$

and on integrating,

$$De^{(-1/(2J\rho_0)+1/2(\ln(2J\rho_0)+O(J\rho_0)))} \sim k_B T_K. \qquad (3.46)$$

or

$$k_B T_K \sim D|2J\rho_0|^{1/2} e^{(-1/(2J\rho_0)+O(J\rho_0))}. \qquad (3.47)$$

Nothing in principle has changed here except that we get a better estimate of T_K. This form for T_K with the prefactor $|2J\rho_0|^{1/2}$ can be obtained by the direct summation of the next leading order logarithmically divergent terms in the calculation of the T matrix, in addition to the leading order ones summed by Abrikosov.

In the general case we can assume a scaling equation of the form,

$$\frac{dJ\rho_0}{d\ln D} = \frac{1}{\Psi'(J\rho_0)}. \qquad (3.48)$$

On integrating this equation can be expressed in the form,

$$\Psi(J\rho_0) = \ln(D/T_K). \qquad (3.49)$$

where Ψ' is the derivative of some function $\Psi(x)$ whose asymptotic form for small x can be deduced from (3.45). The important deduction from this is that $J\rho_0$ is a universal function of D/T_K. Points J, D and $\tilde{J}, \tilde{D}$ on the same trajectory will be related via

$$\Psi(J\rho_0) - \Psi(\tilde{J}\rho_0) = \ln(D/\tilde{D}). \qquad (3.50)$$

We could extend the calculation of $\Psi(J\rho_0)$ to fourth and higher order terms in $J\rho_0$ but these will have little effect on T_K, in the weak coupling limit, $J\rho_0 \ll 1$, which is the physically relevant parameter regime, as they give terms on the exponent of (3.47) which will be negligible compared with the leading order terms which are singular as $J\rho_0 \to 0$.

The scaling method can be used to extend the lowest order perturbational results, and effectively sum the leading order logarithmic terms. The scaling trajectory can be traversed from the initial parameters, D and J, to an effective band width $\tilde{D}$ and effective coupling $\tilde{J}$. For $T \gg T_K$ the perturbational scaling can be extended down to an effective band width $\tilde{D}$ of order T. The effective coupling $\tilde{J}$, which becomes T dependent, can then be deduced from the scaling equation (3.49). With Ψ deduced perturbationally from (3.45) the equation for $\tilde{J}(T)$ becomes

$$\frac{1}{2\tilde{J}\rho_0} - \frac{1}{2}\ln|2\tilde{J}\rho_0| + O((\tilde{J}\rho_0)^0) = \ln(T/T_K), \qquad (3.51)$$

on putting $\tilde{D} = aT$, where a is a positive constant greater than unity.

This can now be used to extend the lowest order perturbational result
for the impurity susceptibility, which for spin $\frac{1}{2}$ is given by

$$\chi_{\mathrm{imp}}(T) = \frac{(g\mu_{\mathrm{B}})^2}{4k_{\mathrm{B}}T}(1 - 2\tilde{J}\rho_0 + \mathrm{O}(\tilde{J}\rho_0)^2). \tag{3.52}$$

Equation (3.51) is solved for $\tilde{J}(T)$ and the result substituted into (3.52)
to give for $T \gg T_{\mathrm{K}}$,

$$\chi_{\mathrm{imp}}(T) = \frac{(g\mu_{\mathrm{B}})^2}{4k_{\mathrm{B}}T}\left\{1 - \frac{1}{\ln(T/T_{\mathrm{K}})} - \frac{\ln(\ln(T/T_{\mathrm{K}}))}{2\ln^2(T/T_{\mathrm{K}})} + \mathrm{O}\left(\frac{1}{\ln^2(T/T_{\mathrm{K}})}\right)\right\} \tag{3.53}$$

with T_{K} given by (3.47). This corresponds to the summation of leading
order and next leading order logarithmic terms. We will show later that
it is asymptotically exact for $\ln(T/T_{\mathrm{K}}) \gg 1$.

What have we learnt from these scaling arguments? First of all that
the energy of the conduction band width can be reduced dramatically for
the calculation of thermodynamic behaviour from its initial value, which
is on an energy scale of electron volts, down to a thermal energy scale
and still be described by a model of the same form but with a renormalized coupling $\tilde{J}$. New types of interaction terms can be generated by
higher terms in the perturbation expansion. These interaction terms,
however, behave as $1/D$ rather than $\ln D$, and so tend to zero rather
than diverge as $D \to \infty$. Such terms are *irrelevant* as they do not affect
the low temperature dynamics in this limit. The relevant $\ln D$ terms
are absorbed into the parameter T_{K}, and $k_{\mathrm{B}}T_{\mathrm{K}}$ is the only energy scale
characterizing the low energy excitations. For the isotropic ferromagnet
the perturbative scaling for $T = 0$ can be continued down to $\tilde{D} \to 0$
giving $\tilde{J} \to 0$, so the impurity spin is uncoupled in this limit. The impurity spin becomes *asymptotically free* giving a Curie law contribution to
the impurity susceptibility which diverges as $T \to 0$. The ferromagnetic
model is essentially solved. For the antiferromagnetic model, however,
which is the more physically interesting case, the conduction band width
can only be reduced to $\tilde{D}$ of the order of $k_{\mathrm{B}}T$ or $k_{\mathrm{B}}T_{\mathrm{K}}$, whichever is the
maximum; the point at which the perturbative approach begins to break
down. If $k_{\mathrm{B}}T_{\mathrm{K}} \sim 10^{-3}D$ then the perturbational scaling can be continued safely down to a band width $\tilde{D} \sim 10^{-2}D$, which is an enormous
reduction in band width. If the lowest order perturbative scaling (3.42)
is pursued beyond this point the coupling $\tilde{J}$ diverges at T_{K} as illustrated
in figure 3.8. It was conjectured in the paper by Anderson et al that
the scaling trajectory should give a finite $\tilde{J}$ for $\tilde{D}$ less than $k_{\mathrm{B}}T_{\mathrm{K}}$ and
diverge only in the limit $\tilde{D} \to 0$.

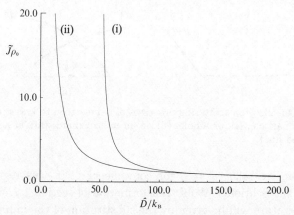

Figure 3.8 Scaling trajectories for the isotropic s-d model based on (i) second order perturbation theory and (ii) third order perturbation theory, with $T_K = 50$ K in each case.

This behaviour in fact occurs in the perturbative scaling to third order, equation (3.51), when pushed beyond its regime of validity as shown in figure 3.8. If the scaling could be continued beyond the perturbative regime a model for describing the lowest lying excitations, and hence the thermodynamic behaviour as $T \to 0$, could be generated. In general such a model will not correspond to an s-d model as further interactions will be generated by the scaling. A non-perturbative way of deriving such a model for the calculation of the low temperature behaviour was devised by Wilson (1975). Wilson's calculation will be described in the next chapter.

It is recommended that at first reading the reader skip the next section, which is best understood when the reader has become acquainted with the solution for the s-d model, and start again at chapter 4.

3.4 Scaling for the Anderson Model

Charge fluctuations of the impurity can occur in the Anderson model in contrast to the s-d model, where spin fluctuations only can occur. We can use the poor man's scaling approach, progressively reducing the band width D, and take these into account (see Haldane, 1978; Jefferson, 1977). If the parameters of the model are such that $\epsilon_d \ll -D$ and $\epsilon_d + U \gg D$, then the scattering processes which involve electrons or holes in the band edges, which take place on an energy scale of the order of D, cannot cause *real* charge fluctuations of the impurity. (Note

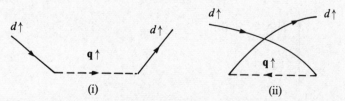

Figure 3.9 One electron scattering processes of a d electron to lowest order in V with a particle, (i), or a hole, (ii), in an intermediate state at a band edge (dashed line).

in this section we take $\epsilon_F = 0$ for convenience.) In this regime we can perform the Schrieffer–Wolff transformation of section 1.7, and the Kondo temperature will be given by the substitution of the induced antiferromagnetic coupling (1.73) into the expression (3.47) for the Kondo temperature,

$$k_B T_K \sim D \left(\frac{\Delta U}{|\epsilon_d||\epsilon_d + U|} \right)^{1/2} e^{-\pi|\epsilon_d||\epsilon_d + U|/2\Delta U}. \tag{3.54}$$

If, however, one or both the impurity levels lie within the conduction band then scattering processes on a scale D can involve real charge fluctuations of the impurity. We can estimate effects, which involve a change of impurity charge and a scattering of a conduction electron to a band edge $D - |\delta D| < |\epsilon_q| < D$, to second order in a perturbation expansion in V. If E_0, E_1 and E_2 are the energies of the three configurations, for zero d , one d and two d electrons respectively, then we can calculate the shift in these energies on reducing the band width by $|\delta D|$ to lowest order in V, using the method applied in the last section. In this case we take for H_{12} and H_{01},

$$H_{12} = \sum_{\mathbf{q},\sigma} V_{\mathbf{q}}^* c_{\mathbf{q},\sigma}^\dagger (X_{0,0:1,\sigma} + \sigma X_{1,\sigma:2}), \tag{3.55}$$

and

$$H_{01} = \sum_{\mathbf{q},\sigma} V_{\mathbf{q}} (X_{1,\sigma:0,0} + \sigma X_{2:1,-\sigma}) c_{\mathbf{q},\sigma}, \tag{3.56}$$

using the X operator notation of section 1.9, which is convenient as we are considering the effects on the different local configurations.

There is a different energy involved in the scattering of a d,σ electron into the band edge depending on whether the d,$-\sigma$ state is occupied or not. Lowest order scatterings of a d,$\uparrow$ electron are shown in figure 3.9, (i) involves an intermediate state with an electron in the upper band edge, and (ii) a hole in the lower band edge. The scattering shown in figure 3.9(i) contributes a term to the effective Hamiltonian in the projected subspace,

$$\sum_{\mathbf{q}} V_{\mathbf{q}} X_{1,\uparrow:0,0} c_{\mathbf{q},\uparrow} (E - H_{22})^{-1} \sum_{\mathbf{q}',\uparrow} V_{\mathbf{q}'}^* c_{\mathbf{q}',\uparrow}^\dagger X_{0,0:1,\uparrow}, \qquad (3.57)$$

which, to lowest order in $V_{\mathbf{q}}$, becomes

$$\sum_{\mathbf{q}} |V_{\mathbf{q}}|^2 (E - H_0 + \epsilon_{\mathrm{d}} - \epsilon_{\mathbf{q}})^{-1} X_{1,\uparrow:1,\uparrow}, \qquad (3.58)$$

on using the contraction $X_{1,\uparrow:0,0} X_{0,0:1,\uparrow} = X_{1,\uparrow:1,\uparrow}$. Summing over $\mathbf{q}$ and using the same approximations as in the previous section, this leads to

$$-\frac{\rho_0 |\delta D| |V|^2}{(D - \epsilon_{\mathrm{d}})} X_{1,\uparrow:1,\uparrow}, \qquad (3.59)$$

which gives a contribution reducing the energy of the state $|1,\uparrow\rangle$ of the impurity. If the other terms of order $|V|^2$ are calculated in the same way we find a Hamiltonian of the same form with modified energies E_0', E_1' and E_2' given by

$$E_0' = E_0 - \frac{2\Delta}{\pi} \frac{|\delta D|}{(D + \epsilon_{\mathrm{d}})}$$

$$E_1' = E_1 - \Delta \frac{|\delta D|}{\pi} \left\{ \frac{1}{(D - \epsilon_{\mathrm{d}})} + \frac{1}{(D + \epsilon_{\mathrm{d}} + U)} \right\}$$

$$E_2' = E_2 - 2\Delta \frac{|\delta D|}{\pi} \left\{ \frac{1}{(D - \epsilon_{\mathrm{d}} - U)} \right\}, \qquad (3.60)$$

using $\rho_0 |V|^2 = \pi \Delta$.

As $\epsilon_{\mathrm{d}} = E_1 - E_0$ and $U = E_2 - 2E_1 + E_0$, equations (3.60) correspond to changes in these parameters,

$$\delta\epsilon_{\mathrm{d}} = \frac{\Delta |\delta D|}{\pi} \left\{ \frac{2}{(D + \epsilon_{\mathrm{d}})} - \frac{1}{(D - \epsilon_{\mathrm{d}})} - \frac{1}{(D + \epsilon_{\mathrm{d}} + U)} \right\},$$

$$\delta U = \frac{2\Delta |\delta D|}{\pi} \left\{ \frac{1}{(D - \epsilon_{\mathrm{d}})} - \frac{1}{(D + \epsilon_{\mathrm{d}})} + \frac{1}{(D + \epsilon_{\mathrm{d}} + U)} - \frac{1}{(D - \epsilon_{\mathrm{d}} - U)} \right\}. \qquad (3.61)$$

We have quite a number of parameter regimes to investigate. If the levels ϵ_{d} and $\epsilon_{\mathrm{d}} + U$ lie well within the conduction band $|\epsilon_{\mathrm{d}}|, |\epsilon_{\mathrm{d}} + U| \ll D$ then $\delta\epsilon_{\mathrm{d}} \sim 0$ and $\delta U \sim 0$, and there is no significant renormalization. This is due to the fact that, on an energy scale D greater than the interaction energy U, the impurity is effectively non-interacting and consequently there are no many-body renormalizations. If, however, U is so large that $\epsilon_{\mathrm{d}} + U \gg D$, but $|\epsilon_{\mathrm{d}}| \leq D$ terms in (3.61) do not cancel and there may be some renormalization of ϵ_{d}. In the regime $D \gg \epsilon_{\mathrm{d}}$, the scaling equation for ϵ_d is approximately given by

$$\frac{d\epsilon_d}{d\ln D} = -\frac{\Delta}{\pi}, \tag{3.62}$$

to order $|V|^2$.

The processes involving charge scattering to the band edges do not lead to any renormalization of the hybridization matrix element $V_{\mathbf{k}}$. These only occur to higher order in V by expanding $(E - H_{22})^{-1}$ in (3.57) in powers of V. The lowest order corrections to the new coupling V' are of order,

$$V' = V\left(1 + O\left(\frac{\Delta}{D^2}\right)\right), \tag{3.63}$$

so that,

$$\frac{d\Delta}{d\ln D} = O\left(\frac{\Delta}{D}\right), \tag{3.64}$$

and hence for $\Delta \ll D$ there is no significant renormalization of Δ. We can therefore integrate (3.62) keeping Δ constant to give the scaling trajectory,

$$\epsilon_d + \frac{\Delta}{\pi}\ln D = \text{const.} \tag{3.65}$$

We choose the scaling invariant as ϵ_d^* so that (3.65) takes the form,

$$\epsilon_d + \frac{\Delta}{\pi}\ln\left(\frac{\pi D}{2\Delta}\right) = \epsilon_d^*. \tag{3.66}$$

Hence, the scaling trajectories are characterized by two scaling invariants Δ and ϵ_d^*. This implies that the impurity susceptibility for the Anderson model in the large U limit can be expressed in an form,

$$\chi_{\text{imp}} = \frac{1}{\Delta}F_A\left(\frac{T}{\Delta}, \frac{T}{\epsilon_d^*}\right), \tag{3.67}$$

in zero field. This also applies if $U + \epsilon_d < D$ but in this case, as the renormalization of ϵ_d does not occur until $\tilde{D} \sim U + \epsilon_d$, then in (3.66), the formula for ϵ_d^*, the upper cut-off D is replaced by $U + \epsilon_d$.

In the local moment regime for $-D \ll \epsilon_d \ll 0$, the conduction band cut-off can be progressively reduced to give a renormalization of ϵ_d until $-\tilde{D}$ is of the order of, but less than, $\tilde{\epsilon}_d$ and the perturbation theory begins to break down. For $\tilde{\epsilon}_d$ such that, $|\tilde{\epsilon}_d|, |(\tilde{\epsilon}_d + U)| \gg \tilde{D}$, real charge fluctuations cannot occur on an energy scale $\tilde{D}$. Virtual charge fluctuations, however, can be taken into account by using the Schrieffer–Wolff transformation to transform to the s-d model. The previous results of the poor man's scaling for this model can be taken over. In this two stage process, the scaling of charge fluctuations in the first stage is terminated when $\tilde{D} \sim -a\tilde{\epsilon}_d$, where a is a positive constant of $O(1)$ but greater than unity, where the renormalized value of the d level $\tilde{\epsilon}_d$ is given by

$$\tilde{\epsilon}_d + \frac{\Delta}{\pi}\ln\left(\frac{a\pi|\tilde{\epsilon}_d|}{2\Delta}\right) = \epsilon_d^*. \qquad (3.68)$$

Then application of the Schrieffer–Wolff transformation of section 1.7 leads to a coupling $\tilde{J}$,

$$\tilde{J} = |V|^2\left(\frac{1}{|\tilde{\epsilon}_d|} + \frac{1}{|U + \tilde{\epsilon}_d|}\right). \qquad (3.69)$$

These values for $\tilde{J}$ and $\tilde{D}$ can then be taken as the initial parameters for the second stage scaling which is that for the s-d model. The value for T_K is therefore obtained by substituting (3.69) into (3.47), together with $D = a|\tilde{\epsilon}_d|$. We consider this for simplicity in the limit $U \to \infty$. The resulting expression for T_K is

$$k_B T_K \sim \Delta e^{\pi\epsilon_d^*/2\Delta}. \qquad (3.70)$$

On using (3.66) for ϵ_d^*,

$$k_B T_K \sim D\left(\frac{\Delta}{D}\right)^{1/2} e^{\pi\epsilon_d/2\Delta}. \qquad (3.71)$$

The prefactor in the expression for T_K in general depends on the relative values of $D, |\epsilon_d|$ and U. For example if $U \to \infty$, and ϵ_d lies below the conduction band ($|\epsilon_d| > D$) from (3.54) we obtain an expression for T_K with a prefactor $(\Delta/|\epsilon_d|)^{1/2}$. Hence for $U \to \infty$ we have

$$k_B T_K \sim D\left(\frac{\Delta}{\max(|\epsilon_d|, D)}\right)^{1/2} e^{\pi\epsilon_d/2\Delta}. \qquad (3.72)$$

The general conclusion we can draw from this is that T_K may be a complicated function of the parameters Δ, $D, |\epsilon_d|$ and U, dependent on their relative values, but if we can transform from our initial model to an effective s-d model for the low energy excitations, then the thermodynamics on this energy scale will be dependent only on the single parameter T_K.

For the Anderson model there are regimes of interest other than the local moment regime. In the mixed valence regime, $\epsilon_d^* < \Delta$, the scaling can be continued until $D \sim \Delta$ or $k_B T$, whichever is the larger. Hence the renormalized level in this regime is given by

$$\epsilon_d' = \epsilon_d + \frac{\Delta}{\pi}\ln\left(\frac{\pi D}{2\max(\Delta, k_B T)}\right), \qquad (3.73)$$

for $U = \infty$, so for $T = 0$, $\epsilon_d' = \epsilon_d^*$. The characteristic feature of the mixed valence regime is that the charge fluctuations of the impurity can be thermally excited and these excitations largely determine the thermodynamic behaviour. As ϵ_d^* increases and passes through the Fermi level the impurity level becomes depopulated. When $\epsilon_d^* \gg \Delta$, $\langle n_d \rangle \sim 0$ thermal fluctuations of the impurity are negligible. In this regime for $U = \infty$ the scaling can be continued until $\tilde{D} = a'\tilde{\epsilon}_d$, where a' is a constant (> 1). The value of $\tilde{\epsilon}_d$ where this occurs, ϵ_d', is given by

$$\epsilon_d' + \frac{\Delta}{\pi}\ln\left(\frac{a'\pi\epsilon_d'}{2\Delta}\right) = \epsilon_d^*. \tag{3.74}$$

In this regime perturbation calculations in powers of Δ/ϵ_d are valid. The impurity susceptibility for the 'bare' model is given by

$$\chi_{\text{imp}} = \frac{(g\mu_B)^2\Delta}{2\pi\epsilon_d{}^2}\left(1 + \frac{2\Delta}{\pi\epsilon_d}\ln\left(\frac{\epsilon_d}{D}\right) + \ldots\right), \tag{3.75}$$

for $\epsilon_d \gg k_B T$ (Haldane, 1978). This can be improved using the scaling results. The band width D is reduced to $\tilde{D} = a'\epsilon_d'$, the perturbation result (3.75) is applied to the new effective Hamiltonian. With $D = a'\epsilon_d'$, $\epsilon_d = \epsilon_d'$ in (3.75) this gives

$$\chi_{\text{imp}} = \frac{(g\mu_B)^2\Delta}{2\pi(\epsilon_d')^2}\left(1 + O\left(\frac{2\Delta}{\pi\epsilon_d'}\right)\right). \tag{3.76}$$

The $\ln D$ terms have now been absorbed in the effective level ϵ_d'. By solving for ϵ_d' in terms of the bare parameters ϵ_d, Δ and D, and iterating in powers of $1/\epsilon_d$, it can be verified that (3.76) generates the terms in (3.75).

The scaling ideas outlined here have not provided a solution for the thermodynamics of the s-d and Anderson models in the local moment regime except in the case of ferromagnetic coupling. This is because the poor man's approach is perturbative and breaks down at $T \sim T_K$ for an antiferromagnetic interaction between a local moment and conduction electrons. Nevertheless they have been useful in extending the leading order perturbational results. More importantly they have provided a conceptual framework, clarifying what is required at the next stage, some non-perturbational scheme for calculating the excitations on an energy scale $T \ll T_K$. The calculational scheme devised by Wilson (1975) to carry out this next stage is described in the next chapter.

4

Renormalization Group Calculations

4.1 The Renormalization Group

The renormalization group approach to problems in condensed matter was largely pioneered by Wilson in the early 1970s. The main area of application initially was to the problem of calculating critical exponents for systems undergoing a second order phase transition (i.e. the values of ν for physical quantities behaving as $(T - T_c)^{\nu}$ near the transition temperature T_c). This approach developed earlier scaling ideas of Kadanoff to give a whole new theoretical framework for considering critical phenomena for phase transitions. Wilson adopted this approach for the $S = \frac{1}{2}$ Kondo problem, building on the scaling ideas of Anderson which we have considered in some detail in the last chapter. For those not familiar with renormalization group methods we shall briefly introduce in this section the key concepts of the approach which we shall require later.

The renormalization group is a mapping R of a Hamiltonian $H(\mathbf{K})$, which is specified by a set of interaction parameters or couplings $\mathbf{K} = (K_1, K_2, \ldots)$ into another Hamiltonian of the same form with a new set of coupling parameters $\mathbf{K}' = (K_1', K_2', \ldots)$. This is expressed formally by

$$R\{H(\mathbf{K})\} = H(\mathbf{K}'), \tag{4.1}$$

or equivalently,

$$R(\mathbf{K}) = \mathbf{K}'.$$

The transformation is in general non-linear. Such transformations were generated in the scaling approach in the last chapter in eliminating higher energy states. The new effective Hamiltonian is then valid over a reduced energy scale. In applications to critical phenomena the new Hamiltonian is obtained by removing short range fluctuations to generate an effective Hamiltonian valid over larger length scales. The transformation is usually characterized by a parameter, say α, which specifies the ratio of the new length or energy scale to the old one. A sequence of transformations,

$$\mathbf{K}' = R_\alpha(\mathbf{K}), \quad \mathbf{K}'' = R_\alpha(\mathbf{K}'), \quad \mathbf{K}''' = R_\alpha(\mathbf{K}''), \quad \text{etc.} \qquad (4.2)$$

generates a sequence of points or, where α is a continuous variable, a *trajectory* in the parameter space $\mathbf{K}$. The transformation is constructed so that it satisfies

$$R_{\alpha'}\{R_\alpha(\mathbf{K})\} = R_{\alpha\alpha'}(\mathbf{K}). \qquad (4.3)$$

In applications, as the transformations are generated either by a reduction in energy scale or a coarse graining of space, inverse transformations do not exist so that, strictly speaking, the transformations constitute a mathematical semi-group rather than a group.

One of the key concepts of the renormalization group is that of a *fixed point*. This is a point $\mathbf{K}^*$ which is invariant under the transformation,

$$R_\alpha(\mathbf{K}^*) = \mathbf{K}^*. \qquad (4.4)$$

The trajectories generated by the repeated application of the renormalization group tend to be drawn towards, or expelled from, the fixed points. The behaviour of the trajectories near a fixed point can usually be determined by linearizing the transformation in the neighbourhood of the fixed point. If in the neighbourhood of a particular fixed point $\mathbf{K} = \mathbf{K}^* + \delta\mathbf{K}$ then, expanding $R_\alpha(\mathbf{K})$ in powers of $\delta\mathbf{K}$,

$$R_\alpha(\mathbf{K}^* + \delta\mathbf{K}) = \mathbf{K}^* + \mathbf{L}_\alpha^* \delta\mathbf{K} + \mathrm{O}(\delta\mathbf{K}^2), \qquad (4.5)$$

where $\mathbf{L}_\alpha^*$ is a linear transformation. If the eigenvectors and eigenvalues of $\mathbf{L}_\alpha^*$ are $\mathbf{O}_n^*(\alpha)$ and $\lambda_n^*(\alpha)$, and if these are complete (there is in general nothing to ensure this, so it is an assumption) then they can be used as a basis for a representation of the vector $\delta\mathbf{K}$,

$$\delta\mathbf{K} = \sum_n \delta K_n \mathbf{O}_n^*(\alpha), \qquad (4.6)$$

where δK_n are the components.

How the trajectories move in the region of a particular fixed point depends on the eigenvalues λ_n^*. If we act m times on a point $\mathbf{K}$ in the neighbourhood of a fixed point $\mathbf{K}^*$ then, from (4.5), we find

$$R_\alpha^m(\mathbf{K}^* + \delta\mathbf{K}) = \mathbf{K}^* + \sum_n \delta K_n \lambda_n^{*m} \mathbf{O}_n^*(\alpha), \qquad (4.7)$$

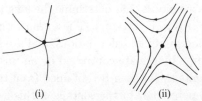

Figure 4.1 Trajectories in the neighbourhood of a fixed point, (i) a stable fixed point, (ii) an unstable fixed point with one relevant eigenvector.

provided all the points generated by the transformation are in the vicinity of the fixed point so that the linear approximation (4.5) remains valid. For eigenvalues $\lambda_n^* > 1$, which are termed *relevant*, the corresponding components of $\delta\mathbf{K}$ in (4.7) increase with m. Those corresponding to eigenvalues $\lambda_n^* < 1$, termed *irrelevant* (this is a technical term and should not be always taken literally), get smaller with m. Eigenvalues $\lambda_n^* = 1$ are termed *marginal* and the corresponding components in (4.7) do not vary with m. The eigenvalues of the linearized equation lead to a classification of the fixed points. *Stable* fixed points have only irrelevant eigenvalues so $\delta\mathbf{K} \to 0$ and the trajectories in their neighbourhood are drawn in towards the fixed point. If there are one or more relevant eigenvalues then the fixed points are *unstable* and the trajectories are eventually driven away from the fixed point in a direction largely determined by the most relevant eigenvector (see figure 4.1). A fixed point is *marginal* if it has *no* relevant eigenvalues and at least one marginal one. In this case the behaviour of the trajectories in the neighbourhood of the fixed point cannot be determined solely from the linearized form (4.5) and the non-linear corrections have to be examined. There may be competitive influences on a trajectory due to several fixed points. The region in which a trajectory passes from the sphere of influence of one fixed point to that of another is known as a *crossover* region.

Applications of the renormalization group to critical phenomena and to the Kondo problem can both be viewed as extensions of earlier scaling approaches. In the real space scaling approach for localized magnetic models, developed by Kadanoff (1966, 1967), blocks of spins of increasing length scale are considered. Transformations are then generated to find how the interaction between blocks varies as the size of the block increases, on the assumption that the interblock interaction can always be described by a model of the same form (at least approximately). The critical exponents are then deduced from these scaling transformations in the limit in which the block size tends to infinity (it is the interactions between the very large blocks that determine the long range fluctuations that determine the critical behaviour). These transformations are essentially renormalization group transformations, and are such that the trajectories for very large blocks are controlled by an unstable fixed point. The critical exponents ν can then be obtained from the relevant eigenvalues of this fixed point. For further details of this type of application we refer the reader to the review of Wilson & Kogut (1974).

The poor man's scaling for the Kondo problem discussed in the last chapter corresponds to a renormalization group approach generated by reducing the energy scale D of the conduction band. In this scaling transformation we map onto the s-d model at each stage so that the parameter space is one dimensional, and consists simply of the coupling J. It follows quite trivially that the transformation has a fixed point at $J = 0$. For a system with a ferromagnetic coupling the scaling trajectories are drawn to this stable fixed point as $D \to 0$. The low temperature behaviour of the ferromagnetic models is determined by the effective Hamiltonian in the neighbourhood of this fixed point, as it is this Hamiltonian which describes the lowest lying excitations of the system.

For an antiferromagnetic interaction we found within the poor man's scaling approach that the coupling strength J increased as the energy scale D decreased. The perturbational derivation, however, became invalid as D approached T_K but the conjecture of Anderson et al (1970) was that J increased indefinitely as $D \to 0$ leading to a spin compensated ground state for $S = \frac{1}{2}$. This is equivalent to the assertion that the renormalization group trajectories approach a $J = \infty$ fixed point as $D \to 0$. This implies that the low temperature behaviour of the system is determined by the effective Hamiltonian in the neighbourhood of this fixed point. Wilson's calculation for the $S = \frac{1}{2}$ s-d model set out to prove that the $J \to \infty$ is the low energy fixed point and to find the effective Hamiltonian near this fixed point to calculate the low temperature thermodynamic behaviour.

In practice renormalization group transformations can be set up in a variety of ways and depend very much on the problem being considered. The perturbational approach was suitable for the ferromagnetic Kondo problem but broke down in the antiferromagnetic case for energy scales of the order of T_K. To transform to lower energy scales Wilson cast the s-d model in the form of a linear chain so that he could use an iterative method based on numerical diagonalization, hence overcoming the limitations of the perturbational approach. In the next section we show how the s-d model can be expressed in a linear chain form, and then see how to modify it for a numerical iterative scheme to work.

4.2 Linear Chain Form for the s-d Model

If we take the $\mathbf{k}$ independent interaction (2.38), or more generally a separable interaction of the form $J_{\mathbf{kk'}} = \alpha_k \alpha_{k'} J$ (which is almost the form generated by the Schrieffer–Wolff transformation (1.73)), then the exchange interaction in the s-d model is between the localized spin S and the spin of an electron in a localized one electron state, $|0\rangle$, with a creation operator $c_{0,\sigma}^\dagger$ given by

$$c_{0,\sigma}^\dagger = \sum_{\mathbf{k}} \alpha_k c_{\mathbf{k},\sigma}^\dagger, \tag{4.8}$$

where α_k is chosen to satisfy the normalization condition $\sum_{\mathbf{k}} |\alpha_k|^2 = 1$. The s-d model (1.64) can then be written in the form,

$$H_{s-d} = 2J\, \mathbf{S} \cdot c_{0,\sigma}^\dagger (\mathbf{s}_0)_{\sigma,\sigma'} c_{0,\sigma'}^\dagger, \tag{4.9}$$

where $\mathbf{s}_0$ is the spin operator for the localized orbital.

The interaction in this form looks simpler but if we keep the conduction electron part of the Hamiltonian unchanged then nothing has been achieved. We can, however, construct a new basis for the conduction electron states starting from the localized state $|0\rangle$. If we denote the conduction electron Hamiltonian (1.2) by H_c then we can construct a new basis for the conduction electron states from the sequence of states, $|0\rangle$, $H_c|0\rangle$, $H_c^2|0\rangle$, $H_c^3|0\rangle$ ... by Schmidt orthogonalization. With respect to this basis the matrix for H_c is tridiagonal and corresponds to a tight-binding chain with nearest neighbour hopping matrix elements. To see this in detail let us construct the state $|1\rangle$ from $|0\rangle$ and $H_c|0\rangle$,

$$|1\rangle = \frac{1}{\gamma_0}(H_c|0\rangle - |0\rangle\langle 0|H_c|0\rangle), \tag{4.10}$$

which is chosen so that $\langle 1|0\rangle = 0$, and γ_0 is chosen so as to normalize $|1\rangle$. Similarly constructing the state $|2\rangle$ from the first three terms in the sequence, so that it is orthogonal to $|0\rangle$ and $|1\rangle$, gives

$$|2\rangle = \frac{1}{\gamma_1}(H_c|1\rangle - |1\rangle\langle 1|H_c|1\rangle - |0\rangle\langle 0|H_c|1\rangle). \tag{4.11}$$

Using the fact that H_c is Hermitian, it is straightforward to prove that the general state $|n+1\rangle$, constructed from the states in the sequence, $H_c^m|0\rangle$ for $m = 0, 1, 2 \ldots n$, is given by

$$|n+1\rangle = \frac{1}{\gamma_n}(H_c|n\rangle - |n\rangle\langle n|H_c|n\rangle - |n-1\rangle\langle n-1|H_c|n\rangle), \tag{4.12}$$

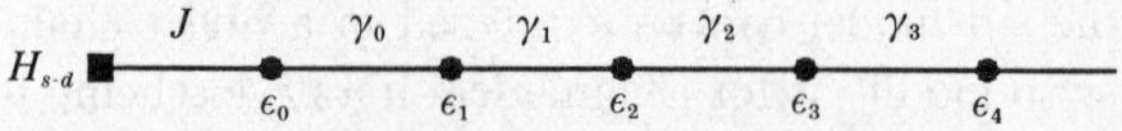

Figure 4.2   Equivalent tight-binding linear chain for the s-d model. The impurity is at the end of the chain. The γ_ns are the nearest neighbour hopping matrix elements between the levels ϵ_n located at sites along the chain (see text).

Multiplying (4.12) on the left by $\langle m|$ gives $\langle m|H_c|n\rangle = 0$ for $m = 0, 1, 2 \ldots n - 2$. It follows that, as H_c is Hermitian, it is tridiagonal in this basis. The off diagonal elements are given by $\langle n+1|H_c|n\rangle = \gamma_n$, which follows from (4.12) on multiplying on the left by $\langle n+1|$. If the diagonal elements $\langle n|H_c|n\rangle = \epsilon_n$ then

$$H_c|n\rangle = \gamma_n|n+1\rangle + \epsilon_n|n\rangle + \gamma_n^*|n-1\rangle. \qquad (4.13)$$

Hence, in second quantized form, the conduction electron Hamiltonian becomes

$$H_c = \sum_{n,\sigma} \epsilon_n c_{n,\sigma}^\dagger c_{n,\sigma} + \sum_{n,\sigma} (\gamma_n c_{n,\sigma}^\dagger c_{n+1,\sigma} + \gamma_n^* c_{n+1,\sigma}^\dagger c_{n,\sigma}). \qquad (4.14)$$

The transformation we have applied corresponds to the well known Lanczos algorithm for converting matrices to a tridiagonal form. What is unusual here is that we applied it to an operator H_c that was initially diagonal, which would seem to be making life more complicated. However, when we combine the form (4.14) with the impurity part (4.9) we have the s-d model in the form of a semi-infinite tight-binding linear chain with the impurity situated at the end; shown schematically in figure 4.2.

We have made no assumptions about the form of the conduction electrons states in deriving (4.14) so we can conclude that the s-d model can be written in a one dimensional form provided the interaction is separable. From any set of conduction electron states $|\mathbf{k}\rangle$, and the initial orbital $|0\rangle$, ϵ_n and γ_n can be calculated by the procedure outlined in equations (4.10) and (4.12). However, as we just want to take a typical density of conduction electrons states, rather than some particular form (we are interested in universal behaviour and not in features arising from a particular choice of conduction electron states), we can *choose* a suitable form for the ϵ_ns and γ_ns. The choice,

$$\epsilon_n = 0, \quad \text{and} \quad \gamma_n = D/2 \quad \text{for all } n, \qquad (4.15)$$

is a suitable one. If the $\alpha_{\mathbf{k}}$s in (4.8) are independent of $\mathbf{k}$ then this choice corresponds to a semi-elliptical density of states for the conduction electrons,

$$\rho_0(\epsilon) = \frac{2}{\pi D^2}(D^2 - \epsilon^2)^{1/2}. \qquad (4.16)$$

We can prove this by calculating the Green's function $G^0_{0,0,\sigma}(\epsilon)$ for the first site of the chain (when uncoupled from the impurity), in two ways. From (4.8) and (1.2) we find,

$$G^0_{0,0,\sigma}(\epsilon) = \sum_{\mathbf{k}} \frac{|\alpha_{\mathbf{k}}|^2}{(\epsilon I - \epsilon_{\mathbf{k}})}. \qquad (4.17)$$

Alternatively, we can calculate it in the form of a recurrence relation by relating the matrix element, $\langle 0|(\epsilon I - H_0)^{-1}|0\rangle$, to the matrix element, $\langle 1|(\epsilon I - H_0')^{-1}|1\rangle$, which is the matrix element for the next site along where H_0' is the Hamiltonian for the chain with the first site removed (we partition the matrix $(\epsilon I - H_0)$ and then invert). This leads to the result,

$$G^0_{0,0,\sigma}(\epsilon) = \frac{1}{(\epsilon - \epsilon_0 - \gamma_0^2 g_1(\epsilon))}, \qquad (4.18)$$

where

$$g_n(\epsilon) = \frac{1}{(\epsilon - \epsilon_n - \gamma_n^2 g_{n+1}(\epsilon))}, \qquad (4.19)$$

and $g_0(\epsilon) = G^0_{0,0,\sigma}(\epsilon)$. Equations (4.18) and (4.19) generate a solution in the form of a continued fraction. For the parameter choice (4.15), $g_0(\epsilon) = g_1(\epsilon)$, so equation (4.19) becomes a quadratic equation for $g_0(\epsilon)$ with the solution,

$$G^0_{0,0,\sigma}(\epsilon) = \frac{2(\epsilon - (\epsilon^2 - D^2)^{1/2})}{D^2}. \qquad (4.20)$$

If $\alpha_{\mathbf{k}}$ is independent of $\mathbf{k}$, then from (4.17), we see that the imaginary part of $G^0_{0,0,\sigma}(\epsilon^+)$ is proportional to $\rho_0(\epsilon)$ so that the normalized density of states calculated from (4.20) is given by (4.16).

A Hamiltonian in this linear chain form, (4.9) and (4.14), suggests a renormalization group type of approach in which the impurity effects are calculated by considering chains of increasing length scale from the impurity. An iterative procedure can be envisaged in which the impurity Hamiltonian (4.9) plus a finite length of chain, say n sites, is diagonalized numerically, and the states of the system examined on progressively lower energy scales by increasing n. There are two main difficulties with this procedure. One problem is that the size of the matrix to be diagonalized increases rapidly with n, and it is difficult to diagonalize the full Hamiltonian for $n > 8$. To obtain energy levels on a scale less than $k_{\mathrm{B}}T$, in order to calculate thermodynamic properties, results are

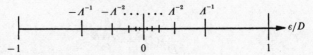

Figure 4.3   Logarithmic discretization of the states of the conduction band. The Fermi energy is zero, and the top and bottom of the conduction band at $\epsilon = \pm D$.

required for very long chains. This problem might be overcome if some form of truncation can be made, keeping only a set of the lowest states at each stage. The problem that arises then is that, as the couplings along the chain do not fall off with increasing n, the procedure does not converge. The fact that the couplings do not fall off with n is not a special feature of (4.15) but is the case for any continuum of conduction electron states. Wilson in his calculation used a form corresponding to a discrete approximation to the density of states,

$$\epsilon_n = 0, \quad \text{and} \quad \gamma_n = \frac{D(1 + \Lambda^{-1})}{2\Lambda^{n/2}} \quad \text{for all } n, \qquad (4.21)$$

with $\Lambda > 1$. We look at the reasoning leading to this choice in the next section.

4.3 Logarithmic Discretization

The problems for which the renormalization group approach was designed are ones in which there is no characteristic energy or length scale, and every energy or length scale makes a contribution. This is the case with the Kondo problem as the integrals that lead to the breakdown of perturbation theory are logarithmic,

$$\int_{k_{\mathrm{B}}T}^{D} \frac{d\epsilon}{\epsilon} \sim \ln\left(\frac{k_{\mathrm{B}}T}{D}\right). \qquad (4.22)$$

As noted earlier, if we cannot take the limit $D \to \infty$ in (4.22), the higher energy scales must be important. To reflect this logarithmic dependence, Wilson divided the energy scales within the band into intervals such that

$$D\Lambda^{-(n+1)} < |\epsilon_{\mathbf{k}} - \epsilon_{\mathrm{F}}| < D\Lambda^{-n} \quad n = 1, 2 \dots . \qquad (4.23)$$

Each subinterval contributes equally to the logarithmic integral (4.22) and Λ is chosen to be greater than unity (see figure 4.3).

A simplification can be made by expanding $\epsilon_{\mathbf{k}}$ in a power series in $k - k_{\mathrm{F}}$, on the assumption that $\epsilon_{\mathbf{k}}$ depends only on $|\mathbf{k}|$ $(= k)$. Then

$$\epsilon_{\mathbf{k}} = \epsilon_{\mathrm{F}} + (k - k_{\mathrm{F}})\frac{d\epsilon_k}{dk} + \mathrm{O}((k - k_{\mathrm{F}})^2), \qquad (4.24)$$

on neglecting terms beyond first order in the expansion, and choosing the units and energy zero so that $k_F = 0$, $(\epsilon_F = 0)$, and $\epsilon = k$. We make a similar simplification to the conduction density of states and keep the constant term only corresponding to the value at the Fermi level which we take to be at the centre of the band. This is equivalent to the flat band density of states,

$$\rho_0(\epsilon) = \frac{1}{2D}, \quad -D < \epsilon < D. \tag{4.25}$$

A complete set of orthogonal functions spanning k-space can be generated by setting up a Fourier series in each of the subintervals,

$$a^\dagger_{m,p,\sigma} = \frac{\Lambda^{m/2}}{D(1 - 1/\Lambda)^{1/2}} \sum_{[k]} e^{i\omega_m pk} c^\dagger_{k,\sigma}, \tag{4.26}$$

where $[k]$ denotes the subinterval m such that $\Lambda^{-(m+1)} < k < \Lambda^{-m}$, p is an integer, $\omega_m = 2\pi\Lambda^{m+1}/(\Lambda - 1)$ and $c^\dagger_{k,\sigma}$ creates an s partial wave state about the impurity site. There is a corresponding set of operators $b^\dagger_{p,m,\sigma}$ for k in the corresponding subinterval in the negative range $(\Lambda^{-(m+1)} < -k < \Lambda^{-m})$. The operators $a^\dagger_{m,p,\sigma}$, $b^\dagger_{m,p,\sigma}$ and their Hermitian conjugates obey the standard fermion anticommutation rules.

For $p = 0$ (4.26) simplifies to

$$a^\dagger_{m,0,\sigma} = \frac{\Lambda^{m/2}}{D(1 - 1/\Lambda)^{1/2}} \sum_{[k]} c^\dagger_{k,\sigma}. \tag{4.27}$$

Hence for $\alpha_\mathbf{k}$ in (4.8) independent of k, $c^\dagger_{0,\sigma}$ can be expressed in terms of the states with $p = 0$,

$$c^\dagger_{0,\sigma} = (1 - \Lambda^{-1})^{1/2} \sum_m \Lambda^{-m/2}(a^\dagger_{m,0,\sigma} + b^\dagger_{m,0,\sigma}). \tag{4.28}$$

The fundamental approximation in Wilson's calculation is made at this point, all the states with $p \neq 0$ are neglected so a single state is selected from each sub-interval. The rejected states are not directly coupled to the impurity and the off-diagonal elements that link them to $p = 0$ states can be shown to be proportional to $(1 - \Lambda^{-1})$, so that the error involved in neglecting them goes to zero in the continuum limit $\Lambda \to 1$. The error in practice in neglecting these can be assessed later. With this assumption, using the tridiagonalization procedure outlined in the previous section, we can put the conduction electron Hamiltonian in the linear form with nearest neighbour hopping (4.14) with

$$\epsilon_n = 0, \quad \text{and} \quad \gamma_n = \frac{D(1 + \Lambda^{-1})\varepsilon_n}{2\Lambda^{n/2}} \quad \text{for all } n, \tag{4.29}$$

where ε_n, as calculated by Wilson, is given by

$$\varepsilon_n = (1 - \Lambda^{-(n+1)})(1 - \Lambda^{-2n+1})^{-1/2}(1 - \Lambda^{-(2n+1)})^{1/2}. \tag{4.30}$$

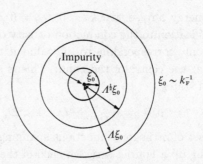

Figure 4.4 Spherical shells about the impurity site indicating the extent of the wavepackets of conduction electrons which are taken into account at successive stages in the numerical renormalization group scheme.

As $\varepsilon_n \to 1$ as $n \to \infty$, this is the same as (4.21) for large n. The conduction density of states corresponding to (4.29) is a discrete approximation to (4.25), due to the neglect of the $p \neq 0$ states, with a limit point at the Fermi level $\epsilon_F = 0$. With this form for the γ_ns it becomes possible to define an iterative procedure which converges as $n \to \infty$. The larger the value of Λ (> 1) the more rapid the convergence, but the greater the error in neglecting the states with non-zero p, so some form of compromise is made in the choice of Λ. Wilson found values of $\Lambda \sim 2$ worked as a suitable compromise in practice. He also found the results for the thermodynamics depended only on the γ_ns for large n so that the term ε_n in (4.30) can be set equal to unity giving the form (4.21).

Before going on to consider the iterative scheme in detail in the next section we should perhaps review the physical as well as the mathematical motivation for Wilson's linear chain form of the s-d model. The basis states of the conduction electrons were chosen so that the impurity is coupled directly to a local orbital, which in turn is coupled to shells or packets of conduction states of increasing length scales, the dimension of the nth shell being approximately proportional to $\Lambda^{n/2}$ (see figure 4.4).

The states most strongly coupled to the impurity are those in shells physically close to the impurity and correspond to small n. These states are built up from conduction states spanning the full band width D. The more distant shells, with larger n, are composed of low energy conduction states, they are built up from states near the Fermi level. Though these are weakly coupled to the impurity they have to be taken into account when calculating the low energy excitations, as these involve states near the Fermi level. Those conduction electrons which are high in energy and physically far removed from the impurity are neglected. As the shells with small and large n are coupled, to calculate the low energy states and thermodynamics of the impurity all energy and length scales are important. In the renormalization group transformation the coupling between the length scales is taken into account progressively, working out from the impurity to the longer length scales and lower energies. We shall see precisely how this scheme is realized in practice in the next section.

4.4 The Numerical Renormalization Group Calculations

In the iterative diagonalization scheme we diagonalize a Hamiltonian H_N, consisting of the impurity spin, which in Wilson's calculation was for $S = \frac{1}{2}$, and N conduction states along the chain, where H_N is given by

$$H_N = D' \sum_{n=0,\sigma}^{N-1} \Lambda^{-n/2}(c_{n,\sigma}^\dagger c_{n+1,\sigma} + c_{n+1,\sigma}^\dagger c_{n,\sigma}) + 2J \sum_{\sigma,\sigma'} \mathbf{S} \cdot c_{0,\sigma}^\dagger (\mathbf{s}_0)_{\sigma\sigma'} c_{0,\sigma'},$$

(4.31)

and $D' = D(1 + \Lambda^{-1})/2$. We have dropped the p index as all states correspond to $p = 0$.

At the next stage, with the addition of a further site along the chain, the Hamiltonian is

$$H_{N+1} = H_N + D' \sum_\sigma \Lambda^{-N/2}(c_{N,\sigma}^\dagger c_{N+1,\sigma} + c_{N+1,\sigma}^\dagger c_{N,\sigma}).$$

(4.32)

The many-body eigenstates of H_N, $|n_e, S, S_z, m\rangle$, can be classified by the quantum numbers, n_e, the total number of electrons, S, the total spin quantum number, S_z the z-component of spin, and an additional number m to identify states with the same set of quantum numbers. We multiply this set of states by the states $|0, 0\rangle$, $|\uparrow, 0\rangle$, $|0, \downarrow\rangle$ and $|\uparrow, \downarrow\rangle$, for the extra site, to construct a basis for the diagonalization of H_{N+1}. When the number of states gets too large for the numerical diagonalization, the lowest N_{st} states (of the order of 1000) are retained at each stage. As the higher energy levels are neglected, the states which are calculated are spread over a decreasing energy scale. To obtain a sequence that has a limit as $N \to \infty$ the energy scale must be adjusted to be comparable with the previous stage. This is done by defining a sequence of Hamiltonians $\overline{H}_N$ by

$$\overline{H}_N = \frac{\Lambda^{(N-1)/2} H_N}{D'}.$$

(4.33)

These Hamiltonians satisfy a recurrence relation,

$$\overline{H}_{N+1} = \Lambda^{1/2}\overline{H}_N + \sum_\sigma (c^\dagger_{N,\sigma}c_{N+1,\sigma} + c^\dagger_{N+1,\sigma}c_{N,\sigma}). \tag{4.34}$$

In the sequence $\overline{H}_N$, the coupling to the extra site is the same for each N, and the energy scale for the lowest states is of the same order. The original Hamiltonian corresponds to the limit,

$$H = \lim_{N\to\infty} (D'\Lambda^{-(N-1)/2}\overline{H}_N). \tag{4.35}$$

A sequence of renormalization group transformations can then be defined by

$$\overline{H}_{N+1} = R(\overline{H}_N). \tag{4.36}$$

If $\overline{H}_N$ is diagonalized numerically and has eigenstates $|\mathbf{m}\rangle$ (an abbreviation for $|n_e, S, S_z, m\rangle$) with a corresponding eigenvalue $E(\mathbf{m})$, then it can be expressed in the form,

$$\overline{H}_N = \sum_{\mathbf{m}} E(\mathbf{m})|\mathbf{m}\rangle\langle\mathbf{m}|. \tag{4.37}$$

The Hamiltonian to be diagonalized at the next stage is

$$\overline{H}_{N+1} = \sum_{\mathbf{m}}^{N_{\mathrm{st}}} \overline{E}(\mathbf{m})|\mathbf{m}\rangle\langle\mathbf{m}| + \sum_{\mathbf{m},\mathbf{m}',\sigma}^{N_{\mathrm{st}}} \{g_\sigma(\mathbf{m},\mathbf{m}')\,|\mathbf{m}\rangle\langle\mathbf{m}'|c_{N+1,\sigma} + \mathrm{h.c.}\}, \tag{4.38}$$

where $g_\sigma(\mathbf{m},\mathbf{m}') = \langle\mathbf{m}|c^\dagger_{N,\sigma}|\mathbf{m}'\rangle$ and $\overline{E}(\mathbf{m}) = \Lambda^{1/2}E(\mathbf{m})$. In retaining a fixed number of states, N_{st}, at each iteration the renormalization group transformation maps onto a Hamiltonian of the same form at each stage, which is specified by the parameter set $\{\overline{E}(\mathbf{m}), g_\sigma(\mathbf{m},\mathbf{m}')\}$. The renormalization group transformation relates these parameters at the N and $N+1$ iterations (the energies are measured relative to the ground state energy in each case). These will be unchanged if the transformation has a fixed point. The iteration procedure is illustrated in figure 4.5.

The conjectured fixed points of the renormalization group transformation for the s-d model correspond to the Hamiltonian for $J = 0$ and $J = \infty$. This can be tested by looking at the form of $\overline{H}_N(J = 0)$ and $\overline{H}_N(J = \infty)$ as $N \to \infty$. The excitations in these two cases are easy to evaluate because they both correspond to a single particle Hamiltonian with the impurity effectively decoupled. For $J = 0$, the single particle levels of the $\overline{H}_N(J = 0)$ are calculated by diagonalization. The excitations about the ground state are described by the single particle Hamiltonian,

$$\overline{H}_N(J = 0) = \sum_n \varepsilon'_n (p^\dagger_{n,\sigma}p_{n,\sigma} + h^\dagger_{n,\sigma}h_{n,\sigma}), \tag{4.39}$$

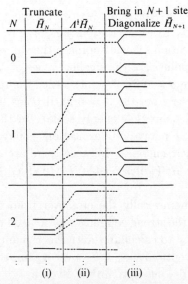

Figure 4.5 Successive stages in the iterative diagonalization: (i) the energy
levels of $\bar{H}_N$; (ii) the rescaling of the levels to correspond to $\Lambda^{1/2}\bar{H}_N$; (iii)
the energy levels obtained by the diagonalization $\bar{H}_{N+1}$, which includes
the next site along the chain. The steps are repeated keeping N_{st} states
corresponding to the lowest levels of $\bar{H}_{N+1}$ at each stage (Krishna-murthy,
1978).

where $p_{n,\sigma}^\dagger$ and $p_{n,\sigma}$ are creation and annihilation operators for the
particle excitations, and $h_{n,\sigma}^\dagger$, $h_{n,\sigma}$ for the holes. The excitation energies
ε_n' calculated by Wilson, for $\Lambda = 2$ and large N are

ε_n'	0	± 1.297	± 2.827	...	$\pm 2^{n-1}\sqrt{2}$	N even
ε_n'		± 0.6555	± 1.976	...	$\pm 2^{n-1}$	N odd

and depend on whether the chain has odd or even number of sites.

For $J = \infty$ the impurity is coupled in a singlet state with the first site
on the chain. As it requires an infinite amount of energy to break up
this state, the impurity and the first site are effectively decoupled from
the system and the low energy excitations of $\overline{H}_N(J = \infty)$ are therefore
the same as for $\overline{H}_{N-1}(J = 0)$.

The hypothesis that a $J = \infty$ fixed point is approached for an antiferromagnetic coupling can now be put to the test. For $\Lambda = 2$ and
$J = 0.009$, Wilson (1975) gives results for the two lowest excitations of
the iterative calculation for various values of N (even),

N	20	22	108	120	130	180
E_1	0.0314	0.0321	0.313	0.363	0.6541	0.6555
E_2	0.0419	-.04428	0.446	0.529	1.3055	1.3110

The first excitation E_1 corresponds to a single particle excitation energy and E_2 to a two particle excitation. For large N it can be seen that the results of the iteration scheme approach the values corresponding to to the $J = \infty$ fixed point. For the smaller values of N these excitations are quite close to the $J = 0$ fixed point. Hence the results demonstrate the complete crossover from the $J = 0$ fixed point (high energy scale) to the $J = \infty$ fixed point (low energy scale). In the crossover region Wilson found that the many-body excitations could not be fitted with additive single particle energies. Because there are different limits for odd and even N it is R^2 (which relates $\overline{H}_N$ to $\overline{H}_{N+2}$) which has a fixed point.

Having achieved numerically the crossover from the weak coupling $J = 0$ fixed point to the strong coupling $J = \infty$ fixed point, the question arises as the best way to calculate the thermodynamic properties, such as the impurity susceptibility $\chi_{\text{imp}}(T)$ over a temperature range down to $T = 0$. The impurity susceptibility is given by

$$
\chi_{\text{imp}}(T) =
$$
$$
\frac{(g\mu_{\text{B}})^2}{k_{\text{B}}T} \left\{ \lim_{N\to\infty} \frac{\text{Tr}\, S_{Nz}^2 e^{-H_N(J)/k_{\text{B}}T}}{\text{Tr}\, e^{-H_N(J)/k_{\text{B}}T}} - \frac{\text{Tr}\, S_{Nz}^2 e^{-H_N(0)/k_{\text{B}}T}}{\text{Tr}\, e^{-H_N(0)/k_{\text{B}}T}} + \frac{1}{4} \right\},
$$

$$(4.40)$$

where S_{Nz} is the z-component of total spin of the electrons for the N site Hamiltonian. The susceptibility of the conduction electrons is subtracted off by using the $H_N(J = 0)$, and the $\frac{1}{4}$ cancels the free spin term in $H_N(J = 0)$. The same g factor has been used for the impurity and conduction electrons.

The lowest energy scale for $\overline{H}_N$ is of the order of unity. If a temperature T_N is defined by $k_{\text{B}}T_N = D'\Lambda^{-N/2}$, then, as $H_N = k_{\text{B}}T_N \overline{H}_N$, the lowest energy excitations of H_N are of the order of $k_{\text{B}}T_N$. Hence the excitation energies of $\overline{H}_N$ are sufficient to calculate $\chi_{\text{imp}}(T)$ for $T \gg T_N$. However, for an accurate evaluation, the excitation energies are required up to energies several times $k_{\text{B}}T$ and it is difficult to keep enough states to span this range. If the coupling matrix elements at the N' site to the rest of the chain are put equal to zero, the contributions to (4.40) cancel for $N > N'$. Hence, the error in neglecting terms beyond $N = N'$ is proportional to this coupling term. This error was estimated in Wilson's calculation by perturbation theory. In this way $\chi_{\text{imp}}(T)$ was calculated numerically in the temperature range $T \sim T_{\text{K}}$, which corresponds to the crossover region. For further details we refer the reader to the original paper.

At high and low temperatures compared with T_K the excitation energies are approximately those of $H_N(J = 0)$ and $H_N(J = \infty)$ for large N. Correction terms to these Hamiltonians can be found by looking at the linearized renormalization group transformation in the neighbourhood of these fixed points. The effective Hamiltonians derived in these regimes can then be used to calculate the high and low temperature behaviour. We shall see how to do this in the next section.

4.5 Effective Hamiltonians near the Fixed Points.

If we denote the Hamiltonians $\overline{H}_N(J = 0)$ and $\overline{H}_N(J = \infty)$, which correspond to the two fixed points in the limit of large N, by $\overline{H}_N^*(J = 0)$ and $\overline{H}_N^*(J = \infty)$ (the * is to indicate that they are associated with a fixed point), then the states of the Hamiltonian $\overline{H}_N(J)$ can be described by an effective Hamiltonian $\overline{H}_N'(J)$,

$$\overline{H}_N' = \overline{H}_N^* + \delta\overline{H}_N, \tag{4.41}$$

where $\delta\overline{H}_N$ is small in the vicinity of the fixed point. The most important contributions to $\delta\overline{H}_N$ are likely to be local interaction terms involving the impurity site and the first few sites along the chain. One can surmise that they should be of the form,

$$\delta\overline{H}_N = D'\Lambda^{(N-1)/2}\sum_m \delta H_m(N), \tag{4.42}$$

where $\delta H_m(N)$ are local interaction terms. The factor $D'\Lambda^{(N-1)/2}$ is introduced to correspond with the definition of $\overline{H}_N$ in equation (4.33). Possible forms for $\delta H_m(N)$ are

$$\delta H_1 = \sum_{\sigma,\sigma'} \mathbf{S}\cdot c_{0,\sigma}^\dagger (\mathbf{s}_0)_{\sigma\sigma'} c_{0,\sigma'}, \qquad \delta H_2 = \left(\sum_\sigma n_{0,\sigma} - 1\right)^2,$$

$$\delta H_3 = \sum_\sigma (c_{0,\sigma}^\dagger c_{1,\sigma} + c_{1,\sigma}^\dagger c_{0,\sigma}), \qquad \delta H_4 = \left(\sum_\sigma n_{0,\sigma} - 1\right)\left(\sum_\sigma n_{1,\sigma} - 1\right),$$

$$\delta H_5 = \left(\sum_\sigma n_{1,\sigma} - 1\right)^2, \qquad \delta H_6 = \sum_\sigma (c_{1,\sigma}^\dagger c_{2,\sigma} + c_{2,\sigma}^\dagger c_{1,\sigma}),$$

$$\tag{4.43}$$

These terms have to be consistent with spin and charge conservation as well as particle–hole symmetry, which restricts the possible forms. The particle–hole symmetry follows from the choice of a symmetric conduction band with the Fermi level at the centre. The question as to which of these terms dominate in the neighbourhood of a particular fixed point can be determined by finding out how these terms transform under R^2

near the fixed point. The transformation R^2 can be applied to $\overline{H}'_N$, as it can be expressed in the form (4.38), so

$$R^2(\overline{H}'_N) = \overline{H}'_{N+2}. \tag{4.44}$$

As $R^2(\overline{H}^*_N) = \overline{H}^*_{N+2}$, then

$$\begin{aligned} \overline{H}'_{N+2} - \overline{H}^*_{N+2} &= R^2(\overline{H}^*_N + (\overline{H}'_N - \overline{H}^*_N)) - R^2\overline{H}^*_N \\ &= L^*(\overline{H}'_N - \overline{H}^*_N), \end{aligned} \tag{4.45}$$

where L^* is the linearized form of R^2 near the fixed point $\overline{H}^*$. If L^* has a complete set of eigenvectors $\mathbf{O}^*_m$ with corresponding eigenvalues λ^*_m, then

$$\delta\overline{H}_N = \overline{H}'_N - \overline{H}^*_N = \sum_m c_m \lambda^{*N/2}_m \mathbf{O}^*_m. \tag{4.46}$$

This can be verified by substitution into equation (4.45).

The terms δH_m in the list (4.43) will not in general correspond to a particular eigenoperator $\mathbf{O}^*_m$, but will be a linear combination of them. How important a particular δH_m will be near a particular fixed point will depend on the largest eigenvalues of the operators in this linear combination. The most important terms will be those that have contributions from the largest eigenvalues of L^* at that fixed point.

The leading eigenoperators in δH_m can be determined by expressing them in terms of the $p^\dagger_{n,\sigma}$, $p_{n,\sigma}$ and $h^\dagger_{n,\sigma}$, $h_{n,\sigma}$, that create the single particle and hole excitations for the corresponding Hamiltonian $\overline{H}^*_N$ (see (4.39)). The argument gets a little technical at this point but we can take a simple example to get the general idea. The operator $c^\dagger_{0,\sigma}$ can be expressed in terms of the operators $p^\dagger_{n,\sigma}$ and $h_{n,\sigma}$ for the Hamiltonian $\overline{H}_N(J = 0)$. Detailed calculations give the result,

$$c^\dagger_{0,\sigma} = \Lambda^{-(N-1)/4} \sum_n \alpha_{0,n}(p^\dagger_{n,\sigma} + h_{n,\sigma}). \tag{4.47}$$

The operator $c^\dagger_{1,\sigma}$ has a similar expansion in the same set of operators, but with a prefactor $\Lambda^{-3(N-1)/4}$. Substituting these, and their Hermitian conjugates, into δH_1, δH_2, and δH_3, and then into (4.42) and comparing with (4.46), we can pick out the eigenvalues from the N dependence. We find that δH_1 has an eigenvalue Λ^0, and hence is a marginal operator. The other two operators, δH_2 and δH_3, have eigenvalues Λ^{-1} and hence are irrelevant operators at the $J = 0$ fixed point. The other terms in the list (4.43) also correspond to irrelevant eigenvalues.

For the $J = 0$ fixed point the effective Hamiltonian has the form,

$$\overline{H}'_N = \Lambda^{(N-1)/2}\Big[\sum_{n=0}^{N-1} \Lambda^{-n/2}(c^\dagger_{n,\sigma}c_{n+1,\sigma} + c^\dagger_{n+1,\sigma}c_{n,\sigma})$$

$$+w(J)\sum_{\sigma,\sigma'}\mathbf{S}\cdot c^\dagger_{0,\sigma}(\mathbf{s}_0)_{\sigma,\sigma'}c_{0,\sigma'}\Big].$$

(4.48)

where the irrelevant operators have been neglected.

For the $J = \infty$ fixed point, the first site in the chain, which is directly coupled to the impurity, is effectively removed, hence terms, $\delta H_1 - \delta H_4$ in (4.43), cannot contribute to the effective Hamiltonian. The last two terms, δH_5 and δH_6, correspond to the eigenvalue Λ^{-1}, which is the leading irrelevant eigenvalue at this fixed point. There are no relevant or marginal eigenvalues so the $J = \infty$ fixed point is a stable one. If we retain the terms corresponding to the leading irrelevant eigenoperators, then the effective Hamiltonian takes the form,

$$\overline{H}'_N = \Lambda^{(N-1)/2}\Big[\sum_{n=1}^{N-1} \Lambda^{-n/2}(c^\dagger_{n,\sigma}c_{n+1,\sigma} + c^\dagger_{n+1,\sigma}c_{n,\sigma})$$

$$+w_1(J)\left(\sum_\sigma c^\dagger_{1,\sigma}c_{1,\sigma} - 1\right)^2 + w_2(J)\sum_\sigma(c^\dagger_{1,\sigma}c_{2,\sigma} + c^\dagger_{1,\sigma}c_{2,\sigma})\Big].$$

(4.49)

Having found effective Hamiltonians for the regions near the two fixed points Wilson was then able to calculate the high and low temperature behaviour. We consider these results in the next section.

4.6 High and Low Temperature Results

The effective Hamiltonian (4.48) near the $J = 0$ fixed point corresponds to that used in the poor man's scaling in the previous chapter, section 3.3. A translation in N in the Wilson scheme is equivalent to a reduction in band width with $D'\Lambda^{-N/2} \sim \tilde{D}$, and $w(J)$ corresponds to the renormalized coupling $\tilde{J}$. The fact that the renormalization group trajectories near the $J = 0$ fixed point are controlled by a marginal operator explains why the effective Hamiltonian in the poor man's scaling retains the same form over a very large change in band width $\tilde{D}$. The slow change was due to a logarithmic dependence on $\tilde{D}$. In a section of his 1975 paper Wilson showed, quite generally, how a marginal fixed point leads to logarithmic terms when scaling trajectories are calculated perturbationally.

In Wilson's calculation, the scaling equation (3.50) takes the form,

$$\Psi_N(J) - \Psi_{\tilde{N}}(\tilde{J}) = \frac{(\tilde{N} - N)}{2} \ln \Lambda, \qquad (4.50)$$

where the right hand side of (4.50) corresponds to the $\ln(D/\tilde{D})$ in (3.50) with $D = D'\Lambda^{-N/2}$ and $\tilde{D} = D'\Lambda^{-\tilde{N}/2}$. The energy levels calculated for small J and N coincide with those for large N (smaller band width) and larger J in the range before the crossover to the strong coupling fixed point $J = \infty$ occurs.

The scaling law for the susceptibility $\chi_{\text{imp}}(T)$ (in units of $(g\mu_B)^2$) can be expressed in the form,

$$\Phi(\chi_{\text{imp}}(T)2k_B T - 1/2) = \ln(T/T_K), \qquad (4.51)$$

where Φ is a universal function.

The high temperature perturbation results are determined by the asymptotic form of Φ for small x,

$$\Phi(x) = \frac{1}{2x} - \frac{1}{2}\ln|2x| + 3.1648x + O(x^2). \qquad (4.52)$$

The first two terms correspond to the sum of the leading order and next leading order logarithmic terms in perturbation theory. The next term, which is linear in x, is a higher order correction calculated by Wilson.

Solving (4.51) and (4.52) gives the result (3.53) derived earlier by poor man's scaling,

$$\chi_{\text{imp}}(T) =$$
$$\frac{(g\mu_B)^2}{4k_B T}\left(1 - \frac{1}{\ln(T/T_K)} - \frac{\ln(\ln(T/T_K))}{2(\ln(T/T_K))^2} + O\left(\frac{1}{(\ln(T/T_K))^3}\right)\right).$$
$$(4.53)$$

The two forms of the scaling law, (4.50) (or (3.50)) for $\tilde{J}$, and (4.51) for $\chi_{\text{imp}}(T)$, are very similar and can be made to correspond precisely by a suitable definition for a band width parameter $\overline{D}$. The perturbation result for $\chi_{\text{imp}}(T)$ from (3.3) is

$$\chi_{\text{imp}}(T) = \frac{1}{4k_B T}(1 - 2J\rho_0 + (J\rho_0)^2\ln(k_B T/\overline{D}) + (J\rho_0)^3 O(\ln(k_B T/\overline{D})))$$
$$(4.54)$$

in units of $(g\mu_B)^2$, where all the non-logarithmic terms, except the term $2J\rho_0$, are absorbed into the definition of $\overline{D}$. Hence $\overline{D} = Dc(J\rho_0)$, where $c(J\rho_0)$ has a power series expansion in $J\rho_0$. Then putting $k_B T = \overline{D}$ in (4.54), the logarithmic terms vanish and we find

$$J\rho_0 = \chi_{\text{imp}}(\overline{D})2\overline{D} - 1/2. \qquad (4.55)$$

Substituting this into (4.51) gives

$$\Phi(J\rho_0) = \ln\left(\frac{\overline{D}}{T_K}\right), \tag{4.56}$$

which we can compare with (3.49), or solving for T_K,

$$T_K = \overline{D}\, e^{-\Phi(J\rho_0)}. \tag{4.57}$$

This defines the Kondo temperature precisely. With this definition T_K is such that there are no terms in $(\ln(T/T_K))^{-2}$ in (4.53).

Calculations for the low temperature range $T \ll T_K$ are based on the effective Hamiltonian $\overline{H}'_N$ near the $J = \infty$ fixed point given in equation (4.49). The two parameters $w_1(J)$ and $w_2(J)$ are determined by fitting the lowest excitations determined by the numerical iteration to the lowest excitations of (4.49). The second correction term in (4.49), the one with the coefficient w_2, when expressed in terms of the single particle operators, $p_{n,\sigma}^\dagger$, $p_{n,\sigma}$ and $h_{n,\sigma}^\dagger$, $h_{n,\sigma}$, is bilinear in these operators. The first correction term in (4.49), the one proportional to w_1, is proportional to $n_{1,\uparrow} n_{1,\downarrow}$ because, due to particle–hole symmetry, $\langle c_{1,\sigma}^\dagger c_{1,\sigma}\rangle = \frac{1}{2}$. This term represents an interaction between the single particle (or hole) excitations. Hence, the coefficient w_2 can be determined by a comparison with the single particle excitation $E_{1,N}$, and w_1 by a comparison with the lowest two particle excitation $E_{2,N}$. Comparison can be made with the energies of the other excitations, which can be determined both from (4.49) and by the numerical iteration, to test the accuracy of $\overline{H}'_N$ in fitting the low lying levels. Having fitted the parameters, w_1 and w_2, for a particular J the results can be extended for other values of J by using the scaling relation (4.50).

As the strong coupling fixed point is approached (larger N) the two correction terms in (4.49), because they correspond to irrelevant operators, become smaller and smaller. At the same time the energy range δE of the excitations of the original full Hamiltonian H, as deduced from $\overline{H}'_N$, becomes smaller. For low enough temperatures, $k_B T \ll \delta E$, the correction terms in $\overline{H}'_N$ can be treated in lowest order perturbation theory. This is the way both the very low temperature susceptibility and specific heat were deduced by Wilson. The contributions arising from $\overline{H}^*_N$ were found to be zero in the continuum limit $\Lambda \to 1$, so these quantities are determined solely from the correction terms. The result for the impurity susceptibility is

$$\chi_{\mathrm{imp}}(T) = \frac{(g\mu_B)^2 (0.4128 \pm .002)}{4k_B T_K}, \tag{4.58}$$

where T_K is defined by (4.57). The fact that this is finite as $T \to 0$

shows that there is no residual local moment, and that the impurity spin is fully compensated. The numerical factor 0.4128 is a universal number for the s-d model, and is known as the *Wilson number*, w. It relates two quite different energy scales for the s-d model, T_K, which is determined from the high temperature perturbative regime, and $\chi_{imp}(0)$, the low temperature susceptibility associated with the strong coupling regime. We shall see later in chapter 6 that, rather remarkably, an exact formula can be derived for this quantity.

Due to the restriction to very low temperatures only the linear term $\gamma_{imp}T$ in the specific heat C_{imp} could be calculated in this way. The ratio R, which was defined in equation (1.61) for $g = 2$, and is essentially the ratio of χ_{imp}/χ_c to γ_{imp}/γ_c (where c denotes the conduction electron values in the absence of the impurity), is also a universal number. In this context, it is usually known as the *Wilson ratio* but more generally as the *Sommerfeld ratio*. Wilson's calculation gave the value,

$$R = \frac{\chi_{imp}/\chi_c}{\gamma_{imp}/\gamma_c} = \frac{4\pi^2 k_B^2}{3(g\mu_B)^2} \frac{\chi_{imp}}{\gamma_{imp}} = 2, \qquad (4.59)$$

to within the numerical accuracy of the calculation. For non-interacting electrons the value is 1. The effective Hamiltonian $\overline{H}'_N$ in terms of the single particle operators, $p^\dagger_{n,\sigma}$, $p_{n,\sigma}$ and $h^\dagger_{n,\sigma}$, $h_{n,\sigma}$, describes the low lying excitations in terms of fermions, which we can identify as quasi-particles (this idea will be developed more fully in the next chapter). The enhancement of (4.59) by a factor of 2 over that for non-interacting particles is due to the quasi-particle interactions.

In the crossover regime the numerical results could be fitted with the formula,

$$\chi_{imp}(T) = \frac{(g\mu_B)^2(0.68)}{4k_B(T + \sqrt{2}T_K)} \qquad 0.5T_K < T < 16T_K. \qquad (4.60)$$

This formula is incorrectly quoted in the original paper and has been corrected (see Wiegmann and Tsvelick (1983) for comments). This Curie–Weiss form corresponds to a reduced moment compared to the free spin form (1.63). Thus the impurity moment, even for $T \sim T_K$, is only of the order of 30% that of the free moment. The residual effects of the screening of the conduction persist to very high temperatures because of the logarithmic dependence on T/T_K. For $T > 16T_K$, the asymptotic formula (4.52) gives results for $\chi_{imp}(T)$ to better than 1%.

The plot of Wilson's results for the inverse susceptibility χ_{imp}^{-1} versus T/T_K is shown in figure 4.6, and the universal plot of $\chi_{imp}(T)T$ against $\log_{10}(T/T_K)$ in figure 4.7.

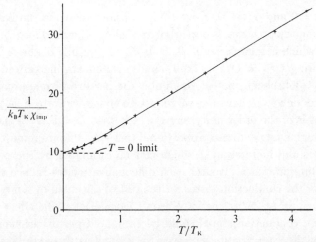

Figure 4.6 The numerical renormalization group results of Wilson (1975) for $\chi_{\text{imp}}(T)$ for the $S = \frac{1}{2}$ s-d model.

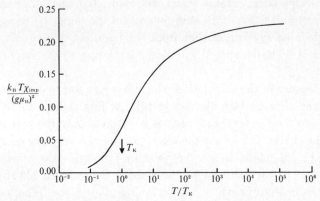

Figure 4.7 The universal susceptibility curve for the $S = \frac{1}{2}$ s-d model (Wilson, 1975).

The very slow asymptotic approach to the Curie form for the susceptibility of a free spin is quite apparent.

Before going on to look at the renormalization group calculations for the Anderson model, there are one or two points in Wilson's calculation that deserve further comment. First of all, why can we expect to make accurate predictions for the model using $\Lambda = 2$ rather than the continuum value $\Lambda = 1$? We can get insight into this by looking at the calculation in a different ways. If we go back to the linear chain form for the s-d model as derived in section 4.2, equations (4.9) and (4.14), with coefficients $\epsilon = 0$ and $\gamma_n = D/2$, we could consider reducing the band width and using the poor man's perturbative scaling. If we reduced D to $\tilde{D} \sim 4k_{\mathrm{B}}T_{\mathrm{K}}$, which is a large factor as $D/k_{\mathrm{B}}T_{\mathrm{K}}$ is typically of the order of 10^3, then we get $\gamma_n = \tilde{D}/2 \sim 2k_{\mathrm{B}}T_{\mathrm{K}}$ and a significant increase in J to $\tilde{J}$ given by (3.42) such that at this point the perturbative approach begins to break down. If, however, we reduce the band width by a factor $\Lambda^{1/2}$, where Λ is of the order unity, say $1 < \Lambda < 2$, so $\tilde{D} = D\,\Lambda^{1/2}$, then a number of high energy states, proportional to $1 - \Lambda^{-1/2}$, are removed, and we get a small increase in J which can be estimated by the perturbative scaling approach. If after each diagonalization we reduce the band width for the conduction states for the rest of the chain by a factor $\Lambda^{1/2}$, we get $\gamma_n = D/2\Lambda^{n/2}$, as in Wilson's calculation. The effect of removing these conduction states could be estimated perturbationally. The effects on the impurity are, however, very small as these states are removed gradually as we move down the chain away from the impurity. The build up of the strong coupling limit is from the states which are retained. The high energy states which are removed at each stage have very little effect on the low lying energy levels of the impurity, because they are high in energy and distant from the impurity, and so can be safely neglected. The important interactions relating to the impurity are retained.

A further feature of the calculation which deserves some comment is the form of the effective Hamiltonian (4.49). At first sight it looks as though the screening is very localized, as it involves only the first few sites along the chain. This is not the case: to reach the very low energy scale required to calculate the low temperature properties, the iteration procedure has been carried out along a chain of the order of 200 sites, and this is the order over which the screening takes place. The local orbitals in (4.59) are built from conduction electrons states over a very much reduced energy range, with ks very close to the Fermi level, so they describe orbitals which are quite extended. Finally, can we get any insight into the origin of the effective interaction $2w_1 n_{1,\uparrow} n_{1,\downarrow}$? If we go to the complete strong coupling limit $w_1 \to 0$ ($\tilde{J} \to \infty$), the impurity is fully screened in a singlet state at the first 'site'. On moving away from this limit effective interactions can be induced at the neighbouring site due to the virtual fluctuations to the triplet state, with an excitation energy of $2\tilde{J}$. This interaction is of the order of $\tilde{D}^2/\tilde{J}$, and is repulsive because the binding energy is lost in the intermediate triplet state.

4.7 The Symmetric Anderson Model

The numerical renormalization group technique can be straightforwardly generalized to the Anderson model (1.66),

$$H = \sum_\sigma \epsilon_d n_{d,\sigma} + U n_{d,\uparrow} n_{d,\downarrow} + \sum_{\mathbf{k},\sigma} \epsilon_{\mathbf{k}} c^\dagger_{\mathbf{k},\sigma} c_{\mathbf{k},\sigma}$$

$$+ \sum_{\mathbf{k},\sigma} (V_{\mathbf{k}} c^\dagger_{d,\sigma} c_{\mathbf{k},\sigma} + V^*_{\mathbf{k}} c^\dagger_{\mathbf{k},\sigma} c_{d,\sigma}).$$

(4.61)

The combination of states which is directly coupled to the impurity, $\sum_{\mathbf{k}} V_{\mathbf{k}} c^\dagger_{\mathbf{k}}$, becomes the orbital associated with the first site in the conduction electron chain (figure 4.2),

$$c^\dagger_{0,\sigma} = \frac{1}{V} \sum_{\mathbf{k}} V_{\mathbf{k}} c^\dagger_{\mathbf{k}},$$

(4.62)

where $V = (\sum_{\mathbf{k}} |V_{\mathbf{k}}|^2 / N_s)^{1/2}$. Using the Lanczos algorithm (4.10)–(4.12) to define the basis states for the rest of the chain, the Hamiltonian (4.61) can be put in the form,

$$H = \sum_\sigma \epsilon_d n_{d,\sigma} + U n_{d,\uparrow} n_{d,\downarrow} + \sum_\sigma (V c^\dagger_{d,\sigma} c_{0,\sigma} + V^* c^\dagger_{0,\sigma} c_{d,\sigma})$$

$$+ \sum_{n,\sigma} \gamma_n (c^\dagger_{n,\sigma} c_{n+1,\sigma} + c^\dagger_{n+1,\sigma} c_{n,\sigma}) + \sum_{n,\sigma} \epsilon_n c^\dagger_{n,\sigma} c_{n,\sigma},$$

(4.63)

where ϵ_n and γ_n depend on $\epsilon_{\mathbf{k}}$ and $V_{\mathbf{k}}$. Using Wilson's discretization scheme for the conduction electrons we can take the forms given in equation (4.29). The same iteration procedure leading to a sequence of Hamiltonians $\bar{H}_N$ can be defined,

$$\bar{H}_N = \Lambda^{(N-1)/2} \left\{ \sum_{n,\sigma} \Lambda^{-n/2} (c^\dagger_{n,\sigma} c_{n+1,\sigma} + c^\dagger_{n+1,\sigma} c_{n,\sigma}) \right.$$

$$\left. + \bar{V} \sum_\sigma (c^\dagger_{d,\sigma} c_{0,\sigma} + c^\dagger_{0,\sigma} c_{d,\sigma}) + \frac{\bar{U}}{2} \left(\sum_\sigma c^\dagger_{d,\sigma} c_{d,\sigma} - 1 \right)^2 \right\},$$

(4.64)

where

$$\bar{U} = \frac{2}{1+\Lambda^{-1}} \frac{U}{D} \quad \bar{V} = \frac{2}{1+\Lambda^{-1}} \frac{V}{D}, \quad \bar{\epsilon}_d = \frac{2}{1+\Lambda^{-1}} \frac{(\epsilon_d + U/2)}{D}. \quad (4.65)$$

The main difference is that the parameter space for this model is larger and more fixed points can come into play. All the fixed points correspond to a special choice of parameters of the original model.

Krishna-murthy, Wilkins and Wilson (1980) have made the most extensive applications of the numerical renormalization group to the Anderson model. The first applications were to the symmetric model, $\epsilon_d = \epsilon_F - U/2$. In this case the impurity levels for $V = 0$ are symmetrically placed about the Fermi level at $E_1 = \epsilon_F - U/2$ and $E_2 = \epsilon_F + U/2$. If the conduction band is symmetric about the Fermi level and half full, the model has complete particle–hole symmetry and the impurity state is always singly occupied, $\langle n_d \rangle = 1$. This model can display the full range of behaviour from non-magnetic for $k_B T, U < \Delta$, to magnetic and Kondo behaviour for $U \gg \Delta$. The later work concerned the asymmetric case with ϵ_d as an independent parameter. A qualitatively different regime, the mixed valence regime with $\epsilon_d \sim \epsilon_F$ and U large, can be explored in this case (see figure 1.9). The fixed point $J = 0$ for the s-d model corresponds to $V = 0$ with $D > U \gg 0$. This is termed the *local moment fixed point* H^*_{LM}. The impurity orbital is occupied by a single electron $(\uparrow,\downarrow)$, and hence has a moment which is uncoupled from the conduction electrons. Using the Schrieffer–Wolff transformation for the symmetric case gives an effective coupling,

$$J = \frac{4|V|^2}{U},\tag{4.66}$$

so $J = 0$ for $V = 0$.

The $J = \infty$ limit, which is termed the *strong coupling fixed point* H^*_{SC}, clearly corresponds to $V \to \infty$, U fixed. In this limit the impurity electron is so strongly coupled to the first site of the chain that it is effectively decoupled so $\bar{H}^*_{\mathrm{N,SC}}$ corresponds to the conduction chain with one site removed. In the symmetric case a third fixed point plays a role. This corresponds to $U = 0$, $V = 0$, in which there is a free orbital at the Fermi level which, because $U = 0$, is independently occupied by $\uparrow$ and $\downarrow$ electrons and has no associated moment. This is termed the *free orbital fixed point* H^*_{FO}. The behaviour of the symmetric model in the whole parameter and temperature regimes can be analysed in terms of these three fixed points.

For $V = 0$ and U finite the susceptibility has the form,

$$\chi_{\mathrm{imp}}(T) = \frac{(g\mu_B)^2}{4k_B T(1 + e^{-U/2k_B T})}.\tag{4.67}$$

Hence on the approach to the local moment fixed point, for which $U \gg k_B T$, $k_B T \chi_{\mathrm{imp}}(T)/(g\mu_B)^2 \to 1/4$. At very high temperatures, $k_B T \gg U$, (4.67) goes over to the susceptibility in the free orbital limit for which $k_B T \chi_{\mathrm{imp}}(T)/(g\mu_B)^2 \to 1/8$. To calculate the correction terms to $\chi_{\mathrm{imp}}(T)$ one needs the leading terms in the effective Hamiltonian from the linearized renormalization group equations near these fixed points. Near the free orbital fixed point the dominant correction terms are

$$\delta H_1 = \sum_\sigma (c_{d,\sigma}^\dagger c_{d,\sigma} - 1)^2, \quad \delta H_2 = \sum_\sigma (c_{d,\sigma}^\dagger c_{0,\sigma} + c_{0,\sigma}^\dagger c_{d,\sigma}), \quad (4.68)$$

which are both relevant operators and correspond to eigenvalues Λ and $\Lambda^{1/2}$ respectively of the linearized renormalization group transformation R^2. Other local operators such as

$$\delta H_3 = \sum_\sigma (c_{d,\sigma}^\dagger c_{0,\sigma} + c_{0,\sigma}^\dagger c_{d,\sigma})^2, \quad \delta H_4 = \left(\sum_\sigma n_{d,\sigma} - 1\right)\left(\sum_\sigma n_{0,\sigma} - 1\right),$$

$$(4.69)$$

can be shown to generate marginal operators.

A general rule can be formulated for selecting the eigenoperators of maximum relevance. They are constructed from the minimum number of operators with the lowest possible site indices n, within the constraints imposed by the overall symmetry.

Using the relevant operators, the effective Hamiltonian near the free orbital fixed point has the form,

$$\bar{H}_N' = \Lambda^{(N-1)/2} \left\{ \sum_{n,\sigma} \Lambda^{-n/2} (c_{n,\sigma}^\dagger c_{n+1,\sigma} + c_{n+1,\sigma}^\dagger c_{n,\sigma}) \right.$$

$$\left. + \tilde{V} \sum_\sigma (c_{d,\sigma}^\dagger c_{0,\sigma} + c_{0,\sigma}^\dagger c_{d,\sigma}) + \frac{\tilde{U}}{2} \left(\sum_\sigma c_{d,\sigma}^\dagger c_{d,\sigma} - 1\right)^2 \right\}.$$

$$(4.70)$$

which is the same as the original model. In terms of the single particle eigenstates of $H_N^*(\text{FO})$, this becomes

$$\bar{H}_N' = \sum_{n,\sigma} \epsilon_n'(p_{n,\sigma}^\dagger p_{n,\sigma} + h_{n,\sigma}^\dagger h_{n,\sigma}) + \tilde{V}\Lambda^{(N-1)/4} \sum_{n,\sigma} \alpha_{0,n}(c_{d,\sigma}^\dagger (p_{n,\sigma}$$

$$+ h_{n,\sigma}^\dagger) + (p_{n,\sigma}^\dagger + h_{n,\sigma})c_{d,\sigma}) + \frac{\tilde{U}}{2}\Lambda^{(N-1)/2} \left(\sum_\sigma c_{d,\sigma}^\dagger c_{d,\sigma} - 1\right)^2,$$

$$(4.71)$$

where $\alpha_{0,n}$ is given by (4.47). In perturbation theory for small $\tilde{\Delta}$ and $\tilde{U}$, the hybridization term gives energy shifts of the same order in $\Lambda^{(N-1)/2}$ as the interaction term $\tilde{U}$ because the lowest order term is proportional to $|\tilde{V}|^2$.

Due to the presence of relevant operators, δH_1 and δH_2, the free orbital fixed point is unstable. The effective Hamiltonian for the local moment fixed point H_{LM}^* is the the same as that found earlier (4.48),

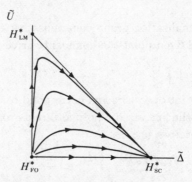

Figure 4.8   A schematic representation of the flow of trajectories generated
by the renormalization group transformation for the symmetric Anderson
model with increasing N (Krishna-murthy et al, 1980).

and similarly the effective Hamiltonian for the strong coupling fixed
point H^*_{SC} is given by (4.49).

Having obtained the effective Hamiltonians near the three fixed points,
H^*_{FO}, H^*_{LM} and H^*_{SC} we are now in a position to get an overall picture of
the behaviour of the symmetric model. (There are, as earlier, two fixed
points in each case corresponding to odd and even N. We can choose
to work with one set only). The true parameter space of the Hamilto-
nian is the set of energy levels $\{\bar{E}(\mathbf{m})\}$ and couplings $\{g_\sigma(\mathbf{m}, \mathbf{m}')\}$ that
appear in the Hamiltonian (4.38). However the effective Hamiltonians
near the fixed points can be described by the Anderson model with effective parameters $\tilde{U}$ and $\tilde{\Delta}$ so we can use these parameters to determine
qualitatively the flow of the trajectories. These trajectories are shown
in figure 4.8.

The trajectories near the line joining H^*_{LM} to H^*_{SC} are those considered
earlier for the s-d model. They move slowly, due to the marginal operator
δH_1 in (4.43), from H^*_{LM} to H^*_{SC} with decreasing $\tilde{U}$ and increasing $\tilde{\Delta}$,
and hence increasing $\tilde{J} \sim \tilde{\Delta}/\tilde{U}$, towards the spin compensated non-
magnetic ground state. The behaviour of the other trajectories falls
into two types. Those starting from points with $\tilde{U} > \tilde{\Delta}$, which are
initially driven away from the free orbital fixed point, due to the relevant
operators δH_1 and δH_2 in (4.68), are drawn towards the local moment
fixed point H^*_{LM}. Then due to the marginal operator in the vicinity of
H^*_{LM} they are eventually pushed towards the strong coupling fixed point.
For $\tilde{\Delta} > \tilde{U}$, systems which are initially categorized as non-magnetic, the
local moment fixed point has little influence and the trajectories are
drawn towards the strong coupling fixed point which is the only stable

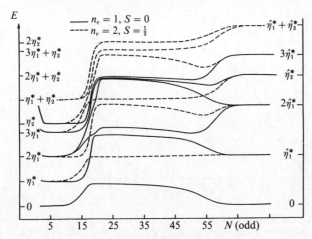

Figure 4.9 The lowest energy levels of $\bar{H}_N$ as a function of N for $U/D = 10^{-3}$ and $U/\pi\Delta = 12.66$ for the symmetric Anderson model (Krishna-murthy et al, 1980).

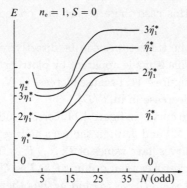

Figure 4.10 The lowest energy levels of $\bar{H}_N$ as a function of N for $U/D = 10^{-3}$, $U/\pi\Delta = 1.013$, for the symmetric Anderson model (Krishna-murthy et al, 1980).

fixed point of the model. Hence all trajectories end up at H^*_{SC}, in a non-magnetic ground state.

This behaviour of the trajectories can be seen more precisely in the plot of the low lying energy levels of $\bar{H}_N$ with N shown in figure 4.9.

The initial parameters correspond to $U/\pi\Delta = 12.66$, close to the free orbital fixed point. For small N they remain close to the free orbital fixed point and then for larger N make a rapid crossover to the local moment fixed point. Due to the marginal operator they remain near the local moment fixed point and only gradually crossover to the strong coupling limit.

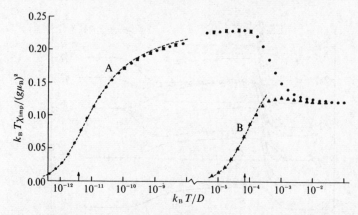

Figure 4.11 Plots of $k_B D\chi_{imp}(T)/(g\mu_B)^2$ vs $\log_{10}(k_B T/D)$ for $U/\pi D = 10^{-2}$. The curve A corresponds to $U/\pi\Delta = 12.66$ and B to $U/\pi\Delta = 1.013$. The dashed curve corresponds to the universal curve shown in figure 4.7 (Krishna-murthy et al, 1980).

Figure 4.10 shows similar low lying levels for a smaller value of $U/\pi\Delta$, $U/\pi\Delta = 1.013$. In this case there is just one crossover from H_{FO}^* to H_{SC}^*.

The behaviour of the trajectories can be directly related to the form of $\chi_{imp}(T)$ with T. This is made apparent by plotting $T\chi_{imp}(T)$ against $\ln(T/D)$ as shown in figure 4.11, because increase in N along a trajectory is proportional to a decrease in $\ln(T/D)$.

If we look at the two curves in figure 4.11, corresponding to the parameters given in figures 4.9 and 4.10, in the *very* high temperature regime $k_B T > U$, both the curves have values of $T\chi_{imp}(T) \sim 1/8$ corresponding to a free orbital. A crossover then occurs at $k_B T \sim U$, in one case to a local moment value $T\chi_{imp}(T) \sim 1/4$, and in the other case to the paramagnetic strong coupling limit at $k_B T \sim U \to 0$. Eventually the local moment case makes a slow crossover at $T \sim T_K$ to the strong coupling limit. The parallels with the behaviour of the energy levels with N is apparent.

4.8 The Asymmetric Anderson Model

For the asymmetric Anderson model there are four independent parameters, U, Δ, ϵ_d and T to consider. If we apply a particle–hole transformation ϵ_d gets replaced by $-(\epsilon_d + U)$, taking $\epsilon_F = 0$, and as a consequence we have the relation,

$$\chi_{imp}(U, \Delta, \epsilon_d, T) = \chi_{imp}(U, \Delta, -(\epsilon_d + U), T). \tag{4.72}$$

This means that it is sufficient to investigate $\epsilon_d > -U/2$. As $\epsilon_d = -U/2$ is the particle–hole symmetric case already considered we can ask what differences occur when ϵ_d increases from this limiting case to larger values.

All the fixed points correspond to a special choice of parameters of the original model. In addition to those of the symmetric model, H^*_{FO}, H^*_{LM}, and H^*_{SC}, there are two further fixed points. The most important of these is the valence fluctuation fixed point, H^*_{VF}, corresponding to $V = 0$, $\epsilon_d = 0$, $U \to \infty$, as it describes qualitatively different behaviour from that found for the symmetric model. The other is the frozen impurity fixed point, H^*_{FI}, for which $V = 0$, $U = 0$, $\epsilon_d \to \infty$. This just leaves the free electron Hamiltonian with no charge on the impurity site. This is less important because it is similar to the strong coupling fixed point and can be directly related to it.

As the valence fluctuation fixed point corresponds to $U = \infty$ the effective Hamiltonian $\bar{H}'_N$ about this point must be written in terms of operators, $d^\dagger_\sigma = c^\dagger_{d,\sigma}(1 - n_{d,-\sigma})$ and $d_\sigma = c_{d,\sigma}(1 - n_{d,-\sigma})$, which project out states of double occupation (or equivalently using Hubbard operators, $X_{\sigma,0}$ and $X_{0,\sigma}$). The leading terms in this effective Hamiltonian are

$$\bar{H}'_N = \Lambda^{(N-1)/2}\Big\{\sum_{n,\sigma}^{N-1} \Lambda^{-n/2}(c^\dagger_{n,\sigma}c_{n+1,\sigma} + c^\dagger_{n+1,\sigma}c_{n,\sigma})$$

$$+\bar{V}\sum_\sigma(d^\dagger_\sigma c_{0,\sigma} + c^\dagger_{0,\sigma}d_\sigma) + \sum_\sigma \bar{\epsilon}_d d^\dagger_\sigma d_\sigma\Big\}.$$

$$(4.73)$$

The term $\bar{\epsilon}_d$ is a relevant operator corresponding to an eigenvalue Λ. The hybridization term $\bar{V}$ is also relevant with an eigenvalue $\Lambda^{1/2}$. There are only two relevant operators in this case. Several marginal operators can be constructed. The presence of a linear term in the occupation number, $\sum_\sigma d^\dagger_\sigma d_\sigma$, reflects the lack of particle–hole symmetry. When the hybridization term is included the loss of particle–hole symmetry affects the conduction electron terms. New operators, which would have been excluded before as lacking particle–hole symmetry, now have to be included in the effective Hamiltonians at all fixed points. To the free electron part of the Hamiltonian can be added a potential scattering term of the conduction electrons at the impurity site,

$$K(\sum_\sigma c^\dagger_{0,\sigma}c_{0,\sigma} - 1).$$

$$(4.74)$$

The Hamiltonians for the N site cluster in the presence of potential

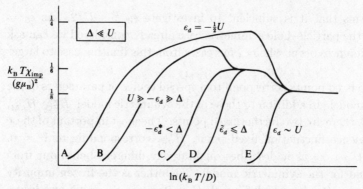

Figure 4.12   A schematic plot of $k_B T \chi_{\text{imp}}(T)/(g\mu_B)^2$ vs $\ln(k_B T/D)$ for $\Delta \ll U$ for the asymmetric Anderson model (Krishna-murthy et al, 1980).

scattering, $\bar{H}_N(K)$, have fixed points of R^2 for each value of K so there is a line of fixed points in each case. The points, $H_{\text{SC}}^*(K)$ and $H_{\text{FI}}^*(K)$, can be related because a large K on the end site is equivalent to an exclusion of charge on the end site plus induced potential scattering, $K' \sim 1/K$, at the next site 1.

The generalization of (4.67) for $\chi_{\text{imp}}(T)$ with $V = 0$ and $\epsilon_d \neq U/2$ is

$$\chi_{\text{imp}}(T) = \frac{(g\mu_B)^2}{2k_B T(2 + e^{-\epsilon_d/k_B T} + e^{-(\epsilon_d+U)/k_B T})}. \tag{4.75}$$

At the valence fluctuation fixed point, $\epsilon_d = 0$ and $U \gg k_B T$, so that $k_B T \chi_{\text{imp}}^{\text{VF}}(T) \to 1/6$.

A schematic plot of the behaviour of $T\chi_{\text{imp}}(T)$ against $\ln(k_B T/D)$ for various parameter regimes is shown in figure 4.12.

The curve A corresponds to the symmetric model, as in figure 4.11. The curves B, C, D, and E correspond to increasing ϵ_d through the valence fluctuation regime to the non-magnetic regime with ϵ_d well above the Fermi level. In curve B a local moment develops and in this region the susceptibility corresponds to the Wilson universal curve until a temperature $k_B T$ comparable with $|\epsilon_d|$. Curve C is in the valence fluctuation regime and deviates from the universal curve at low T as no local moment develops. Curves D and E are in the non-magnetic regime so the the the susceptibility is small at low temperatures and rises rapidly at temperatures $k_B T \sim \epsilon_d$. The results in the appropriate regimes are in agreement with those derived from scaling arguments which we discussed earlier in section 3.4.

A distinctly new feature can be seen in the results for $\chi_{\text{imp}}(T)$ against T as the impurity d level passes through and above the Fermi level, shown in figure 4.13.

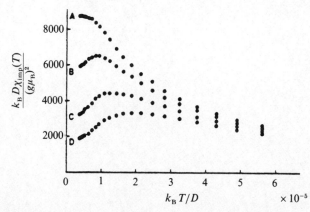

Figure 4.13   A plot of $k_B D\chi_{imp}(T)/(g\mu_B)^2$ vs $k_B T/D$ for $U/D = 10^{-2}$, $\Delta/U = 10^{-3}\pi/2$, for $(\epsilon_d/D)10^5 = -1.57$ (A), $= -0.974$ (B), $= 0$ (C), $= 0.974$ (D), for the asymmetric Anderson model (Krishna-murthy et al, 1980)(see text).

As the curves deviate from the Wilson universal curve they develop a maximum, which arises from the charge fluctuations. Though there is no maximum in the local moment regime of the $S = \frac{1}{2}$ Anderson model, we shall see later a maximum can occur in this regime for models of higher degeneracy so we cannot infer in general that an observation of a maximum implies valence fluctuations. For further results and details of the renormalization group calculations for the Anderson model we refer the reader to the original papers.

5

Fermi Liquid Theories

5.1 Phenomenological Fermi Liquid Theory

The phenomenological Fermi liquid theory of Landau is based on the assumption that the low lying excitations of a system of interacting fermions are in one to one correspondence with the excitations of the system when the interactions are set equal to zero. For non-interacting fermions, in our case electrons, the energy of any excited state E_{ex}, relative to the ground state, can be expressed in the form,

$$E_{\text{ex}} = \sum_{l,\sigma} \epsilon^0_{l,\sigma} \delta n_{l,\sigma}, \qquad (5.1)$$

where the energy $\epsilon^0_{l,\sigma}$ of the state $|l,\sigma\rangle$ is measured with respect to the Fermi energy ϵ^0_{F} and $\delta n_{l,\sigma} = n_{l,\sigma} - n^0_{l,\sigma}$, where $n^0_{l,\sigma}$ is the occupation number in the ground state and $n_{l,\sigma}$ in the excited state. For interacting electrons Landau assumed that this can be generalized to

$$E_{\text{ex}} = \sum_{l,\sigma} \epsilon_{l,\sigma} \delta n_{l,\sigma} + \frac{1}{2} \sum_{l,l',\sigma,\sigma'} f^{l,l'}_{\sigma,\sigma'} \delta n_{l,\sigma} \delta n_{l',\sigma'} + \text{O}(\delta n^3), \qquad (5.2)$$

by assuming a series expansion in $\delta n_{l,\sigma}$, where $\epsilon_{l,\sigma}$ and $f^{l,l'}_{\sigma,\sigma'}$ are coefficients that characterize the low energy excitations of the interacting system. If just a single electron excitation is created in a state $|l,\sigma\rangle$, corresponding to adding an extra particle to the system, its excitation energy is $\epsilon_{l,\sigma}$. If there are two or more excitations then the second term on the right hand side of (5.2) comes into play as an interaction effect. The effective one electron, or quasi-particle energy, $\tilde{\epsilon}_{l,\sigma}$ for the state $|l,\sigma\rangle$, in the presence of the other quasi-particle excitations was defined by Landau as the functional derivative of E_{ex} with respect to $\delta n_{l,\sigma}$,

$$\tilde{\epsilon}_{l,\sigma} = \frac{\partial E_{\text{ex}}}{\partial \delta n_{l,\sigma}} = \epsilon_{l,\sigma} + \sum_{l',\sigma'} f^{l,l'}_{\sigma,\sigma'} \delta n_{l',\sigma'}. \tag{5.3}$$

The energy of this state is modified due to the presence of other excitations. A free energy functional, $F\{\delta n_{l,\sigma}\} = E_{\text{ex}}\{\delta n_{l,\sigma}\} - TS\{\delta n_{l,\sigma}\}$, can be constructed using (5.2) and a Boltzmann expression for the entropy term, $S = k_{\text{B}} \ln W\{\delta n_{l,\sigma}\}$, where W is the number of ways of distributing the excitations over the available states, subject to the Pauli principle. Minimizing $F\{\delta n_{l,\sigma}\}$ with respect to the variational parameters $\delta n_{l,\sigma}$ gives an expression for the free energy and allows the low temperature thermodynamics to be calculated. The expectation value $\langle \delta n_{l,\sigma} \rangle$, which is the value of $\delta n_{l,\sigma}$ which gives the minimum for $F\{\delta n_{l,\sigma}\}$, when substituted into (5.3), results in a temperature dependent quasi-particle energy $\tilde{\epsilon}_{l,\sigma}(T)$,

$$\tilde{\epsilon}_{l,\sigma}(T) = \epsilon_{l,\sigma} + \sum_{l',\sigma'} f^{l,l'}_{\sigma,\sigma'} \langle \delta n_{l',\sigma'} \rangle. \tag{5.4}$$

The linear coefficient of specific heat calculated in this way is equivalent to that of a non-interacting set of quasi-particles of energy $\epsilon_{l,\sigma}$. This is due to the fact that in the interaction term in (5.4) $\langle \delta n_{l,\sigma} \rangle \to 0$ as $T \to 0$ and hence the interaction contributes higher order corrections in T.

The quasi-particle interaction term, however, does play a role in the response to a magnetic field. On switching on a magnetic field H, there is a change in the excitation energy $\epsilon_{l,\sigma}$ linear in H, and also a change in the average distribution of quasi-particles $\delta n_{l,\sigma}$ to the same order. In the phenomenological Fermi liquid theory the interaction parameter $f^{l,l'}_{\sigma,\sigma'}$ is suitably parameterized, taking account of any symmetries of the system, and the parameters deduced by a comparison of the predictions of the theory with experiment. For example, if the system is translationally invariant and has complete rotational symmetry the quantum numbers l can be identified as the wavevectors $\mathbf{k}$ and $f^{\mathbf{k},\mathbf{k}'}_{\sigma,\sigma'}$ depends on the angle θ between $\mathbf{k}$ and $\mathbf{k}'$. As quasi-particles are only well defined in the immediate vicinity of the Fermi level the magnitude of both $\mathbf{k}$ and $\mathbf{k}'$ may be taken to be k_{F}, and an expansion made of $f^{\mathbf{k},\mathbf{k}'}_{\sigma,\sigma'}$ in terms of Legendre polynomials,

$$f^{\mathbf{k},\mathbf{k}'}_{\sigma,\sigma'} = \sum_l (f^l_1 + f^l_2 \, \sigma \cdot \sigma') P_l(\cos \theta), \tag{5.5}$$

where $P_l(x)$ is a Legendre polynomial of degree l. The coefficients f^l_1 and f^l_2 are parameters characterizing the quasi-particle interaction. As, in general, only the low order angular momentum scattering terms in (5.5) corresponding to $l = 0, 1, 2$ are important the interaction can be effectively described by a finite number of parameters, which can be then determined by a comparison with experiment. The main application of this approach has been to helium liquids and we refer the reader to the original papers of Landau (1957, 1958), the book by Nozières (1964) or a review paper such as that of Leggett (1975).

The conclusions of the renormalization group calculations, described in the last chapter, were that the low lying excitations of the s-d model and Anderson model do, indeed, correspond to interacting Fermi quasiparticles. The effective Hamiltonian $\overline{H}'_N$ (4.49) can be identified with Landau theory, and the interaction term w_1 related to the interaction term in (5.4). We can parameterize the effective Hamiltonian (4.49) for the low lying excitations for the s-d model in a slightly different way,

$$H_{\text{eff}} = \sum_{\mathbf{k},\sigma} \epsilon_{\mathbf{k},\sigma} c^\dagger_{\mathbf{k},\sigma} c_{\mathbf{k},\sigma} + \sum_{\mathbf{k},\sigma} (\tilde{V}_{\mathbf{k}} c^\dagger_{1,\sigma} c_{\mathbf{k},\sigma} + \tilde{V}^*_{\mathbf{k}} c^\dagger_{\mathbf{k},\sigma} c_{1,\sigma})$$
$$+ \sum_\sigma \tilde{\epsilon}_{1,\sigma} c^\dagger_{1,\sigma} c_{1,\sigma} + \tilde{U} n_{1,\uparrow} n_{1,\downarrow},$$

(5.6)

with parameters $\tilde{\Delta} = \pi \sum_{\mathbf{k}} |\tilde{V}_{\mathbf{k}}|^2 \delta(\epsilon - \epsilon_{\mathbf{k}})$ and $\tilde{U}$, and $\tilde{\epsilon}_1 = \epsilon_F$, due to particle–hole symmetry. Note that $\tilde{\Delta}$ differs from the definition used in the previous chapter where it was associated with the hybridization between the impurity and the first site along the chain whereas here it is associated with the hybridization of site 1 with the rest of the chain. However with site 1 as the effective impurity this Hamiltonian is identical in form to the Anderson model so we parametrize it in the usual way. If we diagonalize (5.6) for $\tilde{U} = 0$, then

$$H_{\text{eff}}(\tilde{U} = 0) = \sum_{l,\sigma} \epsilon_{l,\sigma} c^\dagger_{l,\sigma} c_{l,\sigma}, \qquad (5.7)$$

where $\epsilon_{l,\sigma}$ are the one electron energies, and $c^\dagger_{l,\sigma}$ and $c_{l,\sigma}$ the corresponding creation and annihilation operators.

For $\tilde{U} \neq 0$, (5.6) can be rewritten using the single particle eigenstates of (5.7) as a basis,

$$H_{\text{eff}} = \sum_{l,\sigma} \epsilon_{l,\sigma} c^\dagger_{l,\sigma} c_{l,\sigma} + \tilde{U} \sum_{l,l'} \sum_{l'',l'''} \alpha^*_l \alpha_{l'} \alpha^*_{l''} \alpha_{l'''} c^\dagger_{l,\uparrow} c_{l',\uparrow} c^\dagger_{l'',\downarrow} c_{l''',\downarrow} \quad (5.8)$$

where

$$c^\dagger_{1,\sigma} = \sum_l \alpha^*_{l,\sigma} c^\dagger_{l,\sigma}. \qquad (5.9)$$

It is straightforward to diagonalize (5.6) for $\tilde{U} = 0$, and to determine the $\epsilon_{l,\sigma}$ and the $\alpha_{l,\sigma}$s. We do not need to do this explicitly because we find later that we only need the result (1.58) for the impurity density of states.

It is the *excitations* of the interacting system from its ground state that are described by the effective Hamiltonian (5.6), so it is appropriate to transform this Hamiltonian to operators which describe the single particle excitations,

$$
\begin{aligned}
c_{l,\sigma}^\dagger = p_{l,\sigma}^\dagger \quad & c_{l,\sigma} = p_{l,\sigma} \quad \epsilon_{l,\sigma} > \epsilon_{\text{F}} \\
c_{l,\sigma}^\dagger = h_{l,\sigma} \quad & c_{l,\sigma} = h_{l,\sigma}^\dagger \quad \epsilon_{l,\sigma} < \epsilon_{\text{F}},
\end{aligned}
\tag{5.10}
$$

where the ground state is such that $p_{l,\sigma}|0\rangle = 0$ and $h_{l,\sigma}|0\rangle = 0$. The interaction term has to be normal ordered in terms of these operators so that $H_{\text{eff}}|0\rangle = 0$, and the Hamiltonian describes interactions only between excitations from the ground state. In the Hartree–Fock or molecular field approximation the expectation values of operators in the interaction terms such as $p_{l,\uparrow}^\dagger p_{l',\uparrow} p_{l'',\downarrow}^\dagger p_{l''',\downarrow}$ are approximated by

$$
\langle p_{l,\uparrow}^\dagger p_{l',\uparrow} p_{l'',\downarrow}^\dagger p_{l''',\downarrow} \rangle = \langle p_{l,\uparrow}^\dagger p_{l,\uparrow} \rangle \langle p_{l'',\downarrow}^\dagger p_{l''',\downarrow} \rangle \delta_{l,l'} \delta_{l'',l'''}.
\tag{5.11}
$$

This decoupling corresponds to Wilson's treatment of the effective interaction in (4.49) in lowest order perturbation theory. This approximation is exact to lowest order as $T \to 0$ as we shall show later.

The quasi-particle energy (5.4) can be identified as the effective one particle energy in this approximation,

$$
\tilde{\epsilon}_{l,\sigma} = \epsilon_{l,\sigma} + \tilde{U}|\alpha_l|^2 \sum_{l'} |\alpha_{l'}|^2 \langle \delta n_{l',-\sigma} \rangle,
\tag{5.12}
$$

where $\langle \delta n_{l',\sigma} \rangle = \langle p_{l',\sigma}^\dagger p_{l',\sigma} \rangle$ for $\epsilon_{l'} > \epsilon_{\text{F}}$ and $\delta n_{l',\sigma} = -\langle h_{l',\sigma}^\dagger h_{l',\sigma} \rangle$ for $\epsilon_{l'} < \epsilon_{\text{F}}$. As $|\alpha_l|^2$ is proportional to $1/N_{\text{s}}$, where N_{s} is the number of sites, because the scattering potential is due to a single impurity, the energy shift in (5.12) is of the order $1/N_{\text{s}}$.

As $\langle p_{l,\sigma}^\dagger p_{l,\sigma} \rangle \to 0$ as $T \to 0$ the quasi-particle interaction does not contribute to the linear term in the specific heat which can be calculated from the non-interacting quasi-particle Hamiltonian (5.7). Using the Sommerfeld expansion (1.6) we find

$$
\gamma_{\text{tot}} = \frac{2\pi^2 k_{\text{B}}^2}{3} \sum_l \delta(\epsilon_{\text{F}} - \epsilon_l) = \frac{2\pi^2 k_{\text{B}}^2}{3} (\rho_0(\epsilon_{\text{F}}) + \tilde{\rho}_{\text{imp}}(\epsilon_{\text{F}})),
\tag{5.13}
$$

where $\tilde{\rho}_{\text{imp}}(\epsilon)$ is the impurity density of states of the quasi-particles for $\tilde{U} = 0$.

In the wide band limit $\tilde{\Delta}(\epsilon)$ is independent of ϵ and for $\tilde{\rho}_{\text{imp}}(\epsilon)$ we can use the result of section 1.4 for the non-interacting Anderson model,

$$\tilde{\rho}_{\text{imp}}(\epsilon) = \frac{1}{\pi} \frac{\tilde{\Delta}}{[(\epsilon - \tilde{\epsilon}_1)^2 + \tilde{\Delta}^2]}, \tag{5.14}$$

with $\tilde{\epsilon}_1 = \epsilon_{\text{F}}$ due to particle–hole symmetry. So the quasi-particle density of states has a resonance at the Fermi level. We are looking at the case for the s-d model which corresponds to the Anderson model with the original impurity level ϵ_{d} well below the Fermi level so that this resonance is quite distinct from the resonance associated with the original level and is a many-body resonance induced through the magnetic scattering at the Fermi level. It is often referred to as the *Kondo resonance* or *Abrikosov–Suhl resonance*.

In calculating the susceptibility the quasi-particle interaction has to be taken into account. In the Hartree–Fock or molecular field approximation for the quasi-particle interaction the total susceptibility in zero field is given by

$$\chi_{\text{tot}} = \frac{(g\mu_{\text{B}})^2}{2} \sum_l \delta(\epsilon_{\text{F}} - \tilde{\epsilon}_l)(1 + |\alpha_l|^2 \tilde{U} \frac{\partial \langle \delta n_{1,-\sigma} \rangle}{\partial h}), \tag{5.15}$$

where $\langle \delta n_{1,-\sigma} \rangle = \sum_l |\alpha_l|^2 \langle \delta n_{l,-\sigma} \rangle$, and $h = g\mu_{\text{B}}H/2$.

In zero field we also have

$$\sum_l \delta(\epsilon - \tilde{\epsilon}_l) = \rho_0(\epsilon) + \tilde{\rho}_{\text{imp}}(\epsilon), \tag{5.16}$$

where $\tilde{\rho}_{\text{imp}}$ is the quasi-particle density of states due to the impurity.

We can forget about the quasi-particle interaction term in calculating $\partial \langle \delta n_{1,-\sigma} \rangle / \partial H$ on the right hand side of (5.15) as it will gives terms of the order of $1/N_{\text{s}}^2$ which can be ignored, so working to order $1/N_{\text{s}}$ we find,

$$\frac{\partial \langle \delta n_{1,-\sigma} \rangle}{\partial h} = \tilde{\rho}_1(\epsilon_{\text{F}}), \tag{5.17}$$

where $\tilde{\rho}_1(\epsilon) = \sum_l |\alpha_l|^2 \delta(\epsilon - \tilde{\epsilon}_l)$. Hence to leading order $1/N_{\text{s}}$ the impurity susceptibility is given by

$$\chi_{\text{imp}} = \frac{(g\mu_{\text{B}})^2}{2} (\tilde{\rho}_{\text{imp}}(\epsilon_{\text{F}}) + \tilde{U}\tilde{\rho}_1^2(\epsilon_{\text{F}})). \tag{5.18}$$

The total charge susceptibility at $T = 0$, $\chi_c = dN_0/d\epsilon_{\text{F}}$, where N_0 is the expectation value of the total number operator for the electrons, can be calculated following precisely the same argument. The result for the impurity contribution is

$$\chi_{c,\text{imp}} = 2(\tilde{\rho}_{\text{imp}}(\epsilon_{\text{F}}) - \tilde{U}\tilde{\rho}_1^2(\epsilon_{\text{F}})). \tag{5.19}$$

Eliminating the term in $\tilde{U}$ between (5.18) and (5.19), and the term in

$\tilde{\rho}_{\mathrm{imp}}(\epsilon_{\mathrm{F}})$ using (5.13), gives the relation,

$$\frac{4\chi_{\mathrm{imp}}}{(g\mu_{\mathrm{B}})^2} + \chi_{\mathrm{c,imp}} = \frac{6\gamma_{\mathrm{imp}}}{\pi^2 k_{\mathrm{B}}^2}. \tag{5.20}$$

In the Kondo limit $\chi_{\mathrm{c,imp}} = 0$ as the occupation of the impurity state is always unity. This is achieved in the quasi-particle picture by the interaction term $\tilde{U}$. The interaction term self-consistently maintains the many-body resonance at the Fermi level so that the impurity occupation is not changed. If we use this result in (5.20) and calculate the Wilson ratio R, as defined by (4.59), we obtain the Wilson result $R = 2$. This was first shown by Nozières (1974, 1975) by an equivalent argument based on assuming a phase shift $\eta(\epsilon, \delta n_{l,-\sigma})$ for electrons of spin σ, expanding to first order in ϵ and $\delta n_{l,-\sigma}$, and then calculating the impurity susceptibility, density of states and specific heat via the Friedel sum rule, using equations (1.29) and (1.30). We shall prove in the next section that the Friedel sum rule, which we proved initially for a non-interacting model, holds when interactions are included.

We can use the fact that $\chi_{\mathrm{c,imp}} = 0$ in the Kondo limit to deduce quasi-particle interaction and the width of the Kondo resonance in terms of T_{K}. We use Wilson's result for γ_{imp} in terms of T_{K} and equate it to the impurity contribution to (5.13) to find the quasi-particle density of states at the Fermi level, and hence the resonance width $\tilde{\Delta}$. If we also take a flat band for the conduction electron density of states so that $\tilde{\Delta}(\epsilon)$ is independent of ϵ, then on calculating $\tilde{\rho}_1(\epsilon)$ using

$$\sum_l |\alpha_l|^2 \delta(\epsilon - \epsilon_l) = -\mathrm{Im}\, G_{1,1}^+(\epsilon)/\pi, \tag{5.21}$$

we can show that $\tilde{\rho}_1(\epsilon) = \tilde{\rho}_{\mathrm{imp}}(\epsilon)$. We can then deduce $\tilde{U}$ from equating (5.19) to zero; the requirement that the Kondo resonance is pinned to the Fermi level. With particle–hole symmetry, $\tilde{\rho}_1(\epsilon_{\mathrm{F}}) = 1/\pi\tilde{\Delta}$, we find

$$\tilde{U} = \pi\tilde{\Delta} = \frac{4k_{\mathrm{B}}T_{\mathrm{K}}}{w}, \tag{5.22}$$

where w is the Wilson number 0.4128 from equation (4.58). Hence we find the width of the quasi-particle resonance and the quasi-particle interaction are of the same order. This is largely to be expected as there is only one energy scale T_{K} in the universal regime $\epsilon, T \ll D$.

Discussion so far has been limited to the case of the Kondo limit of the Anderson model, or equivalently the s-d model for $S = \frac{1}{2}$. If we assume that the effective Hamiltonian (5.6) is applicable more generally for the low temperature excitations of the Anderson model we can use (5.20) to derive a general expression for the Wilson ratio in terms of $\chi_{\text{imp}}/\chi_{\text{c,imp}}$,

$$R = \frac{4\pi^2 k_{\text{B}}^2}{3(g\mu_{\text{B}})^2} \frac{\chi_{\text{imp}}}{\gamma_{\text{imp}}} = \frac{2}{(1 + (g\mu_{\text{B}})^2 \chi_{\text{c,imp}}/4\chi_{\text{imp}})}. \qquad (5.23)$$

This goes over to the correct result for the non-interacting limit $R = 1$ as in this limit $\chi_{\text{c,imp}} = 4\chi_{\text{imp}}/(g\mu_{\text{B}})^2$. We shall prove in section 5.3 that this relation is, in fact, exact.

We can conclude that the low temperature behaviour of the s-d model, or equivalently the Kondo limit of the Anderson model, corresponds precisely to the Fermi liquid picture of Landau, quasi-particle excitations from a ground state with relatively weak interquasi-particle interactions ($\sim k_{\text{B}}T_{\text{K}}$). The renormalization group calculations of Wilson gave the form of the effective Hamiltonian for the quasi-particles with parameters determined by fitting to numerical calculations for the energy levels. This is a rather unusual situation in which the parameters for the low energy excitations can be calculated a priori rather than appealing to experiment. Nozières' argument shows that some results, such as the one for the Wilson ratio $R = 2$, can be obtained under quite general assumptions without a precise knowledge of these parameters. As the energy scale $k_{\text{B}}T_{\text{K}}$ is known from perturbation theory only one number, the Wilson number w, in practice has to be calculated numerically to relate the parameters of the effective Hamiltonians, (1.66) and (5.6), for the high and low energy scales. On the low energy scale the density of states of the quasi-particles has a resonance at the Fermi level, giving significant impurity contributions to the specific heat and magnetic susceptibility. (We shall see later that the symmetry of this resonance about the Fermi level is a special feature of the particle–hole symmetric case and not a general feature.) The situation is similar to the single particle description of an impurity (section 1.2) with a narrow virtual bound state of width of the order of T_{K} at the Fermi level. The model with such a resonance is known as the non-interacting *resonant level model*. It has been used to fit the experimental magnetic isotherms as a function of magnetic field in some Kondo systems, for which it gives a reasonably good fit (see Schotte & Schotte, 1975, and Grüner, 1974). As it does not include the quasi-particle interactions it has a Wilson ratio $R = 1$. The Kondo resonance is, however, a many-body resonance, which includes the quasi-particle interactions, and has an enhanced Wilson ratio $R = 2$.

The phenomenological Fermi liquid approach can be extended to give results for the electrical conductivity as we shall show in section 5.4.

5.2 The Generalized Friedel Sum Rule

As a preliminary to giving a microscopic derivation of the Fermi liquid relation we shall give a proof of the Friedel sum rule for the Anderson model in the presence of interactions. This was originally proved by Langer & Ambegaokar (1961) and Langreth (1966). The proof will follow the pattern of the proof for the non-interacting case given in section 1.2. To do this we will have to introduce many-body Green's functions for dealing with interacting systems. These Green's functions will be the common ones as defined in standard many-body theory. First of all we shall need the thermal or imaginary time Green's functions $G(\tau_1, \tau_2)$. The one electron thermal Green's function is defined by

$$G_{i,j}(\tau_1, \tau_2) = -\langle T_\tau\, c_i(\tau_1) c_j^\dagger(\tau_2)\rangle, \tag{5.24}$$

where $c^{(\dagger)}(\tau) = \exp(\tau H') c^{(\dagger)} \exp(-\tau H')$, with $H' = H - \mu N_0$, $\langle\ldots\rangle$ a thermal average as in (2.24), and T_τ is the 'time' ordering operator defined by

$$\begin{aligned} T_\tau\, c_i(\tau_1) c_j^\dagger(\tau_2) &= \quad c_i(\tau_1) c_j^\dagger(\tau_2) \qquad \tau_1 > \tau_2 \\ &= -\, c_j^\dagger(\tau_2) c_i(\tau_1) \qquad \tau_2 > \tau_1 \end{aligned} \tag{5.25}$$

The Green's function $G_{i,j}(\tau_1, \tau_2)$ can be expressed in the form of a Fourier series,

$$G_{i,j}(\tau_1, \tau_2) = \frac{1}{\beta} \sum_{\omega_n} G_{i,j}(i\omega_n) e^{-i\omega_n(\tau_1 - \tau_2)}, \tag{5.26}$$

where $\omega_n = (2n + 1)\pi/\beta$ and n is an integer.

A perturbation expansion can be developed in powers of the interaction parameter U for the non-degenerate Anderson model (1.66),

$$H = \sum_\sigma \epsilon_d n_{d,\sigma} + U n_{d,\uparrow} n_{d,\downarrow} + \sum_{\mathbf{k},\sigma} \epsilon_\mathbf{k} c_{\mathbf{k},\sigma}^\dagger c_{\mathbf{k},\sigma}$$
$$+ \sum_{\mathbf{k},\sigma} (V_\mathbf{k} c_{d,\sigma}^\dagger c_{\mathbf{k},\sigma} + V_\mathbf{k}^* c_{\mathbf{k},\sigma}^\dagger c_{d,\sigma}), \tag{5.27}$$

and represented by Feynman diagrams with the non-interacting Green's functions $G_{i,j}^0(\tau_1, \tau_2)$ as propagators. The rules for drawing and evaluating diagrams are standard and the reader is referred to one of the many books on many-body perturbation techniques.

It will also be useful to calculate a different type of Green's function, the retarded double-time Green's function $\langle\langle A(t_1) : B(t_2)\rangle\rangle$ for the operators A and B defined by

$$\langle\langle A(t_1) : B(t_2)\rangle\rangle = -i\theta(t_1 - t_2)\langle [A(t_1), B(t_2)]_\eta\rangle, \tag{5.28}$$

where $[A, B]_\eta = AB + \eta BA$, and η can be chosen for convenience with the usual choice $\eta = +1$ for Fermi operators and $\eta = -1$ for Bose operators. The operators are in the Heisenberg representation $A(t) = \exp(iH') A \exp(-iH')$. These Green's functions are essentially the linear response functions $\phi_{AB}(t_1 - t_2)$ defined in appendix B, equation (B.6), with a more general definition for the commutator (we have also put $\hbar = 1$). The Fourier transform $G_{i,j}(\omega)$ of the Green's function $\langle\langle A_i(t_1) : B_j(t_2)\rangle\rangle$ is given by

$$G_{i,j}(\omega^+) = \int_{-\infty}^{\infty} e^{i\omega^+(t_1-t_2)} \langle\langle A_i(t_1) : B_j(t_2)\rangle\rangle \, d(t_1 - t_2), \qquad (5.29)$$

where $\omega^+ = \omega + is$, $s \to +0$. By a suitable choice for A_i and B_j we can deduce frequency dependent response functions such as the electrical conductivity $\sigma(\omega)$ or dynamical susceptibility $\chi(\omega)$. The corresponding spectral density $\rho_{i,j}(\omega)$ is defined by

$$\rho_{i,j}(\omega) = -\frac{1}{\pi} \operatorname{Im} G_{i,j}(\omega^+). \qquad (5.30)$$

The time dependent correlation functions $\langle B_j(t_2) A_i(t_1)\rangle$ can be deduced from the spectral density $\rho_{ij}(\omega)$ using

$$\langle B_j(t_2) A_i(t_1)\rangle = \int_{-\infty}^{\infty} f_\eta(\omega)\rho_{ij}(\omega)e^{-i\omega(t_1-t_2)} d\omega, \qquad (5.31)$$

where $f_\eta(\omega) = 1/(\exp(\beta\omega) + \eta)$. For a derivation of this relation and further discussion of double-time thermal Green's functions we refer the reader to Zubarev (1960).

The total density of states $\rho(\omega)$ for single particle excitations can be deduced from (5.30) using $A_i = c_i$, $B_i = c_i^\dagger$, and $\eta = +1$. The expectation value $\langle c_i^\dagger c_i \rangle$ of the number operator n_i can then be deduced as an integral over this density of states weighted with the Fermi distribution factor $f(\omega)$, using (5.31).

There is a clear correspondence with the retarded Green's functions defined here and the single-particle Green's functions of chapter 1 when there are no two-body interactions present so we shall use the same notation for the Fourier transforms of the two types of functions to bring out the similarities, but distinguish them by using the argument ω for the retarded many-body Green's functions instead of ϵ which we used for the single-particle resolvent functions.

The double-time Green's functions satisfy the equation of motion,

$$\frac{d}{dt_1}\langle\langle A(t_1) : B(t_2)\rangle\rangle =$$
$$\langle [A(t_1), B(t_1)]_\eta\rangle \delta(t_1 - t_2) + \langle\langle [A(t_1), H'(t_1)]_- : B(t_2)\rangle\rangle.$$
$$(5.32)$$

Generating the equations of motion for the single particle Green's functions for the Anderson model using (5.32), and then taking the Fourier transform, leads to the set of equations,

$$(\omega - \epsilon_d)G_{d\sigma,d\sigma}(\omega) - U\Gamma_{d,d}(\omega) = 1 + \sum_{\mathbf{k}} V_{\mathbf{k}}G_{\mathbf{k}\sigma,d\sigma}(\omega), \qquad (5.33)$$

$$(\omega - \epsilon_{\mathbf{k}})G_{\mathbf{k}\sigma,d\sigma}(\omega) = V_{\mathbf{k}}^* G_{d\sigma,d\sigma}(\omega), \qquad (5.34)$$

$$(\omega - \epsilon_{\mathbf{k}})G_{\mathbf{k}\sigma,\mathbf{k}'\sigma}(\omega) = \delta_{\mathbf{k},\mathbf{k}'} + V_{\mathbf{k}}^* G_{d\sigma,\mathbf{k}'\sigma}(\omega), \qquad (5.35)$$

where $\Gamma_{d\sigma,d\sigma}(\omega)$ is the Fourier transform of $\langle\langle c_{d,\sigma}n_{d,-\sigma}(t_1) : c_{d,\sigma}^\dagger(t_2)\rangle\rangle$. The zero of ω has been displaced so that $\omega = \mu$ corresponds to the Fermi level.

For $U = 0$ these equations correspond to the resolvent Green's function equations (1.45), (1.46) and (1.48). Solving for $G_{d\sigma,d\sigma}^0(\omega)$,

$$G_{d\sigma,d\sigma}^0(\omega) = \frac{1}{(\omega - \epsilon_d - \sum_{\mathbf{k}}|V_{\mathbf{k}}|^2/(\omega - \epsilon_{\mathbf{k}}))}. \qquad (5.36)$$

If we differentiate $\langle\langle c_{d,\sigma}(t_1) : c_{\mathbf{k}',\sigma}^\dagger(t_2)\rangle\rangle$ with respect to t_2 and then take the Fourier transform with respect to $t_1 - t_2$, we find

$$(\omega - \epsilon_{\mathbf{k}'})\hat{G}_{d\sigma,\mathbf{k}'\sigma}(\omega) = V_{\mathbf{k}'}G_{d\sigma,d\sigma}(\omega). \qquad (5.37)$$

Solving (5.35) and (5.37) for $G_{\mathbf{k}\sigma,\mathbf{k}'\sigma}(\omega)$ gives

$$G_{\mathbf{k}\sigma,\mathbf{k}'\sigma}(\omega) = \frac{\delta_{\mathbf{k},\mathbf{k}'}}{(\omega - \epsilon_{\mathbf{k}})} + \frac{V_{\mathbf{k}}^*}{(\omega - \epsilon_{\mathbf{k}})}G_{d\sigma,d\sigma}(\omega)\frac{V_{\mathbf{k}'}}{(\omega - \epsilon_{\mathbf{k}'})}. \qquad (5.38)$$

Comparing this with equations (1.50) and (1.51), we can identify the T matrix $T_{\mathbf{k}\sigma,\mathbf{k}'\sigma}(\omega)$ for the many-body scattering,

$$T_{\mathbf{k}\sigma,\mathbf{k}'\sigma}(\omega) = V_{\mathbf{k}}^* G_{d\sigma,d\sigma}(\omega)V_{\mathbf{k}'}. \qquad (5.39)$$

We find, therefore, that $G_{d\sigma,d\sigma}(\omega)$ is the key to calculating the conductivity for the Anderson model.

For $V_{\mathbf{k}} = 0$, equation (5.33) can be solved for $G_{d\sigma,d\sigma}(\omega)$ on generating the equation of motion for $\Gamma_{d\sigma,d\sigma}(t_1,t_2)$ which can be closed by using the operator identity $n_{d,-\sigma}^2 = n_{d,-\sigma}$. The solution is

$$G_{d\sigma,d\sigma}(\omega) = \frac{1 - \langle n_{d,-\sigma}\rangle}{(\omega - \epsilon_d)} + \frac{\langle n_{d,-\sigma}\rangle}{(\omega - \epsilon_d - U)}. \qquad (5.40)$$

This Green's function has poles at the energies ϵ_d and $\epsilon_d + U$, corresponding to the excitation energies on adding an electron to the system in the atomic limit (see figure 1.7).

The equations of motion of the thermal Green's function (imaginary time) are similar in form to those for the double-time Green's functions. The solution for the Fourier coefficient $G_{d\sigma,d\sigma}^0(i\omega_n)$ for $U = 0$ is

$$G_{d\sigma,d\sigma}^0(i\omega_n) = \frac{1}{(i\omega_n + \mu - \epsilon_d - \sum_{\mathbf{k}}|V_{\mathbf{k}}|^2/(i\omega_n + \mu - \epsilon_{\mathbf{k}}))}. \qquad (5.41)$$

It corresponds to (5.38) with $\omega \to i\omega_n + \mu$. This is true generally: the corresponding thermal Green's function can always be deduced from the double-time Green's function in this way. As there are diagrammatic perturbation expansions for the thermal Green's functions but not for the double-time Green's functions, it is of more interest to deduce the double-time functions from the thermal functions. It is possible to do this by analytic continuation (see for instance Fetter and Walecka, 1971, for details). In practice this usually requires the substitution $i\omega_n + \mu \to \omega$ with the check that $G_{i,j}(\omega) \to 0$ as $\omega \to \infty$.

If we use a constant density of states (1.55) and neglect the dependence of $V_{\mathbf{k}}$ on $\mathbf{k}$ we find

$$\sum_{\mathbf{k}} \frac{|V_{\mathbf{k}}|^2}{(i\omega_n + \mu - \epsilon_{\mathbf{k}})} = |V|^2 \rho_0 \int_{-D}^{D} \frac{1}{i\omega_n + \mu - \epsilon} \, d\epsilon$$

$$= \frac{-\Delta}{\pi} \ln\left(\frac{D - i\omega_n - \mu}{-D - i\omega_n - \mu}\right), \qquad (5.42)$$

which tends to $-i\Delta \operatorname{sgn}(\omega_n)$ in the wide band limit $D \to \infty$. Hence in this limit,

$$G^0_{\mathrm{d}\sigma,\mathrm{d}\sigma}(i\omega_n) = \frac{1}{i\omega_n + \mu - \epsilon_{\mathrm{d}} + i\Delta \operatorname{sgn}(\omega_n)}. \qquad (5.43)$$

Perturbation theory in powers of the interaction U can be introduced in the standard way (see Fetter and Walecka, 1971, for example) enabling the interacting Green's function $G_{\mathrm{d}\sigma,\mathrm{d}\sigma}(i\omega_n)$ to be expressed in the form

$$G_{\mathrm{d}\sigma,\mathrm{d}\sigma}(i\omega_n) = \frac{1}{i\omega_n + \mu - \epsilon_{\mathrm{d}} + i\Delta \operatorname{sgn}(\omega_n) - \Sigma_\sigma(i\omega_n)}, \qquad (5.44)$$

where $\Sigma_\sigma(i\omega_n)$ is the proper self-energy. The perturbation series can be used for $\Sigma_\sigma(i\omega_n)$. This approach has been developed in a series of papers by Yamada & Yosida (Yamada 1975, 1976; Yosida & Yamada, 1970, 1975). It is not evident that the perturbation theory is valid for all values of U (we noted earlier that the RPA subset diverges at a critical value of U). We take it as a working hypothesis at this stage. We will be able to justify it later in chapter 6.

We use certain exact results for the self-energy at $T = 0$ in this and subsequent sections. In terms of the self-energy, analytically continued to real frequency ω, these are

(i) $\Sigma_\sigma(\omega \pm is) = \Sigma_\sigma^{\mathrm{R}}(\omega) \mp i\Sigma_\sigma^{I}(\omega)$, where $\Sigma_\sigma^{I}(\omega) \geq 0$ for $s \to +0$.

(ii) $\Sigma_\sigma^{I}(\omega)$ evaluated at the Fermi energy ϵ_{F} is zero.

(iii) $\Sigma_\sigma^{I}(\omega) \propto \omega^2$ as $\omega \to \epsilon_{\mathrm{F}}$.

(iv) $\operatorname{Im} \int_{-\infty}^{\epsilon_{\mathrm{F}}} \partial \Sigma_\sigma(\omega)/\partial\omega \, G_{\sigma,\sigma}(\omega) \, d\omega = 0.$

These have been proved to all orders in perturbation theory for a general class of systems by Luttinger (1960, 1961). The first three have also been proved explicitly for the Anderson model by Yamada and Yosida.

The occupation number for the impurity d level $\langle n_{d,\sigma} \rangle$ can be calculated by integrating the density of states $\rho_{d,\sigma}(\omega)$, which is the spectral density of the Green's function $G_{d\sigma,d\sigma}(\omega)$, using (5.31). Hence, using the analytic continuation of (5.44) for $T = 0$,

$$\langle n_{d,\sigma} \rangle = \int_{-\infty}^{\epsilon_F} \rho_{d,\sigma}(\omega) \, d\omega = -\frac{\mathrm{Im}}{\pi} \int_{-\infty}^{\epsilon_F} \frac{1}{\omega^+ - \epsilon_d + i\Delta - \Sigma_\sigma(\omega^+)} d\omega. \tag{5.45}$$

However, as

$$\frac{1}{\omega^+ - \epsilon_d + i\Delta - \Sigma_\sigma(\omega)} = \frac{\partial}{\partial\omega} \ln\left(\omega^+ - \epsilon_d + i\Delta - \Sigma_\sigma(\omega)\right)$$

$$+ \frac{\partial\Sigma_\sigma(\omega)/\partial\omega}{(\omega^+ - \epsilon_d + i\Delta - \Sigma_\sigma(\omega))}, \tag{5.46}$$

from the result (iv) the second term on the right hand side of (5.46) when integrated over ω gives no contribution to (5.45). The first term when integrated gives

$$\langle n_{d,\sigma} \rangle = \frac{1}{2} - \frac{1}{\pi} \tan^{-1}\left(\frac{\epsilon_d - \epsilon_F + \Sigma_\sigma^R(\epsilon_F)}{\Delta}\right), \tag{5.47}$$

using result (ii) that $\Sigma_\sigma^I(\epsilon_F) = 0$.

As $G_{d\sigma,d\sigma}(\omega)$ is essentially the T matrix the corresponding phase shift, using the definition (1.28), is

$$\eta_\sigma(\omega) = \frac{\pi}{2} - \tan^{-1}\left(\frac{\epsilon_d + \Sigma_\sigma^R(\omega) - \omega}{\Delta + \Sigma_\sigma^I(\omega)}\right). \tag{5.48}$$

Hence, $\langle n_{d,\sigma} \rangle = \eta_\sigma(\epsilon_F)/\pi$, so the Friedel sum rule is applicable for the interacting model in each spin channel. In proving this we took the simplifying case of a wide flat conduction band. This is not a necessary condition and to prove the result more generally we have to include the contribution to $\langle n_{\mathrm{imp},\sigma} \rangle$ from $G_{k\sigma,k\sigma}(\omega)$ as in (1.52). Only in the flat wide band limit does $\langle n_{\mathrm{imp},\sigma} \rangle = \langle n_{d,\sigma} \rangle$.

If $\langle n_{d,\sigma} \rangle$ is known then we can use (5.47) to calculate the impurity density of states at the Fermi level. Calculating $\rho_{d,\sigma}(\omega)$ from (5.44)

$$\rho_{d,\sigma}(\omega) = \frac{(\Delta + \Sigma_\sigma^I(\omega))/\pi}{(\omega - \epsilon_d - \Sigma_\sigma^R(\omega))^2 + (\Delta + \Sigma_\sigma^I(\omega))^2}. \tag{5.49}$$

(Note that in the interacting case, $\pi\rho_{d,\sigma}(\omega) \neq \partial\eta_\sigma(\omega)/\partial\omega$.) Solving for $\Sigma_\sigma^R(\epsilon_F)$ from (5.47), and using $\Sigma_\sigma^I(\epsilon_F) = 0$, gives for the impurity density of states at the Fermi level,

$$\rho_{d,\sigma}(\epsilon_F) = \frac{\sin^2(\pi\langle n_{d,\sigma}\rangle)}{\pi\Delta}, \tag{5.50}$$

which is an exact result. For the symmetric model with particle–hole symmetry $\langle n_{d,\sigma}\rangle = 1/2$ so $\rho_{d,\sigma}(\epsilon_F) = 1/\pi\Delta$, independent of the interaction U.

5.3 Microscopic Fermi Liquid Theory

The results (i)–(iii) given in the previous section can be used to substantiate the quasi-particle concept. If we expand the self-energy $\Sigma(\omega)$ in powers of $\omega - \epsilon_F$ for the retarded Green's function corresponding to (5.44) and keep terms to order $\omega - \epsilon_F$,

$$G_{d\sigma,d\sigma}(\omega) =$$
$$\frac{1}{\omega - \epsilon_d - \Sigma_\sigma^R(\epsilon_F) + i\Delta - (\omega - \epsilon_F)\partial\Sigma_\sigma^R(\epsilon_F)/\partial\omega - iO(\omega^2)}, \tag{5.51}$$

which can be rewritten in the form,

$$G_{d\sigma,d\sigma}(\omega) = \frac{z}{\omega - \epsilon_F - \tilde{\epsilon}_d + i\tilde{\Delta} - iO(\omega^2)}, \tag{5.52}$$

where $\tilde{\Delta} = z\Delta$, $\tilde{\epsilon}_d = z(\epsilon_d - \epsilon_F + \Sigma_\sigma^R(\epsilon_F))$ and $z = 1/(1 - \partial\Sigma_\sigma^R(\epsilon_F)/\partial\omega)$. The z factor is known as the wavefunction renormalization factor, and is omitted in defining a quasi-particle Green's function $\tilde{G}_{d\sigma,d\sigma}(\omega)$,

$$\tilde{G}_{d\sigma,d\sigma}(\omega) = \frac{1}{\omega - \epsilon_F - \tilde{\epsilon}_d + i\tilde{\Delta}}. \tag{5.53}$$

This corresponds to a quasi-particle Hamiltonian of the form (5.6) with $\tilde{U} = 0$ and $\tilde{\epsilon}_1 - \epsilon_F = \tilde{\epsilon}_d$. Hence the corresponding quasi-particle phase shift $\tilde{\eta}_\sigma(\omega)$ and quasi-particle density of states $\tilde{\rho}_{d,\sigma}(\omega)$ are

$$\tilde{\eta}_\sigma(\omega) = \frac{\pi}{2} - \tan^{-1}\frac{\epsilon_F + \tilde{\epsilon}_d - \omega}{\tilde{\Delta}}, \quad \tilde{\rho}_{d,\sigma}(\omega) = \frac{\tilde{\Delta}/\pi}{(\omega - \epsilon_F - \tilde{\epsilon}_d)^2 + \tilde{\Delta}^2}. \tag{5.54}$$

Note that the quasi-particle density of states at the Fermi level differs from the true density of states by the wavefunction renormalization factor z. If the quasi-particle occupation number $\langle\tilde{n}_{d,\sigma}\rangle$ is defined by the application of the Friedel sum rule to the quasi-particle phase shift then

$$\langle\tilde{n}_{d,\sigma}\rangle = \frac{1}{2} - \frac{1}{\pi}\tan^{-1}\left(\frac{\tilde{\epsilon}_d}{\tilde{\Delta}}\right). \tag{5.55}$$

As $\tilde{\epsilon}_d$ and $\tilde{\Delta}$ have a common factor z this cancels out in (5.55) and the right hand side is the same as that in equation (5.47), so $\langle \tilde{n}_{d,\sigma} \rangle = \langle n_{d,\sigma} \rangle$. This establishes the 1:1 correspondence of the particles and quasi-particles which is a general feature of Fermi liquid theory.

We can make use of two further results of microscopic Fermi liquid theory which we can add to the list in the previous section. They are

(v)   The linear coefficient of specific heat $\gamma = 2\pi^2 k_B^2 \tilde{\rho}(\epsilon_F)/3$, where $\tilde{\rho}(\epsilon_F)$ is the density of quasi-particle states at the Fermi level.
(vi)   The zero field $T = 0$ susceptibility is given by $\chi = (g\mu_B)^2 \alpha' \tilde{\rho}(\epsilon_F)/2$, where the quasi-particle energy $\tilde{\epsilon}_\sigma(H)$ is given by $\tilde{\epsilon}_\sigma(h) = \tilde{\epsilon}_\sigma(0) - \alpha' \sigma h$ ($h = g\mu_B H/2$) to first order in h. These have been proved within perturbation theory (see Luttinger 1960, 1961).

The first result (v) is equivalent to the statement that, in the calculation of the specific heat, the interaction between the quasi-particles can be ignored. Applying this to the Anderson model gives

$$\gamma_{\text{imp}} = \frac{2\pi k_B^2}{3} \frac{\tilde{\Delta}}{\tilde{\epsilon}_d^2 + \tilde{\Delta}^2}, \tag{5.56}$$

which in the particle–hole symmetric case $\tilde{\epsilon}_d = 0$ agrees with (5.13) and (5.14). This result is more general, however, as it applies in all parameter regimes. In the Kondo limit $U \gg \Delta$ and $\tilde{\Delta} \sim k_B T_K$, and at the other extreme $U = 0$ and $\tilde{\Delta} = \Delta$.

The second result (vi) differs from that for independent quasi-particles by a factor α', which reflects the fact that the quasi-particles experience a molecular field due to the interaction with the other quasi-particles. Applying this result to the Anderson model gives

$$\chi_{\text{imp}} = \frac{g^2 \mu_B^2}{2\pi} \frac{\alpha' \tilde{\Delta}}{\tilde{\epsilon}_d^2 + \tilde{\Delta}^2}. \tag{5.57}$$

From equation (5.23) we can identify α' as the Wilson ratio R. The quasi-particle energies $\tilde{\epsilon}_\sigma(h)$ are solutions of

$$\tilde{\epsilon}_\sigma(h) - \tilde{\epsilon}_d + \sigma h - \Sigma_\sigma^R(\tilde{\epsilon}_\sigma(h), h) = 0. \tag{5.58}$$

On expanding this to first order in h we find

$$\alpha' = R = \frac{1 - \partial \Sigma_\sigma(\epsilon_F, 0)/\partial h}{1 - \partial \Sigma_\sigma(\epsilon_F, 0)/\partial \omega}. \tag{5.59}$$

The result for χ_{imp} can be verified by calculating the impurity magnetization in a magnetic field, $M_{\text{imp}} = g\mu_B \langle n_{d,\uparrow}(H) - n_{d,\downarrow}(H) \rangle/2$, using the Friedel sum rule (5.47). The α' factor arises from the field dependence of $\Sigma^R(\epsilon_F)$. We can also use this approach to calculate the impurity charge

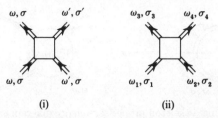

Figure 5.1 Four point vertices $\Gamma_{\sigma,\sigma'}(i\omega, i\omega')$ and $\Gamma_{\sigma_3,\sigma_4}^{\sigma_1,\sigma_2}(i\omega_1, i\omega_2, i\omega_3, i\omega_4)$.

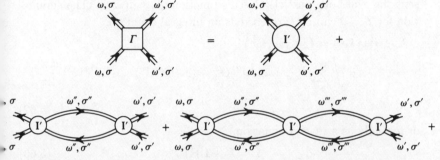

Figure 5.2 Repeated particle–hole diagrams for $\Gamma_{\sigma,\sigma'}(i\omega, i\omega')$ (see equation (5.62))

susceptibility $\chi_{c,imp}$,

$$\chi_{c,imp} = \sum_\sigma \frac{\partial \langle n_{d,\sigma} \rangle}{\partial \mu} = \frac{2\alpha'' \tilde{\Delta}/\pi}{(\epsilon_F - \tilde{\epsilon}_d)^2 + \tilde{\Delta}^2}, \tag{5.60}$$

where α'' is given by

$$\alpha'' = \frac{1 + \partial \Sigma_\sigma(\epsilon_F)/\partial \mu}{1 - \partial \Sigma_\sigma(\epsilon_F)/\partial \omega}. \tag{5.61}$$

To prove (5.20), which relates (5.56), (5.57) and (5.60), we shall need certain Ward identities which relate derivatives of the self-energy to the four point antisymmetrized vertex function $\Gamma_{\sigma,\sigma'}(i\omega, i\omega')$. This vertex represents the sum of all the scattering processes of two particles of spin σ and σ' and is illustrated in figure 5.1(i). It is a special case of the general four point vertex function $\Gamma_{\sigma_3,\sigma_4}^{\sigma_1,\sigma_2}(i\omega_1, i\omega_2, i\omega_3, i\omega_4)$, shown in figure 5.1(ii), with $\sigma_1 = \sigma_2 = \sigma$, $\sigma_3 = \sigma_4 = \sigma'$, $\omega_1 = \omega_2 = \omega$, and $\omega_3 = \omega_4 = \omega'$. The irreducible vertex $I'_{\sigma,\sigma'}(i\omega, i\omega')$ is a similar vertex but contains no contributions that can be separated into two parts by cutting a particle–hole pair. Hence, $\Gamma_{\sigma,\sigma'}(i\omega, i\omega')$ can be obtained from $I'_{\sigma,\sigma'}(i\omega, i\omega')$ by summing the series of particle–hole scattering diagrams shown in figure 5.2.

The diagrams are skeleton diagrams in which the double lines repre-

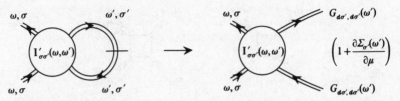

Figure 5.3 Severing a skeleton proper self-energy diagram corresponding to equation (5.66).

sent the exact one particle Green's function $G_{d\sigma,d\sigma}(i\omega_n)$. The summation for $T = 0$ can be expressed as an integral equation,

$$\Gamma_{\sigma,\sigma'}(i\omega, i\omega') = I'_{\sigma,\sigma'}(i\omega, i\omega')$$
$$+ \int \sum_{\sigma''} I'_{\sigma,\sigma''}(i\omega, i\omega'') G^2_{d\sigma'',d\sigma''}(i\omega'') \Gamma_{\sigma'',\sigma'}(i\omega'', i\omega') \frac{d\omega''}{2\pi},$$

$$(5.62)$$

or formally as a matrix equation,

$$\mathbf{\Gamma} = \mathbf{I}' + \mathbf{I}'\mathbf{R}\mathbf{\Gamma}, \qquad (5.63)$$

where $R_{\sigma',\sigma''}(i\omega', i\omega'') = \delta_{\omega',\omega''}\delta_{\sigma',\sigma''}G^2_{d\sigma,d\sigma}(i\omega'')$. The formal solution of (5.63) is

$$\mathbf{\Gamma} = (\mathbf{I} - \mathbf{I}'\mathbf{R})^{-1}\mathbf{I}', \qquad (5.64)$$

which will be used later.

We need to calculate the derivative $\partial\Sigma_\sigma(i\omega)/\partial\mu$ for a proper self-energy diagram as in figure 5.3. Differentiating with respect to μ corresponds to differentiating each propagator in turn in this diagram. This is similar to cutting a single propagator as

$$\frac{dG_{d\sigma,d\sigma}(i\omega)}{d\mu} = G^2_{d\sigma,d\sigma}(i\omega)\left(1 + \frac{\partial\Sigma_\sigma(i\omega)}{\partial\mu}\right). \qquad (5.65)$$

The severed diagram is a contribution to the vertex $I'_{\sigma,\sigma'}(i\omega, i\omega')$.

This follows from the fact that if any two propagators in the diagrammatic series for Γ, shown in figure 5.2, are joined, this does not produce a skeleton diagram for $\Sigma(i\omega)$, except for the first diagram in the series. It can be shown that the sum of these diagrams, with their appropriate weight factors, and after the removal of the external legs, gives the vertex $I'_{\sigma,\sigma'}(i\omega, i\omega')$ (see Nozières, 1964, for further details). Hence the integral equation,

$$\frac{\partial\Sigma_\sigma(i\omega)}{\partial\mu} = \int \sum_{\sigma'} I'_{\sigma,\sigma'}(i\omega, i\omega') G^2_{d\sigma',d\sigma'}(i\omega')\left(1 + \frac{\partial\Sigma_\sigma(i\omega')}{\partial\mu}\right)\frac{d\omega'}{2\pi},$$

$$(5.66)$$

expressed diagrammatically in figure 5.3. Expressing this in the form of a matrix equation,

$$\frac{\partial \boldsymbol{\Sigma}}{\partial \mu} = \mathbf{I}'\mathbf{S} + \mathbf{I}'\mathbf{R}\frac{\partial \boldsymbol{\Sigma}}{\partial \mu}, \qquad (5.67)$$

where $S_{\omega',\sigma'} = G^2_{\mathrm{d}\sigma',\mathrm{d}\sigma'}(i\omega')$. Hence the solution,

$$\frac{\partial \boldsymbol{\Sigma}}{\partial \mu} = (\mathbf{I} - \mathbf{I}'\mathbf{R})^{-1}\mathbf{I}'\mathbf{S} = \boldsymbol{\Gamma}\mathbf{S}, \qquad (5.68)$$

or equivalently,

$$\frac{\partial \Sigma_\sigma(i\omega)}{\partial \mu} = \int \sum_{\sigma'} \Gamma_{\sigma,\sigma'}(i\omega, i\omega') G^2_{\mathrm{d}\sigma',\mathrm{d}\sigma'}(i\omega') \frac{d\omega'}{2\pi}. \qquad (5.69)$$

In a similar way the derivative $\partial \Sigma_\sigma(i\omega)/\partial i\omega$ can be calculated. All the internal lines can be considered to carry the external frequency ω as well as an internal frequency ω' as the frequency is conserved at each vertex. Hence the calculation of $\partial \Sigma_\sigma(\omega)/\partial i\omega$ proceeds in the same way as in the calculation of $\partial \Sigma_\sigma(i\omega)/\partial \mu$ except that

$$\frac{dG_{\mathrm{d}\sigma,\mathrm{d}\sigma}(i\omega)}{di\omega} = \frac{d}{di\omega}\left[\frac{i\omega + \mu - \epsilon_{\mathrm{d},\sigma} - \Sigma_\sigma(i\omega) - i\Delta\mathrm{sgn}\omega}{(i\omega + \mu - \epsilon_{\mathrm{d},\sigma} - \Sigma^{\mathrm{R}}_\sigma(i\omega))^2 + (\Delta - \Sigma^I_\sigma(i\omega))^2}\right]$$

$$= -G^2_{\mathrm{d}\sigma,\mathrm{d}\sigma}(i\omega)\left(1 - \frac{\partial \Sigma_\sigma(i\omega)}{\partial i\omega}\right) - \frac{2\Delta\delta(\omega)}{(\epsilon_{\mathrm{d}} - \mu + \Sigma^{\mathrm{R}}_\sigma(0))^2 + \Delta^2}, \qquad (5.70)$$

which follows straightforwardly on using $\partial(\mathrm{sgn}\,\omega)/\partial\omega = 2\delta(\omega)$. This gives for the derivative of the self-energy,

$$\frac{\partial \Sigma_\sigma(i\omega)}{\partial i\omega} =$$
$$-\int \sum_{\sigma'} \Gamma_{\sigma,\sigma'}(i\omega, i\omega') G^2_{\mathrm{d}\sigma',\mathrm{d}\sigma'}(i\omega') \frac{d\omega'}{2\pi} - \rho_{\mathrm{d},\sigma}(0) \sum_{\sigma'} \Gamma_{\sigma,\sigma'}(i\omega, 0).$$

$$(5.71)$$

It follows on analytically continuing to real frequencies for the self-energy of the retarded function,

$$\frac{\partial \Sigma_\sigma(\omega)}{\partial \omega} + \frac{\partial \Sigma_\sigma(\omega)}{\partial \mu} = -\rho_{\mathrm{d},\sigma}(\epsilon_{\mathrm{F}}) \sum_{\sigma'} \Gamma_{\sigma,\sigma'}(\omega, \epsilon_{\mathrm{F}}). \qquad (5.72)$$

We can generate $\partial \Sigma_\sigma(\omega)/\partial \omega$ also by differentiating only the propagators that carry the external label σ, as spin is conserved at each vertex and propagators with $-\sigma$ must form closed loops and their contributions cancel out. It then follows that the sum of $\partial \Sigma_\sigma(\omega)/\partial \omega$ and $\partial \Sigma_\sigma(\omega)/\partial \mu_\sigma$, where μ_σ is the chemical potential for the spin σ electrons only, are given by an equation of the form (5.71) but with only the terms with $\sigma' = \sigma$

contributing to the right hand side. Hence,

$$\frac{\partial \Sigma_\sigma(\omega)}{\partial \omega} + \frac{\partial \Sigma_\sigma(\omega)}{\partial \mu_\sigma} = -\rho_{\mathrm{d},\sigma}(\epsilon_{\mathrm{F}})\Gamma_{\sigma,\sigma}(\omega, \epsilon_{\mathrm{F}}). \tag{5.73}$$

As

$$\frac{\partial \Sigma_\sigma(\omega)}{\partial \mu} = \sum_{\sigma'} \frac{\partial \Sigma_\sigma(\omega)}{\partial \mu_{\sigma'}}, \tag{5.74}$$

it then follows that

$$\frac{\partial \Sigma_\sigma(\omega)}{\partial \mu_{-\sigma}} = -\rho_{\mathrm{d},\sigma}(\epsilon_{\mathrm{F}})\Gamma_{\sigma,-\sigma}(\omega, \epsilon_{\mathrm{F}}). \tag{5.75}$$

The antisymmetrised vertex for $\omega = \epsilon_{\mathrm{F}}$ must have the form,

$$\Gamma_{\sigma,\sigma'}(\epsilon_{\mathrm{F}}, \epsilon_{\mathrm{F}}) = \Gamma_0(1 - \delta_{\sigma,\sigma'}), \tag{5.76}$$

and hence putting $\omega = \epsilon_{\mathrm{F}}$ in equations (5.72), (5.73) and (5.75) leads to

$$\frac{\partial \Sigma_\sigma(\epsilon_{\mathrm{F}})}{\partial \mu_\sigma} - \frac{\partial \Sigma_\sigma(\epsilon_{\mathrm{F}})}{\partial \mu_{-\sigma}} = \rho_{\mathrm{d},\sigma}(\epsilon_{\mathrm{F}})\Gamma_0 - \frac{\partial \Sigma_\sigma(\epsilon_{\mathrm{F}})}{\partial \omega}, \tag{5.77}$$

and

$$\frac{\partial \Sigma_\sigma(\epsilon_{\mathrm{F}})}{\partial \omega} + \frac{\partial \Sigma_\sigma(\epsilon_{\mathrm{F}})}{\partial \mu} = -\rho_{\mathrm{d},\sigma}(\epsilon_{\mathrm{F}})\Gamma_0. \tag{5.78}$$

As the chemical potential for spin σ in a field H is given by $\mu_\sigma = \mu - \sigma h$,

$$\frac{\partial \Sigma_\sigma(\epsilon_{\mathrm{F}})}{\partial h} = \frac{\partial \Sigma_\sigma(\epsilon_{\mathrm{F}})}{\partial \mu_{-\sigma}} - \frac{\partial \Sigma_\sigma(\epsilon_{\mathrm{F}})}{\partial \mu_\sigma}. \tag{5.79}$$

On eliminating Γ_0 between (5.77) and (5.79), we find a relation between the derivatives of the self-energy evaluated at the Fermi level,

$$\frac{\partial \Sigma_\sigma(\epsilon_{\mathrm{F}})}{\partial h} - \frac{\partial \Sigma_\sigma(\epsilon_{\mathrm{F}})}{\partial \mu} = 2\frac{\partial \Sigma_\sigma(\epsilon_{\mathrm{F}})}{\partial \omega}. \tag{5.80}$$

Using (5.56), (5.57) and (5.60), we get the relation which we conjectured earlier (5.20),

$$\frac{4\chi_{\mathrm{imp}}}{(g\mu_{\mathrm{B}})^2} + \chi_{\mathrm{c,imp}} = \frac{6\gamma_{\mathrm{imp}}}{\pi^2 k_{\mathrm{B}}^2}. \tag{5.81}$$

This verifies the general expression derived earlier for the Wilson ratio (5.23), which is now established rigorously within perturbation theory. Further details can be found in the series of papers by Yamada & Yosida cited earlier.

We can use equations (5.77) and (5.78) to obtain the interaction parameter Γ_0. In the Kondo limit the charge susceptibility must vanish and so $\alpha'' = 0$, or $\partial\Sigma(\epsilon_{\mathrm{F}})/\partial\mu = -1$, hence from (5.77) and (5.56) for particle–hole symmetry $\Gamma_0 = \pi\Delta/z$. The interaction between the quasi-particles is $z^2\Gamma_0$, as there is a $\sqrt{z}$ associated with each of the external legs of the Γ vertex in using the quasi-particle Green's function (5.53) instead of true Green's function (5.52) which carries a z factor

($\sqrt{z}$ is assigned instead to the vertex at the end of each propagator in the quasi-particle case). Using $\Gamma_0 = \pi\Delta/z$ with $z = \tilde{\Delta}/\Delta$, then $z^2\Gamma_0 = \pi\tilde{\Delta} = \tilde{U}$ so confirming (5.22). In the Kondo regime these relations give $z = 4k_BT_K/\pi\Delta w$, $z^2\Gamma_0 = 4k_BT_K/w$. *These exact results confirm completely the results of the more intuitive phenomenological approach of section 5.1.* The connection between the two approaches is made explicit in the 'renormalized perturbation theory' described in appendix L where the expansion parameter is $\tilde{U}$. We now turn our attention to the calculation of transport properties.

5.4 The Electrical Conductivity

If we consider purely s wave scattering we found in chapter 2 that the transport lifetime $\tau_{1\mathbf{k}}(\omega)$ is equal to the scattering lifetime $\tau_{\mathbf{k}}(\omega)$, which in turn is given by the imaginary part of the T matrix,

$$\tau_{\mathbf{k}}^{-1}(\omega) = -2c_{\mathrm{imp}}\mathrm{Im}\,T_{\mathbf{k}\mathbf{k}}(\omega^+),\tag{5.82}$$

where c_{imp} is the impurity concentration. Then using (5.39) we find

$$\tau_{\mathbf{k}}^{-1}(\omega) = 2\pi|V|^2c_{\mathrm{imp}}\rho_{\mathrm{d},\sigma}(\omega),\tag{5.83}$$

so that the temperature dependence at low temperatures arises from the low order expansion of $\rho_{\mathrm{d},\sigma}(\omega)$ in powers of T and ω (in the rest of this chapter we shall take $\epsilon_F = 0$ for convenience). We first of all use the lowest order form for the Green's function in the region of the Fermi level, as given in equation (5.52) which corresponds to non-interacting quasi-particles, and expand to second order in ω^2,

$$\rho_{\mathrm{d},\sigma}(\omega) = \frac{1}{\pi\Delta}\left(1 - \frac{\omega^2}{z^2\Delta^2} + \dots\right),\tag{5.84}$$

where we have restricted our attention to the particle–hole symmetric case $\tilde{\epsilon}_{\mathrm{d}} = 0$. Substituting (5.83) into (2.15),

$$\sigma_{\mathrm{imp}}(T) = -\frac{ne^2}{m}\int\tau(\omega)\frac{\partial f}{\partial\omega}\,d\omega,\tag{5.85}$$

integrating by parts and using the Sommerfeld expansion (1.6), we find

$$\sigma_{\mathrm{imp}}(T) = \sigma_0\left\{1 + \frac{\pi^2}{3}\left(\frac{k_BT}{z\Delta}\right)^2 + \mathrm{O}(T^4)\right\},\tag{5.86}$$

with $\sigma_0 = ne^2\pi\rho_0(0)/2mc_{\mathrm{imp}}$, which corresponds to the 'unitarity' limit of (2.20) in which the s phase shift at the Fermi level $\eta(0)$ is equal to $\pi/2$. In the Kondo regime $z = 4k_BT_K/\pi\Delta w$, and hence

$$\sigma_{\mathrm{imp}}(T) = \sigma_0\left\{1 + \frac{\pi^4w^2}{48}\left(\frac{T}{T_K}\right)^2 + \mathrm{O}(T^4)\right\}.\tag{5.87}$$

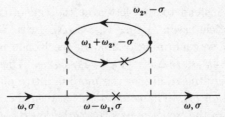

Figure 5.4 Second order self-energy diagram for $\Sigma_\sigma(i\omega)$. The crosses indicate the replacement of the propagators $G^0_\sigma(\omega)$ by $2\delta(\omega)/\Delta$ for the calculation of $\partial^2\Sigma_\sigma(0)/\partial(i\omega)^2$.

This form corresponds to a resonance in the relaxation function $\tau(\omega)$ of width $\tilde{\Delta}$. Such a form was used by experimentalists to interpret the low temperature resistivity and other data (see Grüner, 1974) and is qualitatively correct. If we have a resistance minimum (conductivity maximum) we expect $\sigma_{\text{imp}}(T)$ to increase with T. The coefficient of the T^2 term, however, is not exact. For the correct T^2 coefficient the density of states $\rho_{\text{d},\sigma}(\omega)$ is required to order ω^2 and to order T^2. In the lowest order expansion of the self-energy to derive (5.52), terms of this order, which contribute to the imaginary part of $\Sigma_\sigma(i\omega)$, were neglected. These are inelastic terms which give the quasi-particle lifetime.

To calculate the spectral weight, $\rho_{\text{d},\sigma}(\omega)$, exactly to order ω^2 we need to evaluate $\partial^2\Sigma_\sigma(i\omega)/\partial(i\omega)^2$. Let us calculate the contribution to this from the second order diagram shown in figure 5.4. At $T = 0$ this diagram gives

$$-\frac{U^2}{(2\pi)^2} \int\int \frac{\partial^2 G^0_\sigma(i\omega - i\omega_1)}{\partial(i\omega)^2} G^0_{-\sigma}(i\omega_1 + i\omega_2) G^0_{-\sigma}(i\omega_2)\, d\omega_1 d\omega_2 \quad (5.88)$$

where we have made an obvious simplification of notation and all propagators are d Green's functions. In calculating this we need the second derivative of the Green's function $G^0_{\text{d}\sigma,\text{d}\sigma}(i\omega)$,

$$\frac{\partial^2 G^0_\sigma(i\omega)}{\partial(i\omega)^2} = 2(G^0_\sigma(i\omega))^3 - \frac{2}{i\Delta}\delta'(\omega), \quad (5.89)$$

where $\delta'(\omega)$ is the derivative of the delta function, and we have assumed particle–hole symmetry. Substituting this into (5.88) we get two terms,

$$= \frac{-iU^2}{2\pi^2\Delta} \int\int \delta'(\omega - \omega_1) G^0_{-\sigma}(i\omega_1 + i\omega_2) G^0_{-\sigma}(i\omega_2)\, d\omega_1 d\omega_2$$

$$-\frac{U^2}{2\pi^2} \int\int (G^0_\sigma(i\omega - i\omega_1))^3 G^0_{-\sigma}(i\omega_1 + i\omega_2) G^0_{-\sigma}(i\omega_2)\, d\omega_1 d\omega_2.$$

$$(5.90)$$

For $\omega = 0$ the second integral vanishes because $G^0_\sigma(i\omega)$ for particle–hole

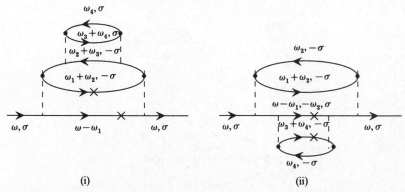

(i)   (ii)

Figure 5.5 Diagrams leading to possible contributions to $\partial^2 \Sigma_\sigma(0)/\partial(i\omega)^2$.

symmetry is an odd function of ω, and the integral contains an odd number of Green's functions. The first integral gives

$$\frac{U^2}{2\pi^2 \Delta} \int_{-\infty}^{\infty} \frac{dG^0_{-\sigma}(i\omega + i\omega_2)}{di\omega} G^0_{-\sigma}(i\omega_2) \, d\omega_2. \qquad (5.91)$$

Using (5.70) for $dG^0_{-\sigma}(i\omega + i\omega_2)/di\omega$, and then substituting into (5.91), generates two terms, a term involving three Green's functions which again vanishes due to odd symmetry, and a delta function term,

$$-\left(\frac{U}{\pi\Delta}\right)^2 \int \delta(\omega + \omega_2) G^0_{-\sigma}(i\omega_2) \, d\omega_2, \qquad (5.92)$$

which in the limit $\omega \to 0$ gives

$$-i \left(\frac{U}{\pi\Delta}\right)^2 \frac{\mathrm{sgn}(\omega)}{\Delta}. \qquad (5.93)$$

In this evaluation the non-zero contribution arises from a term in which two of the propagators are replaced by delta functions. This can be shown quite generally, the remaining terms vanish due to odd symmetry as they involve an odd number of Green's functions. We look at higher order diagrams of the type shown in figure 5.5, where a cross x indicates that this propagator has been replaced by a delta function.

For example, figure 5.5(i) has a contribution after integration over the delta functions proportional to

$$\int\int G^0_\sigma(i\omega_3 - i\omega) G^0_{-\sigma}(i\omega_3 + i\omega_4) G^0_{-\sigma}(i\omega_4) \, d\omega_3 d\omega_4, \qquad (5.94)$$

which vanishes for $\omega = 0$. Similarly the contribution from the diagram in figure 5.5(ii) also vanishes. There is another contribution from a diagram of the type in 5.5(ii) with the cross on the propagator $G^0_{-\sigma}(i\omega_3 + i\omega_4)$ transferred to $G^0_{-\sigma}(i\omega_1 + i\omega_2)$ but this also vanishes. Generalizing this

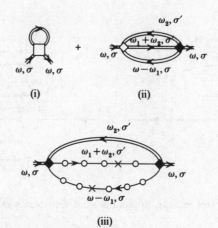

(iii)

Figure 5.6 (i) (ii) Total skeleton diagrams for $\Sigma_\sigma(i\omega)$, and (iii) form of diagrams giving non-zero contributions to $\partial^2 \Sigma_\sigma(0)/\partial(i\omega)^2$, where the open circles represent proper self-energy insertions. The filled vertices correspond to the four point vertex 5.1(i).

argument it can be shown that the only non-zero contributions arise when there are three intermediate propagators, with two of them crossed.

The exact self-energy $\Sigma_\sigma(i\omega)$ is given by the diagrams of figure 5.6(i) and (ii), where the full four point vertex $\Gamma_{\sigma,\sigma',\sigma',\sigma}(i\omega_1, i\omega_2, i\omega_3, i\omega_4)$ is used. The open square represents the bare vertex which is given by

$$\Gamma^0_{\sigma,\sigma',\sigma',\sigma}(i\omega_1, i\omega_2, i\omega_3, i\omega_4) = U(1 - \delta_{\sigma,\sigma'}). \qquad (5.95)$$

The only non-zero contribution arises from diagram 5.6(iii). The two vertices are full vertices unlike diagram 5.6(ii) because vertex corrections can be uniquely assigned because of the crosses. The contribution to $\partial^2 \Sigma_\sigma(i\omega)/\partial(i\omega)^2$ from this diagram is the integral of

$$\frac{-\Gamma^2_{\sigma,\sigma',\sigma',\sigma}(i\omega_1, i\omega_2, i\omega_1 + i\omega_2, i\omega - i\omega_1) G_{-\sigma}(i\omega_2)\delta(\omega_1 + \omega_2)\delta(\omega - \omega_1)}{(1 - G^0_\sigma(i\omega - i\omega_1)\Sigma_\sigma(i\omega - i\omega_1))^2 (1 - G^0_\sigma(i\omega_1 + i\omega_2)\Sigma_\sigma(i\omega_1 + i\omega_2))^2} \qquad (5.96)$$

with respect to ω_1 and ω_2, divided by $(\pi\Delta)^2$.

As $\Sigma_\sigma(0) = 0$, and $G_\sigma(-i\omega) \to -1/i\Delta\,\mathrm{sgn}(\omega)$, we find

$$\frac{\partial^2 \Sigma_\sigma(i\omega)}{\partial(i\omega)^2} = -i\frac{\mathrm{sgn}(\omega)}{\Delta}\frac{\Gamma^2_{\sigma,-\sigma}(0,0)}{(\pi\Delta)^2}, \qquad (5.97)$$

using the fact that $\Gamma_{\sigma,\sigma}(0,0) = 0$.

To calculate the T^2 coefficient of $\Sigma_\sigma(0,T)$ we use the relation,

$$\frac{1}{\beta}\sum_n F(i\omega_n) = \int_{-\infty}^{\infty} F(i\omega)\frac{d\omega}{2\pi} + \frac{\pi k_B^2 T^2}{12}\left(\frac{dF(0^+)}{di\omega_n} - \frac{dF(0^-)}{di\omega_n}\right) + \mathrm{O}(T^4), \qquad (5.98)$$

which is the equivalent of the Sommerfeld expansion (1.6), for the ω_n sums for a given diagram for $\Sigma_\sigma(i\omega)$. For example, if we consider the second order diagram shown in figure 5.4, we have the factor,

$$G_\sigma^0(i\omega_n - i\omega_{n_1})G_{-\sigma}^0(i\omega_{n_1} + i\omega_{n_2})G_{-\sigma}^0(i\omega_{n_2}). \tag{5.99}$$

If we write this in the form $G_{-\sigma}^0(i\omega_{n_2})R(i\omega_{n_2}, i\omega_n)$ in considering the sum over ω_{n_2}, then

$$\lim_{\omega \to 0} \left\{ \frac{d}{di\omega_2}[G_{-\sigma}^0(i\omega_2)R(i\omega_2, i\omega)]_{0+} - \frac{d}{di\omega_2}[G_{-\sigma}^0(i\omega_2)R(i\omega_2, i\omega)]_{0-} \right\}$$

$$= \frac{1}{i\Delta} \left\{ \frac{d}{di\omega_2}[R(i\omega_2, i\omega)]_{0+} + \frac{d}{di\omega_2}[R(i\omega_2, i\omega)]_{0-} \right\}, \tag{5.100}$$

where we have used $R(\omega_2, 0) = R(-\omega_2, 0)$. This term arises from the discontinuity of the propagator $G_{-\sigma}^0(i\omega_2)$ at $\omega_2 = 0$. The result is equivalent to the substitution of $-\delta'(\omega_2)/\Delta i$ for $G_{-\sigma}^0(i\omega_2)$ in $T = 0$ evaluation of the diagram 5.4 for the self-energy. Hence the calculation proceeds on exactly the same lines as that for the first term in (5.90), which was the only term from this diagram which gave a finite result for $\partial^2 \Sigma_\sigma(i\omega)/\partial(i\omega)^2$. It is equivalent to placing crosses on two of the propagators in diagram 5.4. Contributions of the form (5.100) arise from the discontinuities in the other propagators, and there is an extra factor of 3 when both the ω_{n_1} and the ω_{n_2} are taken into account. This corresponds to the three distinct ways of placing two crosses on the three propagators. This diagram then contributes a T^2 term,

$$-i\pi^2(k_B T)^2 \left(\frac{U}{\pi\Delta} \right)^2 \frac{\operatorname{sgn}(\omega)}{2\Delta}, \tag{5.101}$$

to $\Sigma_\sigma(0, T)$ on using (5.98). This can be generalized to give an exact result to order T^2 following the same argument as in the evaluation of $\partial^2 \Sigma_\sigma(i\omega)/\partial(i\omega)^2$. The result to lowest order in T and ω for the self-energy of the retarded Green's function is

$$\Sigma_\sigma(\omega^+) = \frac{-i\Gamma_{\sigma,-\sigma}^2(0,0)}{2\Delta(\pi\Delta)^2} \left(\omega^2 + \pi^2(k_B T)^2 \right). \tag{5.102}$$

Hence the exact result for $\rho_{d,\sigma}(\omega)$ to second order in T and ω in the case of particle–hole symmetry,

$$\rho_{d,\sigma}(\omega) =$$

$$\frac{1}{\pi\Delta} \left\{ 1 - (2 + (R-1)^2)\frac{\omega^2}{2z^2\Delta^2} - (R-1)^2\frac{\pi^2(k_B T)^2}{2z^2\Delta^2} \cdots \right\}, \tag{5.103}$$

on using (5.78) and (5.80) for $\Gamma_{\sigma,-\sigma}(0,0)$. Substituting this into (5.83) and (5.85) gives for the conductivity,

$$\sigma_{\text{imp}}(T) = \sigma_0 \left\{ 1 + \frac{\pi^2}{3} \left(\frac{k_{\text{B}}T}{z\Delta} \right)^2 (1 + 2(R-1)^2) + \text{O}(T^4) \right\}. \quad (5.104)$$

This result, which is exact to order T^2 for particle–hole symmetry, was derived by Yamada (1975), using the arguments we have outlined. For U large, $R \to 2$, $z \to 4k_{\text{B}}T_{\text{K}}/\pi\Delta w$, and hence in this limit, the Kondo limit, (5.104) becomes

$$\sigma_{\text{imp}}(T) = \sigma_0 \left\{ 1 + \frac{\pi^4 w^2}{16} \left(\frac{T}{T_{\text{K}}} \right)^2 + \text{O}(T^4) \right\}, \quad (5.105)$$

and differs from (5.87) by an extra factor of three in the T^2 coefficient. This result was first derived by Nozières (1974) using phenomenological Fermi liquid theory. Nozières' calculation is equivalent to using the quasi-particle Hamiltonian (5.6), to calculate the quasi-particle phase shift, and then including the inelastic contributions from the quasi-particle scattering diagram 5.4 (without the crosses) with $U = \tilde{U}$ and $\Delta = \tilde{\Delta}$, and the relation (5.22), which applies in the Kondo limit. The result is exact to order T^2.

Two further exact relations, one due to Shiba (1975) which is known as the *Korringa relation* which relates the imaginary part of the frequency susceptibility $\chi_{\text{imp}}(\omega)$ divided by ω to χ_{imp}^2, the other due to Houghton, Read & Won (1987) and Kawakami, Usuki & Okiji (1987) which relates the linear coefficient of the thermopower $S_{\text{imp}}(T)$ to γ_{imp}, are given in appendix E with an outline of their proofs.

5.5 Finite Order Perturbation Results

The perturbation approach in powers of U has been very fruitful in providing exact relations. However, as mentioned in chapter 3, attempts to find approximate solutions based on a partial summation of diagrams have not been successful. The Hartree–Fock solution, as discussed in chapter 1, becomes unstable for values of U exceeding a critical value U_{c} ($U_{\text{c}} = \pi\Delta$ in the particle–hole symmetric case) and favours a unphysical broken symmetry state with $\langle n_{\text{d},\sigma} \rangle \neq \langle n_{\text{d},-\sigma} \rangle$. This is reflected in the RPA sum of repeated particle–hole scattering which develops a singularity for $U = U_{\text{c}}$. The most successful application of the perturbation approach has been that of Yosida & Yamada. Apart from the exact relations already mentioned, all diagrams for the free energy and one particle Green's function were calculated to fourth order in $U/\pi\Delta$ for the symmetric model about the non-magnetic Hartree–Fock solution

$(\langle n_{\mathrm{d},\sigma} \rangle = \langle n_{\mathrm{d},-\sigma} \rangle = 1/2)$. The result for the ground state energy $E_{\mathrm{gs}}(U)$ (Yamada, 1975) is

$$E_{\mathrm{gs}}(U) = E_{\mathrm{gs}}(0) - U \left\{ 0.25 + 0.0369 \left(\frac{U}{\pi\Delta} \right) - 0.0008 \left(\frac{U}{\pi\Delta} \right)^3 \dots \right\}. \tag{5.106}$$

As the leading coefficients are very small this gives good results for U up to $\pi\Delta$. Results for χ_{imp}, γ_{imp} and $\chi_{\mathrm{c,imp}}$ at $T = 0$ are

$$\chi_{\mathrm{imp}} = \frac{(g\mu_{\mathrm{B}})^2}{2\pi\Delta} \left\{ 1 + \frac{U}{\pi\Delta} + \left(3 - \frac{\pi^2}{4} \right) \left(\frac{U}{\pi\Delta} \right)^2 + \left(15 - \frac{3\pi^2}{2} \right) \left(\frac{U}{\pi\Delta} \right)^3 \right.$$
$$\left. + \left(105 - \frac{45\pi^2}{4} + \frac{\pi^4}{16} \right) \left(\frac{U}{\pi\Delta} \right)^4 \dots \right\}, \tag{5.107}$$

$$\gamma_{\mathrm{imp}} = \frac{2\pi k_{\mathrm{B}}^2}{3\Delta} \left\{ 1 + \left(3 - \frac{\pi^2}{4} \right) \left(\frac{U}{\pi\Delta} \right)^2 \right.$$
$$\left. + \left(105 - \frac{45\pi^2}{4} + \frac{\pi^4}{16} \right) \left(\frac{U}{\pi\Delta} \right)^4 \dots \right\}, \tag{5.108}$$

and

$$\chi_{\mathrm{c,imp}} = \frac{2}{\pi\Delta} \left\{ 1 - \frac{U}{\pi\Delta} + \left(3 - \frac{\pi^2}{4} \right) \left(\frac{U}{\pi\Delta} \right)^2 - \left(15 - \frac{3\pi^2}{2} \right) \left(\frac{U}{\pi\Delta} \right)^3 \right.$$
$$\left. + \left(105 - \frac{45\pi^2}{4} + \frac{\pi^4}{16} \right) \left(\frac{U}{\pi\Delta} \right)^4 \dots \right\}. \tag{5.109}$$

The Fermi liquid relation (5.20), or equivalently (5.81), can be verified explicitly to fourth order in $U/\pi\Delta$. The Wilson ratio R (5.23) calculated from these results approaches the value 2 until $U/\pi\Delta \sim 2$, and then decreases as the perturbation approximation breaks down.

The development of the narrow many-body resonance at the Fermi level is seen clearly in the results for the spectral density of the one electron Green's function shown in figure 5.7.

For $U = 0$ there is a single resonance of width Δ at the Fermi level. For $U/\pi\Delta$ between 1 and 2 the resonance at the Fermi level narrows markedly and develops shoulders which emerge as distinct peaks at $\omega \sim \pm U/2$. These peaks occur in the atomic limit $\Delta = 0$ at $\omega = \pm U/2$, as can be seen from the Green's function (5.40) $\epsilon_{\mathrm{d}} = -U/2$, $\langle n_{\mathrm{d},\sigma} \rangle = 1/2$. With this substitution (5.40) can be written in the form,

$$[G_{\mathrm{d},\sigma,\mathrm{d},\sigma}(\omega)]^{-1} = \omega - \frac{U^2}{4\omega}. \tag{5.110}$$

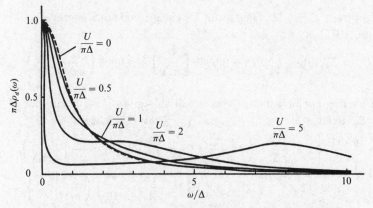

Figure 5.7 Fourth order perturbational results (Yamada, 1975) for the spectral density $\pi\Delta\rho_d(\omega)$ for the symmetric Anderson model.

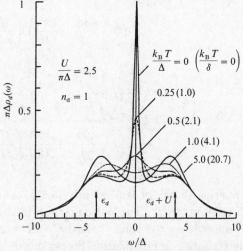

Figure 5.8 Results for the spectral density $\pi\Delta\rho_d(\omega)$ at various temperatures (Horvatić et al, 1987) for the symmetric Anderson model with $U/\pi\Delta = 2.5$ and $\langle n_d \rangle = 1$. The arrows indicate the energies of the excitations for $\Delta = 0$. The temperature is indicated in units of Δ and also in terms of the half width of the central resonance $\delta \sim T_K$.

It is evident that in this case second order perturbation theory is exact in the atomic limit. This is a special feature of the symmetric model and is not true in general. The case $U/\pi\Delta = 5$ shown in figure 5.7 is beyond the region of applicability of this theory $U/\pi\Delta < 2.5$, and gives a Kondo resonance which narrows algebraically in U rather than exponentially as

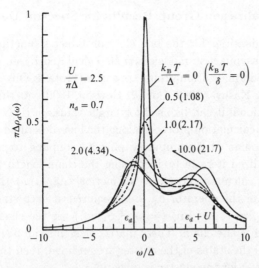

Figure 5.9 Results for the spectral density $\pi\Delta\rho_d(\omega)$ at various temperatures (Horvatić et al, 1987) for the asymmetric Anderson model with $U/\pi\Delta = 2.5$ and $\langle n_d(T=0)\rangle = 0.7$. The arrows indicate the energies of the excitations for $\Delta = 0$. The temperature is indicated in units of Δ and also in terms of the half width of the lower resonance.

in the exact result. Nevertheless the features are qualitatively correct.

These calculations have been extended by Horvatić, Šokčević & Zlatić (1987) to cases without particle–hole symmetry. They have also included the temperature dependence. Results for the temperature dependence in the particle–hole symmetric case are shown in figure 5.8.

The height of the Kondo peak is reduced as it broadens with temperature but only disappears as a discernible peak for $T > 4T_K$.

A case without particle–hole symmetry for $n_d = 0.7$ is shown in figure 5.9 over the same temperature range. The Kondo peak gets displaced relative to the Fermi level. This is to be expected as the quasi-particle phase shift (5.54) satisfies the Friedel sum rule and the occupied states within the quasi-particle peak give the impurity occupation number $\langle n_d \rangle$. When $\langle n_d \rangle < 1$ the level is less than half-full and the peak lies above the Fermi level and for $\langle n_d \rangle > 1$ the peak lies below. This can have consequences for predictions for the leading term in the thermopower which depends on the derivative of the density of states at the Fermi level. It differs in sign in these two situations and is zero for the symmetric model (see appendix E).

5.6 Renormalization Group Results for Spectral Densities

The spectral densities for the one electron Green's function for the symmetric and asymmetric models can be calculated by an extension of the Wilson renormalization group approach (Frota & Oliveira, 1986; Sakai, Shimizu & Kasuya, 1989; Costi & Hewson, 1990). As this method is non-perturbational it can be used for larger values of U/Δ including $U/\Delta \to \infty$. The calculation proceeds along the lines described in section 4.4; the Hamiltonian of the impurity plus its coupling to an N site chain is diagonalized iteratively, increasing the chain length by one at each stage. To calculate the Green's function $G_{d\sigma,d\sigma}(\omega)$ the matrix elements $V^\sigma_{\mathbf{m},\mathbf{m}'}$ of the operator $c^\dagger_{\mathrm{d},\sigma}$ are required at each stage, where $V^\sigma_{\mathbf{m},\mathbf{m}'} = \langle \mathbf{m}|c^\dagger_{\mathrm{d},\sigma}|\mathbf{m}'\rangle$, and $|\mathbf{m}\rangle$ $(= |n, S, S_z, m\rangle)$ are the eigenstates of the impurity and N site chain. These matrix elements can be calculated recursively. If all the states of the cluster are retained, then the Green's function $G_{d\sigma,d\sigma}(\omega)$ for the cluster is given by

$$G_{d\sigma,d\sigma}(\omega) = \sum_{\mathbf{m},\mathbf{m}'} \frac{|V^\sigma_{\mathbf{m},\mathbf{m}'}|^2 \langle X_{\mathbf{m},\mathbf{m}} + X_{\mathbf{m}',\mathbf{m}'}\rangle}{(\omega - E_{\mathbf{m}'} + E_{\mathbf{m}})}. \tag{5.111}$$

where $X_{\mathbf{m},\mathbf{m}} = |\mathbf{m}\rangle\langle\mathbf{m}|$ is the Hubbard projection operator for the state $|\mathbf{m}\rangle$. The corresponding spectral density is a set of delta functions,

$$\rho_{\mathrm{d},\sigma}(\omega) = \sum_{\mathbf{m},\mathbf{m}'} |V^\sigma_{\mathbf{m},\mathbf{m}'}|^2 \langle X_{\mathbf{m},\mathbf{m}} + X_{\mathbf{m}',\mathbf{m}'}\rangle \delta(\omega - E_{\mathbf{m}'} + E_{\mathbf{m}}), \tag{5.112}$$

As, in practice, only a finite number (N_{st}) of the lowest states is kept and the matrix is truncated the levels are calculated on a logarithmically decreasing energy scale. For a chain with N_{s} sites the spectral density is evaluated for $\omega \sim \omega_N$, where $\omega_N = k_{\mathrm{B}}T_N = D'\Lambda^{-N/2}$. This is because information about the higher energy states $\omega \gg \omega_N$ has been lost through earlier truncations, and because the low energy excitations $\omega \ll \omega_N$ have not been calculated accurately enough due to the finite cluster size. Hence the spectral density is evaluated at a sequence of points ω on a logarithmically decreasing energy scale. The delta functions in (5.112) are broadened by convolution with a Gaussian having a width which decreases according to the energy scale. This procedure can give higher energy peaks that are a little too broad and which have some slight asymmetry where the points are more widely spaced but the overall weights of these peaks are correct.

Results for the spectral density for the symmetric model at $T = 0$ are shown in figure 5.10 for $U/\pi\Delta = 0, 1.5, 4$ and 6. This method gives accurate predictions for the Kondo peak where the points are closely

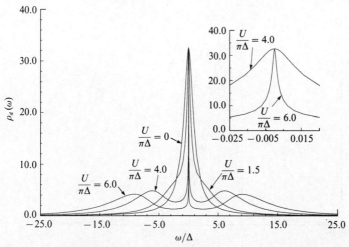

Figure 5.10 Results for the spectral density $\rho_d(\omega)$ calculated via the numerical renormalization group method for the symmetric Anderson model for $\langle n_d \rangle = 1$ and different values of $U/\pi\Delta$. The inset shows the region near $\omega = 0$ for the two largest values of $U/\pi\Delta$ (Costi & Hewson, 1990).

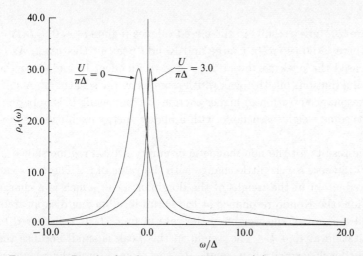

Figure 5.11 Results for the spectral density $\rho_d(\omega)$ calculated via the numerical renormalization group method for the asymmetric Anderson model in the mixed valence regime $\epsilon_d = -\Delta$ for different values of $U/\pi\Delta$ (Costi & Hewson, 1990).

spaced and the exponential decrease in width of this resonance is apparent in the inset of figure 5.10 which shows details of the resonance on a reduced energy scale about $\omega = 0$ for the cases $U/\pi\Delta = 4$ and 6.

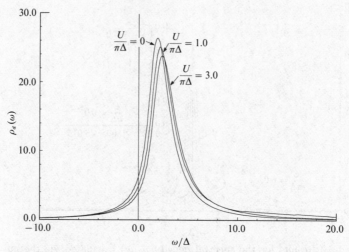

Figure 5.12 Results for the spectral density $\rho_d(\omega)$ calculated via the numerical renormalization group method for the asymmetric Anderson model in the empty orbital regime $\epsilon_d = 2\Delta$ for different values of $U/\pi\Delta$ (Costi & Hewson, 1990).

Figure 5.11 gives results in the mixed valence regime $\epsilon_d^* \sim 0$, $U/\pi\Delta = 0, 3$. There is no separate charge and Kondo peak in this case. As U is increased the peak narrows a little due to its effect in inhibiting the charge fluctuations but the peak width remains on the scale of Δ. A shift in the resonance position to higher energies on increasing U is apparent. There is some weight associated with a higher energy excitation at $\omega \sim \epsilon_d + U$.

In the results for the non-magnetic or empty orbital regime shown in figure 5.12 there is very little change with the value of U. There is some slight reduction in the weight of the dominant peak which is a charge excitation, the Kondo resonance at the Fermi level having disappeared. A small amount of the weight in the case of finite U is transferred to an excitation at $\epsilon_d + U$. The weight in this peak is small because the impurity level occupation is small. In the atomic limit the weight is equal to $\langle n_d \rangle/2$ (see equation (5.40)). The weight of the peak at $\epsilon_d + U$ will increase when the temperature is such that $k_B T \sim \epsilon_d$ and the level ϵ_d becomes thermally populated. The magnitude of $\rho_d(\omega)$ at the Fermi level is in agreement with the Friedel sum rule in all cases. This is a good test of the method as it relates n_d, obtained by an integration over all energy scales of $\rho_d(\omega)$, to $\rho_d(\epsilon_F)$.

The renormalization group approach has been extended to the calculation of other spectra such as the dynamic charge susceptibility $\chi_{c,imp}(\omega)$ and the dynamic spin susceptibility $\chi_{imp}(\omega)$ (Sakai et al, 1989). It has also been extended to models with degeneracy (Sakai et al, 1989), and to the Anderson model with a screened Coulomb interaction U_{dc} with the conduction electrons (Costi & Hewson, 1992a).

6

Exact Solutions and the Bethe Ansatz

6.1 The Linear Dispersion s-d Model

In 1980 exact solutions for the s-d model were found independently
by Andrei (1980) and Wiegmann (1980), and later also exact solutions
for the Anderson model (Wiegmann, 1981; Kawakami & Okiji, 1981).
This was a rather surprising development, more particularly, because the
methods of solution were based on the ansatz used by Bethe as early as
1931 in his solution of the one dimensional Heisenberg model. Though
a quantitative understanding of the Kondo problem was obtained in the
70s using the renormalization group and Fermi liquid approaches, the
exact solutions have produced new results, analytic formulae for the behaviour in weak and strong magnetic fields, the form of the electronic
specific heat over the full temperature range. The results give complete confirmation of the physical picture that emerged with the earlier
work. The approach has proved to be generalizable to some of the more
physically relevant models, the s-d model with spin greater than $\frac{1}{2}$, the
degenerate models for rare earth impurities (we will discuss these in the
next chapter), and models including crystal fields, providing a greater
range of predictions for comparison with experiment. The solutions
have proved to be immensely valuable in testing some of the approximate methods, ones which can also be used to calculate dynamic as well
as thermodynamic properties (it has not proved possible to calculate
dynamics directly from the Bethe ansatz), methods which may prove
useful in tackling multiple impurity and lattice problems.

The solution via the Bethe ansatz requires a modification of the original s-d model. Wilson for simplicity used a linear dispersion for the

conduction electrons, $\epsilon(\mathbf{k}) = k$, where k is the wavevector (assuming spherical symmetry) measured relative to the Fermi wavevector k_F (see section 4.3). A symmetric cut-off D for the conduction band was assumed, $|k| < D$, allowing the s-d model to be written in the form,

$$H_{\text{sd}} = \sum_\sigma \int_{-D}^{D} k c_{k,\sigma}^\dagger c_{k,\sigma} \, dk + 2J \int_{-D}^{D} \int_{-D}^{D} \sum_{\sigma,\sigma'} c_{k,\sigma}^\dagger \mathbf{S}_{\sigma,\sigma'} c_{k,\sigma'} \cdot \mathbf{S}_o \, dk \, dk',$$

(6.1)

where J is now a dimensionless coupling constant $(2\pi J\rho_0)$. The modification required to use the Bethe ansatz is to assume linear dispersion for all energies and let $D \to \infty$. This is only a temporary measure and we will restore a cut-off later. The scaling theory implies that the cut-off only appears in the low temperature behaviour through the Kondo temperature T_K. The motivation for removing the cut-off here is mathematical so that we can make a Fourier transform to operators,

$$c_\sigma(x) = \int_{-\infty}^{\infty} e^{ikx} c_{k,\sigma} \, dk,$$

(6.2)

with respect to a variable x. In terms of the transformed variable the Hamiltonian (6.1) in the limit $D \to \infty$ becomes

$$H_{\text{sd}} = \int \Big\{ -i \sum_\sigma c_\sigma^\dagger(x) \frac{\partial c_\sigma(x)}{\partial x} + 2J\delta(x) \sum_{\sigma,\sigma'} c_\sigma^\dagger(x) \mathbf{S}_{\sigma,\sigma'} c_{\sigma'}(x) \cdot \mathbf{S}_o \Big\} dx.$$

(6.3)

It will be convenient to use the Hamiltonian in the first quantized form and work with the Schrödinger equation,

$$H_{\text{sd}}\Psi = \Big\{ \sum_{j=1}^{N_e} -i\frac{\partial}{\partial x_j} + 2J \sum_{j=1}^{N_e} \delta(x_j)\mathbf{S}_j \cdot \mathbf{S}_0 \Big\} \Psi = E\Psi,$$

(6.4)

for the wavefunction Ψ and N_e electrons, where the impurity spin is now denoted by $\mathbf{S}_0$. This one dimensional model differs from the semi-infinite linear chain form used in the renormalization group calculations. The x-coordinate has no clear interpretation in terms of the original s-d model. It is helpful at this stage to consider this as a model in its own right. It corresponds to conduction electrons, with spatial coordinate x_j, moving in one dimension and interacting via a contact exchange interaction with an impurity spin at the origin. To solve the Schrödinger equation boundary conditions need to be imposed. These can be taken to be periodic by taking the system to be a ring of length L.

There is a special feature of the linear dispersion model. The kinetic energy of the conduction electrons depends only on the momentum of the centre of mass as $\sum_j -i\partial/\partial x_j$ is simply $-iN_e\partial/\partial x_{\text{cm}}$, where

$x_{\text{cm}} = \sum_j x_j/N_e$ is the centre of mass coordinate. As a result the usual constraint, that the wavefunction must be continuous for $x_i = x_j$ for the kinetic energy to be bounded, is no longer necessary. As a consequence this gives more flexibility in constructing a suitable complete set of states. We can illustrate this for the non-interacting model $J = 0$. Consider two electrons with coordinates x_1, x_2 in eigenstates of the one electron Hamiltonian, plane waves $\exp(ik_1x_1), \exp(ik_2x_2)$, with a total z-component of spin S_z equal to zero. If continuity at $x_i = x_j$ is imposed together with antisymmetry with respect to electron interchange then the wavefunction has the form,

$$\Psi_{s,t}(x_1, x_2) = \frac{1}{2}(e^{ik_1x_1}e^{ik_2x_2} \pm e^{ik_2x_1}e^{ik_1x_2})(\alpha_1\beta_2 \mp \beta_1\alpha_2), \quad (6.5)$$

where $\alpha(\uparrow)$ and $\beta(\downarrow)$ are the spin wavefunctions. The upper sign corresponds to a singlet $(S = 0)$ and the lower sign a triplet $(S = 1)$. If continuity at $x_i = x_j$ is not necessary, then spin singlet and spin triplets can also be constructed via

$$\Psi'_{s,t}(x_1, x_2) = \frac{1}{2}(e^{ik_1x_1}e^{ik_2x_2} \mp e^{ik_2x_1}e^{ik_1x_2})$$
$$(\alpha_1\beta_2 \mp \beta_1\alpha_2)(\theta(x_1 - x_2) - \theta(x_2 - x_1)).$$

$$(6.6)$$

It is possible to choose a set such that charge part of the wavefunction is antisymmetric irrespective of the spin state,

$$\Psi''_{s,t}(x_1, x_2) = \frac{1}{2}(e^{ik_1x_1}e^{ik_2x_2} - e^{ik_2x_1}e^{ik_1x_2})$$
$$(\alpha_1\beta_2 \mp \beta_1\alpha_2)(\theta(x_1 - x_2) \mp \theta(x_2 - x_1)).$$

$$(6.7)$$

With the wavefunction in this form there is a phase shift of π in the singlet case when the particles pass one another, though the particles are non-interacting.

It will be useful to use the set (6.7) as a basis for finding a solution via the Bethe ansatz. In this basis the wavefunction factorizes into charge and spin with the charge part of the wavefunction corresponding to spinless fermions. It is possible to change the signs of the first and last factors in (6.7) so that the charge part becomes symmetric corresponding to Bose symmetry. Complete sets of states can be built from either set, there is an analogy here with a half range Fourier series expansion using either sines (antisymmetric) or cosines (symmetric). We will use the set with the charge part antisymmetric.

We now consider the interacting model restricting attention for the moment to two conduction electrons and the impurity spin. We consider a wavefunction with the charge and spin factorized,

$$\Psi(x_1, x_2) = (e^{ik_1x_1}e^{ik_2x_2} - e^{ik_2x_1}e^{ik_1x_2})$$

$$\sum_{P,\sigma's} A_{\sigma_0,\sigma_1,\sigma_2}[P]\chi_0(\sigma_0)\chi_1(\sigma_1)\chi_2(\sigma_2)\theta(P),$$

(6.8)

where $\chi_i(\sigma_i)$ is the spin wavefunction, $\chi(\uparrow) = \alpha$ and $\chi(\downarrow) = \beta$, P denotes a permutation of $[0, 1, 2]$ which designates a region $x_{P0} > x_{P1} > x_{P2}$, within which $\theta(P) = 1$ and is zero otherwise. This wavefunction satisfies the Schrödinger equation *within* any region P, provided $E = k_1 + k_2$. The assumption that the eigenfunctions can be constructed from a given set of wavevectors (k_1, k_2) is the essential feature of the Bethe ansatz. Due to the delta function interaction with the impurity the wavefunction must have a discontinuity at $x_j = 0$. To calculate this we divide by Ψ and integrate (6.4) over x_j from 0^- to 0^+. The discontinuity as x_j passes the origin is then given by

$$\Psi(x_1, x_2, ...x_j = 0^+, ...x_{N_e}) = \mathbf{R}_j\,\Psi(x_1, x_2, ...x_j = 0^-, ...x_{N_e}), \quad (6.9)$$

where

$$\mathbf{R}_j = \frac{e^{-iJ/2}(\mathbf{I}\cos J - i\mathbf{P}_{j,0}\sin J)}{(\cos J - i\sin J)}, \quad (6.10)$$

on using the identity $2\mathbf{S_j}\cdot\mathbf{S_0} = 2\mathbf{P}_{j,0} - 1$, where $\mathbf{P}_{j,0}$ is the permutation operator which exchanges the spin states of the impurity and the j conduction electron. The operator $\mathbf{R}_j$ is unitary, $\mathbf{R}_j^\dagger\mathbf{R}_j = \mathbf{I}$, and operates only on the spin components.

Due to the discontinuity there is some ambiguity in integrating over the delta function, as $\theta(x)\delta(x)$ is not uniquely defined. Other choices, such as taking the wavefunction at $x_j = 0$ to be the mean of the values at 0^- and 0^+, give the same results to lowest order in J, so there is no ambiguity in weak coupling $J \ll 1$, which is the physically relevant situation, and in general differences can be absorbed by a renormalization of J.

We now apply the periodic boundary condition $x_1 \to x_1 + L$ and $x_2 \to x_2 + L$, corresponding to $x_{\text{cm}} \to x_{\text{cm}} + L$, starting in the region $[012]$ $(x_0 > x_1 > x_2)$ (see figure 6.1).

In generating $\Psi(x_1 + L, x_2 + L)$ from $[012]$, electrons 1 and 2 are scattered by the impurity in turn, so that,

$$\Psi(x_1 + L, x_2 + L) = e^{i(k_1+k_2)L}(e^{ik_1x_1}e^{ik_2x_2} - e^{ik_2x_1}e^{ik_1x_2})\times$$

$$\mathbf{R}_2\mathbf{R}_1\sum_{\sigma's} A_{\sigma_0,\sigma_1,\sigma_2}[012]\chi_0(\sigma_0)\chi_1(\sigma_1)\chi_2(\sigma_2).$$

(6.11)

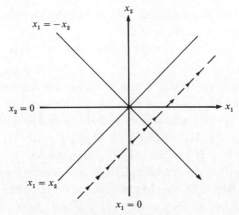

Figure 6.1 Coordinate regions for two conduction electrons plus impurity which is at the origin. The dashed line indicates the trajectory in applying the boundary condition $x_{\rm cm} \to x_{\rm cm} + L$ to the centre of mass coordinate, $x_{\rm cm} = (x_1 + x_2)/2$, starting in the region [012], and the crosses indicate the interaction of the conduction electrons with the impurity.

Equating this to $\Psi(x_1, x_2)$ gives the equation in matrix form,

$$\mathbf{R}_2\mathbf{R}_1\mathbf{A}[012] = e^{-i(k_1+k_2)L}\mathbf{A}[012], \qquad (6.12)$$

where $\mathbf{A}[012]$ is a vector in spin space with components $A_{\sigma_0,\sigma_1,\sigma_2}[012]$. The eigenvalues of the operator $\mathbf{R}_2\mathbf{R}_1$, which operates entirely in the spin space, determine the total energy $k_1 + k_2$. If the eigenvalue of $\mathbf{R}_2\mathbf{R}_1$ is $\exp(-iE_{sp}L)$, it can be taken to be of this form because it is a unitary operator, then $E = k_1 + k_2 = E_{\rm sp} + 2\pi n/L$, where n is an integer. The term $2\pi n/L$ must be the total kinetic energy contribution of the conduction electrons.

We could have set up the boundary condition $x_1 \to x_1 + L$ and $x_2 \to x_2 + L$, starting in some other region, say [021], then the equation corresponding to (6.12) takes the form,

$$\mathbf{R}_1\mathbf{R}_2\mathbf{A}[021] = e^{-i(k_1+k_2)L}\mathbf{A}[021]. \qquad (6.13)$$

As $\mathbf{R}_1\mathbf{R}_2 \neq \mathbf{R}_2\mathbf{R}_1$ then it is not obvious that this is compatible with (6.12). However, due to the symmetry condition on the spin wavefunction, for overall antisymmetry of Ψ with respect to an interchange of the conduction electrons 1 and 2, $\mathbf{A}[021] = \mathbf{P}_{1,2}\mathbf{A}[012]$, where $\mathbf{P}_{1,2}$ permutes the spin components of 1 and 2. As $\mathbf{P}_{1,2}\mathbf{R}_1\mathbf{R}_2\mathbf{P}_{1,2} = \mathbf{R}_2\mathbf{R}_1$, irrespective of the definition of $\mathbf{R}_j$, then the two equations are equivalent.

We only have equations so far that determine the sum of the wavevectors k_1 and k_2, and not the individual values. We can determine these using one more boundary condition. We start in the region [012] and let $x_1 \to x_1 + L$ to determine $\Psi(x_1 + L, x_2)$,

$$\Psi(x_1 + L, x_2) = \{e^{ik_1 L} e^{ik_1 x_1 + ik_2 x_2} - e^{ik_2 L} e^{ik_2 x_1 + ik_1 x_2}\} \mathbf{P}_{1,2} \mathbf{R}_1 \mathbf{A}[012]. \tag{6.14}$$

The factor $\mathbf{R}_1$ is the scattering when x_1 passes the impurity, and $\mathbf{P}_{1,2}$ is the permutation when x_1 passes x_2. A similar equation can be generated for $\Psi(x_1, x_2 + L)$ which corresponds to (6.14) with k_1 and k_2 interchanged, and $\mathbf{P}_{1,2}\mathbf{R}_1$ replaced by $\mathbf{R}_2\mathbf{P}_{1,2}$. Let $\mathbf{M}_1 = \mathbf{P}_{1,2}\mathbf{R}_1$, $\mathbf{M}_2 = \mathbf{R}_2\mathbf{P}_{1,2}$, then $\mathbf{R}_1 = \mathbf{P}_{1,2}\mathbf{R}_2\mathbf{P}_{1,2}$ implies $\mathbf{M}_1 = \mathbf{M}_2$, and as $\mathbf{R}_2\mathbf{R}_1 = \mathbf{M}_1\mathbf{M}_2$, then the transformation that diagonalizes $\mathbf{R}_2\mathbf{R}_1$ also diagonalizes the Ms. As the Ms are unitary then their eigenvalues can be expressed in the form $\exp(-i\eta L)$, where $2\eta = E_{\mathrm{sp}}$ (modulo $2\pi/L$). We see that in the diagonal representation the boundary condition $\Psi(x_1 + L, x_2) = \Psi(x_1, x_2)$ is satisfied if $k_j = \eta + 2\pi n_j/L$ where n_j is an integer, as we can see by substituting these values for k_1, k_2 into (6.14) with $\mathbf{M}_1 = \exp(-i\eta L)$. The wavevectors k_1, k_2 are modified by a phase shift η due to the impurity scattering.

To summarize the two particle problem, eigenfunctions of the form (6.8) can be found which satisfy periodic boundary conditions and the antisymmetry condition under particle exchange. These require the diagonalization of the subsidiary spin problem (6.12). The diagonalization of this spin problem will be considered after we have established that this form of solution can be generalized to an arbitrary number of particles.

6.2 Diagonalization of the s-d Model

Generalizing the wavefunction we considered in the previous section, we assume a form,

$$\Psi(x_1, x_2, ... x_{N_e}) = \Phi_{k_1, k_2, ... k_{N_e}}(x_1, x_2, ... x_{N_e})$$

$$\sum_{P, \sigma's} A_{\sigma_1, \sigma_2, ... \sigma_{N_e}}[P] \prod_{j=0}^{N_e} \chi_j(\sigma_j)\theta(P), \tag{6.15}$$

where $\Phi_{k_1, k_2, ... k_{N_e}}(x_1, x_2, ... x_{N_e})$ is the Slater determinant for spinless fermions constructed from plane waves with wavevectors $k_1, k_2, ... k_{N_e}$. This satisfies the Schrödinger equation (6.4) within each region P for $E = \sum_{j=1}^{N_e} k_j$. Starting in the region $P = I = [0, 1, 2, ... N_e]$ the periodic boundary condition,

$$\Psi(x_1 + L, x_2 + L, ... x_{N_e} + L) = \Psi(x_1, x_2, ... x_{N_e}), \tag{6.16}$$

generates the equation,

$$\mathbf{R}_{N_e}...\mathbf{R}_2\mathbf{R}_1\mathbf{A}[I] = e^{-i(k_1+k_2+...k_{N_e})L}\mathbf{A}[I]. \qquad (6.17)$$

where $[I]$ denotes the region $[0, 1, 2, ...N_e]$. So the diagonalization of $\mathbf{R}_{N_e}...\mathbf{R}_2\mathbf{R}_1$, which we shall denote by $\mathbf{T}_1$, determines the total energy modulo $2\pi/L$.

Starting in any of the other regions P and applying the same boundary condition generates a matrix equivalent to $\mathbf{T}_1$. This follows straightforwardly generalizing the argument for the two electron case. The boundary condition, $\Psi(x_1, x_2, ...x_j+L, ...x_{N_e}) = \Psi(x_1, x_2, ...x_j, ...x_{N_e})$, applied to the region $P = I$ generates the operator,

$$\mathbf{M}_j = \mathbf{P}_{j+1,1}...\mathbf{P}_{N_e,j}\mathbf{R}_j\mathbf{P}_{1,j}...\mathbf{P}_{j-2,j}\mathbf{P}_{j-1,j}, \qquad (6.18)$$

which describes the spin scattering of the jth electron as it is moved round the ring. Again we can generalize the argument for the two electron problem to show that the $\mathbf{M}_j$s are all equal. They are diagonalized by the same transformation that diagonalizes $\mathbf{T}_1$ as $\mathbf{T}_1 = \mathbf{M}_{N_e}\mathbf{M}_{N_e-1}...\mathbf{M}_1$. As the $\mathbf{M}_j$s are unitary their eigenvalues can be expressed in the form $e^{-i\eta L}$, giving an eigenvalue for $\mathbf{T}_1$ of $e^{-iN_e\eta L}$. The boundary condition, $\Psi(x_1, ...x_j + L, ...x_{N_e}) = \Psi(x_1, ...x_j, ...x_{N_e})$, is satisfied for $k_j = \eta + 2\pi n_j/L$, where n_j is an integer. Hence we have established that the argument given in section 6.1 can be generalized to N_e electrons.

The complete diagonalization of the linear dispersion s-d model now reduces to the diagonalization of the subsidiary spin problem (6.17). Fortunately this can be also diagonalized by a further Bethe ansatz, as was shown originally by Yang (1967) in considering the problem of a one dimensional electron gas interacting via a delta function interaction. Yang reduced that problem to a similar subsidiary spin problem to (6.17); in fact (6.17) is a special case of Yang's problem and the solution is given in Yang's paper. The same type of problem occurs in quite a different context, the calculation of the classical partition function for a two dimensional lattice which has directed bonds between lattice sites (generalizations of the two dimensional 'ice' problem). Different distributions of bonds at a particular lattice site or vertex have different energies in general, and when the problem is expressed in terms of transfer matrices it reduces to diagonalization of matrices of the form $\mathbf{T}_1$ in special cases (Baxter, 1982). The correspondence with transfer matrix problems and the one dimensional Heisenberg model can be made clearer by writing (6.17) in terms of spin operators, using the identity, $\mathbf{P}_{i,j} = (1 + 2\mathbf{S}_i \cdot \mathbf{S}_j)/2$, in (6.10).

The state with all the spins aligned, say all up, is clearly an eigenstate of $\mathbf{T}_1$, and though not the lowest eigenstate is a convenient reference state. States with one spin reversed, or two or more spins reversed, can be generated from this with associated wavevectors λ_r, similar to spin waves. In this context the wavevectors are often known as *rapidities*. There is an elegant method of generating these solutions based on the transfer matrix approach of Baxter (1982) and the quantum inverse scattering method (Sklyanin & Faddeev, 1978), known as the *algebraic Bethe ansatz* (for a review see Takhatajan, 1985). We give an outline of this approach in appendix F. Here we quote only the results, which can also be deduced directly from Yang's 1967 paper. The set of equations determining the λ_rs is

$$N_e\theta(2\lambda_r) + \theta(2\lambda_r + 2/c) = \sum_{s=1}^{M}\theta(\lambda_r - \lambda_s) + 2\pi I_r, \qquad (6.19)$$

for $r, s = 1, 2, ...M$, where M is the number of reversed spins, $\theta(\lambda) = 2\tan^{-1}\lambda$ and $c = \tan J$. The I_rs are integers or half integers depending on whether $N - M$ is odd or even. The k_js are given by

$$k_j = \frac{2\pi n_j}{L} - \frac{1}{L}\sum_{s=1}^{M}\left[\theta(2\lambda_s) + \pi\right], \qquad (6.20)$$

where the second term on the right hand side is the η associated with the eigenvalues of the $\mathbf{M}_j$s. The total energy eigenvalues E for the model are

$$E = \sum_{j}^{N_e}\frac{2\pi n_j}{L} - \frac{N_e}{L}\sum_{r=1}^{M}\left[\theta(2\lambda_r) + \pi\right], \qquad (6.21)$$

for states with z-component of spin, $S_z = (N_e + 1 - 2M)/2$. These states are characterized by the set of quantum numbers $\{n_j, I_r\}$, the n_j are associated with the charge excitations and I_r with the spin excitations. No two n_js can be equal or the wavefunction vanishes, as is evident from (6.15). The I_rs also have to be distinct to generate a solution to the spin problem. This set of quantum numbers also applies to the linear dispersion electron gas (we simply put the interaction to zero in this case), if we choose to work with the basis set which is discontinuous at $x_i = x_j$, rather than the continuous basis set which would be specified by the conventional set $\{n_{j\uparrow}, n_{j\downarrow}\}$, an integer for each spin type. This approach to the non-interacting electron gas is very complicated as it requires the solution of the set of equations (6.19) for the λ_rs. Nevertheless it has been proved that these lead to the standard results for an electron gas for $k_B T \ll D'$ (Filyov, Tsvelick & Wiegmann, 1981).

To apply these results to calculate the ground state impurity properties and the thermodynamics we need to reinstate some form of cut-off
on the charge excitations, as this spectrum is unbounded below. The
charge quantum numbers are required to lie within a band width D',
$|2\pi n_j/L| < D'$. The cut-off D' is similar to the band width D of the
original model. However, they must be distinguished as D acts as a
cut-off at all stages in calculations on the conventional model, whereas
D' is a cut-off imposed after diagonalization on the *solution* of the linear
dispersion model. We found from scaling arguments in section 3.3 that
the low energy spectrum depended only on a single parameter T_K. We
should obtain, therefore, the same behaviour as the original model in
the universal regime, $\mu_B H, k_B T \ll D$; the way the cut-off is imposed
should only affect the equation for T_K.

The ground state for the charge excitations clearly corresponds to
to taking consecutive values of n_j from the lowest value, $-D'L/2\pi$, to
the Fermi level, which we take as zero, corresponding to the centre of
the conduction band. It can be shown that the ground state of the spin
energy corresponds to consecutive filling of the I_r values from the lowest
value $-(N_e - M)/2$, which we take to correspond to $\lambda = -B$, to $(3M -
N_e)/2$ for which $\lambda = \infty$. In the continuum limit, $N_e, M \to \infty$ and $L \to
\infty$, such that $\pi N_e/L = D'$, the equations that determine the λ values
go over to integral equations. If $\sigma(\lambda)$ is the density of states the number
of solutions in the interval $d\lambda$ is $N_e\sigma(\lambda)d\lambda$. An equation for $\sigma_{con}(\lambda)$,
the density of states for the λ excitations of the conduction electrons in
the absence of the impurity, can be derived by taking consecutive values
of I_r in (6.19), subtracting the two equations and then taking the limit
$N_e \to \infty$, giving the integral equation

$$\sigma_{con}(\lambda) = \frac{2/\pi}{(1 + 4\lambda^2)} - \frac{1}{\pi}\int_{-B}^{\infty} K(\lambda - \lambda')\sigma_{con}(\lambda')d\lambda', \qquad (6.22)$$

using $\sigma_{con}(\lambda) = \text{limit}\{1/N_e(\lambda_{r+1} - \lambda_r)\}$ as $N_e \to \infty$, where $K(\lambda) =
(1 + \lambda^2)^{-1}$.

There is an extra contribution to the density of states $\sigma_{imp}(\lambda)/N_e$ of
order $1/N_e$ to the density of states $\sigma(\lambda)$ in the presence of the impurity,
arising from the second term on the left hand side of (6.19). This extra
contribution is given by an integral equation of the same form as (6.22)
but with a different driving term,

$$\sigma_{imp}(\lambda) = \frac{2/\pi}{(1 + 4(\lambda + 1/c)^2)} - \frac{1}{\pi}\int_{-B}^{\infty} K(\lambda - \lambda')\sigma_{imp}(\lambda')d\lambda'. \quad (6.23)$$

The magnetization of the electron gas, M_{con}, and the impurity contribution, M_{imp}, are given by

$$M_\alpha = g\mu_{\text{B}} \left(\frac{1}{2} - \int_{-B}^{\infty} \sigma_\alpha(\lambda')\, d\lambda' \right), \qquad (6.24)$$

where $\alpha = \text{con}, \text{imp}$. It can be readily demonstrated that the ground state of the electron gas in zero magnetic field corresponds to $B \to \infty$. For $B = \infty$ (6.22) can be solved by Fourier transform to give

$$\sigma_{\text{con}}^0(\lambda) = \frac{1}{2\cosh(\pi\lambda)}, \qquad (6.25)$$

which on substitution into (6.21) and (6.24) gives

$$E_{\text{con}}^{\text{gs}} = -\frac{\pi N D'}{2}, \qquad M_{\text{con}}^{\text{gs}} = 0, \qquad (6.26)$$

for the ground state of the electron gas.

With this value for B the equation for the impurity contribution σ_{imp}^0 to the density of states in the ground state can be calculated, and corresponds to (6.25) with λ replaced by $(\lambda + 1/c)$. This leads to zero impurity magnetization in the ground state, $M_{\text{imp}}^{\text{gs}} = 0$, when substituted into (6.24). To establish that this corresponds to a fully compensated state we need to calculate the impurity susceptibility. This requires the solution of (6.23) in the presence of an applied field H for which $B \neq \infty$. To determine B we calculate the field induced magnetization of the conduction electron gas in the absence of the impurity from (6.22), and then equate it to the known result, calculated by conventional methods. Once B is known in terms of the applied field H, the induced impurity magnetization can deduced by substituting the solution of (6.23) into (6.24).

In carrying out this calculation it will be convenient to cast equation (6.22) into a slightly different form,

$$\sigma_{\text{con}}(\lambda) = \sigma_{\text{con}}^0(\lambda) + \int_{-\infty}^{-B} R(\lambda - \lambda')\sigma_{\text{con}}(\lambda')\, d\lambda', \qquad (6.27)$$

by applying a Fourier transformation. The kernel $R(\lambda)$ of the equation in this form is given by

$$R(\lambda) = \frac{1}{4\pi} \int_{-\infty}^{\infty} \frac{e^{ip\lambda}e^{-|p|/2}}{\cosh(p/2)}\, dp. \qquad (6.28)$$

To switch the B dependence to the driving term the λ is displaced so that $\lambda \to \lambda - B$, and a new function, $\rho_{\text{con}}(\lambda)$, is introduced via $\rho_{\text{con}}(\lambda) = \sigma_{\text{con}}(\lambda - B)$. Precisely the same transformations can be applied to (6.23), so that both (6.22) and (6.23) can be written in the form,

$$\rho_\alpha(\lambda) = \rho_\alpha^0(\lambda) + \int_{-\infty}^{0} R(\lambda - \lambda')\rho_\alpha(\lambda')\, d\lambda', \qquad (6.29)$$

where $\alpha = \text{con}, \text{imp}$.

With this form for the equation we can deduce the weak field magnetization and field independent susceptibility of the impurity without explicitly solving (6.29). In the weak field limit $H \to 0$ we need only the asymptotic form for the driving term in (6.29) as $B \to \infty$. For the conduction electrons this can be deduced from (6.25), and also for the impurity term by replacing λ by $\lambda + 1/c$. Hence we find leading terms,

$$\rho_{\text{con}}^0(\lambda) = e^{(-\pi B + \pi \lambda)}, \quad \rho_{\text{imp}}^0(\lambda) = e^{(-\pi B + \pi \lambda)} e^{\pi/c}. \qquad (6.30)$$

These driving terms are the same apart from a constant factor $e^{\pi/c}$. As the equations are linear their solutions will differ by the same factor. Consequently the impurity and conduction electron magnetizations in weak field are in the ratio $e^{\pi/c}$, so

$$\chi_{\text{imp}}/\chi_{\text{con}} = e^{\pi/c} \quad \text{or} \quad \chi_{\text{imp}} = \frac{(g\mu_{\text{B}})^2 e^{\pi/c}}{4D'}, \qquad (6.31)$$

where χ_{con} is the susceptibility per electron for the conduction electrons. We can identify $D' e^{-\pi/c}/k_{\text{B}}$ as the Kondo temperature apart from a constant factor. For the moment we shall denote this quantity by T_0 and relate it more precisely to T_{K} later.

To calculate the relation between B and H, and hence calculate the induced impurity magnetization in higher fields, the explicit solution is required. The equation can be solved analytically by the Wiener–Hopf technique and we give an outline of the solution in appendix G. Using this solution for the conduction electrons alone and equating the induced magnetization linear in field to the standard result, gives the relation between B and H,

$$e^{-\pi B} = H(2\pi e)^{1/2} \chi_{\text{con}}/g\mu_{\text{B}} = g\mu_{\text{B}}(\pi e/8)^{1/2} H/D'. \qquad (6.32)$$

Only the weak field response need be calculated for the conduction electrons as the correction terms are of order $(\mu_{\text{B}} H/D')^2$, involving the ratio of the Zeeman energy to the Fermi energy and hence are completely negligible. For the impurity contribution, however, this factor gets replaced by $(\mu_{\text{B}} H/k_{\text{B}} T_0)^2$, and as the impurity response to fields comparable with the Kondo energy is of physical interest the higher order terms are important. From the Wiener–Hopf solution the induced impurity magnetization can be expressed as a power series in $g\mu_{\text{B}} H/k_{\text{B}} T_1$, where $T_1 = (8/\pi e)^{1/2} T_0$,

$$M_{\text{imp}} = \frac{1}{\sqrt{\pi}} \sum_{n=0}^{\infty} \frac{(-1)^n}{n!} \left(n + \frac{1}{2}\right)^{n-1/2} e^{-(n+1/2)} \left(\frac{g\mu_{\text{B}} H}{k_{\text{B}} T_1}\right)^{2n+1},$$

$$(6.33)$$

which is valid for $g\mu_{\text{B}} H/k_{\text{B}} T_1 \leq 1$. For very large fields $g\mu_{\text{B}} H/k_{\text{B}} T_1 \gg 1$,

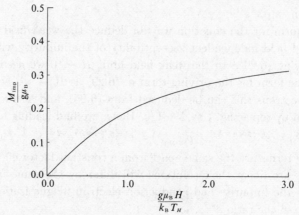

Figure 6.2 A plot of the impurity moment $M_{\mathrm{imp}}(H)$ versus magnetic field H. The slow asymptotic approach to the saturation value $g\mu_{\mathrm{B}}/2$ is apparent.

M_{imp} tends to the saturated value $g\mu_{\mathrm{B}}/2$ but there are significant logarithmic corrections. As $H \to \infty$ the asymptotic form is

$$M_{\mathrm{imp}} = \frac{g\mu_{\mathrm{B}}}{2}\left\{1 - \frac{1}{2\ln(g\mu_{\mathrm{B}}H/k_{\mathrm{B}}T_{\mathrm{H}})} - \frac{\ln\ln(g\mu_{\mathrm{B}}H/k_{\mathrm{B}}T_{\mathrm{H}})}{4(\ln(g\mu_{\mathrm{B}}H/k_{\mathrm{B}}T_{\mathrm{H}}))^2}\right.$$

$$\left. + O(\ln(g\mu_{\mathrm{B}}H/k_{\mathrm{B}}T_{\mathrm{H}}))^{-3}\right\}, \quad (6.34)$$

where $T_{\mathrm{H}} = T_1/\sqrt{2}$ and has been defined so that there is no term in $(\ln(g\mu_{\mathrm{B}}H/k_{\mathrm{B}}T_{\mathrm{H}}))^{-2}$ in the expansion.

The full magnetization curve for the impurity as a function of field is shown in figure 6.2.

The curvature of $M_{\mathrm{imp}}(H)$ with H is always negative, which implies that the field dependent impurity magnetization $\chi_{\mathrm{imp}}(H)$ monotonically decreases with increasing H (we shall see later this does not apply to all models). We also note the very slow approach to the saturated value due to the logarithmic terms. For fields of the order, $g\mu_{\mathrm{B}}H \sim k_{\mathrm{B}}T_H$, $M_{\mathrm{imp}}(H)$ is only about 40% of its saturation value, and is within 10% of its saturated value only for fields as large as $g\mu_{\mathrm{B}}H \sim 150k_{\mathrm{B}}T_H$.

6.3 Excitations

So far we have only considered the ground state solution. The full excitation spectrum is required to calculate the thermodynamic behaviour. As the charges are decoupled from the spins and non-interacting their excitations are simply independent particle–hole excitations from the

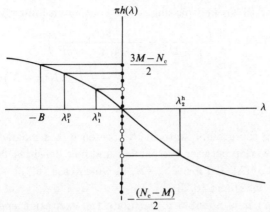

Figure 6.3 A schematic representation of the assignment of wavevectors λ^p and λ^h for particle and hole states for an excited state. The allowed set of integer and half integers is shown on the vertical axis from $-(N_e - M)/2$ to $(3M - N_e)/2$ (see equation (6.35)).

Fermi sea. Similar excitations occur for the spins but due to interactions these are not independent. For the ground state in the absence of a magnetic field, the allowed values of $\{I_r\}$ are used. We can consider these as occupied or particle states. In this case there can be no further particle states but 'holes' can be created by omitting integers from the allowed set $\{I_r\}$. In the presence of a magnetic field $M < N_e/2$ in the ground state so some of the set $\{I_r\}$ are available and particle-like excited states can also be created. Following Yang & Yang (1969), a particle and hole density of states can be introduced. Consider a subset $\{I_r\}'$ for which solutions of (6.19), $\lambda_1, \lambda_2, ...\lambda_M$, are found. These constitute the particle states. Let $h(\lambda)$ be defined by

$$\pi h(\lambda) = N_e\theta(\lambda) + \theta(2\lambda + 2/c) - \sum_{s=1}^{M} \theta(\lambda - \lambda_s), \qquad (6.35)$$

then $h(\lambda_r) = I_r \in \{I_r\}'$, where $\{I_r\}'$ denotes the 'occupied' set of allowed integers. For values, J_r contained in the allowed set $\{I_r\}$ but not included in $\{I_r\}'$ hole wavevectors λ_r^h can be defined by $h(\lambda_r) = J_r$. This is indicated schematically in figure 6.3.

As $N_e, M \to \infty$ it is possible to define a density for the distribution of these states, $N_e\sigma(\lambda)d\lambda$ being the number of particle states in the range $d\lambda$, and $N_e\sigma^h(\lambda)d\lambda$ being the number of hole states. As $\delta h(\lambda)$ gives the total number of states in the range $\delta\lambda$, then

$$\frac{dh(\lambda)}{d\lambda} = N_e(\sigma(\lambda) + \sigma^h(\lambda)). \qquad (6.36)$$

Differentiating (6.35) and passing to the continuum limit gives

$$(\sigma(\lambda) + \sigma^{\mathrm{h}}(\lambda)) = \frac{2}{\pi}\left(\frac{1}{1+4\lambda^2} + \frac{1/N_{\mathrm{e}}}{1+4(\lambda+1/c)^2}\right)$$
$$-\frac{1}{\pi}\int_{-\infty}^{\infty} K(\lambda-\lambda')\sigma(\lambda')d\lambda'.$$

(6.37)

In calculating the ground state in the presence of a magnetic field we assumed no overlap between the particle and hole densities, $\sigma(\lambda) = 0$ for $\lambda < -B$ and $\sigma^{\mathrm{h}}(\lambda) = 0$ for $\lambda > -B$, in which case (6.37) is equivalent to equations (6.22) and (6.23).

Using (6.37) it is possible to calculate the excitation energies associated with a hole at λ^{h} which corresponds to a hole density $\sigma^{\mathrm{h}} = \delta(\lambda - \lambda^{\mathrm{h}})/N_{\mathrm{e}}$. When this is substituted into (6.37) it gives an equation for the particle density $\sigma(\lambda)$. This changes from the ground state value due to the adjustment of all the other states to the presence of the hole. The equation can be solved by Fourier transformation, and on substituting the solution for $\sigma(\lambda)$ into (6.24), the value of M is found to decrease corresponding to a reduction in the value of S_z of $\frac{1}{2}$. This purely spin excitation has been termed a *spinon*, and corresponds to an excitation energy ,

$$\epsilon(\lambda^{\mathrm{h}}) = \frac{2D'}{\pi}\tan^{-1}e^{\pi\lambda^{\mathrm{h}}}.$$

(6.38)

If λ^{h} corresponds to an integer n_h, a density of states for spinon excitations can be defined via

$$D_s(\epsilon) = \frac{\partial n_h}{\partial\epsilon(\lambda^{\mathrm{h}})} = \sigma^0(\lambda^{\mathrm{h}})\left(\frac{\partial\epsilon(\lambda^{\mathrm{h}})}{\partial\lambda}\right)^{-1}.$$

(6.39)

Solving for λ^{h} in terms of ϵ from (6.38), on the assumption of low lying excitations such that $e^{\pi\lambda^{\mathrm{h}}} \ll 1$ so $\tan^{-1}(e^{\pi\lambda^{\mathrm{h}}})$ can be replaced by $e^{\pi\lambda^{\mathrm{h}}}$, gives for the impurity contribution,

$$D_s(\epsilon) = \frac{1}{\pi}\frac{(2k_{\mathrm{B}}T_0/\pi)}{\epsilon^2 + (2k_{\mathrm{B}}T_0/\pi)^2}.$$

(6.40)

This has the form of the Kondo resonance found earlier for the quasiparticle density of states. The width of the resonance, however, differs by a factor of 2 from (5.14), (5.22).

A spinon excitation cannot be excited without adding or removing an electron, in which case there is an additional energy, $2\pi n/L$, due to the change of charge. For two holes λ_1^{h}, λ_2^{h}, the change in S_z is -1, and this is a possible excitation with a fixed number of particles. Solution of the equations in this case gives a total excitation energy, ΔE given by

$$\Delta E = \epsilon(\lambda_1^{\text{h}}) + \epsilon(\lambda_2^{\text{h}}).\tag{6.41}$$

For λ^{h} large and negative these excitation energies are rather small and this is the reason why in the presence of a magnetic field hole states corresponding to the large negative values in $\{I_r\}$ were created. It was shown by Andrei, Furuya & Lowenstein (1983) that this excitation is a triplet.

Other excitations can occur in the spin system corresponding to complex values of λ, as were found in Bethe's original work on the Heisenberg model. The simplest of these excitations is a string, $\lambda^{\pm} = \bar{\lambda} \pm i/2$, which can occur if two holes, λ_1^{h}, λ_2^{h}, are present where $2\bar{\lambda} = \lambda_1^{\text{h}} + \lambda_2^{\text{h}}$. Including these in the sum over the solutions on the right hand side of (6.35) and using the relation $\theta(\lambda - \lambda^+) + \theta(\lambda - \lambda^-) = \theta(2(\lambda - \bar{\lambda})) + \theta((2\lambda - \bar{\lambda})/3)$, then passing to the continuum limit, we can derive a new equation for $\sigma(\lambda)$ in the presence of the two holes. This has an additional driving term due to the string in the form,

$$\frac{-3}{(3/2)^2 + (\lambda - \bar{\lambda})^2} + \frac{-1}{(1/2)^2 + (\lambda - \bar{\lambda})^2}.\tag{6.42}$$

When the equations are solved in the presence of the string the two hole excitation becomes a singlet with $\Delta S_z = 0$, and the energy, given by (6.41), is unchanged (Andrei et al, 1983).

The problem of determining all possible excitations, with both real and complex wavevectors, is in general very difficult. Fortunately there is a rather simple conjecture, given in the original paper by Bethe (1931), that as $N_e \to \infty$ the complex solutions are all of the form,

$$\lambda_r^{(n,j)} = \lambda_r^n + i\left(\frac{n+1}{2} - j\right), \qquad j = 1,...n,\tag{6.43}$$

where λ_r^n is real. Such a set is known as an n-string and is associated with an n spin bound state. The real solutions considered earlier correspond to a 1-string. A proof of this hypothesis is given in the review of Tsvelick & Wiegmann (1983), who also clarify the conditions for its validity. The limit $N_e \to \infty$ is not a sufficient condition for the hypothesis to hold. An additional assumption must be made that there is a macroscopic number of holes present. This means it can be used in the thermodynamic limit.

When the string solutions are included in (6.19), and the sum over j is taken for each string, the equation for the real part of an n-string becomes

$$N_e \theta \left(\frac{2\lambda_r^{(n)}}{n} \right) + \theta \left(\frac{2\lambda_r^{(n)} + 2/c}{n} \right) = \sum_{m \neq n, s} \theta_{nm} \left(\lambda_r^{(n)} - \lambda_s^{(m)} \right) + 2\pi I_r^{(n)},$$

(6.44)

where $\theta_{nm}(x)$ is given by

$$\theta_{nm}(x) = \theta \left(\frac{2x}{|n-m|} \right) + \theta \left(\frac{2x}{n+m} \right) + 2 \sum_{k}^{\min(m,n)-1} \theta \left(\frac{2x}{|n-m|+2k} \right),$$

(6.45)

with the first term omitted for $m = n$. Using (6.45) $h(\lambda)$ can be generalized to define a corresponding quantity $h_n(\lambda)$ for each string so that $h_n(\lambda_r^{(n)}) = I_r^{(n)}$ is a particle solution for an n-string, and $h_n(\lambda_r^{(n)h}) = J_r^{(n)}$ for a hole, where $J_r^{(n)}$ is an integer in the admissible set $\{I_r^{(n)}\}$ which does not correspond to a particle solution. Similarly particle and hole densities, $\sigma_n(\lambda)$ and $\sigma_n^h(\lambda)$ can be introduced for each string, which are required to satisfy a coupled set of integral equations similar to (6.37),

$$\sigma_n^h(\lambda) + \sum_{m=1}^{\infty} A_{nm} \sigma_n(\lambda) = f_n(\lambda),$$

(6.46)

where A_{nm} are integral operators defined by

$$A_{nm} = [|n-m|] + 2[|n-m|+2] + \ldots 2[n+m-2] + [n+m],$$

(6.47)

$$[k]a(\lambda) = \frac{1}{\pi} \int_{-\infty}^{\infty} \frac{ka(\lambda')/2}{(k/2)^2 + (\lambda - \lambda')^2} \, d\lambda',$$

(6.48)

and

$$f_n(\lambda) = \frac{n}{2\pi} \left\{ \frac{1}{(n/2)^2 + (\lambda)^2} + \frac{1/N_e}{(n/2)^2 + (\lambda + 1/c)^2} \right\}.$$

(6.49)

The total spin energy is given by

$$E_{sp} = \frac{2\pi}{L} \sum_{j=1}^{N_e} n_j + \frac{N_e}{L} \sum_{n=1}^{\infty} \left[\theta \left(\frac{2\lambda}{n} \right) - \pi \right] \sigma_n(\lambda) \, d\lambda,$$

(6.50)

and the z-component of spin,

$$S_z = \frac{1}{2} - \sum_{n=1}^{\infty} \int_{-\infty}^{\infty} n \sigma_n(\lambda) \, d\lambda.$$

(6.51)

The solutions of these equations form the basis for the calculation of the thermodynamics.

6.4 Thermodynamics of the s-d Model for $S = \frac{1}{2}$

As the spin and charge excitations are decoupled in the Bethe ansatz solution the total free energy is the independent sum of spin and charge contributions. As the charge excitations are non-interacting their contribution F_{ch} is very simple to calculate,

$$
\begin{aligned}
F_{\mathrm{ch}} &= -\frac{N_e k_{\mathrm{B}} T}{2D'} \int_{-D'}^{D'} \ln(1 + e^{\epsilon/k_{\mathrm{B}}T})\, d\epsilon - \frac{D'}{4} \\
&= -\frac{\pi^2 k_{\mathrm{B}}^2 T^2}{12 D'} + \mathrm{O}(e^{-D'/k_{\mathrm{B}}T}),
\end{aligned}
\tag{6.52}
$$

measured relative to the ground state energy.

The free energy of the spin excitations can be found by forming a free energy functional $F_{\mathrm{sp}}(\{\sigma_n, \sigma_n^{\mathrm{h}}\})$ in terms of the string particle and hole densities,

$$
F_{\mathrm{sp}}(\{\sigma_n, \sigma_n^{\mathrm{h}}\}) = E_{\mathrm{sp}} - hS_z(\{\sigma_n, \sigma_n^{\mathrm{h}}\}) - k_{\mathrm{B}} T S(\{\sigma_n, \sigma_n^{\mathrm{h}}\}),
\tag{6.53}
$$

where S is the entropy term. The calculation then follows along the lines originally devised by Yang & Yang (1969) for deriving the thermodynamics of the one dimensional Bose gas subject to a repulsive delta function interaction. This is the simplest model solvable by the Bethe ansatz and has no string excitations. The approach was generalized to the one dimensional Heisenberg model by including strings in a series of papers by Takahashi (1970–4). The entropy term is expressed as the logarithm of the number of ways of distributing the particles and holes for a given energy. The numbers of particles and holes over a small interval $d\lambda$ are $N_e \sigma_n d\lambda$ and $N_e \sigma_n^{\mathrm{h}} d\lambda$ so the total number of ways of distributing these over $d\lambda$ is

$$
\frac{N_e(\sigma_n + \sigma_n^{\mathrm{h}}) d\lambda!}{N_e \sigma_n d\lambda!\, N_e \sigma_n^{\mathrm{h}} d\lambda!}.
\tag{6.54}
$$

Taking logarithms and using Stirling's formula gives

$$
S = N_e \sum_n \int \{(\sigma_n + \sigma_n^{\mathrm{h}})\ln(\sigma_n + \sigma_n^{\mathrm{h}}) - \sigma_n \ln(\sigma_n) - \sigma_n^{\mathrm{h}} \ln(\sigma_n^{\mathrm{h}})\}\, d\lambda.
\tag{6.55}
$$

The spin free energy is now deduced from (6.53), with E_{sp} given by (6.50), S_z by (6.51) and S by (6.55), by minimizing with respect to σ_n and σ_n^{h} subject to the constraint (6.46). The form of (6.55), reflecting the single or zero occupancy of any given allowed quantum number, leads to Fermi statistics for the excitations. After some lengthy manipulations the result for $F_{\rm sp}$ can be expressed in the form,

$$F_{\rm sp} = N_{\rm e} E_{\rm sp}^{\rm gs} - k_{\rm B} T \int \sigma_0(\lambda) \ln(1 + e^{\epsilon_1(\lambda)/k_{\rm B}T}) \, d\lambda, \qquad (6.56)$$

where $e^{\epsilon_1(\lambda)/k_{\rm B}T} = \sigma_1^{\rm h}(\lambda)/\sigma_1(\lambda)$, $E_{\rm sp}^{\rm gs}$ is the ground state energy, and $\sigma_0(\lambda)$ is defined by

$$\sigma_0(\lambda) = \frac{1}{2} \left[\frac{N_{\rm e}}{\cosh \pi \lambda} + \frac{1}{\cosh \pi(\lambda + 1/c)} \right] \qquad (6.57)$$

The equilibrium string particle and hole densities are determined by the set of equations,

$$\epsilon_1 = -2D' \tan^{-1} e^{-\pi \lambda} + k_{\rm B} T \mathbf{G} \ln(1 + e^{\epsilon_2(\lambda)/k_{\rm B}T}), \qquad (6.58)$$

and

$$\epsilon_n(\lambda) = \mathbf{G} \left[\ln(1 + e^{\epsilon_{n+1}(\lambda)/k_{\rm B}T}) + \ln(1 + e^{\epsilon_{n-1}(\lambda)/k_{\rm B}T}) \right], \qquad (6.59)$$

where $e^{\epsilon_n(\lambda)/k_{\rm B}T} = \sigma_n^{\rm h}/\sigma_n$, and $\mathbf{G}$ is an integral operator defined by

$$\mathbf{G} f(\lambda) = \frac{1}{2} \int \frac{1}{\cosh \pi(\lambda - \lambda')} f(\lambda') \, d\lambda', \qquad (6.60)$$

together with the boundary conditions as $n \to \infty$,

$$\lim_{n \to \infty} \left\{ \frac{\epsilon_n}{n} \right\} = g\mu_{\rm B} H. \qquad (6.61)$$

For details of the derivations of these equations we refer the reader to the original papers and the reviews of Andrei et al (1983) and Tsvelick & Wiegmann (1983).

This infinite set of non-linear integral equations has to be solved numerically to deduce the thermodynamics over the full temperature range. Analytic results can be deduced, however, for the low and high temperature limits. If we set the impurity contribution to $\sigma_0(\lambda)$ equal to zero, then we obtain a set of equations which determine the contribution of the spin excitations to the free energy of the non-interacting electron gas with linear dispersion. The free energy of the conduction electrons alone is, however, known ($k_{\rm B} T \ll D'$). It is twice that given by (6.52), the extra factor being due to the spin degeneracy. Rather than solving the difficult set of integral equations we can use this known result to deduce the spin contribution. From the total free energy minus the charge contribution (6.52), we find

$$F_{\rm sp}^{\rm con} = -N_{\rm e} \left\{ \frac{\pi^2 k_{\rm B}^2 T^2}{12D'} + \frac{(g\mu_{\rm B} H)^2}{8D'} + \frac{D'}{4} \right\}, \qquad (6.62)$$

for $k_{\rm B} T \ll D'$.

We can deduce the free energy of the impurity from this in the limit of low temperatures and low fields. We first of all establish our earlier scaling results, such as (3.43), that for $k_B T \ll D'$ the free energy of the impurity is a universal function of T/T_0 and H/T. To demonstrate this we transform to new variables λ' and $\tilde{\epsilon}_n$, where $\lambda = \lambda' - \ln(k_B T/2D')/\pi$ and $\tilde{\epsilon}_n(\lambda') = \epsilon_n(\lambda' - \ln(k_B T/2D')/\pi)$. The equations for $F_{\rm sp}^{\rm con}$ and $F_{\rm sp}^{\rm imp}$ become

$$F_{\rm sp}^{\rm con} = -\frac{k_B T N_e}{2} \int_{-\infty}^{\infty} \frac{\ln(1 + e^{-\tilde{\epsilon}_1(\lambda')/k_B T})}{\cosh[\pi\lambda' - \ln(k_B T/2D')]} d\lambda', \qquad (6.63)$$

$$F_{\rm sp}^{\rm imp} = -\frac{k_B T}{2} \int_{-\infty}^{\infty} \frac{\ln(1 + e^{-\tilde{\epsilon}_1(\lambda')/k_B T})}{\cosh[\pi\lambda' - \ln(T/2T_0)]} d\lambda'. \qquad (6.64)$$

The driving term in the equations for $\tilde{\epsilon}_n(\lambda')$ is modified to

$$-2D'\tan^{-1}\left(\frac{k_B T}{2D'}e^{-\pi\lambda'}\right). \qquad (6.65)$$

In the expression for the impurity spin free energy the significant contributions arise from $\lambda' \sim \ln(T/2T_0)/\pi$. For λ' in this range we can simplify the driving term (6.65) to give $-e^{-\pi\lambda'}$ (replacing $\tan^{-1}x$ by x for $x \ll 1$). Then D' cancels out in (6.65) so the equations that determine $\tilde{\epsilon}_1(\lambda')$ now depend only on H/T. Consequently in this regime

$$F_{\rm sp}^{\rm imp} = -Tf\left(\frac{T}{T_0}, \frac{H}{T}\right). \qquad (6.66)$$

From (6.63) and (6.64) we see that the same universal function describes the spin contribution to the free energy of the conduction electrons,

$$F_{\rm sp}^{\rm con} = -Tf\left(\frac{k_B T}{D'}, \frac{H}{T}\right). \qquad (6.67)$$

The form of this function for $k_B T \ll D'$, however, can be deduced from (6.62) without explicit evaluation of (6.63). We then find for the impurity free energy,

$$F^{\rm imp} = -\left\{\frac{\pi^2 k_B T^2}{12T_0} + \frac{(g\mu_B H)^2}{8k_B T_0} + \frac{k_B T_0}{4}\right\} + O(T^4, H^4, T^2 H^2), \qquad (6.68)$$

which is entirely due to spin excitations. Using this to calculate the $T = 0$ susceptibility gives the result (6.31) derived earlier. However, we are now in a position to calculate the specific heat coefficient $\gamma_{\rm imp}$ for the impurity. From (6.68) we obtain

$$\frac{\gamma_{\rm imp}}{\gamma_{\rm con}} = \frac{D'}{2k_B T_0} = \frac{e^{\pi/c}}{2}, \qquad (6.69)$$

where $\gamma_{\rm con}$ is the specific heat coefficient per conduction electron. The extra factor of 2 compared to (6.31) occurs because the specific heat for the impurity arises purely from the spin excitations, whereas that for the conduction electrons has equal contributions from the spin and charge

excitations. From (6.31) and (6.69) we can derive the earlier results for
the χ/γ ratio, $R = 2$. The Bethe ansatz solution does lead also to new
results. Tsvelick & Wiegmann (1983) have shown that this ratio holds
in the presence of an arbitrary magnetic field H so

$$\frac{4\pi^2 k_B^2}{3(g\mu_B)^2} \frac{\chi_{\text{imp}}(H)}{\gamma_{\text{imp}}(H)} = 2. \tag{6.70}$$

We will use this result a little later to deduce the T^2 coefficient in the
low temperature expansion for $\chi_{\text{imp}}(T)$.

At high temperatures analytic results can be extracted for $\chi_{\text{imp}}(T)$
which confirm the perturbational result (3.53),

$$\chi_{\text{imp}}(T) = \frac{(g\mu_B)^2}{4k_B T}\left\{1 - \frac{1}{\ln(T/T_K)} - \frac{\ln(\ln(T/T_K))}{2\ln^2(T/T_K)} + O\left(\frac{1}{\ln^2(T/T_K)}\right)\right\}$$

$$\tag{6.71}$$

where $T_K = \alpha T_0$, and α is chosen to eliminate the term in $\ln(T/T_K)^{-2}$
in (6.71), corresponding to the same procedure used by Wilson which
determines T_K precisely. The coefficient of the $\ln(T/T_K)^{-2}$ cannot be
deduced analytically from the equations (6.56)–(6.60), so it is not possible to calculate α and hence deduce the Wilson number w, as defined
in (4.58), analytically by this method.

One unexpected feature of the Bethe ansatz expression for T_K is the
absence of the square root factor $(J\rho_0)^{1/2}$ found in the earlier expression
(3.47) for T_K. As the cut-off procedure adopted to derive the Bethe
ansatz solution differs from that for conventional calculations it is not
surprising that the form for T_K is not precisely the same. It can be shown
that two couplings J and J' which can be related by an analytic function,
$J = F(J')$, lead to the same expressions for T_K in weak coupling, and
differ only in the terms of order $J\rho_0$ or higher in the exponential of (3.47).
The couplings in the Bethe ansatz expression for T_K and that for the
conventional model, (3.47), cannot be related analytically so there was
some doubt as to whether or not the linear dispersion s-d model belongs
to the same universality class as the conventional model. This led Andrei
& Lowenstein (1981) to devise an indirect way of calculating the Wilson
number w (defined in (4.58)) to compare with Wilson's result. The
Wilson number is a universal ratio relating the behaviour of the model
on quite distinct energy scales $T \ll T_K$ and $T \gg T_K$ and so can only
be deduced on the basis of a complete solution. The close agreement of
the value found by Andrei & Lowenstein with the Wilson's numerical
result established beyond doubt that the models belong to the same

universality class, so dispelling any lingering doubt as to whether or not the Bethe ansatz results give the solution to the Kondo problem.

The argument of Andrei & Lowenstein, leading to an analytic expression for w, was based on re-expressing this quantity in terms of two other universal ratios, T_L/T_H and T_H/T_K, via

$$w = \frac{T_K}{T_L} = \frac{T_H}{T_L}\frac{T_K}{T_H}, \qquad (6.72)$$

where T_L is defined by $\chi_{imp}(0) = (g\mu)^2/4k_B T_L$ and T_H by (6.34). From (6.31) we find $T_L = T_0$, and from (6.34) $T_H = 2T_0/(\pi e)^{1/2}$, and hence

$$\frac{T_H}{T_L} = \frac{2}{\sqrt{\pi e}}. \qquad (6.73)$$

This is the difficult ratio to calculate because like w it relates different regimes, the low temperature strong coupling regime through T_L and the high field perturbative regime through T_H. The other ratio in (6.72) is much simpler to estimate as it relates quantities which can be calculated from perturbative regimes, T_K from the high temperature expansion and T_H from the high field expansion. For this ratio it is more convenient to use the perturbational results for the conventional model given in section 3.1. Following the Wilson procedure for defining T_K precisely and a similar procedure for T_H, we obtain from (3.3) and (3.6), with c_2 and c_2' evaluated for a flat band,

$$\frac{T_K}{T_H} = \frac{e^{(C+3/4)}}{(2\pi)^{1/2}}. \qquad (6.74)$$

where C is Euler's constant which has the value 0.577216. This gives an exact expression for w,

$$w = \frac{e^{(C+1/4)}}{\pi^{3/2}} = 0.41071.... \qquad (6.75)$$

which agrees with Wilson's numerical result $w = 0.4128 \pm .002$ to within the numerical accuracy quoted. Using $T_0 = T_K/w$ and $T_H = 2T_K/w\sqrt{e\pi}$, we can express results (6.33), (6.34) and (6.68), in terms of the Kondo temperature as defined by Wilson.

From the thermodynamic relation,

$$\frac{\partial^2 C(T,H)}{\partial H^2} = T\frac{\partial^2 \chi(T,H)}{\partial T^2}, \qquad (6.76)$$

where $C(T,H)$ is the specific heat, we can deduce the T^2 term in the expansion of $\chi_{imp}(T,0)$ for $T \ll T_K$. From (6.76) we find

$$\left(\frac{\partial^2 \chi(T,H)}{\partial T^2}\right)_{0,0} = \left(\frac{\partial^2 \gamma(H)}{\partial H^2}\right)_0. \qquad (6.77)$$

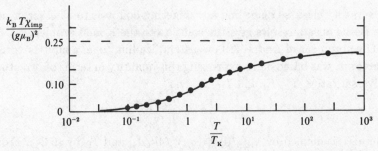

Figure 6.4 A plot of $T\chi_{\text{imp}}(T)$ versus the temperature T where the dots represent the renormalization group results (see figure 4.7) (Andrei et al, 1983).

Using the independence of the χ/γ ratio of the magnetic field,

$$\frac{\partial^2 \gamma_{\text{imp}}(H)}{\partial H^2} = \frac{4\pi^2 k_{\text{B}}^2}{3(g\mu_{\text{B}})^2} \frac{\partial^2 \chi_{\text{imp}}(0,H)}{\partial H^2}. \tag{6.78}$$

The right hand side of (6.78) can be deduced from (6.33). Substituting the result for $H \to 0$, we find

$$\chi_{\text{imp}}(T) = \chi_{\text{imp}}(0) \left\{ 1 - \frac{\sqrt{3}\pi^3 w^2}{4} \left(\frac{T}{T_{\text{K}}}\right)^2 + \text{O}\left(\frac{T}{T_{\text{K}}}\right)^4 \right\}. \tag{6.79}$$

The fact that $\chi_{\text{imp}}(T)$ initially decreases with T follows from the negative H^3 coefficient in the expansion for $M_{\text{imp}}(H)$.

To determine $\chi_{\text{imp}}(T)$ and $C_{\text{imp}}(T)$ over the complete temperature range equations (6.56)–(6.61) have to be solved numerically. This has been done by several groups. There are well controlled approximations for truncating the equations and the results from the different groups agree to within 1%.

Results for $\chi_{\text{imp}}(T)$ and $C_{\text{imp}}(T)$ are shown in figures 6.4 and 6.5. They confirm the the numerical renormalization group results of Wilson for $\chi_{\text{imp}}(T)$, and those of Oliviera & Wilkins (1981) for $C_{\text{imp}}(T)$ (these latter calculations were not strictly speaking for the s-d model, due to the computational problem of retaining sufficient states. They used a mapping to a spinless interacting resonant level model which reduced the computation to a manageable level).

6.5 Results for the s-d Model ($S > \frac{1}{2}$)

The Bethe ansatz method can be generalized to give exact results for the ground state properties and thermodynamics of many other models of magnetic impurities. Results for the general spin s-d model can be

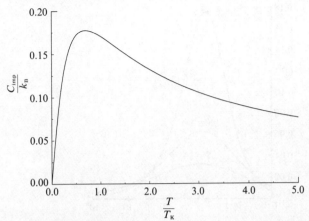

Figure 6.5 The specific heat $C_{\mathrm{imp}}(T)$ as a function of T/T_{K} using the results of Desgranges and Schotte (1982).

derived along the lines outlined for the $S = \frac{1}{2}$ case. There is an important difference for $S > \frac{1}{2}$ in the low temperature behaviour due to incomplete compensation of the impurity spin at $T = 0$. The ground state is characterized by a total spin $S - \frac{1}{2}$, as was originally proved by Mattis (1967). At high fields and $T = 0$, the impurity magnetization approaches its saturation value $g\mu_{\mathrm{B}}S$,

$$M_{\mathrm{mp}}(H) = g\mu_{\mathrm{B}}S\left\{ 1 - \frac{1}{2\ln(g\mu_{\mathrm{B}}H/k_{\mathrm{B}}T_{\mathrm{H}})} - \frac{\ln\ln(g\mu_{\mathrm{B}}H/k_{\mathrm{B}}T_{\mathrm{H}})}{4(\ln(g\mu_{\mathrm{B}}H/k_{\mathrm{B}}T_{\mathrm{H}}))^2} \right.$$
$$\left. +O(\ln(g\mu_{\mathrm{B}}H/k_{\mathrm{B}}T_{\mathrm{H}}))^{-3} \right\},$$

(6.80)

for $g\mu_{\mathrm{B}}H/k_{\mathrm{B}}T_{\mathrm{H}} \gg 1$ as in the $S = \frac{1}{2}$ case. As the field is reduced there is a crossover to the strong coupling regime and a partially reduced moment $g\mu_{\mathrm{B}}(S - \frac{1}{2})$ for $g\mu_{\mathrm{B}}H/k_{\mathrm{B}}T_{\mathrm{H}} \ll 1$. The approach to this value is logarithmic, in contrast to the the low field behaviour of the $S = \frac{1}{2}$ model,

$$M_{\mathrm{imp}}(H) = g\mu_{\mathrm{B}}\left(S - \frac{1}{2}\right)\left\{ 1 - \frac{1}{2\ln(g\mu_{\mathrm{B}}H/k_{\mathrm{B}}T_{\mathrm{H}})} - \frac{\ln\ln(g\mu_{\mathrm{B}}H/k_{\mathrm{B}}T_{\mathrm{H}})}{4(\ln(g\mu_{\mathrm{B}}H/k_{\mathrm{B}}T_{\mathrm{H}}))^2} \right.$$
$$\left. +O(\ln(g\mu_{\mathrm{B}}H/k_{\mathrm{B}}T_{\mathrm{H}}))^{-3} \right\},$$

(6.81)

The energy scale $k_{\mathrm{B}}T_{\mathrm{H}}$ is chosen so as to eliminate any term of order $\ln(g\mu_{\mathrm{B}}H/k_{\mathrm{B}}T_{\mathrm{H}})^{-2}$ and is the same in the low and high field results. The

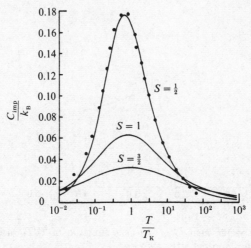

Figure 6.6　The specific heat $C_{\mathrm{imp}}(T)$ as a function of $\log_{10} T$ for spin values $\frac{1}{2}$, 1, $\frac{3}{2}$ (Rajan et al, 1982).

logarithmic approach in the low field case can be explained by a reduced spin $S - \frac{1}{2}$ having a ferromagnetic interaction with the conduction electrons. The approach to the saturated value as $H \to 0$ is then similar to the $S = \frac{1}{2}$ model with ferromagnetic coupling which is also logarithmic.

Results for the temperature dependence of the specific heat $C_{\mathrm{imp}}(T)$ are shown in figure 6.6 on a $\log_{10} T$ scale for $S = \frac{3}{2}, 1$ and $\frac{1}{2}$. There is a clear broadening and reduction of the specific heat peak with increasing S. For further details of the results of the $S > \frac{1}{2}$ model we refer the reader to the reviews of Andrei et al (1983), and Tsvelick & Wiegmann (1983), which contain references to the original papers.

The N-fold degenerate Coqblin–Schrieffer model, which has application to rare earth impurity systems, can also be diagonalized by the Bethe ansatz. This again follows very much along the lines outlined for the $S = \frac{1}{2}$ s-d model. The Coqblin–Schrieffer model has SU(N) symmetry and the s-d model corresponds to the case $N = 2$. The diagonalization of the subsidiary spin problem (6.17) to the SU(N) case was generalized by Sutherland (1968) and the solution appropriate to the Coqblin–Schrieffer model can be deduced from Sutherland's result. There are also generalizations of the inverse scattering method of solution (see appendix F) by Kulish & Reshetikhin (1981). The next chapter is devoted entirely to the degenerate impurity models, so we discuss the results there. We now consider the application of the Bethe ansatz to the Anderson model.

6.6 Integrability of the Anderson Model

Wiegmann (1981) showed that the Bethe ansatz approach could be extended to give exact results for the ground state and thermodynamic properties of the Anderson model. The assumptions made for the applicability of the method are basically the same as those for the s-d model, linear dispersion for the conduction electrons and point interactions. With these assumptions the Anderson model can be expressed in the form,

$$H_A = \int \left\{ -i \sum_\sigma c_\sigma^\dagger(x) \frac{\partial c_\sigma(x)}{\partial x} + V\delta(x)(c_\sigma^\dagger(x)c_{d,\sigma} + c_{d,\sigma}^\dagger c_\sigma(x)) \right\} dx$$
$$+ \sum_\sigma \epsilon_d n_{d,\sigma} + U n_{d,\uparrow} n_{d,\downarrow},$$

$$(6.82)$$

where V is the hybridization parameter. To prove integrability we follow similar arguments to those used in section 6.1 and look at situations with only a few particles present.

Consider first of all the one particle case. We look for an eigenstate of energy $E = k$ of the form,

$$|k,\sigma\rangle = a_{k,\sigma}^\dagger |0\rangle = \left\{ \int_{-\infty}^\infty g_k(x) c_\sigma^\dagger(x) \, dx + e_k c_{d,\sigma}^\dagger \right\} |0\rangle, \qquad (6.83)$$

where $|0\rangle$ is the vacuum state. As we are using a mixed basis set we continue using the second quantized form rather than the first quantized form used earlier. Substituting (6.83) into the Schrödinger equation gives

$$\left(-i \frac{\partial}{\partial x} \right) g_k(x) + V\delta(x) e_k = 0,$$

$$(\epsilon_d - k)e_k + V g_k(0) = 0, \qquad (6.84)$$

where $g_k(0) = (g_k(0^+) + g_k(0^-))/2$.

The solution is

$$g_k(x) = e^{ikx}(e^{i\delta(k)}\theta(x) + e^{-i\delta(k)}\theta(-x)),$$

$$e_k = -\frac{2}{V} \sin \delta(k), \qquad (6.85)$$

where

$$\delta(k) = \tan^{-1}\left(\frac{V^2}{2(\epsilon_d - k)} \right). \qquad (6.86)$$

This describes the scattering of a state in which the particle experiences a phase shift $2\delta(k)$ on passing the origin.

For $U = 0$ the two particle solution is $a^\dagger_{k_1,\sigma_1} a^\dagger_{k_2,\sigma_2}|0\rangle$ is an eigenstate with energy $k_1 + k_2$. This will also be a solution for $\sigma_1 = \uparrow, \sigma_2 = \uparrow$, and $\sigma_1 = \downarrow, \sigma_2 = \downarrow$, as the interaction is between particles of opposite spin. For $S_z = 0$ and $U = 0$ the general two particle state must be of the form,

$$|\Psi\rangle = \left\{ \int g(x_1, x_2) c^\dagger_\uparrow(x_1) c^\dagger_\downarrow(x_2)\, dx_1 dx_2 \right.$$

$$\left. + \int e(x)(c^\dagger_\uparrow(x) c^\dagger_{d,\downarrow} - c^\dagger_{d,\uparrow} c^\dagger_\downarrow(x))\, dx + f c^\dagger_{d,\uparrow} c^\dagger_{d,\downarrow} \right\}|0\rangle.$$

$$(6.87)$$

To satisfy the Schrödinger equation for energy E,

$$\left\{ -i\left(\frac{\partial}{\partial x_1} + \frac{\partial}{\partial x_1} \right) - E \right\} g(x_1, x_2) + V(\delta(x_1)e(x_2) + \delta(x_2)e(x_1)) = 0,$$

$$\left(-i\frac{\partial}{\partial x} - E - \epsilon_d) \right) e(x) + V g(0, x) + V\delta(x)f = 0,$$

$$(U + 2\epsilon_d - E)f + 2Ve(0) = 0. \qquad (6.88)$$

For $U = 0$ there is a singlet solution corresponding to

$$g_{k_1,k_2}(x_1, x_2) = g_{k_1}(x_1)g_{k_2}(x_2) + g_{k_1}(x_2)g_{k_2}(x_1),$$

$$e_{k_1,k_2}(x) = g_{k_1}(x)e_{k_2} + g_{k_2}(x)e_{k_1}, \qquad f_{k_1,k_2} = 2e_{k_1}e_{k_2}. \qquad (6.89)$$

In the spirit of the Bethe ansatz we require a solution $(U \neq 0)$ for all states as linear combinations of products of single particle states for a given set of ks. This is possible in this model, as for the s-d model, by exploiting the fact that for linear dispersion we are not required to impose continuity conditions at $x_1 = x_2$. We can include terms of the form,

$$r(k_1, k_2)\left\{(g_{k_1}(x_1)g_{k_2}(x_2) - g_{k_1}(x_2)g_{k_2}(x_1))(\theta(x_1 - x_2) - \theta(x_2 - x_1)\right\},$$

$$(6.90)$$

with those corresponding to $U = 0$ to form a more general ansatz for $g(x_1, x_2)$ for the two particle singlet state, and similarly for $e(x)$ include

$$r(k_1, k_2)(e_{k_1}g_{k_2}(x) - e_{k_2}g_{k_1}(x))(\theta(x) - \theta(-x)). \qquad (6.91)$$

The generalized ansatz for the singlet state satisfies (6.88) if

$$r(k_1, k_2) = \frac{i}{2(\alpha(k_1) - \alpha(k_2))}, \qquad \alpha(k) = \frac{(k - \epsilon_d - U/2)^2}{2U\Delta}, \qquad (6.92)$$

with $E = k_1 + k_2$. This solution for $g(x_1, x_2)$ has a factor,

$$\left(1 - \frac{i}{2(\alpha(k_1) - \alpha(k_2))} \right)\theta(x_1 - x_2) + \left(1 + \frac{i}{2(\alpha(k_1) - \alpha(k_2))} \right)\theta(x_2 - x_1),$$

$$(6.93)$$

so that the wavefunction changes by a factor,

$$\frac{2(\alpha(k_1) - \alpha(k_2)) - i}{2(\alpha(k_1) - \alpha(k_2)) + i} = e^{i\delta_{12}}, \qquad (6.94)$$

when the particles pass one another, corresponding to a phase shift δ_{12}. This can be described by a unitary scattering matrix $\mathbf{S}_{j,j'}$,

$$\mathbf{S}_{j,j'} = \frac{2(\alpha(k_j) - \alpha(k_{j'}))\mathbf{I} + i\mathbf{P}_{j,j'}}{2(\alpha(k_j) - \alpha(k_{j'})) + i}, \qquad (6.95)$$

with $j = 1$, $j' = 2$, where the operator $\mathbf{P}_{j,j'}$ permutes the spin indices. This is included to cover both the singlet ($\mathbf{P}_{1,2} = -\mathbf{I}$ and the triplet ($\mathbf{P}_{1,2} = \mathbf{I}$) cases, reflecting the fact that the $r(k_1, k_2)$ is only needed in the singlet case so $\mathbf{S} = \mathbf{I}$ in the triplet case. Having built up a solution from products of single particle solutions, using just two wavevectors k_1, k_2, we are now in a position to make an ansatz for a general N_e particle solution. The ansatz for $g_{\sigma_1, \dots \sigma_{N_e}}(x_1, x_2, \dots x_{N_e})$ for the solution constructed from a set of wavevectors, $\{k_j\}$ $j = 1, 2, \dots N_e$, is

$$g_{\sigma_1 \dots \sigma_{N_e}}(x_1, x_2, \dots x_{N_e}) = \sum_{P, Q, \sigma's} A_{\sigma_1, \dots \sigma_{N_e}}(P|Q) \prod_{j=1}^{N_e} g_{k_Q}(x_j)\theta(P), \qquad (6.96)$$

where P and Q are permutations of the integers $[1, 2, \dots N_e]$, P denotes the spatial region (as earlier), Q the set of momentum indices of the gs at $x_1, x_2, \dots x_{N_e}$, and the σs the spin states. A similar ansatz for $e_{\sigma_1 \dots \sigma_{N_e}}(x_1, x_2, \dots x_{N_e - 1})$ can be made. (Note that earlier for the s-d model we were able to assume As independent of the ks, and hence independent of Q except for the antisymmetry requirement, so we could completely factorize the spin and spatial parts of the wavefunction. This was due to the fact that the phase shifts were independent of k, and it does not apply here.)

It can be shown that the wavefunction is antisymmetric if the coefficients satisfy

$$A_{\sigma_1, \dots \sigma_{N_e}}(P|Q) = (-1)^q A_{\sigma'_1, \dots \sigma'_{N_e}}(PQ), \qquad (6.97)$$

where PQ is the product of the permutations P and Q, $[\sigma'_1, \dots \sigma'_{N_e}] = Q[\sigma_1, \dots \sigma_{N_e}]$ and q is the parity of the Q permutation.

The coefficients for the different regions are linked through the Schrödinger equation. If $x_j = x'_j$ is the boundary between the regions P and P', then the coefficients in the two regions are related by the matrix $\mathbf{S}_{j,j'}$ defined in (6.95) as in the two particle case. A consistency problem now arises as it is possible to relate coefficients by different routes. For example, for three particles the regions, $P = [123]$ and $P' = [321]$, can be related by either $123 \to 231 \to 213 \to 123$ or $321 \to 312 \to 132 \to 123$. Consistency is achieved only if $\mathbf{S}_{j,j'}$ satisfies

$$\mathbf{S}_{1,2}\mathbf{S}_{1,3}\mathbf{S}_{2,3} = \mathbf{S}_{2,3}\mathbf{S}_{1,3}\mathbf{S}_{1,2}. \qquad (6.98)$$

This is the well known Yang–Baxter relation (see appendix F, equation (F.5)) for integrability and a Bethe ansatz solution. It is a sufficient condition for the two particle factorizability of the many particle scattering $\mathbf{S}$ matrix. It ensures consistency of the equations for the coefficients in the general N_e particle case. For $\mathbf{S}_{j,j'} = a_{jj'}\mathbf{I} + b_{jj'}\mathbf{P}_{j,j'}$, where $a_{jj'} + b_{jj'} = 1$, to satisfy (6.98) the coefficients must be related via

$$\frac{a_{21}}{b_{21}} + \frac{a_{32}}{b_{32}} = \frac{a_{31}}{b_{31}}. \qquad (6.99)$$

As $a_{jj'}/b_{jj'} \propto (\alpha_j - \alpha_{j'})$, (6.98) is satisfied, and the model is integrable.

If now periodic boundary conditions are imposed the wavevectors k_j are required to satisfy

$$\mathbf{S}_{j,j+1}...\mathbf{S}_{j,N_e}\mathbf{S}_{j,1}...\mathbf{S}_{j,j-1}\mathbf{A}(\mathbf{I}) = e^{(ik_j L + 2\delta(k_j))}\mathbf{A}(\mathbf{I}). \qquad (6.100)$$

This is generated by $x_j \to x_j + L$ starting in the region $\mathbf{P} = \mathbf{I}$ so there is a scattering matrix as the jth particle passes each of the particles on the ring in turn. As $\mathbf{S}_{j,j'}$ acts only on the spin indices we have a subsidiary spin problem similar in form to (6.17). It is now, in fact, the same as that of Yang (1967). The same condition (6.98) is required to be able to solve this problem via the algebraic Bethe ansatz. The solution is given in appendix F. Integrability at the first stage resulted in a set of conserved quantities, the set of ks, the wavevectors associated with the charge excitations. Integrability of the subsidiary spin problem leads to a conserved set of wavevectors $\{\lambda_r\}$ $r = 1, 2, ...M$, associated with the spin excitations, where M is the total number of reversed spins from the fully aligned state. The equations which determine these wavevectors are

$$Lk_j + 2\delta(k_j) = 2\pi n_j - \sum_{s=1}^{M}(\theta(2\lambda_s - 2\alpha(k_j)) + \pi), \qquad (6.101)$$

$$\sum_{j=1}^{N_e}\theta(2\lambda_r - 2\alpha(k_j)) = 2\pi I_r + \sum_{s=1}^{M}\theta(\lambda_r - \lambda_s). \qquad (6.102)$$

The total energy is

$$E = \sum_{j=1}^{N_e} k_j. \qquad (6.103)$$

A new feature of these equations compared to the earlier ones for the s-d model, is that there are solutions with complex ks. In the ground

state solution, as $N_e, M \to \infty$, the λs are all real and distinct as for the
s-d model, but there are complex ks of the form,

$$\alpha(k_r^\pm) = \lambda_r \pm \frac{i}{2}, \qquad (6.104)$$

associated with real wavevectors λ_r of the spins (Kawakami & Okiji,
1981). These correspond to bound singlet pairs (complex ks are found
in one dimensional models with attractive interactions associated with
bound states). The energy is reduced by these bound singlet pairs, and
the ground state has the maximum number M, one for each λ_r, leaving
$N_e - 2M$ real ks associated with unbound charge excitations. The total
energy then takes the form,

$$E = \sum_{j=1}^{N_e - 2M} k_j + \sum_{r=1}^{M} x(\lambda_r), \qquad (6.105)$$

where $x(\lambda_r)$ is the real part of $k_r^\pm$ given by (6.104). In the absence of
a magnetic field the ground state is a singlet constructed purely from
bound state pairs, corresponding to consecutive values of $\{I_r\}$.

In the continuum limit linear integral equations can be derived for
the densities of states, $\rho(k)$ and $\sigma(\lambda)$, for the two types of excitations.
Wiener–Hopf solutions of these equations lead to exact expressions for
the d level occupation, n_d, and the impurity susceptibility χ_{imp}. We
shall discuss these results in the next two sections. For the details of
their derivation we refer the reader to Wiegmann & Tsvelick (1983) and
the review of Tsvelick & Wiegmann (1983) and the references therein.

It is possible to define spinon and holon excitations from the ground
state. Removing one bound state of wavevector λ^h from the ground
state, and calculating the new distribution $\sigma(\lambda)$ (dependent on λ^h), leads
to an expression for the energy $\epsilon(\lambda^h)$ of this excitation. The new distribution is such that there is a change of charge e but no change of spin,
so this excitation is termed a *holon*. Using (6.39) a density of states
$D_c(\epsilon)$ of this excitation can be defined. Adding an electron with quasi-
momentum k to the ground state leads to a new particle distribution
$\sigma(\lambda)$. On solving for $\sigma(\lambda)$ this excitation is found to carry no charge
but has a spin $\frac{1}{2}$, and so is a spinon. The density of states $D_s(\epsilon)$ can
be calculated as earlier using (6.39). The densities of states of these two
types of excitations have been calculated numerically by Kawakami &
Okiji (1990) for the symmetric model, and their results are shown in
figure 6.7 for various values of U.

They clearly illustrate that the sharp resonance at the Fermi level for
larger U is due to the spin excitations, while the charge peak, which is

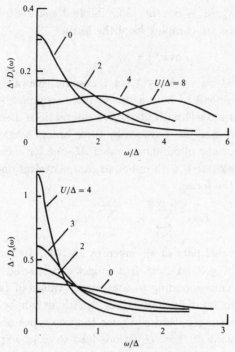

Figure 6.7 Excitation spectra for the local holon $D_c(\omega)$ and spinon $D_s(\omega)$ for the symmetric Anderson model (Kawakami & Okiji, 1990).

at the Fermi level for $U = 0$, remains broad and is shifted to energies $\epsilon \sim U/2$. There is a clear correspondence with the results of Yosida & Yamada which we showed earlier for the spectral density of the one electron Green's function in which both features were present (see figure 5.7). In the Kondo limit the expression of Kawakami & Okiji for $D_s(\epsilon)$ agrees with that for the s-d model (6.40).

For the thermodynamics all possible types of excitations have to be taken into account. As well as the n-string complex solutions, $\lambda_r^{(n,j)}$, associated with bound spins considered earlier, there are charge complexes involving n particles. These have complex ks and are associated with an n-string solution $\lambda_r^{(n,j)}$, their wavevectors $k_r^{\pm(n,j)}$ are given by an equation of the form (6.104) with λ_r replaced by $\lambda_r^{(n,j)}$. The quantum numbers associated with each set of excitations is distinct. Particle densities for the unbound charge excitations, $\rho(k)$, the n-charge complex, $\sigma'(\lambda')$ and the n-string excitations $\sigma(\lambda)$, and densities of the corresponding holes, $\rho^h(k)$, $\sigma'^h(\lambda')$ and $\sigma^h(\lambda)$, can be defined. Setting up a free energy functional and minimizing with respect to these, subject to constraints, leads to an infinite set of non-linear integral equations which determine the equilibrium densities of states, from which the thermodynamic behaviour can be calculated as in the s-d model. We outline some of the results in the next two sections.

6.7 Results for the Symmetric Anderson Model

In the particle–hole symmetric model $\epsilon_d = U/2$, and $n_d = 1$, independent of U. The susceptibility, however, can vary from 'non-magnetic' behaviour for small U/Δ to 'magnetic' Kondo behaviour for $U/\Delta \gg 1$. The full expression for χ_{imp} is given by Wiegmann & Tsvelick (1983) and can be expressed in various forms. For small U/Δ it is most usefully expressed in the form of a power series,

$$\chi_{\text{imp}} = \frac{(g\mu_B)^2}{2\pi\Delta} \sum_{n=0}^{\infty} C_n \left(\frac{U}{\pi\Delta}\right)^n, \qquad (6.106)$$

where the coefficients satisfy the recurrence relation,

$$C_n = (2n-1)C_{n-1} - (\pi/2)^2 C_{n-2}, \qquad (6.107)$$

with $C_0 = C_1 = 1$. The coefficients up to C_4 correspond to the fourth order perturbations results of Yosida & Yamada quoted earlier. The coefficients fall off rapidly with n which explains why the perturbational results to this order can be used for values of $U/\pi\Delta \sim 2$, in the strong coupling regime. For large n the ratio of the coefficients C_{n+1}/C_n tends to $(\pi/2)^2/(2n+3)$, and hence the series is absolutely convergent for $U < \infty$ (Horvatić & Zlatić, 1985). This gives an a postiori justification for the validity of the perturbation theory in powers of U used in chapter 5.

Though (6.106) holds for arbitrarily large U it is more useful to express χ_{imp} in the form,

$$\chi_{\text{imp}} = \frac{(g\mu_B)^2}{4k_B T_L} \left(1 + \frac{1}{\sqrt{\pi}} \int_0^{\pi\Delta/2U} \frac{e^{x - \pi^2/16x}}{\sqrt{x}} \, dx\right), \qquad (6.108)$$

where

$$k_B T_L = U \left(\frac{\Delta}{2U}\right)^{1/2} e^{-\pi U/8\Delta + \pi\Delta/2U}. \qquad (6.109)$$

As U is large χ_{imp} is clearly dominated by the first term as the integral vanishes as $U \to \infty$, and is small for $U/\pi\Delta > 1$. Hence T_L can be identified as the Kondo temperature apart from a constant. The first term in the exponent corresponds to the usual expression for T_K with $J = 4V^2/U$, as given by the Schrieffer–Wolff transformation (1.73). The second term in the exponent is small for U large and corresponds to a term $O(J)$ in the definition of T_K in equation (3.47). The prefactor differs in that D is replaced by U. This is to be expected because in the linear dispersion model we are dealing with the regime $D \gg U$. As noted earlier in the poor man's scaling, no scaling or renormalization occurs as $\tilde{D}$ is reduced until $\tilde{D} \sim U$, because on an energy scale much greater than U the impurity is effectively non-interacting. Hence the relevant cut-off in this case is U rather than D.

When the numerical prefactor in (6.109) is compared with that for T_K deduced from the renormalization group calculations of Krishna-murthy et al (1980), $T_K/T_L = 4 \times 0.182/\sqrt{\pi} = 0.4107$, in precise agreement with the Wilson number for the s-d model (6.75).

In the large U limit (6.108) can be rewritten as a Kondo term plus corrections in Δ/U,

$$\chi_{\text{imp}} = \frac{(g\mu_B)^2}{4k_B T_L} + \frac{(g\mu_B)^2 \Delta}{\pi U^2} \left\{ 1 - \frac{6}{\pi} \left(\frac{2\Delta}{U} \right) + \left(\frac{60}{\pi^2} - 1 \right) \left(\frac{2\Delta}{U} \right)^2 ... \right\}. \tag{6.110}$$

The charge susceptibility $\chi_{c,\text{imp}}$ and the linear coefficient of specific heat γ_{imp} can also be expressed in similar forms,

$$\chi_{c,\text{imp}} = \frac{2}{\pi \Delta} \sum_{n=0}^{\infty} C_n (-1)^n \left(\frac{U}{\pi \Delta} \right)^n, \tag{6.111}$$

and

$$\gamma_{\text{imp}} = \frac{2\pi k_B^2}{3\Delta} \sum_{n=0}^{\infty} C_{2n} \left(\frac{U}{\pi \Delta} \right)^{2n}, \tag{6.112}$$

where the coefficients are given by (6.107). The Fermi liquid relation relating these to χ_{imp} can be verified directly from (6.106), (6.111) and (6.112).

The charge susceptibility tends to zero for large U as Δ/U^2, and has the asymptotic expansion,

$$\chi_{c,\text{imp}} = \frac{8\Delta}{\pi U^2} \left\{ 1 - \frac{6}{\pi} \left(\frac{2\Delta}{U} \right) + \left(\frac{60}{\pi^2} - 1 \right) \left(\frac{2\Delta}{U} \right)^2 ... \right\}. \tag{6.113}$$

The specific heat coefficient γ_{imp} is dominated by the Kondo term arising from the spin fluctuations for large U,

$$\gamma_{\text{imp}} = \frac{\pi^2 k_B}{6T_L} + \frac{8\pi k_B^2 \Delta}{3U^2} \left\{ 1 + \left(\frac{60}{\pi^2} - 1 \right) \left(\frac{2\Delta}{U} \right)^2 ... \right\}, \tag{6.114}$$

with correction terms of the order Δ/U^2. The crossover from the non-magnetic to the Kondo behaviour occurs rather rapidly for $U/\pi\Delta > 2$.

For varying magnetic field strength, $T = 0$, there are a number

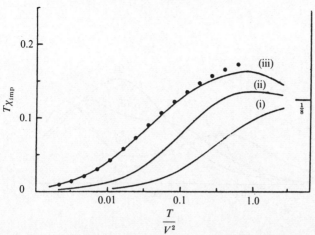

Figure 6.8   Plots of $T\chi_{\mathrm{imp}}(T)$ versus $\log_{10}(T/V^2)$ for the symmetric Anderson model where (i) $U/V^2 = 0$, (ii) 2.0 and (iii) 4.0. The points are the results for the renormalization group calculation for the s-d model (see figure 4.7) (Okiji & Kawakami, 1983).

of regimes. These regimes are related to the renormalization group fixed points, with smoothly varying behaviour from one regime to another. For $U \gg \Delta$, and $H < H_0$, where $\mu_{\mathrm{B}}H_0 \sim (U\Delta)^{1/2}$, the behaviour is identical to that of the s-d model, a strong coupling regime for $\mu_{\mathrm{B}}H < k_{\mathrm{B}}T_{\mathrm{K}}$, in which the impurity moment is compensated with a large magnetic response, $\chi_{\mathrm{imp}} \sim \mu_{\mathrm{B}}^2/T_{\mathrm{K}}$, passing over to a localized moment for $\mu_{\mathrm{B}}H_0 > \mu_{\mathrm{B}}H > k_{\mathrm{B}}T_{\mathrm{K}}$ with $M_{\mathrm{imp}} = g\mu_{\mathrm{B}}/2$ with logarithmic corrections. Eventually for $H \gg H_0$ there is a free orbital regime in which the corrections to $M_{\mathrm{imp}} = g\mu_{\mathrm{B}}/2$ are of order $\Delta/\mu_{\mathrm{B}}H$.

For $U < \Delta$, $H < H_0$ there is only a weak magnetic response with $\chi_{\mathrm{imp}} \sim \mu_{\mathrm{B}}^2/\Delta$, which passes over to a free orbital regime for $H > H_0$.

The susceptibility has been calculated as a function of temperature by Okiji & Kawakami (1983), and the results are shown in figure 6.8.

One sees here that the behaviour as a function of temperature is similar to that as a function of field. There is an enhancement of the low temperature response with increasing U/Δ, agreeing with that for the s-d model for large U, and an approach to free orbital response for $k_{\mathrm{B}}T > \Delta$.

The corresponding curves for the specific heat are shown in figure 6.9.

With increasing U a many-body peak associated with the spin fluctuations emerges on the low temperature side, agreeing with that for the

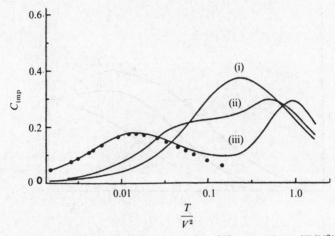

Figure 6.9   Plots of the specific heat $C_{\text{imp}}(T)$ versus $\log_{10}(T/V^2)$ for the symmetric Anderson model where (i) $U/V^2 = 0$, (ii) 2.0 and (iii) 4.0. The points are the values for the s-d model (see figure 6.5) (Okiji & Kawakami, 1983).

s-d model for larger U. The peak associated with the charge fluctuations gets pushed to higher temperatures.

6.8 Results for the Asymmetric Anderson Model

If we now consider the asymmetric model for large fixed U, and move away from the symmetric case $\epsilon_d = -U/2$, increasing ϵ_d up to the Fermi level, we move out of the Kondo regime into the mixed valence regime $\epsilon_d \sim 0$.

The consequent change in susceptibility is clearly shown in figure 6.10, where $\chi_{\text{imp}}(T)$ is shown for various values of ϵ_d. The large susceptibility seen in the Kondo regime is suppressed for $k_B T < \Delta$, and the results at $\epsilon_d \sim 0$ look qualitatively similar to those of the symmetric model for $U = 0$. The results for the specific heat as a function of temperature shown in figure 6.11 show the same qualitative trend as $\epsilon_d \to 0$. The two peak structure in the Kondo regime merges into a single peak, qualitatively similar to that of the symmetric model at $U = 0$.

Results for the induced magnetic moment as a function of field are shown in figure 6.12 for various values of U and ϵ_d.

In the regime which is dominated by Kondo behaviour T_L is given by

$$T_{\text{L}} = U \left(\frac{\Delta}{2U} \right)^{1/2} e^{-\pi |\epsilon_d| |\epsilon_d + U| / 2U\Delta}. \tag{6.115}$$

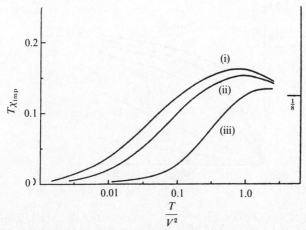

Figure 6.10   Plots of $T\chi_{\text{imp}}(T)$ versus $\log_{10}(T/V^2)$ for the asymmetric Anderson model for $U/V^2 = 4.0$, and values of ϵ_{d}/V^2 of (i) -2.0, (ii) -1.0 and (iii) 0.0 (Okiji & Kawakami, 1983).

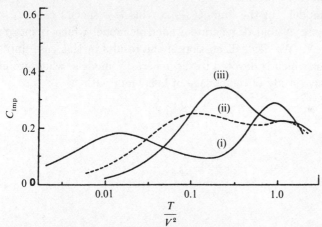

Figure 6.11   Plots of the specific heat $C_{\text{imp}}(T)$ versus $\log_{10}(T/V^2)$ for the asymmetric Anderson model for $U/V^2 = 4.0$, and values of ϵ_{d}/V^2 of (i) -2.0, (ii) -1.0 and (iii) 0.0 (Okiji & Kawakami, 1983).

The exponent is of the form expected from the Schrieffer–Wolff transformation (1.73) with $D = U$.

In the mixed valence regime the scaling results derived in chapter 3 are verified. The effective d level in this regime, ϵ_{d}^*, is given by (3.66) with $D = U$. The mixed valence regime corresponds to $\epsilon_{\text{d}}^* \sim 0$.

The solution we have considered here is for the regime $D \gg U, \epsilon_{\text{d}}$.

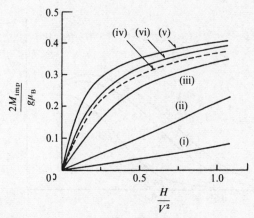

Figure 6.12 Plots of the induced magnetic moment of $M_{\mathrm{imp}}(H)/(g\mu_{\mathrm{B}}/2)$ as a function of magnetic field H for the for $\epsilon_{\mathrm{d}}/V^2 = -1.0$ and values of U/V^2 of (i) -0.0, (ii) 0.7, (iii) 1.5, (iv) 2.0, (v) 6.5 and (vi) 15.0 (Okiji and Kawakami, 1982).

There is another parameter regime $U \gg D \gg \epsilon_{\mathrm{d}}$ for the asymmetric Anderson model. In the limit $U \to \infty$ this is a special case, $N = 2$, of the $U = \infty$ N-fold degenerate Anderson model which is integrable for general N. We defer discussion of the results in this case until the next chapter which is devoted to the general N models, where we make a comparative study of the change of behaviour with N.

7

N-fold Degenerate Models I

7.1 Introduction

The approaches developed so far, the renormalization group, Fermi liquid and the Bethe ansatz solutions, have led to rather a complete picture of the ground state and thermodynamic behaviour of the s-d model for spin $\frac{1}{2}$ and the non-degenerate Anderson model. We now turn our attention to the N-fold degenerate models, the Coqblin–Schrieffer model (1.89), and the N-fold degenerate Anderson model with $U = \infty$ (1.84). The Hamiltonians for these models are

$$H = \sum_{k,m} \epsilon_k c^\dagger_{k,m} c_{k,m} + \sum_{k,k',m,m'} J_{k,k'} X_{m,m'} c^\dagger_{k',m'} c_{k,m}, \tag{7.1}$$

for the Coqblin–Schrieffer model of a localized magnetic moment, arising from an N-fold degenerate spin–orbit split multiplet, interacting via an exchange interaction with the conduction electrons, and

$$H = E_0 |0,0\rangle\langle 0,0| + \sum_m E_{1,m} |1,m\rangle\langle 1,m|$$

$$+ \sum_{k,m} \epsilon_k c^\dagger_{k,m} c_{k,m} + \sum_{k,m} (V_k |1,m\rangle\langle 0,0| c_{k,m} + V_k^* c^\dagger_{k,m} |0,0\rangle\langle 1,m|),$$

$$\tag{7.2}$$

for the corresponding Anderson model ($U = \infty$), where there is an explicit hybridization term V between the multiplet and conduction electrons so that charge fluctuations of the impurity can occur in this model. The notation was explained in chapter 1.

These models are directly applicable to rare earth impurity systems, particularly those containing cerium and ytterbium impurities. The techniques already developed, such as the Fermi liquid and Bethe ansatz approaches, are applicable to this more general class of models, and provide exact results for the ground state and thermodynamics. However, with the introduction of the new variable N, the degeneracy factor, some new techniques become available. These are $1/N$ expansions and large N methods which have been successfully applied in other areas of physics such as critical phenomena and field theory. They will be useful in this context, going beyond the range of the exact techniques, because they will provide ways of calculating the dynamic response functions, such as the one electron Green's functions and dynamic susceptibility. This will make possible a more extensive comparison between the theoretical predictions and the experimental results. The large N approaches give new insights from the simplifications that are found in the large N limit. There is a mean field theory which becomes exact in this limit, which will be described in section 7.6. There are also ways of calculating corrections to this mean field limit which we shall describe in the next chapter. These simple approximation schemes work very well for values of N that are of interest in applications to rare earth impurities. For example, the lowest spin-orbit split multiplet for $Ce\ 4f^1$ has $j = \frac{5}{2}$, corresponding to $N = 6$, and for $Yb\ 4f^{13}$, $j = \frac{7}{2}$, corresponding to $N = 8$. Even for $N = 2$ one can get good semi-quantitative results, which can be accurate to within a few percent for some quantities.

In the next section we introduce a diagrammatic perturbation expansion in powers of the hybridization V. We can classify terms in this by the dependence on N, and sum subseries to obtain leading terms in a $1/N$ expansion. We then look at some of the exact results derived via the Bethe ansatz, and make a comparison between these and the predictions based on the $1/N$ expansion to see how well the approach works in practice. We then look at the Fermi liquid methods developed in chapter 5 and find that they give an interpretation of the results in terms of a resonance in the quasi-particle density of states above the Fermi level. In the last section we look at a mean field approach based on 'slave bosons'. This approach gives the Fermi liquid picture in the large N limit with explicit expressions for the quasi-particle parameters. Further developments of these techniques, particularly with a view to the calculation of response functions, are dealt with in chapter 8.

7.2 Perturbation Theory and the 1/N Expansion

The form of perturbation theory we considered earlier in chapter 5 was an expansion in powers of the Coulomb interaction U, and hence an expansion about the non-interacting limit $U = 0$, so that the standard Feynman diagram method based on Wick's theorem could be used. The main disadvantage of this approach is that, in systems of physical interest, U is usually one of the larger energy scales (~ 5 eV) in the problem, so that the results of perturbation theory to finite order are in general not good enough (though see section 5.5). Summation of the RPA diagrams results in a divergence at $U = U_c$, and no-one has so far managed to sum an infinite subsequence of diagrams to give reliable results in the strong correlation regime. The hybridization parameter V would seem to be a much more useful expansion parameter, as in the systems of interest, it is usually the smallest of the relevant energy scales. The main disadvantage is that $V = 0$ corresponds to the 'atomic' limit,

$$H_0 = \sum E_\alpha |\alpha\rangle\langle\alpha| + \sum_{k,m} \epsilon_k c_{k,m}^\dagger c_{k,m}, \qquad (7.3)$$

where the sum over $|\alpha\rangle$ includes the states $|0,0\rangle$ and $|1,m\rangle$. These many-body states include the Coulomb interaction, so the standard diagrammatic methods which use Wick's theorem are not applicable. The standard methods are based on the commutation rule $[c, c^\dagger]_\pm = 1$, and this does not hold for the Hubbard operators, (1.87), used to describe the atomic limit. Nevertheless we can usefully develop an earlier form of many-body perturbation theory, that of Bloch (1965) and Balian & de Dominicis (1971), which is applicable, and can be somewhat simplified in situations we are dealing with here where the two-body interaction is solely on the impurity site (Keiter & Kimball, 1971; Kuromoto, 1983).

Our effort will be initially concentrated on calculating the partition function Z which can be written in the form,

$$Z = \mathrm{Tr}_f \mathrm{Tr}_c \{e^{-\beta H}\} = \frac{1}{2\pi i} \int_\Gamma e^{-\beta z} \mathrm{Tr}_f \mathrm{Tr}_c (z - H)^{-1} \, dz, \qquad (7.4)$$

where Γ encircles all the singularities of the integrand. The subscripts c and f denote a trace over the conduction and impurity states respectively (in this chapter we will use f rather than d to denote the localized states as the applications will be mainly to 4f rather than 3d systems). It will

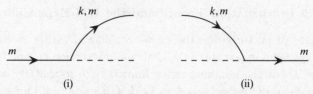

(i)   (ii)

Figure 7.1 Hybridization vertices at which, (i) an electron in state $|1,m\rangle$ is annihilated and a conduction electron k, m is created leaving the state $|0,0\rangle$, and (ii) the reverse process in which a conduction electron k, m is annihilated and the state $|1,m\rangle$ created.

be useful to introduce the resolvent operator,

$$R(z) = (z - H)^{-1} = (z - H_0)^{-1} \sum_{n=0}^{\infty} [H_I(z - H_0)^{-1}]^n, \qquad (7.5)$$

where $H = H_0 + H_I$. The difference between the resolvent Green's function which was introduced in chapter I and (7.5) is that here we are dealing with a system with two-body interactions so we use a different notation. Then matrix elements of $R(z)$ are given by

$$\langle \alpha|\langle c|R(z)|c\rangle|\alpha\rangle = (z - E_c - E_\alpha)^{-1}\langle \alpha|\langle c| \sum_{n=0}^{\infty} [H_I(z - H_0)^{-1}]^n|c\rangle|\alpha\rangle, \qquad (7.6)$$

where $|c\rangle$ is a conduction state of energy E_c. On changing the variable z to $z + E_c$ for each term in the trace, we can write Z as

$$Z/Z_c = \frac{1}{2\pi i} \int_\Gamma e^{-\beta z} \mathrm{Tr}_f\, R_f(z)\, dz, \qquad (7.7)$$

with $R_f(z) = \sum_\alpha |\alpha\rangle R_\alpha(z)\langle \alpha|$, where R_α is known as the reduced resolvent because it includes a sum over conduction electron states,

$$R_\alpha = \frac{1}{Z_c} \sum_c e^{-\beta E_c}(z - E_\alpha)^{-1}\langle \alpha|\langle c| \sum_{n=0}^{\infty} [H_I(z + E_c - H_0)^{-1}]^n|c\rangle|\alpha\rangle, \qquad (7.8)$$

where $|\alpha\rangle = |n_f, 0\rangle$, or $|n_f + 1, m\rangle$.

The diagrammatic expansion will be developed for $R_\alpha(z)$. To zero order it is given by

$$R_\alpha^0(z) = \frac{1}{(z - E_\alpha)}. \qquad (7.9)$$

In applying the method for the Anderson model (7.2), the form for H_I will be

$$H_I = \sum_{k,m}(V_k|1,m\rangle\langle 0,0|c_{k,m} + V_k^* c_{k,m}^\dagger|0,0\rangle\langle 1,m|). \qquad (7.10)$$

Vertices can be associated with the interaction term as indicated in

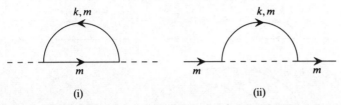

Figure 7.2 The lowest order diagrams for the resolvents: diagram (i) for the resolvent $R_0(z)$ with a sum over m, and diagram (ii) for the resolvent $R_{1,m}(z)$.

figure 7.1. The lowest order non-vanishing terms in the expansion are second order in V and represented by diagrams (i) and (ii) in figure 7.2, corresponding to $R_0(z)$ and $R_{1,m}(z)$ respectively. The contribution corresponding to that of 7.2(i) is

$$\sum_{k,m} \frac{|V_k|^2}{(z - E_0)^2} \frac{1}{(z - E_{1m} + \epsilon_k)} \sum_c \frac{e^{-\beta E_c}}{Z_c} \langle c|c_{k,m}^\dagger c_{k',m'}|c\rangle. \qquad (7.11)$$

The last part of the expression corresponds to the thermal average $\langle c|c_{k,m}^\dagger c_{k',m'}|c\rangle_c$ with respect to the non-interacting conduction states, and hence equal to $f(\epsilon_{k,m})\delta_{k,k'}\delta_{m,m'}$, the Fermi–Dirac distribution function. Similarly, figure 7.2(ii) corresponds to

$$\sum_k \frac{|V_k|^2}{(z - E_{1m})^2} \frac{(1 - f(\epsilon_k))}{(z - E_0 + \epsilon_k)}. \qquad (7.12)$$

Higher order terms are generated by alternate combinations of the vertices (i) and (ii) of figure 7.1. As the conduction electrons are non-interacting the expectation value of the conduction electron terms can be factorized using Wick's theorem. We can then join the conduction electron propagators in all possible ways consistent with their directions.

Figure 7.3 gives all the possible third order diagrams. A complete set of rules for drawing and evaluating the diagrams is given in appendix H. The conduction electron propagators give factors $f(\epsilon_k)$ or $(1 - f(\epsilon_k))$ dependent on whether the arrow is directed to the left or to the right as in (7.11) and (7.12). Contributions from diagrams with non-overlapping conduction electrons, such as figure 7.3(i), (ii) and (iii), factorize. They can be taken into account by the introduction of a resolvent self-energy $\Sigma_\alpha(z)$,

$$R_\alpha(z) = \frac{1}{z - E_\alpha - \Sigma_\alpha(z)}, \qquad (7.13)$$

which is similar to the conventional Feynman diagram self-energy. Diagrams of the type 7.3(i) are taken into account if the second order self-

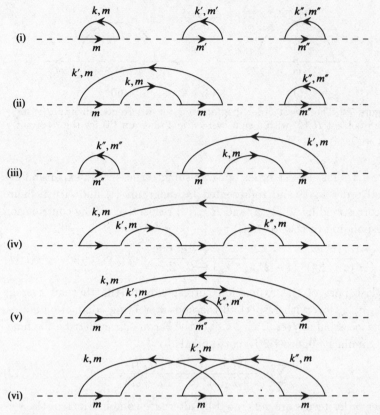

Figure 7.3 All the diagrams of order $|V|^6$ contributing to the resolvent $R_0(z)$.

energy diagram (figure 7.2(i) with the external legs removed) is taken into account. This contribution to the self-energy $\Sigma_0^{(2)}(z)$ is given by

$$\Sigma_0^{(2)}(z) = \sum_{k,m} |V_k|^2 \frac{f(\epsilon_k)}{(z - E_{1m} + \epsilon_k)}. \tag{7.14}$$

The first part of diagram 7.3(ii) and the last part of diagram 7.3(iii) represent a fourth order self-energy term.

Diagrams may be classified by their dependence on the degeneracy factor N. If the hybridization matrix element is scaled by a factor $1/\sqrt{N}$, $V_k \to V_k' = V_k/\sqrt{N}$, then the diagrams can be classified by their dependence on $(1/N)^r$, $r \geq 0$. This scaling is required, as we shall see later, to enable a physically meaningful $N \to \infty$ limit to be taken. For example the contributions to $R_0(z)$ shown in figure 7.3(i) are of order $(1/N)^0$, figure 7.3(ii), (iii) and (v) of order $(1/N)^1$, as they have only

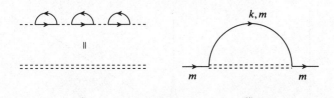

Figure 7.4 The leading order diagrams for $R_0(z)$ and $R_{1,m}(z)$ in a $1/N$ expansion. The double dashed line in figure (ii) represents the sum of bubble diagrams as indicated in figure (i).

two independent m summations, and diagrams (iv) and (vi) have only one free m and are of order $(1/N)^2$.

The order $(1/N)^0$ diagrams for $R_0(z)$ are summed by taking into account the second order self-energy (7.14). The lowest order diagrams for $R_{1,m}(z)$ are of order $(1/N)^1$. The diagram in figure 7.2(ii) is of this order as there is no internal sum over m. To calculate the partition function to order $(1/N)^0$ we need $R_{1,m}(z)$ to order $(1/N)^1$ as there is a final sum over m for this term in Z contributing an extra factor of N. All the diagrams for $R_{1,m}(z)$ to order $(1/N)^1$ are obtained by dressing the internal propagator $R_0(z + \epsilon_k)$ with the second order self-energy (7.14). This is indicated diagrammatically in figure 7.4.

Evaluation of the second order self-energy (7.14) using a flat band density of states gives, as $T \to 0$

$$\Sigma_0^{(2)}(z) = \frac{N\Delta}{\pi}\ln\left|\frac{z - E_1}{z - E_1 + D}\right|, \tag{7.15}$$

for $\epsilon_F = 0$. The resolvent $R_0(z)$ has a pole when z satisfies the equation,

$$z - E_0 - \frac{N\Delta}{\pi}\ln\left|\frac{z - E_1}{z - E_1 + D}\right| = 0. \tag{7.16}$$

This equation has three solutions in general (see figure 7.5) but only one of them lies below E_1 below the cut $E_1 < z < D - E_1$ on the real axis and so is undamped at $T = 0$. If we denote this solution by $z = E_1 - k_BT_A$, then T_A satisfies

$$k_BT_A = \epsilon_f - \frac{N\Delta}{\pi}\ln\left|\frac{k_BT_A}{D}\right|, \tag{7.17}$$

where $\epsilon_f = E_1 - E_0$, and we have assumed $k_BT_A \ll D$. This gives an isolated pole contribution to $R_0(z)$,

$$R_0(z) = \frac{(1 + N\Delta/\pi k_BT_A)^{-1}}{(z - E_1 + k_BT_A)}, \tag{7.18}$$

at $T = 0$. A similar contribution arises from the diagram 7.4 for $R_{1,m}(z)$ which, after summing over m and combining with (7.18), has a weight

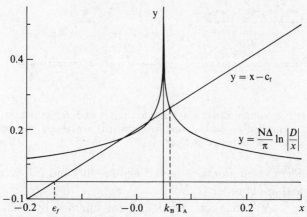

Figure 7.5   The graphical solution of equation (7.17) for $k_B T_A$ for $\epsilon_f = -0.2$, $N\Delta = 0.15$ and $D = 1.0$.

unity. This pole contribution dominates the partition function as $T \to 0$ and gives the ground state energy of the impurity system as $N \to \infty$, such that $N\Delta$ is finite,

$$E_{gs}^{(0)} = E_1 - k_B T_A. \tag{7.19}$$

The occupation number of the impurity level $n_f^{(0)}$ to this order is given by

$$n_f^{(0)} = \frac{\partial E_{gs}^{(0)}}{\partial E_1} = \frac{N\Delta/\pi k_B T_A}{(1 + N\Delta/\pi k_B T_A)}. \tag{7.20}$$

The magnetic susceptibility $\chi_{imp}^{(0)}$ can be similarly calculated. If the pole contribution is denoted by $E^{(0)}(T, H)$ at finite temperature and finite field, and for simplicity we assume the same g factor for the f electrons and the conduction electrons, then

$$\chi_{imp}^{(0)} = -\frac{\partial^2 E^{(0)}(0, H)}{\partial H^2}, \tag{7.21}$$

and,

$$-\frac{\partial E^{(0)}}{\partial H} = \frac{\Delta}{\pi} \frac{\partial E^{(0)}}{\partial H} \sum_m \int_{-D}^{D} \frac{f(\epsilon - mg\mu_B H)}{(E^{(0)} - \epsilon_f + \epsilon)^2} \, d\epsilon$$

$$+ \frac{\Delta}{\pi} \sum_m \int_{-D}^{D} \frac{\partial f(\epsilon - mg\mu_B H)}{\partial H} \frac{d\epsilon}{(E^{(0)} - \epsilon_f + \epsilon)}$$

$$= \frac{N\Delta}{\pi k_B T_A} \frac{\partial E^{(0)}}{\partial H} + (g\mu_B)^2 \frac{H\Delta}{\pi k_B^2 T_A^2} \sum_m m^2, \tag{7.22}$$

to first order in H at $T = 0$, so finally,

$$\chi_{\text{imp}}^{(0)} = \frac{\pi(g\mu_{\text{B}})^2 j(j+1)}{3N\Delta} \frac{(n_{\text{f}}^{(0)})^2}{(1 - n_{\text{f}}^{(0)})} = (g\mu_{\text{B}})^2 \frac{n_{\text{f}}^{(0)} j(j+1)}{3k_{\text{B}}T_{\text{A}}}. \qquad (7.23)$$

In the localized limit $n_{\text{f}}^{(0)} \to 1$ and $T_{\text{A}} \to T_{\text{L}}^{(0)}$, where $T_{\text{L}}^{(0)}$ is given by

$$T_{\text{L}}^{(0)} = D e^{-\pi |\epsilon_{\text{f}}| / N\Delta}. \qquad (7.24)$$

This can be identified as the Kondo temperature, to within a constant. It is clear that the $N \to \infty$ limit has to be taken so that $N\Delta$ remains finite for a finite Kondo temperature.

In the approach to the localized limit the leading correction to $1 - n_{\text{f}}^{(0)}$ is proportional to $T_{\text{L}}^{(0)}$,

$$n_{\text{f}}^{(0)} = 1 - \frac{\pi T_{\text{L}}^{(0)}}{N\Delta} + O\left(\frac{T_{\text{L}}^{(0)}}{N\Delta}\right)^2. \qquad (7.25)$$

The charge susceptibility $\chi_{\text{c,imp}}^{(0)}$ is obtained by taking a further derivative of (7.20) with respect to E_1, which gives

$$\chi_{\text{c,imp}}^{(0)} = \frac{\pi(n_{\text{f}}^{(0)})^2}{N\Delta}(1 - n_{\text{f}}^{(0)}). \qquad (7.26)$$

At temperatures $T \ll T_{\text{A}}$, the pole contributions give the dominant contribution to the specific heat. As

$$-\frac{\partial E^{(0)}(T)}{\partial T} = \frac{N\Delta}{\pi} \int_{-D}^{D} -\frac{\partial f(\epsilon)}{\partial T} \frac{1}{(E^{(0)} - \epsilon_{\text{f}} + \epsilon)}$$

$$+ \frac{f(\epsilon)}{(E^{(0)} - \epsilon_{\text{f}} + \epsilon)^2} \frac{\partial E^{(0)}}{\partial T} d\epsilon, \qquad (7.27)$$

then to lowest order in T,

$$= \frac{N\Delta\pi}{3} \frac{T}{k_{\text{B}}^2 T_{\text{A}}^2} + \frac{N\Delta}{\pi k_{\text{B}} T_{\text{A}}} \frac{\partial E^{(0)}}{\partial T}, \qquad (7.28)$$

and so,

$$\gamma_{\text{imp}}^{(0)} = \frac{\pi^2 n_{\text{f}}^{(0)}}{3k_{\text{B}}T_{\text{A}}} \qquad (7.29)$$

Deduction of the Wilson or χ/γ ratio (see (4.59)) from (7.23) and (7.29) gives $R = 1$ in this case, as for non-interacting particles.

To summarize these calculations: we have summed up a subset of diagrams corresponding to the leading term $(1/N)^0$ in a $1/N$ expansion, and evaluated expressions for n_{f}, χ_{imp}, $\chi_{\text{c,imp}}$ and γ_{imp} at $T = 0$. Finite temperature results are given by Ramakrishnan & Sur (1982). We leave aside the question of how useful these $N \to \infty$ results are for finite N

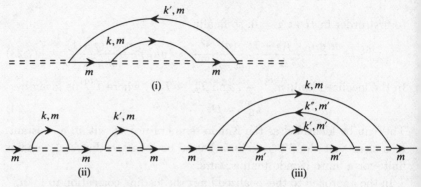

Figure 7.6 The diagrams required for the evaluation of the partition function to order $(1/N)^1$.

systems, we look at that question in the next section. Instead we shall look briefly to see what is involved in calculating the next term in the series, the diagrams of order $(1/N)^1$. The diagrams to be evaluated are shown in figure 7.6.

Diagram 7.6(i) is the $(1/N)^1$ contribution to $R_0(z)$, and diagrams 7.6(ii), (iii) the corresponding contribution to $R_{1,m}(z)$. Expressions for χ_{imp} and γ_{imp} to order $1/N$ were evaluated by Rasul & Hewson (1984) and Brandt, Keiter & Liu (1985). As these are rather lengthy we give them in appendix I.

For further results and discussion of this perturbational technique we refer the reader to the reviews of Bickers (1987) and Bickers, Cox & Wilkins (1987).

7.3 Exact Results

Both the Coqblin–Schrieffer and the $U = \infty$ Anderson models can be diagonalized by the Bethe ansatz, on the assumption of linear dispersion for the conduction electrons, using the arguments outlined in sections 6.2 and 6.8. Exact analytic expressions for arbitrary N can be derived for the $T = 0$ susceptibility χ_{imp}, the impurity level occupation number n_{f}, and the specific heat coefficient γ_{imp}. In the large N limit the results for $\chi_{\mathrm{imp}}^{(0)}$ and $n_{\mathrm{f}}^{(0)}$ are

$$n_{\mathrm{f}}^{(0)} = \int_0^\infty \frac{x^{-x} e^{-x \ln t'}}{\Gamma(1-x)} \, dx, \quad \chi_{\mathrm{imp}}^{(0)} = \frac{(g\mu_{\mathrm{B}})^2 \pi}{3N\Delta} \int_0^\infty \frac{x^{-(x-1)} e^{-x \ln t'}}{\Gamma(2-x)} \, dx, \tag{7.30}$$

for $t'e \geq 1$, where $t' = (D'/e\Delta)e^{\pi \epsilon_{\mathrm{f}}/N\Delta}$ and D' is the lower charge excitation cut-off introduced after diagonalization (Ogievetskii, Tsvelick

& Wiegmann, 1983). For $t'e < 1$ the corresponding expressions are

$$n_f^{(0)} = 1 - \sum_{r=1}^{\infty} \frac{(-1)^r}{r!} r^r (t')^r, \qquad (7.31)$$

$$\chi_{\text{imp}}^{(0)} = \frac{(g\mu_B)^2 \pi}{3N\Delta} \left[\frac{1}{t'} - \sum_{r=1}^{\infty} \frac{r^{(r+1)}}{r+1!} (-t')^r \right]. \qquad (7.32)$$

To relate these to the expressions derived in the previous section we write (7.20) in the form,

$$n_f^{(0)} = (1+s)^{-1}, \quad s = \frac{\pi k_B T_A}{N\Delta}. \qquad (7.33)$$

The equation for T_A becomes with this notation $s = te^{-s}$, where $t = (\pi D/N\Delta)e^{\pi \epsilon_f/N\Delta}$. Expanding (7.33) in powers of t using Lagrange's theorem gives a result identical to (7.31) if $t = t'$ or equivalently $D' = e\pi D/N$. This gives a relation between the conduction band cut-off D' in the Bethe ansatz solution and the conduction band width D of the flat band model in the large N limit. Using this relation it can be shown that the results (7.30) are equivalent to (7.20) and (7.23) (Rasul & Hewson, 1984).

It is possible to find a relation between D' and D for general N by calculating n_f or χ_{imp} for both models in the non-magnetic limit $\epsilon_f \to \infty$ and then comparing them. Perturbation theory for the conventional model gives the asymptotic forms for these as

$$n_f = \frac{N\Delta}{\pi \epsilon_f^*} - \left(\frac{N\Delta}{\pi \epsilon_f^*}\right)^2 \left(1 - \frac{1}{N}\right)\left(1 + \ln\left|\frac{N\Delta}{\pi \epsilon_f^*}\right|\right) + O\left(\frac{1}{\epsilon_f}\right)^3, \qquad (7.34)$$

$$\chi_{\text{imp}} = \frac{(g\mu_B)^2 \pi j(j+1)}{3N\Delta} \left(\frac{N\Delta}{\pi \epsilon_f^*}\right)^2 \times$$
$$\left\{1 - \left(\frac{N\Delta}{\pi \epsilon_f^*}\right)\left[\left(1 - \frac{3}{N}\right) + 2\left(1 - \frac{1}{N}\right)\ln\left|\frac{\pi \epsilon_f^*}{N\Delta}\right|\right]\right\} + O\left(\frac{1}{\epsilon_f}\right)^3 \qquad (7.35)$$

(see appendix H for a derivation), with

$$\epsilon_f^* = \epsilon_f + \frac{(N-1)\Delta}{\pi} \ln\left(\frac{\pi D}{N\Delta}\right), \qquad (7.36)$$

The renormalized level ϵ_f^* agrees with the Haldane scaling result (3.66) for $N = 2$. These results are exactly equivalent to those derived from the Bethe ansatz if

$$D' = \frac{\pi e D}{N^{N/N-1}}, \qquad (7.37)$$

which holds for general N. With this relation it is possible to translate the Bethe ansatz results into exact results for the flat band model. Non-analytic terms in $1/N$, such as $\ln N$, disappear from the Bethe ansatz results once this is done, and it is then possible to expand them in a $1/N$ expansion. The exact results for n_{f} and χ_{imp} for general N are

$$\chi_{\mathrm{imp}} = \frac{(g\mu_{\mathrm{B}})^2 j(j+1)}{3}$$
$$\times \left\{ \frac{e^{-a}}{\Gamma(1+1/N)} + \frac{1}{2\pi} \int_{-\infty}^{\infty} \frac{\Gamma(1+i\omega)(-i\omega+0^+)e^{-i\omega(N-1)/N}}{\Gamma(1+i\omega/N)(1-i\omega)e^{(-i\omega a - \pi|\omega|/N)}} \, d\omega \right\},$$
(7.38)

$$n_{\mathrm{f}} = 1 - \frac{i}{2\pi} \int_{-\infty}^{\infty} \frac{\Gamma(1+i\omega)(-i\omega+0^+)e^{-i\omega(N-1)/N}e^{(-i\omega a - \pi|\omega|/N)}}{\Gamma(1+i\omega/N)(\omega+i0^+)} \, d\omega,$$
(7.39)

where $a = \pi\epsilon_{\mathrm{f}}^*/N\Delta$. These can be expressed as an asymptotic expansion in the limits $\epsilon_{\mathrm{f}} \to \pm\infty$. For $\epsilon_{\mathrm{f}} \to \infty$ $(n_{\mathrm{f}} \to 0)$ we obtain (7.34) and (7.35). For the Kondo regime $\epsilon_{\mathrm{f}} \to -\infty$ $(n_{\mathrm{f}} \to 1)$, we find

$$n_{\mathrm{f}} = 1 - \frac{1}{N}\left|\frac{N\Delta}{\pi\epsilon_{\mathrm{f}}^*}\right| - \frac{1}{N}\left(\frac{N\Delta}{\pi\epsilon_{\mathrm{f}}^*}\right)^2 \left(1-\frac{1}{N}\right)\left(1-\ln\left|\frac{N\Delta}{\pi\epsilon_{\mathrm{f}}^*}\right|\right) + O\left(\frac{1}{\epsilon_{\mathrm{f}}^{*3}}\right).$$
(7.40)

$$\ln\left(\frac{3\chi_{\mathrm{imp}}N\Delta}{(g\mu_{\mathrm{B}})^2 j(j+1)\pi}\right) = \frac{-\pi\epsilon_{\mathrm{f}}^*}{N\Delta} - \Gamma\left(1+\frac{1}{N}\right) + O\left(\left|\frac{N\Delta}{\pi\epsilon_{\mathrm{f}}^*}\right|e^{\pi\epsilon_{\mathrm{f}}^*/N\Delta}\right).$$
(7.41)

If we express χ_{imp} in the Kondo regime in terms of the high temperature Kondo temperature T_{K}, as defined by Wilson, using equation (I.12) in appendix I (see Rasul & Hewson, 1984), we find

$$\chi_{\mathrm{imp}} = \frac{(g\mu_{\mathrm{B}})^2 j(j+1)w_N}{3k_{\mathrm{B}}T_{\mathrm{K}}},$$
(7.42)

where w_N, which is a generalization of the Wilson number (4.58), is given by

$$w_N = \frac{e^{(1+C-3/2N)}}{2\pi\Gamma(1+1/N)},$$
(7.43)

which agrees with (6.75) for $N = 2$. It also agrees with the results for the Coqblin–Schrieffer model (Hewson & Rasul, 1983; Schlottmann, 1984a). The Kondo temperature for the flat band model, width $2D$ with the Fermi level at the centre, is given by

$$k_{\mathrm{B}}T_{\mathrm{K}} = \frac{De^{(1+C-3/2N)}}{2\pi}\left(\frac{N\Delta}{\pi\max(eD,|\epsilon_{\mathrm{f}}|)}\right)^{1/N} e^{\pi\epsilon_{\mathrm{f}}/N\Delta}.$$
(7.44)

The prefactor agrees with that found by scaling arguments (see appendix

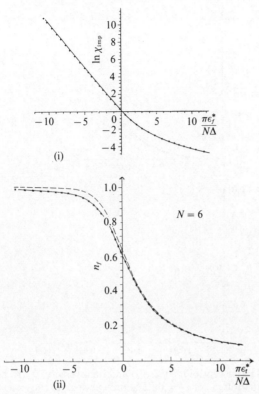

Figure 7.7 Plots of (i) $\ln\chi_{imp}$ in units of $\pi(g\mu_B)^2 j(j+1)/3$ and (ii) the f occupation n_f, as a function of $\pi\epsilon_f^*/N\Delta$ for $N = 6$. The full curve gives the exact results, the broken curve the leading order results in a $1/N$ expansion, and the crosses the results of the $1/N$ expansion to order $(1/N)^1$ (Rasul & Hewson, 1984).

C) for the Coqblin–Schrieffer model (1.89), in the regime where the Schrieffer–Wolff transformation is valid $|\epsilon_f| \gg D$.

We are now in a position to answer the question how good an approximation to the exact results are the first terms in the expansion at $T = 0$? We can see from (7.34) and (7.35) retaining the first two terms in the expansion is certainly a good approximation in the non-magnetic regime $\epsilon_f \to \infty$. The $(1/N)^0$ terms in the expressions for n_f, χ_{imp} (and also $\chi_{c,imp}$) are asymptotically exact in this limit, and the $(1/N)^1$ terms are the next order corrections. In the Kondo limit $\epsilon_f \to -\infty$, to order $(1/N)^0$, the expression for $\ln\chi_{imp}$ is asymptotically exact, but not the asymptotic form of n_f as $n_f \to 1$. The leading order correction here is proportional to $(N\Delta/\pi|\epsilon_f|)$, and not exponential as in (7.25).

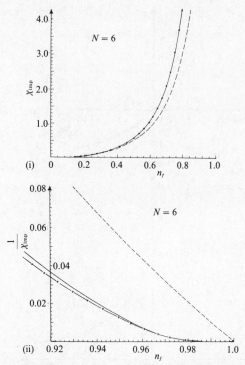

Figure 7.8 The susceptibility χ_{imp} (i) and its inverse χ_{imp}^{-1} (ii) as a function of the f occupation n_{f} for $N = 6$. The full curve gives the exact results, the broken curve the leading order results in a $1/N$ expansion, and the crosses the results of the $1/N$ expansion to order $(1/N)^1$ (Rasul & Hewson, 1984).

However, the coefficient of the correction term is of order $(1/N)^1$ so, if we take both the leading and next leading order terms into account, the results for n_{f} and $\ln\chi_{\text{imp}}$ are asymptotically exact in this limit also.

A general numerical comparison between the exact, the leading order and next leading order results of the $1/N$ expansion over the whole parameter range is shown in figures 7.7 and 7.8 for $N = 6$. We see the very close agreement over the whole range. The main deviation is in the asymptotic form of χ_{imp} against n_{f} to leading order $(1/N)^0$ as $n_{\text{f}} \to 1$. The exact results behaves as

$$\chi_{\text{imp}} \sim \frac{(g\mu_{\text{B}})^2 \pi j(j+1)}{3N\Delta} e^{-1/N(1-n_{\text{f}})}, \qquad (7.45)$$

for finite N as $n_{\text{f}} \to 1$, and not as predicted by (7.23). The difference stems from the order $(1/N)^1$ correction term in (7.40). When this is corrected by including the next leading order term the agreement is also good in this limit as seen in figure 7.8(ii).

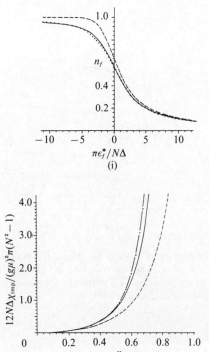

Figure 7.9 (i) The f occupation $n_{\rm f}$, as a function of $\pi\epsilon_f^*/N\Delta$ for $N = 2$ with the full curve the exact results, the dashed curve the leading order $1/N$ results and the crosses the $1/N$ expansion to order $(1/N)^1$, (ii) $\chi_{\rm imp}$ as a function of $n_{\rm f}$ for $N = 2$ with the full curve the exact results, the dashed curve the leading order $1/N$ results and the curve with the crosses the $1/N$ expansion to order $(1/N)^1$ (Rasul & Hewson, 1984).

In figure 7.9 we compare the results for $n_{\rm f}$ and $\chi_{\rm imp}$ for $N = 2$ with the exact results. We see that even in this case there is reasonably good overall agreement.

The finite temperature results to order $(1/N)^1$ have been compared with the exact results by Zhang & Lee (1983). They find it constitutes a good approximation over the whole temperature range.

The electronic specific heat coefficient $\gamma_{\rm imp}$ deduced from the thermodynamic equations in the Coqblin–Schrieffer limit is given by

$$\gamma_{\rm imp} = \frac{\pi^2}{3k_{\rm B}T_{\rm L}}\frac{(N-1)}{N}, \tag{7.46}$$

where $T_{\rm L} = T_{\rm K}/w_N$.

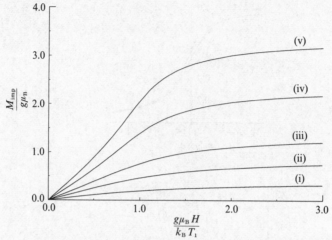

Figure 7.10   Plots of the exact results for $M_{\text{imp}}(H)$ versus $g\mu_{\text{B}}H/T_1$ for various values of N: (i) $N = 2$, (ii) $N = 3$, (iii) $N = 4$, (iv) $N = 6$ and (v) $N = 8$, where T_1 is proportional to T_{H} with a proportionality factor which depends upon N (Hewson & Rasul, 1983).

The Wilson ratio in this limit is given by

$$R = \frac{\chi_{\text{imp}}/\chi_{\text{c}}}{\gamma_{\text{imp}}/\gamma_{\text{c}}} = \frac{\pi^2 k_{\text{B}}^2}{j(j+1)(g\mu_{\text{B}})^2} \frac{\chi_{\text{imp}}}{\gamma_{\text{imp}}} = \frac{N}{N-1}, \qquad (7.47)$$

hence $R = 2$ for $N = 2$, and $R = 1$ in the limit $N \to \infty$, in agreement with earlier results. More generally there is a generalization of the Fermi liquid relation (5.20),

$$\frac{3(N-1)\chi_{\text{imp}}}{j(j+1)(g\mu_{\text{B}})^2} + \chi_{\text{c,imp}} = \frac{3N\gamma_{\text{imp}}}{\pi^2 k_{\text{B}}^2}. \qquad (7.48)$$

and hence,

$$R = \frac{N}{N-1+j(j+1)(g\mu_{\text{B}})^2 \chi_{\text{c,imp}}/3\chi_{\text{imp}}}, \qquad (7.49)$$

which reduces to (7.47) in the localized limit $\chi_{\text{c,imp}} \to 0$.

The ground state impurity magnetization $M_{\text{imp}}(H)$ can be calculated as a function of magnetic field. The equations have to be solved numerically because for $N > 3$ simultaneous equations are involved and an exact analytic solution using the Wiener–Hopf technique is no longer possible.

Results are shown in figure 7.10 for the Coqblin–Schrieffer model for several values of N. We note the appearance of a new feature, a positive curvature for small H and for $N > 3$. This shows up clearly as a peak

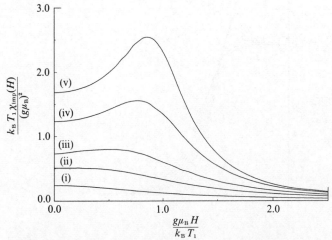

Figure 7.11 Plots of the exact results for $\chi_{\text{imp}}(H)$ versus $g\mu_B H/T_1$ for values of N: (i) $N = 2$, (ii) $N = 3$, (iii) $N = 4$, (iv) $N = 6$ and (v) $N = 8$, where T_1 is proportional to T_H with a proportionality factor which depends upon N (Hewson & Rasul, 1983).

in $\chi_{\text{imp}}(H)$, see figure 7.11, the peak developing steadily with increasing N. The s-d model with $S = \frac{1}{2}$ corresponds to the case $N = 2$.

As in equation (6.33), for weak fields M_{imp} can be expanded in a power series in $g\mu_B H/k_B T_K$, but for $N > 3$ the coefficients have to be calculated numerically. The H^3 coefficient is zero for $N = 3$, and then becomes positive for $N > 3$. For large fields, the asymptotic form for M_{imp} is

$$M_{\text{imp}}(H) = g\mu_B j$$
$$\times \left\{ 1 - \frac{1}{N}\frac{1}{\ln(\alpha_B H/T_H)} - \frac{1}{N^2}\frac{\ln(\ln(\alpha_B H/T_H))}{(\ln(\alpha_B H/T_H))^2} + O\left(\frac{1}{(\ln(\alpha_B H/T_H))^2}\right) \right\},$$
$$(7.50)$$

in the Coqblin–Schrieffer regime for $g\mu_B H \gg T_H$, with $T_H = T_K/\lambda$, where $\lambda = e^{(1+C-1/2N)}\delta$, $\ln\delta = 2\sum_{N=1}^{N-1} n\ln n/N(N-1)$ and $\alpha_B = g\mu_B/k_B$.

A peak also occurs in the Coqblin–Schrieffer model in $\chi_{\text{imp}}(T)$ for $N > 3$, as can be seen in the numerical results shown in figure 7.12.

This peak could have been anticipated from the results for $\chi_{\text{imp}}(H)$, because, from the thermodynamic relation (6.76) and the independence of the Wilson ratio (7.47) on H, the T^2 coefficient in the expansion of $\chi_{\text{imp}}(T)$ can be related to the H^2 coefficient in $\chi_{\text{imp}}(H)$ and has the same sign. Hence for $N > 3$ when this coefficient is positive $\chi_{\text{imp}}(T)$

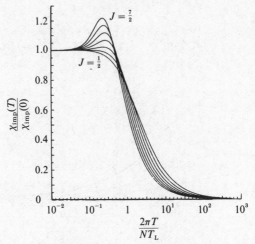

Figure 7.12   Plots of the exact results for $\chi_{\mathrm{imp}}(T)$ versus $2\pi T/NT_{\mathrm{L}}$ for $N = 2, 3, 4, 5, 6, 7, 8$ (Rajan, 1983).

initially increases and develops a maximum. The asymptotic form of $\chi_{\mathrm{imp}}(T)$ for $T \gg T_{\mathrm{K}}$ is given by

$$\chi_{\mathrm{imp}}(T) = \frac{(g\mu_{\mathrm{B}})^2 j(j+1)}{3 k_{\mathrm{B}} T}$$

$$\times \left\{ 1 - \frac{2}{N\ln(T/T_{\mathrm{K}})} - \frac{2\ln(\ln(T/T_{\mathrm{K}}))}{N^2(\ln(T/T_{\mathrm{K}}))^2} + \mathrm{O}\left(\frac{1}{(\ln(T/T_{\mathrm{K}}))^2}\right) \right\},$$

$$(7.51)$$

which generalizes the earlier result for $N = 2$. The peak in $\chi_{\mathrm{imp}}(H, T)$ as a function of H is temperature dependent and vanishes for $T > T_{\mathrm{max}}$ where T_{max} is the temperature at which $\chi_{\mathrm{imp}}(0, T)$ has its maximum. (see Cox, 1987a).

We now move away from looking exclusively at the Coqblin–Schrieffer regime, and look more generally at the results for n_{f}, χ_{imp} and the specific heat C_{imp}, as a function of T. These quantities are shown in figure 7.13 for $N = 6$ and for values of the bare parameter ϵ_{f} ranging from the non-magnetic through the mixed valence to the Coqblin–Schrieffer regime.

The peak in χ_{imp} in the non-magnetic and mixed valence regimes is due to the broad resonance above the Fermi level which sharpens into a narrow Kondo resonance as $n_{\mathrm{f}} \to 1$. (For $n_{\mathrm{f}} \to 1$ there is a broad resonance well below the Fermi level at $\omega \sim \epsilon_{\mathrm{f}}^*$ but this has little direct effect on the thermodynamic behaviour.) The results are scaled to $\chi_{\mathrm{imp}}(0)$,

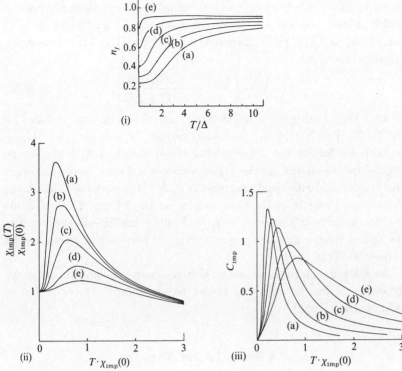

Figure 7.13 Plots of (i) $n_f(T)$, (ii) $\chi_{\text{imp}}(T)$, (iii) $C_{\text{imp}}(T)$ for $N = 6$. The curves correspond to (a) $\epsilon_f = 6N\Delta$, (b) $\epsilon_f = 3N\Delta$, (c) $\epsilon_f = 0$, (d) $\epsilon_f = -3N\Delta$, (e) $\epsilon_f = -6N\Delta$, with $D = 45N\Delta$. The susceptibility and specific heat curves have been scaled with the zero temperature susceptibility $\chi_{\text{imp}}(0)$ (Schlottmann, 1989).

which shows the peak is relatively larger in the non-magnetic and the mixed valence regimes, when ϵ_f is not too far above the Fermi level. However, $\chi_{\text{imp}}(0)$ is enhanced by a factor of the order of $N\Delta/k_B T_K$, in passing from the mixed valence to the Kondo regime, so in absolute terms the peak in the Kondo regime is larger. For general N the trend in the results is for $\chi_{\text{imp}}(T)$ to increase with T at low T and to have a maximum. The exception is the case $N = 2$ in the Kondo regime, where $\chi_{\text{imp}}(T)$ decreases and has no maximum, and $N = 3$ in the same regime, where $\chi_{\text{imp}}(T)$ is very flat and again has no maximum. This also applies to the specific heat, which at low T has the form, $C_{\text{imp}}(T) = \gamma_{\text{imp}}T + \delta_{\text{imp}}T^3$, where δ_{imp} is in general positive, except in the Kondo regime for $N = 2$, $\delta_{\text{imp}} < 0$ and for $N = 3$ when $\delta_{\text{imp}} = 0$.

These results can be given a qualitatively correct naive interpretation in terms of an effective one-body resonance in the density of states of width $\tilde{\Delta}$ and position $\tilde{\epsilon}_f$. If we calculate the T^2 coefficient in $\chi(T)$ from (1.9) and the T^3 coefficient in $C(T)$ from (1.7) for a Lorentzian resonance, we find

$$\left(\frac{\rho'}{\rho}\right)^2 - \left(\frac{\rho''}{\rho}\right) = -\frac{2(\tilde{\epsilon}_f^2 - \tilde{\Delta}^2)}{(\tilde{\epsilon}_f^2 + \tilde{\Delta}^2)^2}, \qquad (7.52)$$

so that they both have a negative sign for $|\tilde{\epsilon}_f| < \tilde{\Delta}$, and positive for $|\tilde{\epsilon}_f| > \tilde{\Delta}$. For $N = 2$ in the Kondo regime, $\tilde{\epsilon}_f \sim 0$ and $\tilde{\Delta} \sim k_B T_K$, so both coefficients are negative. For larger values of N in the Kondo regime the resonance moves higher above the Fermi level (to satisfy the Friedel sum rule) and narrows ($\tilde{\Delta} \propto N^{-1}$), we will see this in the Fermi liquid results in the next section, so the T^2 and T^3 coefficients become positive and increase with N. Positive coefficients also occur in the mixed valence and non-magnetic regimes because the resonance lies above the Fermi level.

For further exact results on the thermodynamic behaviour of the degenerate models we refer the reader to the comprehensive review of Schlottmann (1989).

7.4 Fermi Liquid Theories

We can get further insight into the Bethe ansatz and large N results from Fermi liquid theory. Both the phenomenological and the microscopic approaches described in chapter 5 are applicable and can be straightforwardly generalized to the N-fold degenerate models. Looking first of all at the phenomenological theory, we can conjecture an effective Hamiltonian for the low lying excitations about the $T = 0$ fixed point of the form,

$$H_{\text{eff}} = \sum_{k,m} \epsilon_{k,m} c_{k,m}^\dagger c_{k,m} + \sum_{k,m} (\tilde{V}_k c_{1,m}^\dagger c_{k,m} + \tilde{V}_k^* c_{k,m}^\dagger c_{1,m})$$
$$+ \sum_m \tilde{\epsilon}_{1,m} c_{1,m}^\dagger c_{1,m} + \sum_{m<m'} \tilde{U} n_{1,m} n_{1,m'}, \qquad (7.53)$$

as a natural generalization of (5.6). Making a molecular field approximation for the *excitations* and then calculating χ_{imp}, $\chi_{c,\text{imp}}$ and γ_{imp} at $T = 0$, as in section 5.1, we find

$$\chi_{\text{imp}} = \frac{(g\mu_B)^2 j(j+1)}{3} N \tilde{\rho}_{\text{imp}}(\epsilon_F)(1 + \tilde{U} \tilde{\rho}_{\text{imp}}(\epsilon_F)), \qquad (7.54)$$

$$\chi_{c,\text{imp}} = N \tilde{\rho}_{\text{imp}}(\epsilon_F)(1 - (N-1) \tilde{U} \tilde{\rho}_{\text{imp}}(\epsilon_F)) \qquad (7.55)$$

and

$$\gamma_{\text{imp}} = \frac{\pi^2 k_{\text{B}}^2}{3} N \tilde{\rho}_{\text{imp}}(\epsilon_{\text{F}}), \tag{7.56}$$

where $\tilde{\rho}_{\text{imp}}(\epsilon)$ is the quasi-particle density of states (we have assumed a flat band so $\tilde{\rho}_1(\epsilon)$ of (5.17) is $\tilde{\rho}_{\text{imp}}(\epsilon)$; from this point onwards we will take $\epsilon_{\text{F}} = 0$). Eliminating the density of states gives the relation (7.48) between these quantities. In the local moment limit $\chi_{\text{c,imp}} \to 0$, and we obtain (7.47) for the Wilson ratio $R = N/(N - 1)$.

At $T = 0$, $H = 0$, there are no excitations so the interaction term plays no role, and the quasi-particle density of states is that of a non-interacting Anderson model,

$$\tilde{\rho}_{\text{imp}}(\omega) = \frac{\tilde{\Delta}/\pi}{((\omega - \tilde{\epsilon}_{\text{f}})^2 + \tilde{\Delta}^2)}, \tag{7.57}$$

where we have identified the localized level in (7.53) as an effective f level. We can calculate the occupation number n_{f} by assuming the Friedel sum rule to apply so that

$$n_{\text{f}} = \frac{N}{\pi} \tan^{-1} \left(\frac{\tilde{\Delta}}{\tilde{\epsilon}_{\text{f}}} \right). \tag{7.58}$$

This is equivalent to integrating the quasi-particle density of states up to the Fermi level.

If we assume n_{f} as given, we can write (7.58) as a relation between $\tilde{\epsilon}_{\text{f}}$ and $\tilde{\Delta}$,

$$\tilde{\epsilon}_{\text{f}} = \tilde{\Delta} \cot \left(\frac{\pi n_{\text{f}}}{N} \right). \tag{7.59}$$

The quasi-particle density of states (7.57) evaluated at the Fermi level can then be expressed in the form,

$$\tilde{\rho}_{\text{imp}}(0) = \frac{\sin^2(\pi n_{\text{f}}/N)}{\pi \tilde{\Delta}}. \tag{7.60}$$

In the localized limit $n_{\text{f}} \to 1$, $\chi_{\text{c,imp}} \to 0$ and we can derive expressions for $\tilde{\Delta}$, $\tilde{\epsilon}_{\text{f}}$ and $\tilde{U}$ in terms of a single energy scale, the Kondo temperature. If we assume χ_{imp} to be given by $\chi_{\text{imp}} = (g\mu_{\text{B}})^2 j(j+1)/3k_{\text{B}}T_{\text{L}}$, where T_{L} is proportional to the Kondo temperature ($= T_{\text{K}}/w_N$ with Wilson's definition of T_{K}), we can calculate the density of quasi-particle states at the Fermi level from (7.56) using the result for the Wilson ratio $R = N/(N - 1)$,

$$\tilde{\rho}_{\text{imp}}(0) = \frac{N - 1}{N^2 k_{\text{B}} T_{\text{L}}}. \tag{7.61}$$

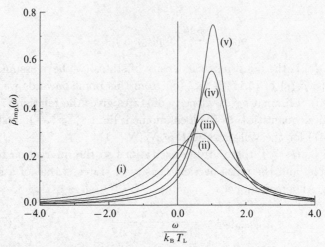

Figure 7.14 The quasi-particle density of states $\tilde{\rho}_{\mathrm{imp}}(\omega)$ in the Kondo limit for various values of N, (i) $N = 2$, (ii) $N = 3$, (iii) $N = 4$, (iv) $N = 6$ and (v) $N = 8$, as a function of $\omega/k_{\mathrm{B}}T_{\mathrm{L}}$.

Then, using (7.60) for $n_{\mathrm{f}} = 1$, gives for $\tilde{\Delta}$,

$$\tilde{\Delta} = k_{\mathrm{B}}T_{\mathrm{L}}\frac{N^2\sin^2(\pi/N)}{\pi(N-1)}, \tag{7.62}$$

and for $\tilde{\epsilon}_{\mathrm{f}}$ from (7.59),

$$\tilde{\epsilon}_{\mathrm{f}} = k_{\mathrm{B}}T_{\mathrm{L}}\frac{N^2\sin(\pi/N)\cos(\pi/N)}{\pi(N-1)}. \tag{7.63}$$

As $\chi_{\mathrm{c,imp}} = 0$ in the localized limit we can deduce $\tilde{U}$ from (7.55). We find

$$\tilde{U} = \left(\frac{N}{N-1}\right)^2 k_{\mathrm{B}}T_{\mathrm{L}}. \tag{7.64}$$

These expressions agree with the earlier results in chapter 5 for $N = 2$.

In figure 7.14 we give plots of the quasi-particle density of states (7.57) using (7.62) and (7.63) for various values of N and a given T_{L} in the localized limit $n_{\mathrm{f}} = 1$. They show clearly that on increasing N from $N = 2$ the resonance becomes more and more asymmetric with respect to the Fermi level so as to satisfy the Friedel sum rule. The integrated density of states up to the Fermi level must be such as to accommodate $1/N$ of an electron per channel.

In the large N limit (7.62)–(7.64) simplify to

$$\tilde{\Delta} = \frac{\pi}{N}k_{\mathrm{B}}T_{\mathrm{L}}, \quad \tilde{\epsilon}_{\mathrm{f}} = k_{\mathrm{B}}T_{\mathrm{L}}, \quad \tilde{U} = k_{\mathrm{B}}T_{\mathrm{L}} \tag{7.65}$$

and $R = 1$. This gives for the quasi-particle density of states for large N,

$$\tilde{\rho}_{\text{imp}}(\omega) = \frac{(k_B T_L/N)}{(\omega - k_B T_L)^2 + (\pi k_B T_L/N)^2},$$ (7.66)

which as $N \to \infty$ gives a delta function at $\omega = k_B T_L$. This is a surprisingly simple result. We will confirm it by explicit calculation in the next chapter.

We can also look at the large N limit quite apart from the localized limit $n_f = 1$. We take $\chi_{\text{imp}} = (g\mu_B)^2 j(j + 1)n_f/3k_B T_A$, which defines an energy scale $k_B T_A$, the n_f is included so $\chi_{\text{imp}} \to 0$ as $n_f \to 0$ in the non-magnetic limit. The Friedel sum rule (7.58) and the quasi-particle density of states in this limit become

$$n_f^{(0)} = \frac{N\tilde{\Delta}}{\pi \tilde{\epsilon}_f}, \qquad \tilde{\rho}_{\text{imp}}(0) = \frac{\tilde{\Delta}}{\pi \tilde{\epsilon}_f^2} = \frac{n_f^{(0)}}{N\tilde{\epsilon}_f},$$ (7.67)

as $\tilde{\Delta} \to 0$ as $N \to \infty$. Substituting $N\tilde{\rho}_{\text{imp}}(0)$ from (7.67) into (7.54) and (7.59) gives

$$\tilde{\epsilon}_f = k_B T_A, \qquad \tilde{\Delta} = \frac{\pi n_f^{(0)} k_B T_A}{N},$$ (7.68)

and from (7.57) for large N,

$$\tilde{\rho}_{\text{imp}}(\omega) = \frac{(n_f^{(0)} k_B T_A/N)}{(\omega - k_B T_A)^2 + (n_f^{(0)} \pi k_B T_A/N)^2}.$$ (7.69)

Hence we find more generally a delta function form for the density of quasi-particle states in the large N limit. In this case, however, we cannot deduce $\tilde{U}$ without having more information about the charge susceptibility $\chi_{\text{c,imp}}$, which we only know in the localized limit.

The Fermi liquid theory for $\tilde{U} = 0$ is equivalent to the local Fermi liquid theory of Newns & Hewson (1980) who used it to interpret experimental data on rare earth compounds. It corresponds to the non-interacting resonant level model generalized to arbitrary N.

All these results can be confirmed by generalizing the microscopic theory of sections 5.2 and 5.3 (Yoshimori, 1976; Schlottman, 1983). Formally the perturbation theory is in terms of U so we consider a finite U version of (7.2), which can be written in terms of the standard creation and annihilation operators,

$$H = \sum_m \epsilon_f n_{f,m} + U \sum_{m<m'} n_{f,m} n_{f,m'} + \sum_{k,m} \epsilon_k c_{k,m}^\dagger c_{k,m}$$
$$+ \sum_{k,m} (V_k c_{f,m}^\dagger c_{k,m} + V_k^* c_{k,m}^\dagger c_{f,m}).$$

(7.70)

This Hamiltonian corresponds formally to replacing the spin component σ by the z-component of angular momentum m. Apart from this, the approach is exactly the same as that described in chapter 5. The key relations (5.72) and (5.75) involving the self-energy derivatives are generalized to

$$\frac{\partial \Sigma_m(\omega)}{\partial \omega} + \frac{\partial \Sigma_m(\omega)}{\partial \mu} = -\rho_{\mathrm{imp}}(0) \sum_{m'} \Gamma_{m,m'}(\omega, 0), \tag{7.71}$$

$$\frac{\partial \Sigma_m(\omega)}{\partial \mu_{m'}} = -\rho_{\mathrm{imp}}(0)\Gamma_{m,m'}(\omega, 0), \tag{7.72}$$

for $m \neq m'$ in zero field. The antisymmetrized vertex at $\omega = 0$ must have the form,

$$\Gamma_{m,m'}(0,0) = \Gamma_0(1 - \delta_{m,m'}), \tag{7.73}$$

due to the SU(N) symmetry of the model.

We can write the results for γ_{imp}, $\chi_{\mathrm{c,imp}}$ and χ_{imp} in the form,

$$\gamma_{\mathrm{imp}} = \frac{\pi^2 k_{\mathrm{B}}^2}{3} \gamma^* N \rho_{\mathrm{imp}}(0), \quad \chi_{\mathrm{c,imp}} = \chi_{\mathrm{c}}^* N \rho_{\mathrm{imp}}(0),$$

$$\chi_{\mathrm{imp}} = \frac{(g\mu_{\mathrm{B}})^2 j(j+1)}{3} \chi^* N \rho_{\mathrm{imp}}(0), \tag{7.74}$$

where $\rho_{\mathrm{imp}}(0)$ is the density of states of the one electron Green's function at the Fermi level, and the starred quantities are many-body enhancement factors, $\gamma^* = 1/z = (1 - \partial\Sigma/\partial\omega)$, $\chi_{\mathrm{c}}^* = (1 - \partial\Sigma/\partial\mu)$, and

$$\chi^* = 1 + \frac{1}{j(j+1)} \sum_{m,m'} mm' \frac{\partial \Sigma_{1,m}(0)}{\partial \mu_{m'}}. \tag{7.75}$$

Using relations (7.71) and (7.72) for $\omega = 0$ and (7.73) we find

$$\gamma^* - \chi_{\mathrm{c}}^* = \rho_{\mathrm{imp}}(0)(N - 1)\Gamma_0, \tag{7.76}$$

and

$$\chi^* - \gamma^* = \rho_{\mathrm{imp}}(0)\Gamma_0. \tag{7.77}$$

Eliminating Γ_0 gives the relation,

$$(N - 1)\chi^* + \chi_{\mathrm{c}}^* = N\gamma^*, \tag{7.78}$$

which is equivalent to (7.48).

We can prove the Friedel sum rule using the arguments of section 5.2 which, for arbitrary N, becomes

$$n_{\mathrm{f}} = \frac{N}{\pi}\tan^{-1}\left(\frac{\Delta}{\epsilon_{\mathrm{f}} + \Sigma(0)}\right). \tag{7.79}$$

This is equivalent to (7.58) using $\tilde{\Delta} = \Delta/\gamma^*$, $\tilde{\epsilon}_{\mathrm{f}} = (\epsilon_{\mathrm{f}} + \Sigma(0))/\gamma^*$ as the wavefunction renormalization factor $z = 1/\gamma^*$ cancels out. Solving for

$\epsilon_f + \Sigma(0)$, we can derive an exact expression for the density of states in the immediate neighbourhood of the Fermi level in terms of n_f,

$$\rho_{\text{imp}}(\omega) = \frac{\sin^2(\pi n_f/N)}{\pi\Delta}\left(1 + \frac{6\gamma_{\text{imp}}\omega}{\pi N k_B^2}\cot\left(\frac{\pi n_f}{N}\right) + O(\omega^2)\right), \quad (7.80)$$

generalizing (5.50) to arbitrary N. The coefficient of the term linear in ω is related to the linear in T term in the low temperature thermopower (see appendix E)

In the localized moment limit $U \to \infty$, $n_f \to 1$, we have $\chi^*_{c,\text{imp}} = 0$, and we can express χ_{imp} in terms of T_L as earlier (T_L is a commonly used alternative definition of the Kondo temperature and differs from the Wilson definition only by the numerical factor $1/w_N$). Solving for γ^*,

$$\gamma^* = \frac{\pi\Delta}{k_B T_L}\frac{(N-1)}{N^2\sin^2(\pi/N)}, \quad (7.81)$$

we can calculate $\tilde{\epsilon}_f$ and $\tilde{\Delta}$, and confirm (7.62) and (7.63). We can also confirm the expression for the quasi-particle density of states (7.66) using the arguments of section 5.3.

We can calculate Γ_0 in the localized limit. Solving (7.76) and (7.77) with $\chi^*_c = 0$, and using (7.80) with $n_f = 1$, we find

$$\Gamma_0 = \frac{\pi^2\Delta^2\text{cosec}^4(\pi/N)}{N(N-1)T_L}. \quad (7.82)$$

Then on identifying $\tilde{U}$ as $z^2\Gamma_0$, or equivalently $\Gamma_0/(\gamma^*)^2$, we can confirm equation (7.64) for $\tilde{U}$.

We can also rigorously establish results (7.67)–(7.69) for arbitrary n_f, in the limit $U \to \infty$ and $N \to \infty$. We write the susceptibility in the form $\chi_{\text{imp}} = (g\mu_B)^2 j(j+1)n_f/3k_B T_A$ as earlier. We find, using $R = 1$ and (7.80) for $N\rho_{\text{imp}}(0)$,

$$\gamma^* = \frac{N\Delta}{\pi k_B T_A}\frac{1}{n_f^{(0)}}, \quad (7.83)$$

which leads to (7.68) for $\tilde{\epsilon}_f$ and $\tilde{\Delta}$, and (7.69) for the quasi-particle density of states. We thus have a proof of all the results of the phenomenological approach.

Though we cannot calculate Γ_0 and $\tilde{U}$ other than in the local moment limit (where we know $\chi_{c,\text{imp}} = 0$), we can estimate them in the large N limit using the results for $\chi_{c,\text{imp}}$ in section 7.2. Using (7.23) and (7.26) we find

$$\frac{\chi^*_c}{\chi^*} = (1 - n_f^{(0)})^2. \quad (7.84)$$

Solving (7.76) and (7.77) for Γ_0 and γ^*, and using $\chi^*/\gamma^* = 1$, we find

$$\Gamma_0 = \frac{(2 - n_{\mathrm{f}}^{(0)})}{(n_{\mathrm{f}}^{(0)})^2} \frac{N^2 \Delta^2}{\pi^2 k_{\mathrm{B}} T_{\mathrm{A}}}, \quad \tilde{U} = (2 - n_{\mathrm{f}}^{(0)}) k_{\mathrm{B}} T_{\mathrm{A}}. \qquad (7.85)$$

There is always some effective short range interaction $\tilde{U}$ between the quasi-particles, even for $N \to \infty$, due to the constraint arising from the strong Coulomb interaction $U \to \infty$, which suppresses the charge susceptibility and enhances the spin susceptibility compared with a non-interacting electrons. However, there is some simplification in the large N limit due to the fact that $\tilde{\rho}_{\mathrm{imp}}(\omega) \propto 1/N$. This implies that in the frequency and temperature range $k_{\mathrm{B}}T, \omega \ll k_{\mathrm{B}}T_{\mathrm{L}}$, the quasi-particle propagator $\tilde{G}_{\mathrm{f}}^0(\omega)$ will tend to zero as $N \to \infty$ as $1/N$. The only diagram of order $(1/N)^0$ in a perturbation expansion in powers of $\tilde{U}$ is then the Hartree bubble corresponding to the mean field approximation. The diagram for the inelastic quasi-particle scattering, figure 5.4, which gives the quasi-particles a lifetime proportional to $1/\omega^2$ away from the Fermi surface, is down by a factor $1/N$. This is due to the fact that the particle–hole pair carry the same m label so there is one less sum over m than the number of Green's functions so one of the $1/N$ factors remains uncancelled. Thus we can conjecture that, for $N \to \infty$ only the Hartree–Fock bubble has to be taken into account and the mean field treatment of the effective Hamiltonian becomes exact for $k_{\mathrm{B}}T, \omega \ll k_{\mathrm{B}}T_{\mathrm{L}}$. For general N the Hartree–Fock bubble gives exact results for the thermodynamics as $\omega, k_{\mathrm{B}}T \to 0$ corresponding to Landau Fermi liquid theory. In the large N limit the condition for exact results is less restrictive requiring only $\omega, k_{\mathrm{B}}T \ll k_{\mathrm{B}}T_{\mathrm{L}}$.

We only conjectured the form of the effective Hamiltonian (7.53) and, except for $N = 2$, it has not been derived from a first principles renormalization group calculation. Such a first principles calculation would be required to obtain expressions for $\tilde{\epsilon}_{\mathrm{f}}$, $\tilde{\Delta}$ and $\tilde{U}$, in terms of bare the parameters, ϵ_{f}, Δ. In the next section we consider an alternative approach via slave bosons which leads to an independent quasi-particle picture in the large N limit for $k_{\mathrm{B}}T, \omega \ll k_{\mathrm{B}}T_{\mathrm{L}}$. This theory will have the advantage of giving us explicit expressions for the renormalized parameters, $\tilde{\epsilon}_{\mathrm{f}}$, $\tilde{\Delta}$ in terms of the bare parameters ϵ_{f}, Δ.

7.5 Slave Bosons and Mean Field Theory

The $U = \infty$ degenerate Anderson model (7.2) can be expressed in terms of Hubbard operators, $|1, m\rangle\langle 0, 0| = X_{m,0}$, $|0, 0\rangle\langle 1, m| = X_{0,m}$,

$|0,0\rangle\langle0,0| = X_{0,0}$ and $|1,m\rangle\langle1,m| = X_{m,m}$. These operators obey commutation relations $[X_{p,q}, X_{q',p'}]_{\pm} = X_{p,p'}\delta_{q,q'} \pm X_{q',q}\delta_{p',p}$, so that diagrammatic methods based on Wick's theorem are not applicable. Nevertheless we showed in section 7.2 an earlier form of perturbation theory, not dependent on Wick's theorem, can be developed and applied successfully to this model. This form of perturbation theory, known as the Keiter–Kimball perturbation theory in this context, has the disadvantage that it becomes difficult to apply to many-site Anderson models. Models of this type are used to describe rare earth lattice systems, the model which has f electrons hybridizing with conduction electrons at every site being known as the *periodic Anderson model*. This will be considered as a model of heavy fermion systems in chapter 10.

An alternative way of developing perturbation theory, with a view to finding an approach suitable for the lattice problems, is to find a representation of the X-operators in terms of conventional boson and fermion creation and annihilation operators (Coleman, 1984). The representation,

$$X_{m,0} = f_m^\dagger b, \qquad X_{0,m} = f_m b^\dagger,$$
$$X_{0,0} = b^\dagger b, \qquad X_{m,m} = f_m^\dagger f_m, \tag{7.86}$$

is a possible representation where $[b, b^\dagger]_- = 1$, and $[f_m, f_m^\dagger]_+ = \delta_{m,m'}$ (corresponding to Bose and Fermi particles), provided that we work in the subspace such that

$$Q = b^\dagger b + \sum_m f_m^\dagger f_m = I. \tag{7.87}$$

In terms of this representation the Anderson model becomes

$$H = \sum_m \epsilon_f f_m^\dagger f_m + \sum_{k,m}(V_k f_m^\dagger c_{k,m} b + V_k^* c_{k,m}^\dagger f_m b^\dagger) + \sum_{k,m}\epsilon_k c_{k,m}^\dagger c_{k,m}. \tag{7.88}$$

The operator Q commutes with this Hamiltonian and is a conserved quantity.

We can think of the f operators here as describing the f electron and the Bose operators describing the hole in the f configuration when the f electron leaves the impurity site. We seem to have the great advantage that we are now dealing with operators for which the standard Wick's theorem is applicable. However, we still have to apply the constraint $Q = I$. If we use perturbation theory in V and apply the constraint to each order, projecting out all contributions outside the physical subspace (7.87), then we simply reproduce the perturbation theory of section 7.2, and there is no real advantage. There is an alternative way, however, of imposing the constraint by an auxiliary field λ (Read & Newns, 1983a). We can use the representation for the Krönecker delta function,

$$\frac{\beta}{2\pi} \int_{-\pi/\beta}^{\pi/\beta} e^{-i\lambda\beta n} \, d\lambda = \Delta_{n,0}, \qquad (7.89)$$

where $\Delta_{n,0} = 1$ for $n = 0$ and is zero for any other integer. The partition function for (7.88) subject to the constraint (7.87), can be expressed in the form,

$$Z = \frac{\beta}{2\pi} \int_{-\pi/\beta}^{\pi/\beta} e^{-\beta H(\lambda)} \, d\lambda, \qquad (7.90)$$

where

$$H(\lambda) = \sum_m \epsilon_f f_m^\dagger f_m + \sum_{k,m} (V_k f_m^\dagger c_{k,m} b + V_k^* c_{k,m}^\dagger f_m b^\dagger)$$

$$+ \sum_{k,m} \epsilon_k c_{k,m}^\dagger c_{k,m} + i\lambda (b^\dagger b + \sum_m f_m^\dagger f_m - 1).$$

$$(7.91)$$

Conventional perturbation theory can be developed for $H(\lambda)$, and then the constraint is imposed by a final integration over λ. We could do this term by term but we would then simply generate the perturbation theory of section 7.2. An alternative strategy, which appears quite natural if we represent Z by a functional integral over the Bose and Fermi fields as well as the λ field, is to carry out an approximate evaluation of the integrals using a saddle point evaluation. This corresponds to a mean field type of approximation, which we can carry out without using the functional integral representation for Z. The functional integral formulation does have some advantages but we will use the Hamiltonian formulation so as not to introduce too many different techniques.

In the mean field approximation on (7.91) we replace the Bose operators $b^\dagger$ and b by their expectation values, $b^\dagger \to \langle b^\dagger \rangle = r$, and $b \to \langle b \rangle = r$ (Read & Newns, 1983b). The mean fields r and λ are then determined by minimization of the free energy with respect to these variables. It is straightforward to calculate the free energy because the Hamiltonian is of the form

$$H(\lambda) = \sum_m \tilde{\epsilon}_f f_m^\dagger f_m + \sum_{k,m} (\tilde{V}_k f_m^\dagger c_{k,m} + \tilde{V}_k^* c_{k,m}^\dagger f_m)$$

$$+ \sum_{k,m} \epsilon_k c_{k,m}^\dagger c_{k,m} + (\tilde{\epsilon}_f - \epsilon_f)(r^2 - 1),$$

$$(7.92)$$

corresponding to a non-interacting Anderson or resonant level model with renormalized parameters, $\tilde{V}_k = rV_k$ and $\tilde{\epsilon}_f = \epsilon_f + i\lambda$.

As there is no two-body interaction term in (7.92) we can calculate the free energy F from the partition function given by (1.5),

$$F = -\frac{1}{\beta}\ln Z = -\frac{N}{\beta}\int_{-\infty}^{\infty}\ln(1 + e^{-\beta(\omega-\mu)})\bar{\rho}_f(\omega)\,d\omega, \qquad (7.93)$$

where $\bar{\rho}_f(\omega) = \sum_n \delta(\omega - \epsilon_n)$, and ϵ_n are the one-electron energies of (7.92). These energies can be calculated from the poles of the one-electron Green's function (1.47) or (5.36), so $\bar{\rho}_f(\omega)$ is given by

$$\bar{\rho}_f(\omega) = \frac{\text{Im}}{\pi}\left[\frac{1 + \sum_k |\tilde{V}_k|^2/(\omega^+ - \epsilon_k)^2}{\omega^+ - \tilde{\epsilon}_f - \sum_k |\tilde{V}_k|^2/(\omega^+ - \epsilon_k)}\right] = \frac{\text{Im}}{\pi}\frac{\partial}{\partial\omega}\ln\tilde{G}_f(\omega^+),$$
$$(7.94)$$

where

$$\tilde{G}_f(\omega^+) = \frac{1}{\omega^+ - \tilde{\epsilon}_f - \sum_k |\tilde{V}_k|^2/(\omega^+ - \epsilon_k)}. \qquad (7.95)$$

Substituting $\bar{\rho}_f(\omega)$ into (7.93), and integrating by parts, gives for the change in free energy due to the impurity,

$$F_{\text{MF}} = \frac{N}{\pi}\int_{-\infty}^{\infty}f(\omega)\text{Im}\ln\tilde{G}_f(\omega^+)\,d\omega + (\tilde{\epsilon}_f - \epsilon_f)(r^2 - 1). \qquad (7.96)$$

For a wide flat conduction band of width $2D$ this becomes

$$F_{\text{MF}} = -\frac{N}{\pi}\int_{-D}^{D}f(\omega)\tan^{-1}\left(\frac{\tilde{\Delta}}{\tilde{\epsilon}_f - \omega}\right)\,d\omega + (\tilde{\epsilon}_f - \epsilon_f)(r^2 - 1), \qquad (7.97)$$

where $\tilde{\Delta} = r^2\Delta$. For $T = 0$, we find on evaluating the integral,

$$E_{\text{gs}} = -\frac{N\tilde{\Delta}}{\pi} + \frac{N\tilde{\epsilon}_f}{\pi}\tan^{-1}\left(\frac{\tilde{\Delta}}{\tilde{\epsilon}_f}\right) + \frac{N\tilde{\Delta}}{2\pi}\ln\left(\frac{\tilde{\epsilon}_f^2 + \tilde{\Delta}^2}{D^2}\right) + (\tilde{\epsilon}_f - \epsilon_f)(\frac{\tilde{\Delta}}{\Delta} - 1).$$
$$(7.98)$$

Minimizing E_{gs} with respect to λ and r is equivalent to minimization with respect to the renormalized parameters $\tilde{\epsilon}_f$ and $\tilde{\Delta}$,

$$\frac{\partial E_{\text{gs}}}{\partial\tilde{\epsilon}_f} = \frac{\tilde{\Delta}}{\Delta} - 1 + \frac{N}{\pi}\tan^{-1}\left(\frac{\tilde{\Delta}}{\tilde{\epsilon}_f}\right) = 0, \qquad (7.99)$$

$$\frac{\partial E_{\text{gs}}}{\partial\tilde{\Delta}} = \frac{1}{\Delta}(\tilde{\epsilon}_f - \epsilon_f) + \frac{N}{\pi}\ln\left(\frac{(\tilde{\epsilon}_f^2 + \tilde{\Delta}^2)^{1/2}}{D}\right) = 0. \qquad (7.100)$$

As the f occupation is given by

$$n_f = \frac{N}{\pi}\tan^{-1}\left(\frac{\tilde{\Delta}}{\tilde{\epsilon}_f}\right), \qquad (7.101)$$

and $\tilde{\Delta}/\Delta = \langle b^\dagger b \rangle$, then (7.99) corresponds to an average of the constraint equation (7.87), $\langle b^\dagger b \rangle + \langle n_f \rangle = 1$.

We can make an immediate connection between these mean field equations and the large N results in section 7.2. As $N \to \infty$, $\tilde{\Delta} \to 0$, equation (7.100) becomes

$$\tilde{\epsilon}_f = \epsilon_f - \frac{N\Delta}{\pi}\ln\left(\frac{\tilde{\epsilon}_f}{D}\right), \qquad (7.102)$$

which is identical to (7.17) with $\tilde{\epsilon}_f = k_B T_A$. In this limit (7.101) becomes

$$n_f = \frac{N\tilde{\Delta}}{\pi\tilde{\epsilon}_f}. \qquad (7.103)$$

Using this result and the constraint, $\tilde{\Delta}/\Delta = 1 - n_f$, we can eliminate $\tilde{\Delta}$, to derive an expression for n_f,

$$n_f = \frac{N\Delta/\pi k_B T_A}{(1 + N\Delta/\pi k_B T_A)}, \qquad (7.104)$$

which is equivalent to (7.20). Hence the mean field approximation for (7.91) gives exact results as $N \to \infty$. The renormalized parameters $\tilde{\epsilon}_f$ and $\tilde{\Delta}$ agree with those derived from Fermi liquid theory in section 7.4, which are given by equations (7.62) and (7.63). The spectral density of the quasi-particle Green's function $\tilde{\rho}_f(\omega)$ gives the resonance (7.57).

The susceptibility χ_{imp} can be calculated by including a magnetic field and then taking the second derivative of the free energy $F(H)$ with respect to H. This gives

$$\chi_{\text{imp}} = \frac{(g\mu_B)^2 j(j+1)}{3} N\tilde{\rho}_f(0), \qquad (7.105)$$

and agrees with (7.23) and (7.54) in the large N limit. The coefficient of specific heat can be deduced by using the Sommerfeld expansion,

$$f(\epsilon) = \theta(-\epsilon) - \frac{\pi^2}{6}(k_B T)^2 \delta'(\epsilon) + O(T^4), \qquad (7.106)$$

in equation (7.97) and differentiating twice with respect to T. This gives

$$\gamma_{\text{imp}} = \frac{\pi^2}{3} k_B^2 N\tilde{\rho}_f(0), \qquad (7.107)$$

in agreement with (7.56) and (7.29) for large N. As the quasi-particles are non-interacting in this approximation the Wilson ratio $R = 1$.

The impurity charge susceptibility can be calculated from $\chi_{c,\text{imp}} = -\partial n_f/\partial \epsilon_f$, and using (7.101) and (7.100), we find

$$\chi_{c,\text{imp}} = \frac{\pi}{N\tilde{\Delta}(1 - n_f)}\left(\frac{\pi^2}{N^2} + \left(\frac{1}{1 - n_f} + \frac{\pi}{N}\cot\frac{\pi n_f}{N}\right)^2\right)^{-1}, \qquad (7.108)$$

which simplifies to the expression given in equation (7.26) in the large N limit.

In this representation the f electron Green's function $G_{f,m}(\tau)$ for $U = \infty$ is given by

$$\langle T X_{0,m}(\tau) X_{m,0}(0) \rangle = \langle T b^\dagger(\tau) f_m(\tau) f_m^\dagger(0) b(0) \rangle. \qquad (7.109)$$

This simplifies in the mean field approximation to

$$G_{f,m}(\tau) = |\langle b \rangle|^2 \langle T f_m(\tau) f_m^\dagger(0) \rangle, \qquad (7.110)$$

which is $|\langle b \rangle|^2$ multiplied by the quasi-particle Green's function (7.95). The Fourier transform of the corresponding retarded Green's function $G_{f,m}(\omega)$ calculated by analytic continuation from (7.110) is

$$G_{f,m}(\omega) = \frac{|\langle b \rangle|^2}{\omega^- \tilde{\epsilon}_f - \sum_k |\tilde{V}_k|^2/(\omega^+ - \epsilon_k)}, \qquad (7.111)$$

so $|\langle b \rangle|^2 \, (= r^2)$ is the wavefunction renormalization factor z. In the mean field approximation $|\langle b \rangle|^2 = (1 - n_f^{(0)})$ from the constraint equation so $z = (1 - n_f^{(0)})$. As $z = 1/\gamma^*$ this agrees with the Fermi liquid relation (7.83) on using the large N relation (7.20) between T_A and $n_f^{(0)}$.

The temperature dependence of $\tilde{\epsilon}_f(T)$ and $\tilde{\Delta}(T)$ can be calculated from the mean field equations. For large N and low T compared to T_A they are given by

$$\tilde{\epsilon}_f(T) = \tilde{\epsilon}_f(0) \left(1 + \frac{\pi^2}{6} \left(\frac{T}{T_A} \right)^2 n_f^{(0)} + O(T^4) \right), \qquad (7.112)$$

$$\tilde{\Delta}(T) = \tilde{\Delta}(0) \left(1 - \frac{\pi^2}{6} \left(\frac{T}{T_A} \right)^2 n_f^{(0)}(2 - n_f^{(0)}) + O(T^4) \right), \qquad (7.113)$$

so the resonance narrows with increase of temperature.

These give corrections to $n_f(T)$,

$$n_f(T) = n_f(0) \left(1 + \frac{\pi^2}{6} \left(\frac{T}{T_A} \right)^2 (1 - n_f^{(0)})(2 - n_f^{(0)}) + O(T^4) \right), \qquad (7.114)$$

so the population of the f state increases with temperature due to the peak in the resonance above the Fermi level (Read, 1986). In the Kondo limit, $T_A \to T_L$, $n_f \to 1$, so the T^2 coefficient is very small, though on an energy scale $k_B T_L$, due to the $U = \infty$ constraint. In the non-magnetic regime $n_f \to 0$, $k_B T_A \to \epsilon_f$, and the coefficient in this case is very little affected by the constraint.

The values of $\tilde{\epsilon}_f$ and $\tilde{\Delta}$ are also H dependent. In the Kondo regime for large N and small fields they are given by

$$\tilde{\epsilon}_f(H) = \tilde{\epsilon}_f(0) \left(1 + \frac{j(j+1)}{6} \left(\frac{g\mu_B H}{k_B T_L} \right)^2 + O(H^4) \right), \qquad (7.115)$$

$$\tilde{\Delta}(H) = \tilde{\Delta}(0) \left(1 - \frac{j(j+1)}{6} \left(\frac{g\mu_B H}{k_B T_L} \right)^2 + O(H^4) \right), \qquad (7.116)$$

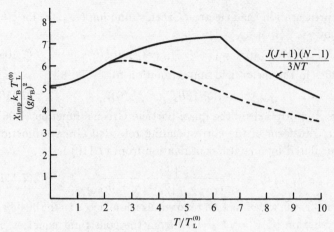

Figure 7.15 A plot of the susceptibility $\chi_{\text{imp}}(T)$ versus T/T_{L} for $N = 8$. The full curve is the prediction of mean field theory and the dot-dashed curve the exact result (Newns & Read, 1987).

and so they change with field in a similar way to their change with temperature.

For further results of the mean field approach we refer the reader to the review article of Newns & Read (1987) and Read's thesis (1986).

Considering the simplicity of the approach the mean field gives remarkably good results. It suffers from the same defect as the leading order in $1/N$ result in that the corrections to $1 - n_{\text{f}}$ in the Kondo regime are exponential rather than proportional to $1/|\epsilon_{\text{f}}|$ for finite N. The mean field results get less good with increasing temperature due thermal fluctuations which violate the constraint condition, as this is only imposed as an average in the mean field theory. The approximation eventually breaks down at a temperature $T_{\text{c}} \sim T_{\text{L}}/\ln N$, when $\tilde{\Delta} = 0$. Above this temperature the susceptibility has a Curie law behaviour,

$$\chi_{\text{imp}}(T) = \frac{(g\mu_{\text{B}})^2 j(j+1)}{3k_{\text{B}}T} \frac{N-1}{N}, \tag{7.117}$$

which is exact in the atomic limit for $N \to \infty$. The mean field theory, which is only correct for large N, does not contain the logarithmic terms, $1/\ln(T/T_{\text{K}})$, as these are of order $1/N$ (see equation (7.51)). These logarithmic terms are responsible for the slow approach of χ_{imp} to the free spin value at high temperatures. At the transition temperature T_{c} there is a discontinuous change of gradient of $\chi_{\text{imp}}(T)$ with T as shown in figure 7.15.

Though this behaviour is unphysical the general form is what one

might expect for large N, the initial rise in $\chi_{\text{imp}}(T)$ with T becomes
more pronounced with increase of N, and then for $T > T_c$ falls off as
a Curie law. There is the same trend as a function of magnetic field.
For $T = 0$ there is a critical field $H_c \sim g\mu_{\text{B}}/T_{\text{L}}\ln N$, at which M_{imp} for
finite N reaches its saturation value, and there is a solution which is
independent of field for $H > H_c$. Again this is what one might expect
for large N as the initial slope of $M_{\text{imp}}(H)$ with H increases with N,
and the logarithmic terms at higher fields, which give the approach to
saturation, are missing as these are of order $1/N$ (see equation (7.50)).

The value of T_c can be understood roughly as a competition between
a singlet state, with an energy lowering of $k_{\text{B}}T_{\text{L}}$, and a free spin with
degeneracy N so, crudely speaking, we have

$$Z \sim e^{-\beta(-k_{\text{B}}T_{\text{L}})} + Ne^0. \tag{7.118}$$

The transition occurs when these two terms become comparable at a
temperature $T_c \sim T_{\text{L}}/\ln N$.

As the mean field only imposes the constraint as an average, there
are fluctuations which violate the constraint. The constraint field λ
plays the role of a Lagrange multiplier when treated in the saddle point
approximation and there are fluctuations about this mean field to take
into account. There are also fluctuations in the Bose field about the
mean field average r. The success of the mean field solution for large N
and low T must be due to the fact that these fluctuations can be ignored
in this regime. In the next chapter we will consider how to take some
account of these fluctuations and obtain systematic corrections in $1/N$.

8

N-Fold Degenerate Models II

8.1 Introduction

In the last chapter we introduced some new techniques of calculation which, though approximate, gave asymptotically exact results in
the limit $N \to \infty$. This is a chapter of advanced topics in which we
develop these techniques further, particularly with a view to calculating
dynamic quantities and response functions. We first of all extend the
perturbational scheme of section 7.2 to a self-consistent scheme which
takes into account all non-crossing diagrams, and known generally as the
Non-Crossing Approximation (NCA). This is used to calculate dynamic
response functions at finite temperatures. It has been used extensively
for making predictions to compare with experiment for anomalous rare
earth systems. Following this we also reconsider the mean field slave
boson approach. The aim here is to go beyond the mean field theory
and take some account of fluctuations. It is possible to generalize the
approach so that the corrections to the mean field can be systematically treated in a $1/N$ expansion. We shall also develop approximations
so that the full spectrum of the one particle Green's function can be
calculated, and not just the quasi-particle contribution. Finally we discuss briefly yet another formulation of a $1/N$ expansion, the variational
approach. This has been developed for the calculation of the spectral
density of Green's functions and response functions for comparison with
several types of photoemission experiments, and also for one electron
absorption spectra. It has been used extensively in the interpretation of
data on Ce and other rare earth compounds.

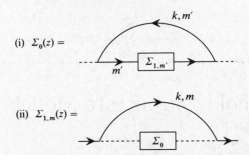

(i) $\Sigma_0(z) =$

(ii) $\Sigma_{1,m}(z) =$

Figure 8.1 A schematic representation of the integral equation for the sum of the non-crossing diagrams for the self-energies, (i) $\Sigma_0(z)$ and (ii) $\Sigma_{1,m}(z)$.

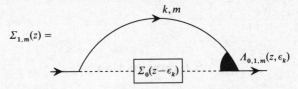

$\Sigma_{1,m}(z) =$

Figure 8.2 A schematic representation of the complete sum of diagrams for the self-energy, $\Sigma_{1,m}(z)$, where $\Lambda_{0,1m}(z,\epsilon_k)$ takes account of all vertex parts.

8.2 The Non-Crossing Approximation (NCA)

In section 7.2 we looked at calculations based on the summation of subsets of diagrams to low order in $1/N$. We now consider ways of taking more diagrams into account by setting up self-consistent calculations for the two lowest order diagrams 7.1(i) and (ii). We insert a self-energy into the internal propagators of these two diagrams.

This then gives coupled integral equations for the two self-energies, $\Sigma_0(z)$ and $\Sigma_{1,m}(z)$ (see figure 8.1). The iteration of these equations generates a set of diagrams which includes all non-crossing diagrams but does not correspond to any specific order in the $1/N$ expansion. The coupled equations for the self-energies are

$$\Sigma_0(z) = \frac{N\Delta}{\pi} \int_{-D}^{D} \frac{f(\epsilon)}{(z - E_{1,m} + \epsilon - \Sigma_{1,m}(z+\epsilon))} \, d\epsilon, \qquad (8.1)$$

and

$$\Sigma_{1,m}(z) = \frac{\Delta}{\pi} \int_{-D}^{D} \frac{(1 - f(\epsilon))}{(z - E_0 - \epsilon - \Sigma_0(z-\epsilon))} \, d\epsilon, \qquad (8.2)$$

taking a flat conduction band of width $2D$ for the conduction electrons. The set of diagrams summed by these equations includes all the terms of order $(1/N)^0$ and $(1/N)^1$ and a subset of contributions from the higher

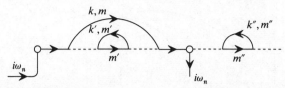

Figure 8.3 A non-crossing contribution to the Green's function $G_m^f(i\omega_n)$.

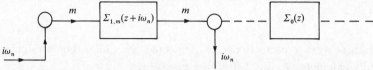

Figure 8.4 A schematic representation of the sum of all non-crossing diagrams for the Green's function $G_m^f(i\omega_n)$ where the self-energies, $\Sigma_0(z)$ and $\Sigma_{1,m}(z)$, are solutions of the integral equation illustrated in figure 8.1.

order terms. The lowest order diagram which is not included for $R_0(z)$ is that in figure 7.3(vi), which is of order $(1/N)^2$. There is a similar diagram for $R_{1,m}(z)$ of the same order which is not included.

In principle, equations (8.1) and (8.2) can be made exact by including vertex parts, $\Lambda_{1,0}(z,\epsilon)$ and $\Lambda_{0,1}(z,\epsilon)$, in the integrals to take into account the crossing diagrams (see figure 8.2). As the lowest order contributions to these vertex parts, such as diagram (vi) in figure 7.3, are of order $(1/N)^2$ or higher, it seems reasonable to neglect them for large values of N.

We will need to extend the formalism to calculate the one electron Green's function, $G_m^f(\tau)$, as defined in (7.109). The Fourier coefficient, $G_m^f(i\omega_n)$, can be calculated within the framework of the perturbation theory of section 7.2 and represented by diagrams which have two extra vertices, represented by open circles, in which an f electron external line carrying a frequency $i\omega_n$ enters and leaves. The first vertex is the one in which an f electron of frequency $i\omega_n$ enters, which can be subsequently removed at any later stage. The full rules for drawing and evaluating these diagrams are given in appendix G.

We can develop an approximation for $G_m^f(i\omega_n)$ based on the non-crossing diagrams for the resolvents. A diagram of this type is shown in figure 8.3. The summation of all non-crossing diagrams is shown schematically in figure 8.4.

Diagrams of the type shown in figure 8.5 have conduction electron propagators over the external vertex. These can be represented as contributions dressing the external vertex, as in the perturbation theory considered in chapter 2, section 2.3.

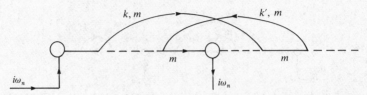

Figure 8.5 A crossing diagram of order $(1/N)^2$ for the Green's function $G^f_m(i\omega_n)$ which can be viewed as a vertex correction to the NCA sum shown in figure 8.4.

If these vertex corrections are neglected we can rather straightforwardly calculate $G^f_m(i\omega_n)$ from the resolvents.

It is useful at this stage to introduce a spectral density $\rho_\alpha(\omega)$ for the two resolvents $R_\alpha(\omega)$ defined by

$$\rho_\alpha(\omega) = -\frac{\mathrm{Im}}{\pi} R_\alpha(\omega + is), \tag{8.3}$$

for $s \to +0$. In terms of these the partition function of the impurity (7.7) can be expressed in the form,

$$Z/Z_c = Z_f = \int_{-\infty}^{\infty} e^{-\beta\epsilon}(\rho_0(\epsilon) + \sum_m \rho_{1,m}(\epsilon))\, d\epsilon, \tag{8.4}$$

by distorting the contour Γ used in equation (7.7) to just above and just below the real axis (note the singularities of the resolvent lie only on the real axis).

The Green's function $G^f_m(i\omega_n)$ is given by

$$G^f_m(i\omega_n) = \frac{1}{2\pi Z_f} \int_\Gamma e^{-\beta z} R_0(z)R_{1,m}(z + i\omega_n)\, dz, \tag{8.5}$$

if the corrections to the external vertex are neglected, where Γ encloses all the singularities of the integrand in a counter-clockwise sense. The lowest order external vertex correction is of order $(1/N)^2$, the same order as the vertex corrections for the resolvent. The neglect of these terms for the external vertex again seems reasonable for large N, and is consistent with the approximation used in calculating the resolvents. By distorting the contour Γ to surround the poles of the resolvents in (8.5), which lie on the real axis $z = 0$ and the line $z = -i\omega_n$, (8.5) can be expressed in terms of the spectral densities of the resolvents. The corresponding retarded Green's function can be calculated by analytic continuation. We can then obtain an expression for the spectral density $\rho_{f,m}(\omega)$ of the retarded Green's function in the form,

$$\rho_{f,m}(\omega) = \frac{1}{Z_f}(1 + e^{-\beta\omega}) \int_{-\infty}^{\infty} e^{-\beta\epsilon}\rho_0(\epsilon)\rho_{1,m}(\epsilon + \omega)\, d\epsilon. \tag{8.6}$$

To leading order in the $1/N$ expansion, $\rho_0(\omega)$ and $\rho_{1,m}(\omega)$, are given by

$$\rho_0^{(0)}(\omega) = (1 - n_f^{(0)})\delta(\omega - E_1 - k_B T_A), \qquad \rho_{1,m}^{(0)}(\omega) = \delta(\omega - E_1), \quad (8.7)$$

for $T = 0$ in the absence of a field. Then the spectral density for the retarded one electron Green's function deduced from (8.6) is given by

$$\rho_f^{(0)}(\omega) = (1 - n_f^{(0)})\delta(\omega - k_B T_A). \tag{8.8}$$

This is consistent with the results of the previous chapter in the limit $N \to \infty$ where we found a delta function form for the spectral weight of the quasi-particle density of states, and a wavefunction renormalization factor $(1 - n_f^{(0)})$. The spectral weight is consistent with the sum rule,

$$\int_{-\infty}^{\infty} \rho_f(\omega)\, d\omega = \langle X_{0,0} + X_{m,m} \rangle = 1 - n_f + \frac{n_f}{N}, \tag{8.9}$$

in the limit $N \to \infty$ (using $\langle X_{m,m} \rangle = n_f/N$ in the absence of a magnetic field). Using the standard results for calculating correlation functions from double time thermal Green's functions (Zubarev, 1960), we find

$$\int_{-\infty}^{\infty} \rho_f(\omega)(1 - f(\omega))\, d\omega = 1 - n_f, \tag{8.10}$$

and

$$\int_{-\infty}^{\infty} \rho_f(\omega) f(\omega)\, d\omega = \frac{n_f}{N}, \tag{8.11}$$

where $f(\omega)$ is the Fermi–Dirac distribution function. The sum rule (8.9) follows from the addition of (8.10) and (8.11). We can deduce from (8.11) that the spectral weight below the Fermi level for $T = 0$ is of order $(1/N)$, and hence zero in the limit $N \to \infty$. However, for *finite* N, and in the Kondo regime $n_f \to 1$, the bulk of the spectral weight is below the Fermi level, in a broad peak at $\omega \sim \epsilon_f^*$ (7.36). The width of the peak is of order $N\Delta$. The Kondo peak above the Fermi level has very little weight in this limit, as it is predominantly due to spin fluctuations rather than charge fluctuations. It is very narrow for large N as it is of order T_K/N.

These features are shown clearly in the calculations of $\rho_f(\omega)$ for $N = 6$ shown in figure 8.6. The Kondo resonance broadens and is reduced in height with increase of temperature as shown in figure 8.7. It is still visible, however, up to temperatures of the order of $10T_L$.

In the mixed valence and non-magnetic regimes the Kondo peak disappears. As ϵ_f increases from the Kondo regime to the mixed valence regime the two peaks merge as $\epsilon_f^* \to 0$. In these regimes there is no distinct energy scale associated with the spin fluctuations, and they have

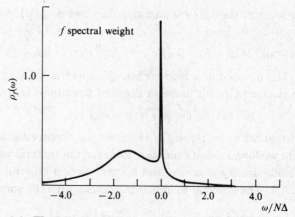

Figure 8.6 The spectral density of $\rho_f(\omega)$ calculated within the NCA in the Kondo regime for $N = 6$ with $\epsilon_f = -2.0N\Delta$, and a conduction band width of $10N\Delta$ (Coleman, 1984).

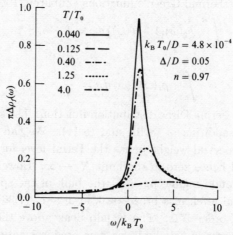

Figure 8.7 The temperature dependence of the Kondo resonance in the f density of states as calculated within the NCA (Bickers, Cox & Wilkins, 1987). The temperature T_0 is defined to correspond to the position of the peak, ω_{pk}, in the Kondo resonance at $T = 0$, $\omega_{pk} = k_B T_0$.

a lifetime of the same order as that for the charge fluctuations, $1/N\Delta$.

There are analytic solutions of the NCA equations for a flat conduction band at $T = 0$ (Kuromoto & Müller-Hartmann, 1985). These solutions reveal some of the short-comings of the self-consistent approach. There is a small spurious non-analytic cusp in $\rho_f(\omega)$ at the Fermi level $\omega = 0$. As a result the Fermi liquid relations are not satisfied. The frequency

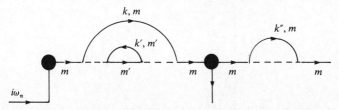

Figure 8.8 A low order contribution to the dynamic susceptibility.

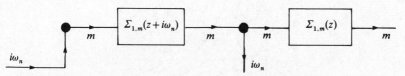

Figure 8.9 The NCA diagrams for the dynamic susceptibility. The self-energies are determined from the equations shown in figure 8.1.

and temperature scale for this spurious behaviour is of the order,

$$\omega, k_{\mathrm{B}}T \sim \frac{k_{\mathrm{B}}T_{\mathrm{K}}}{(N+1)} \left(\frac{\pi k_{\mathrm{B}}T_{\mathrm{K}}}{\Delta} \right)^{(N+1)/(N-1)}. \tag{8.12}$$

This is only a small part of the low energy and temperature range of interest, $\omega, k_{\mathrm{B}}T \le k_{\mathrm{B}}T_{\mathrm{K}}$. With this limitation apart, the NCA gives a flexible way of calculating the behaviour of the dynamic response functions with temperature. The agreement of the results for thermodynamic quantities, such as the specific heat as a function of temperature, with the exact Bethe ansatz results gives a quantitative test of the reliability of the approximation.

Kojima, Kuramoto & Tachiki (1984) and Bickers, Cox & Wilkins (1987) have calculated the dynamic susceptibility $\chi_{\mathrm{imp}}(\omega)$, which is the Fourier transform of the response function,

$$\chi_{\mathrm{imp}}(t) = i\theta(t) \langle [M_z(t), M_z(0)]_- \rangle, \tag{8.13}$$

where $M_z = g\mu_{\mathrm{B}} \sum_m m n_{\mathrm{f},m}$. This can be calculated perturbationally in a similar way to the one electron Green's function, and described by diagrams with external vertices to represent the operators, $X_{m,m}$ and $X_{m',m'}$, represented by full circles, and external lines carrying a frequency ω. The rules for drawing and evaluating the diagrams contributing to $\chi_{\mathrm{imp}}(\omega)$ are outlined in appendix G.

An example is shown in figure 8.8. A diagrammatic sum based on the NCA for the resolvents, with the neglect of any diagrams which involve corrections to the external vertex, is shown in figure 8.9.

Using this approximation and making similar manipulations to those

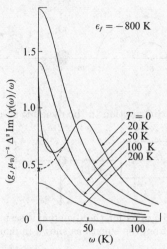

Figure 8.10 A plot of $\text{Im}(\chi_{\text{imp}}(\omega)/\omega)$ as a function of ω at various temperatures as calculated within the NCA for $N = 6$, $\Delta = 30$ K, $\epsilon_{\text{f}} = -800$ K, and a conduction band width 2×10^4 K. The cross indicates the Fermi liquid value at $\omega = 0$ and the dashed line an extrapolation from the $T = 0$ curve to this point (Kuramoto & Kojima, 1984).

used in deriving (8.5) and (8.6), we obtain an expression for $\text{Im}\,\chi_{\text{imp}}(\omega)$,

$$
\text{Im}\,\chi_{\text{imp}}(\omega) = \frac{(g\mu_{\text{B}})^2 j(j+1)N\pi(1 - e^{-\beta\omega})}{3Z_{\text{f}}} \int_{-\infty}^{\infty} e^{-\beta\epsilon}\rho_{1,m}(\epsilon)\rho_{1,m}(\epsilon + \omega)\, d\epsilon,
$$

$$(8.14)$$

and for the static susceptibility $\chi_{\text{imp}}(0)$,

$$
\chi_{\text{imp}}(0) = \frac{(g\mu_{\text{B}})^2 j(j+1)N\pi}{3Z_{\text{f}}} \int_{-\infty}^{\infty} e^{-\beta\epsilon}\rho_{1,m}^2(\epsilon)\, d\epsilon. \tag{8.15}
$$

The temperature dependence of $\chi_{\text{imp}}(0)$ can be checked against the Bethe ansatz results.

Results for $\text{Im}(\chi_{\text{imp}}(\omega)/\omega)$ are shown in figure 8.10 in the Kondo regime (Kuromoto & Kojima, 1984). This quantity is a symmetric function of ω and only the region $\omega > 0$ is shown. At very low temperatures there are inelastic peaks at $\omega \sim \pm k_{\text{B}}T_{\text{L}}$. These peaks can be interpreted as an excitation of the localized f electron between the singlet ground state with the conduction electrons, energy $\sim E_1 - k_{\text{B}}T_{\text{L}}$, and an excited state, energy $\sim E_1$. Consequently the peak position for positive ω is about the same as that in $\rho_{\text{f}}(\omega)$, although the excitation is purely magnetic in this case and does not involve removing or adding an electron. (We will see in the next chapter results of numerical renormalization group calculations which show that for $N = 2$ there is a single peak at the origin.) The spurious low frequency results at $T = 0$ show up as a singularity at $\omega = 0$ of the form $\text{Im}(\chi_{\text{imp}}(\omega)/\omega) \sim |\omega|^{-2(N+1)}$. This means that the Fermi liquid relation (E.10), proved for $N = 2$ by Shiba (1975) and generalized by Yoshimori & Zawadowski (1982), the Korringa relation,

$$\frac{(g\mu_{\text{B}})^2 j(j+1)N\pi}{3} \lim_{\omega \to 0} \text{Im}\left\{\frac{\chi_{\text{imp}}(\omega)}{\omega \chi_{\text{imp}}^2}\right\} = \frac{\pi}{N}, \qquad (8.16)$$

is not satisfied. The higher frequency results, however, do plausibly extrapolate to this value as shown in figure 8.10.

At higher temperatures the peaks merge into a single peak at the origin and $\text{Im}(\chi_{\text{imp}}(\omega)/\omega)$ is well represented by the Lorentzian

$$\text{Im}\left\{\frac{\chi_{\text{imp}}(\omega)}{\omega \chi_{\text{imp}}(T)}\right\} = \frac{\Gamma(T)}{\omega^2 + \Gamma^2(T)}, \qquad (8.17)$$

corresponding to a quasi-elastic line with a temperature dependent line width $\Gamma(T)$. In the next chapter we will examine how well these predictions correspond to data from neutron scattering experiments.

Bickers, Cox & Wilkins (1987) have calculated the temperature dependence of the resistivity and other transport coefficients within the NCA. If the scattering occurs entirely within one orbital channel the conductivity can be expressed in terms of the single particle lifetime $\tau(\omega)$, which can be related to $\rho_{\text{f}}(\omega)$ by generalizing (5.82) and (5.83). We leave discussion of their results to the next chapter where a comparison is made with experimental data.

8.3 Beyond Mean Field Theory

In this section we return to the slave boson description for the degenerate Anderson model, section 7.5, equation (7.91),

$$H(\lambda) = \sum_m \epsilon_{\text{f}} f_m^\dagger f_m + \sum_{k,m}(V_k f_m^\dagger c_{k,m} b + V_k^* c_{k,m}^\dagger f_m b^\dagger) + \sum_{k,m} \epsilon_k c_{k,m}^\dagger c_{k,m}$$

$$+ i\lambda(b^\dagger b + \sum_m f_m^\dagger f_m - Q), \qquad (8.18)$$

where we leave Q (7.87) as arbitrary for the moment. In the mean field solution b and $b^\dagger$ were replaced by their expectation values r and the constraint field $i\lambda$ by $\tilde{\epsilon}_{\text{f}} - \epsilon_{\text{f}}$. To go beyond mean field theory we make

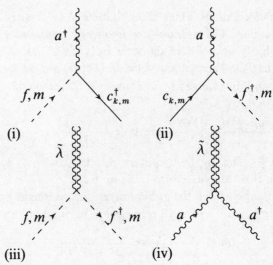

Figure 8.11 Vertices associated with terms in the interaction Hamiltonian (8.21).

the transformation,

$$b = a + r, \quad b^\dagger = a^\dagger + r, \quad i\lambda = \tilde{\epsilon}_f - \epsilon_f + \tilde{\lambda}, \tag{8.19}$$

where a and $a^\dagger$ satisfy standard Bose commutation relations. In transforming to these variables we can separate the Hamiltonian (8.18) into two parts. The first part, H_0, is given by

$$H_0 = \sum_m \tilde{\epsilon}_f f_m^\dagger f_m + \sum_{k,m} (\tilde{V}_k f_m^\dagger c_{k,m} + \tilde{V}_k^* c_{k,m}^\dagger f_m) + \sum_{k,m} \epsilon_k c_{k,m}^\dagger c_{k,m}$$
$$+ (\tilde{\epsilon}_f - \epsilon_f)(a^\dagger a + r^2 - Q), \tag{8.20}$$

which, with the appropriate choice of parameters $\tilde{V}$ and $\tilde{\epsilon}_f$, corresponds to the mean field theory. The second part of the Hamiltonian, H_{int}, which is given by

$$H_{\text{int}} = \sum_{k,m} (V_k f_m^\dagger c_{k,m} a + V_k^* c_{k,m}^\dagger f_m a^\dagger)$$
$$+ i\tilde{\lambda}(\sum_m \epsilon_f f_m^\dagger f_m + a^\dagger a + r(a^\dagger + a) + r^2 - Q) + r(\tilde{\epsilon}_f - \epsilon_f)(a^\dagger + a), \tag{8.21}$$

describes the interactions not taken into account within the mean field theory. This will be taken into account using standard many-body perturbation theory.

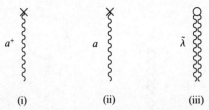

(i)   (ii)   (iii)

Figure 8.12 A representation of the terms in the interaction Hamiltonian (8.21) corresponding to coupling with external fields, $r(i\tilde{\lambda} + \tilde{\epsilon}_f - \epsilon_f)$ (**x**) and $i(r^2 - Q)$ (**O**).

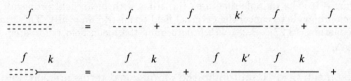

Figure 8.13 The series (in powers of the hybridization $\tilde{V}_k$) corresponding to the mean field propagators $\tilde{G}_f^{(MF)}(i\omega_n)$ and $\tilde{G}_{f,k}^{(MF)}(i\omega_n)$.

There are four interaction vertices which describe the scattering of particles which are shown in figure 8.11. The first two, in figure 8.11(i) and (ii), describe the scattering of an f electron in or out of the conduction band with the annihilation or creation of a Bose particle. The second two vertices, in figure 8.11(iii) and (iv), describe the interaction of the $\tilde{\lambda}$ field with the f electron and Bose field respectively. Note that the $\tilde{\lambda}$ field, which is a constraint field, is time independent and has a frequency of zero. The other vertices shown in figure 8.12 describe the interactions with external fields.

The one electron Green's functions with respect to the mean field Hamiltonian are

$$\tilde{G}_f^{(MF)}(i\omega_n) = \frac{1}{i\omega_n - \tilde{\epsilon}_f - \sum_k |\tilde{V}_k|^2/(i\omega_n - \epsilon_k)}, \qquad (8.22)$$

for the thermal f Green's function, and

$$\tilde{G}_{f,k}^{(MF)}(i\omega_n) = \frac{\tilde{V}_k}{(i\omega_n - \epsilon_k)(i\omega_n - \tilde{\epsilon}_f - \sum_k |\tilde{V}_k|^2/(i\omega_n - \epsilon_k))}, \qquad (8.23)$$

for the Fourier coefficient of the Green's function $\langle T c_{k,m}(\tau) f_m^\dagger(0) \rangle$, and can be derived from the equation of motion with H_{MF} or by summing the series shown in figure 8.13.

As the Bose particles are not conserved in general there are four Bose Green's functions which can be written in matrix form,

$$\mathbf{D}(\tau) = \begin{pmatrix} \langle T a(\tau) a^\dagger(0) \rangle & \langle T a(\tau) a(0) \rangle \\ \langle T a^\dagger(\tau) a^\dagger(0) \rangle & \langle T a^\dagger(\tau) a(0) \rangle \end{pmatrix}. \qquad (8.24)$$

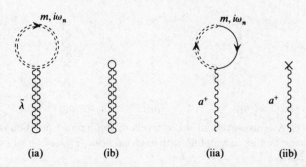

(ia)　　　　(ib)　　　　(iia)　　　　(iib)

Figure 8.14 The tadpole diagrams (ia), (iia) with mean field propagators which are made to cancel the external field terms (ib) and (iib). This leads to equations (8.27), (8.28) which determine the mean field parameters $\tilde{\epsilon}_f$ and $\tilde{\Delta}$.

For the mean field Hamiltonian the off diagonal terms are zero and the Fourier components of the diagonal terms are given by

$$D_{i,i}^{(0)}(i\omega_n) = \frac{1}{i\omega_n - (\tilde{\epsilon}_f - \epsilon_f)(\delta_{i,1} - \delta_{i,2})}, \tag{8.25}$$

where $i = 1, 2$. There is also a Green's function, $\langle T R(\tau) R(0) \rangle$, associated with the pair of Bose operators, $R = a + a^\dagger$, which appear in H_{int}. The Fourier coefficient of this Green's function is given by

$$D_R^{(0)}(i\omega_n) = \frac{2(\tilde{\epsilon}_f - \epsilon_f)}{(i\omega_n)^2 - (\tilde{\epsilon}_f - \epsilon_f)^2}. \tag{8.26}$$

The mean field equations can be derived by choosing the external fields r and $(\tilde{\epsilon}_f - \epsilon_f)$ so that they cancel the tadpole diagrams as indicated in figure 8.14(i) and (ii).

The requirement is equivalent to putting the expectation value of the coefficients of the linear terms in H_{int}, $\tilde{\lambda}$, $a^\dagger$ and a, to zero. The first equation is

$$r^2 - Q + \sum_{m,n} \tilde{G}_f^{(\text{MF})}(i\omega_n) = 0, \tag{8.27}$$

which is the condition that the constraint (7.87) is satisfied on the average. The second condition equates the linear terms in the Bose field to zero,

$$r(\tilde{\epsilon}_f - \epsilon_f) + \sum_{m,n} V_k \tilde{G}_{k,f}^{(\text{MF})}(i\omega_n) = 0. \tag{8.28}$$

Using equations (8.22) and (8.23), and carrying out the frequency summation and k integration for a flat band, gives

$$\frac{\tilde{\Delta}}{\Delta} - Q + \frac{N}{\pi}\tan^{-1}\left(\frac{\tilde{\Delta}}{\tilde{\epsilon}_f}\right) = 0, \tag{8.29}$$

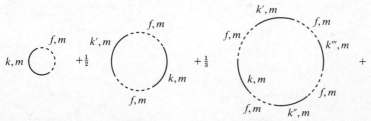

Figure 8.15 The ring diagrams in an expansion in powers of the hybridization $\tilde{V}_k$ which sum to give the mean field free energy for the impurity in equation (8.31).

and

$$r(\tilde{\epsilon}_f - \epsilon_f) + \frac{N\tilde{\Delta}}{2\pi r}\ln\left(\frac{(\tilde{\epsilon}_f^2 + \tilde{\Delta}^2)}{D^2}\right) = 0, \qquad (8.30)$$

which, for $Q = 1$ and $\tilde{\Delta} = r^2\Delta$, are equivalent to the mean field equations (7.99), (7.100), derived earlier. The impurity free energy in the mean field solution can be calculated by summing the ring diagrams shown in figure 8.15.

This gives the result,

$$F_{\mathrm{MF}} = -\frac{N}{\pi}\sum_n \ln(-i\omega_n + \tilde{\epsilon}_f + \sum_k |\tilde{V}_k|^2/(i\omega_n - \epsilon_k)) + (\tilde{\epsilon}_f - \epsilon_f)\left(\frac{\tilde{\Delta}}{\Delta} - 1\right).$$

$$(8.31)$$

On carrying out the sum over n in the standard way by contour integration we recover the earlier result (7.96). Minimizing F_{MF} with respect to $\tilde{\Delta}$, $\tilde{\epsilon}_f$, as we did in section 7.5, would be an alternative way of deriving (8.27) and (8.28) for the mean field parameters.

The lowest order corrections to the mean field theory arise from the quadratic terms in the fields a, $a^\dagger$ and $\tilde{\lambda}$. In the Bose propagator these are taken into account within second order perturbation theory for the self-energy Σ, which is a matrix defined by

$$\mathbf{D}^{-1}(i\omega_n) = (\mathbf{D}^{(0)}(i\omega_n))^{-1} - \Sigma(i\omega_n). \qquad (8.32)$$

These second order self-energy diagrams are indicated in figure 8.16. There are second order terms in $\tilde{\lambda}$ and R, which are only needed in the static case, $i\omega_n = 0$.

Read (1985) has shown that the $1/N$ corrections to the physical quantities (χ_{imp}, n_f etc) can be calculated from the contribution to the free energy from the sum of the RPA-like diagrams shown in figure 8.17.

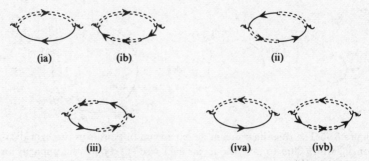

Figure 8.16 The matrix of second order self-energies $\Sigma^{(2)}$ for the Bose propagator.

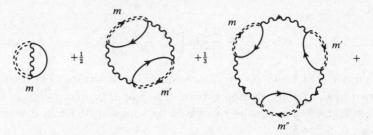

Figure 8.17 RPA-like correction terms to the mean field free energy.

This set sums to

$$\Delta F = \frac{1}{\beta} \sum_n \ln[-i\omega_n + \tilde{\epsilon}_f - \epsilon_f + \Gamma(i\omega_n)], \qquad (8.33)$$

where $\Gamma(i\omega_n)$ is the self-energy $\Sigma^{(2)}_{(ia)}(i\omega_n)$ (see figure 8.16), and is given by

$$\Gamma(i\omega_n) = \frac{N|V|^2}{\beta} \sum_{k,\nu} \frac{1}{(i\omega_\nu - i\omega_n - \epsilon_k)} \frac{1}{(i\omega_\nu - \tilde{\epsilon}_f + i\tilde{\Delta}\mathrm{sgn}(\omega_\nu))}, \qquad (8.34)$$

where $\omega_\nu = (2\nu + 1)\pi/\beta$, $\omega_n = 2\pi n/\beta$ and ν, n are integers. The ω_ν integration can be re-expressed as a contour integral in the standard way. On integrating over a flat conduction band with a constant density of states ρ_0 of width $2D$ (half-filled) this gives

$$\Gamma(i\omega_n) = \frac{N\Delta}{2} \int_C f(\epsilon) \frac{\theta(D-\epsilon)\theta(D+\epsilon)\mathrm{sgn}(\mathrm{Im}\epsilon - \omega_n)}{[\epsilon - \tilde{\epsilon}_f + i\tilde{\Delta}\mathrm{sgn}(\mathrm{Im}\epsilon)]} d\epsilon, \qquad (8.35)$$

where the contour C surrounds the poles on the imaginary axis as shown in figure 8.18. The integral can be evaluated analytically at $T = 0$ by distorting the contour to C' to surround the cuts (see figure 8.18), and

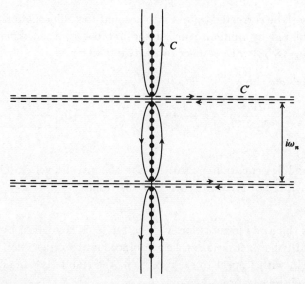

Figure 8.18 The contour C used for the integral (8.35). To evaluate analytically the contour is distorted to C' (dotted lines) above and below the cuts (heavy lines).

gives

$$\Gamma(i\omega_n) = \frac{N\Delta}{\pi}\ln\left(\frac{i\omega_n - \tilde{\epsilon}_f + i\tilde{\Delta}\mathrm{sgn}(\omega_n)}{-\tilde{\epsilon}_f + i\tilde{\Delta}\mathrm{sgn}(\omega_n)}\right) + \frac{N\Delta}{2\pi}\ln\left(\frac{(\tilde{\epsilon}_f^2 + \tilde{\Delta}^2)}{D^2}\right).$$

(8.36)

On substituting this result into (8.33) and carrying out the sum over the Bose frequencies ω_n in a similar way, but in this case distorting the contour to surround the negative real axis, we obtain for ΔF, $T = 0$ and $|\epsilon_f| \ll D$,

$$\Delta F =$$
$$\frac{-1}{\pi}\int_{-D}^{0}\tan^{-1}\left[\frac{N\Delta/\pi(\tan^{-1}(\tilde{\Delta}/(\tilde{\epsilon}_f - \epsilon)) - \tan^{-1}(\tilde{\Delta}/\tilde{\epsilon}_f))}{\epsilon - (\tilde{\epsilon}_f - \epsilon_f + N\Delta/\pi\ln\{[(\epsilon - \tilde{\epsilon}_f)^2 + \tilde{\Delta}^2]^{1/2}/D\})}\right]d\epsilon.$$

(8.37)

To leading order in $1/N$ this can be simplified to give

$$\Delta F =$$
$$\frac{-1}{\pi}\int_{-D}^{0}\frac{\tilde{\Delta}}{\pi}\left(\frac{1}{\epsilon - \tilde{\epsilon}_f} + \frac{1}{\tilde{\epsilon}_f}\right)\left[\epsilon - \left(\tilde{\epsilon}_f - \epsilon_f + \frac{N\Delta}{\pi}\ln\left|\frac{\epsilon - \tilde{\epsilon}_f}{D}\right|\right)\right]^{-1}d\epsilon.$$

(8.38)

The mean field parameters, $\tilde{\epsilon}_f$ and $\tilde{\Delta}$, which are found self-consistently, have now to be corrected to take into account these fluctuations. If these are calculated by minimization of the free energy then differentiating (8.31) plus (8.38) with respect to $\tilde{\Delta}$ gives

$$\tilde{\epsilon}_{\mathrm{f}} - \epsilon_{\mathrm{f}} + \frac{N\Delta}{\pi}\ln\frac{\tilde{\epsilon}_{\mathrm{f}}}{D}$$
$$+ \frac{(N\Delta)^2}{N\pi}\int_{-D}^{0}\left(\frac{1}{\epsilon - \tilde{\epsilon}_{\mathrm{f}}} + \frac{1}{\tilde{\epsilon}_{\mathrm{f}}}\right)\left(\epsilon - \frac{N\Delta}{\pi}\ln\left|\frac{\epsilon - \tilde{\epsilon}_{\mathrm{f}}}{\tilde{\epsilon}_{\mathrm{f}}}\right|\right)^{-1} d\epsilon = 0,$$

$$(8.39)$$

which modifies equation (7.100) for the effective f level $\tilde{\epsilon}_{\mathrm{f}}$. Combining (8.38) with (7.98) gives to order $1/N$ the result

$$E_{\mathrm{gs}} = \epsilon_{\mathrm{f}} - \tilde{\epsilon}_{\mathrm{f}}. \tag{8.40}$$

This has the same form as for $N \to \infty$ but $\tilde{\epsilon}_{\mathrm{f}}$ is now given by equation (8.39) with the fluctuations taken into account to order $1/N$. Differentiating E_{gs} with respect to ϵ_{f} gives for n_{f} the result (I.5) in appendix I, which is correct to order $1/N$.

On taking the magnetic field and temperature dependence into account Read (1985) has shown that the susceptibility χ_{imp} and the specific heat coefficient γ_{imp} can be calculated to order $1/N$ in a similar way, in agreement with the results (I.3) and (I.4) in appendix I. The calculations are not completely straightforward, however, due to the presence of infrared divergent terms. For example if we differentiate (8.31) plus (8.38) with respect to $\tilde{\epsilon}_{\mathrm{f}}$ to derive a new equation for $\tilde{\Delta}$ to replace the mean field equation (7.99), then an infrared divergent integral appears. The resulting equation can be written in the form,

$$\frac{\tilde{\Delta}}{\Delta}(\mu + 1) - 1 + \left(n_{\mathrm{f}} - \frac{\mu}{1 + \mu}\right)(1 + \mu) + I = 0, \tag{8.41}$$

where $\mu = N\Delta/\pi T_{\mathrm{A}}$, and n_{f} is given by (I.5), and I is the integral,

$$I = \frac{\mu^2}{N}\int_{\Lambda}^{D/T_{\mathrm{A}}}\frac{x}{1+x}\frac{1}{[x + \mu\ln(1+x)]^2}\, dx. \tag{8.42}$$

As I is divergent a lower cut-off Λ has been introduced. The quantity $\tilde{\Delta}$ cannot be calculated to order $1/N$ due to the infrared divergent term. However, $\tilde{\Delta}$ is not a physical observable. If physical observables are calculated, such as χ_{imp} and γ_{imp}, without intermediate evaluation of $\tilde{\Delta}$ then the divergent integrals cancel. These problems arise due to the breaking of symmetry in the mean field solution. The symmetry involved is a local gauge invariance. If the Bose fields are changed by a phase factor, $b \to be^{i\theta}$, $b^{\dagger} \to b^{\dagger}e^{-i\theta}$, with a simultaneous change of the fermi operators, $f \to fe^{i\theta}$, $f^{\dagger} \to f^{\dagger}e^{-i\theta}$, this leaves the Hamiltonian invariant. It is this symmetry which is broken in the mean field solution when $\langle b \rangle$ and $\langle b^{\dagger} \rangle$ have non-zero values. Symmetry breaking is possible in the limit $N \to \infty$, but for finite N there is a general theorem that a local gauge symmetry cannot be broken (Elitzur's theorem), hence the symmetry must be restored by the fluctuations. The divergent part of I behaves asymptotically as $I \sim -n_f^2/N\ln\Lambda$. It has been argued by Read (1985), in analogy with a similar situation considered by Witten (1978), that higher order terms of this type sum to an exponential so that

$$|b|^2 \sim \Lambda^{n_f^2/N}, \tag{8.43}$$

and the symmetry will be restored for $\Lambda \to 0$ with N finite.

It is possible to avoid the infrared divergent terms using the functional integral formulation for the partition function. The divergences arise essentially because there is no constraint on the dynamics of the phase of the Bose mean field. By transforming to a radial gauge, the phase of the Bose field can be absorbed into a new time dependent field $\lambda'(\tau)$, where $\lambda'(\tau) = \lambda - d\theta(\tau)/d\tau$. The $1/N$ corrections can then be calculated by considering the quadratic fluctuations in the Bose and λ' fields without the complication of the infrared divergences (see Read & Newns, 1983a, who introduced this transformation for the Coqblin-Schrieffer model).

It has been shown by explicit calculation that the mean field approximation gives the exact results in the limit $N \to \infty$, and that taking the Gaussian fluctuations into account gives $1/N$ corrections. The question naturally arises as to whether there is a systematic way of generating the diagrams to a specific order in the $1/N$ expansion. If Q is made extensive with N, $Q = q_0 N$, each Bose propagator gives a factor $1/N$ and each fermion loop a factor N due to the sum over m. The mean field free energy is then proportional to N and the correction terms from diagrams with L is the number of independent boson loops are of order N^{1-L} (this would mean that all the diagrams in figure 8.17 are of order unity). Coleman & Andrei (1986) have found exact solutions via the Bethe ansatz for the arbitrary Q model. They have looked at the results for a sequence of values of N with $q_0 = Q/N$, the filling factor, fixed at the value $1/2$. Some of their results are shown in figure 8.19.

It was proved that the $N \to \infty$ result for $T = 0$ corresponds to the mean field solution of this model. Only the curve with $N = 2$ corresponds to the original model with $Q = 1$. It is possible, however, to regard the $1/N$ expansion as a formal mathematical device for making systematic approximations, the model itself does not have to be the actual physical model but has to provide good approximations to the

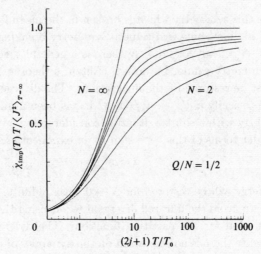

Figure 8.19 Exact results for $k_B T \chi_{\text{imp}}(T)$ for the Coqblin-Schrieffer model with general Q. The sequence of curves is for $Q/N = \frac{1}{2}$, $N = 2$ (lowest curve) towards the large N limit for $N = 2, 4, 6, 8, 12, \infty$ (Coleman & Andrei, 1986).

physically relevant model in the parameter regime of interest. Coleman (1985) and Millis & Lee (1987) have used the model with $Q = q_0 N$, fixed q_0, to classify the terms in a $1/N$ expansion and then to make the physical choice $q_0 = 1/N$ at the end of the calculation. Houghton, Read & Won (1987) on the other hand have kept strictly to the $Q = 1$ model at all stages and have extended the earlier work of Read (1985) by including the fluctuations of the zero frequency $\tilde{\lambda}$ field. They conclude that, on taking into account cancellations, diagrams with up to L independent boson loops are required to calculate physical quantities to order $1/N^L$. They prove this explicitly for $L = 2$ and give arguments that it holds for general L.

The mean field theory, as we found in the last chapter, is restricted to the Fermi liquid regime $T \ll T_K$ and breaks down at higher temperatures. Including second order fluctuations does not extend the temperature range, and the results are still restricted to $T \ll T_K$. The difficulty lies in the approximations used for the λ field when thermal fluctuations become important. It is possible to avoid this problem in the slave boson approach by introducing a chemical potential $-\mu'$ associated with $(Q - 1)$, and then to project on to the physical subspace by taking the limit $\mu' \to \infty$. This is essentially similar to the technique used by Abrikosov (1965) which we described briefly in section 3.1, and was used in the original slave boson paper of Coleman (1984). This method of imposing the constraint, combined with the $1/N$ classification of terms, has been used to calculate the thermodynamics over the full temperature range and also to calculate the spectral density $\rho_f(\omega)$ (Jin & Kuroda, 1988; Jin, Matsuura & Kuroda, 1991). As there is no breaking of symmetry in this approach the complications due to the infrared divergences are avoided.

The importance of the slave boson approach is that, unlike many of the other techniques we have considered, it can be generalized and applied to lattice problems where a strong correlation constraint is imposed. The mean field theory is easy to generate and, though fluctuations are harder to take into account, calculations to second order are quite feasible. We will look at the mean field theory for the lattice problem in some detail in the next chapter.

8.4 The Variational 1/N Expansion

There is yet another way of carrying out the $1/N$ expansion for the degenerate Anderson model. This is the variational method developed by Gunnarsson & Schönhammer (1983), which they have used extensively for calculating photoemission and absorption spectra for comparison with experiments on rare earth and actinide compounds. We shall first of all illustrate this approach for the calculation of ground state properties. It is a generalization of the variational approach described in chapter 3, section 3.2. Though we will be rederiving results obtained earlier the approach is sufficiently different that it will lead us on to a new way of calculating the spectra for dynamical response functions. Being a variational method the calculations are limited to $T = 0$.

As we know the exact many-body ground state of the Anderson model is a singlet, an appropriate starting point for a variational calculation is the full Fermi sphere $|\Omega\rangle$ with the impurity state empty, which is also a singlet. From this state we can construct other singlet states by acting on $|\Omega\rangle$ with the hybridization term, and use these as a basis for constructing a suitable variational wavefunction. Operating once on the state $|\Omega\rangle$ with the hybridization term in (7.2) generates states of the form $X_{m,0}c_{k,m}|\Omega\rangle = |\epsilon_f \epsilon_k^h m\rangle$, in which the impurity state is singly occupied and a hole is created in the Fermi sphere (the superscript h on ϵ_k^h indicates a hole state). We can show that we obtain the leading order in $1/N$ results for the ground state energy using linear combinations of

$|\epsilon_f \epsilon_k^h m\rangle$ and $|\Omega\rangle$ only. The trial wavefunction is

$$|\phi^{(0)}\rangle = A\{|\Omega\rangle + \frac{1}{\sqrt{N}}\sum_{k,m}^{\text{occ}}\alpha_k|\epsilon_f \epsilon_k^h m\rangle\}, \qquad (8.44)$$

where the sum is over occupied k states only and α_k are variational parameters. These are determined by minimization of the function

$$\langle\phi^{(0)}|H|\phi^{(0)}\rangle - E^{(0)}(\langle\phi^{(0)}|\phi^{(0)}\rangle - 1), \qquad (8.45)$$

where $E^{(0)}$ is a Lagrange multiplier to ensure normalization. On minimizing with respect to A we find $E^{(0)}$ is the ground state energy,

$$E^{(0)} = \frac{\langle\phi^{(0)}|H|\phi^{(0)}\rangle}{\langle\phi^{(0)}|\phi^{(0)}\rangle}. \qquad (8.46)$$

On evaluating (8.46) using (8.44) and (8.45) we find,

$$A^2[\sum_{k,m}^{\text{occ}}\alpha_k^2(\epsilon_f - \epsilon_k) + 2\sqrt{N}V\sum_{k,m}^{\text{occ}}\alpha_k] - E^{(0)}[A^2(1 + \sum_{k,m}^{\text{occ}}\alpha_k^2) - 1]. \quad (8.47)$$

On minimizing with respect to α_k we find

$$\alpha_k = \frac{\sqrt{N}V}{E^{(0)} - \epsilon_f + \epsilon_k}. \qquad (8.48)$$

Using this result in equation (8.47) we obtain an equation for the ground state energy $E^{(0)}$

$$E^{(0)} = \frac{N\Delta}{\pi}\ln\left|\frac{\epsilon_f - E^{(0)}}{D}\right|, \qquad (8.49)$$

using a flat density of states of width $2D$ with the Fermi level at the centre. Writing $E^{(0)} = \epsilon_f - k_B T_A$, (8.49) can be seen to correspond to the ground state energy to leading order in $1/N$ found in section 7.2, equation (7.17). The occupancy of the f level $n_f^{(0)}$ can be calculated via

$$n_f^{(0)} = \sum_{k,m}^{\text{occ}}|\langle\epsilon_f \epsilon_k^h m|\phi^{(0)}\rangle|^2 = \frac{N\Delta/\pi k_B T_A}{1 + N\Delta/\pi k_B T_A}, \qquad (8.50)$$

which agrees with (7.20). Our earlier expression for the susceptibility (7.23) follows straightforwardly on including a magnetic field and taking the second derivative of $E^{(0)}(H)$ with respect to H.

The ground state solution (8.49) for $V \neq 0$ is non-perturbative and does not evolve from the ground state for $V = 0$, $\epsilon_f < 0$ ($|\Omega\rangle$ plus an f electron in the impurity state) which is not a singlet and hence has the wrong symmetry. The parallel in the perturbative approach developed in section 7.2 is that there the new ground state is found as a pole in the resolvent $R_0(z)$ in the singlet channel rather than in the resolvent for the magnetic channel $R_1(z)$, which gives the ground state only for $V = 0$.

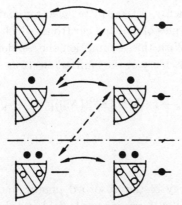

Figure 8.20 A schematic representation of the basis states used by Gunnarsson & Schönhammer (1987) in the variational calculation of the ground state energy. Solid circles show electrons and open circles holes in the conduction band and localized f level (horizontal line). The arrow indicates the states which are coupled by the one-body hybridization; those with the full line by V and those with the dashed line by $V/\sqrt{N}$.

The fact that the ground state wavefunction can be constructed entirely from the states, $|\Omega\rangle$ and $|\epsilon_f \epsilon_k^h m\rangle$, for $N \to \infty$ is because in this limit $V \to 0$ as $N\Delta$ remains finite. As a consequence the higher order states constructed by the application of the hybridization term in the Hamiltonian on the state $|\Omega\rangle$ beyond the first are of higher order in $1/N$.

These states are shown schematically in figure 8.20, and the order in $1/N$ to which they contribute to the ground state energy is indicated. The states which must be included for calculations to order $1/N$ are $|\epsilon_{k_1}^p \epsilon_{k_2}^h m\rangle$ (where p indicates a particle state) and $|(\epsilon_f \epsilon_k^h m)(\epsilon_{k_1}^p \epsilon_{k_2}^h m')\rangle$, corresponding to the second line in figure 8.20. Generalizing the ansatz (8.44) for the ground state wavefunction to include these states gives the ground state energy to order $1/N$ (see Bickers, 1987, for details).

As mentioned earlier, the main application of the variational approach has been to the calculation of spectra. In valence photoemission experiments on rare earth and actinide compounds absorption of a high energy photon can cause an f electron to be emitted. If the system is initially in the ground state $|E^{(0)}(N_e)\rangle$ for N_e electrons the photon induces a transition of the system to a state $|E^{(n)}(N_e - 1)\rangle|\mathbf{k}\rangle$, where the state $|\mathbf{k}\rangle$ describes the emitted electron of energy ϵ and the state $|E^{(n)}(N_e - 1)\rangle$ is a state of the system with one less electron. In the 'sudden' approximation it is assumed that the emitted electron has no interaction with the remaining $N_e - 1$ electrons. This approximation gets better the higher the kinetic energy of the emitted electron and is a reasonable approximation for energies of emitted electrons, which are typically in the range 50-1000 eV. With this assumption the Golden Rule can be used to estimate of the intensity of the emitted electrons of energy ϵ. This is proportional to

$$P(\epsilon) = \sum_{\mathbf{k},n} |\langle E^{(n)}(N_{\mathrm{e}} - 1)|\langle \mathbf{k}|H_{\mathrm{I}}|E^{(0)}(N_{\mathrm{e}})\rangle\rangle|^2 \delta(\epsilon - \epsilon_{\mathbf{k}})$$

$$\delta(\epsilon + E^{(n)}(N_{\mathrm{e}} - 1) - \omega - E^{(0)}(N_{\mathrm{e}})),$$

(8.51)

where ω is the energy of the absorbed photon and H_{I} is the dipolar interaction between the electromagnetic field and the electronic system. For processes in which an f electron is emitted the relevant term in the interaction has the form,

$$H_{\mathrm{I}} = \sum_{\mathbf{k},m} \langle \mathbf{k}|\mathbf{d}|f, m\rangle c_{\mathbf{k}}^\dagger X_{0,m},$$

(8.52)

where $\langle \mathbf{k}|\mathbf{d}|f, m\rangle$ is the dipole matrix element. If the state $|\mathbf{k}\rangle$ is a high energy scattering state then it will have small weight in the initial state $|E^{(0)}(N_{\mathrm{e}})\rangle$ so we can assume $c_{\mathbf{k}}|E^{(0)}(N_{\mathrm{e}})\rangle = 0$. With these assumptions (8.51) can be written in the form,

$$P(\epsilon) = \sum_{\mathbf{k}} |\langle \mathbf{k}|\mathbf{d}|f, m\rangle|^2 \delta(\epsilon - \epsilon_{\mathbf{k}}) \mathrm{Im} G_{\mathrm{f}}^<(\epsilon - \omega - is)/\pi,$$

(8.53)

for $s \to +0$, where $G_{\mathrm{f}}^<(\omega)$ is given by

$$G_{\mathrm{f}}^<(\omega) = \langle E^{(0)}(N_{\mathrm{e}})|X_{m,0}(\omega + H - E^{(0)}(N_{\mathrm{e}}))^{-1} X_{0,m}|E^{(0)}(N_{\mathrm{e}})\rangle,$$ (8.54)

where H is the Hamiltonian describing the system. We will assume for the high energy excitations we are considering here that it is reasonable to neglect the intersite correlation of the f states and use the impurity Anderson model. Over the energy range of the spectrum it is usually reasonable to neglect the ϵ dependence of the matrix element in (8.53) and then $P(\epsilon)$ is proportional to $\rho_{\mathrm{f}}^<(\epsilon - \omega)$, where

$$\rho_{\mathrm{f}}^<(\epsilon - \omega) = \mathrm{Im} G_{\mathrm{f}}^<(\epsilon - \omega - is)/\pi,$$

(8.55)

and is the density of the occupied f states.

The density of the unoccupied f states above the Fermi level can be measured by bremsstrahlung isochromat spectroscopy (BIS). In these experiments a system is exposed to an electron beam and the emission of photons induced by the electronic transitions is studied. Under similar assumptions to the derivation of (8.53), the energy distribution of the photons emitted due to the absorption of an electron in an f state is proportional to $\rho_f^>(E - \omega)$, where E is the energy of the incoming electrons and $\rho_f^>(\omega)$ is given by

$$\rho_f^>(\omega) = \mathrm{Im} G_f^>(\omega - is)/\pi, \tag{8.56}$$

where $G_f^>(\omega)$ is given by

$$G_f^>(\omega) = \langle E^{(0)}(N_e)|X_{0,m}(\omega - H + E^{(0)}(N_e))^{-1}X_{m,0}|E^{(0)}(N_e)\rangle. \tag{8.57}$$

The two response functions, $G_f^<(\omega)$ and $G_f^>(\omega)$, are the two components of the retarded Green's function $G_f(\omega)$ at $T = 0$ which are associated with the emission and absorption of a f electron. It follows straightforwardly from (7.109) that

$$G_f(\omega) = G_f^<(\omega) + G_f^>(\omega). \tag{8.58}$$

The calculation of $G_f^<(\omega)$ or $G_f^>(\omega)$ is performed by introducing a complete set of many particle states states for the system with $N_e - 1$ electrons so, for example,

$$G_f^<(\omega) = \sum_{n,n'} \langle E^{(0)}(N_e)|X_{m,0}|E^{(n)}(N_e - 1)\rangle\langle E^{(n)}(N_e - 1)|$$

$$(\omega + H - E^{(0)}(N_e))^{-1}|E^{(n')}(N_e - 1)\rangle\langle E^{(n')}(N_e - 1)|X_{0,m}|E^{(0)}(N_e)\rangle,$$

$$\tag{8.59}$$

which requires the inverse of the matrix,

$$\bar{H}_{n,n'} = \langle E^{(n)}(N_e - 1)|(\omega + H - E^{(0)}(N_e))|E^{(n')}(N_e - 1)\rangle, \tag{8.60}$$

to be calculated. This calculation is made tractable by invoking the $1/N$ expansion which allows the basis states $|E^{(n)}(N_e - 1)\rangle$, which have non-zero matrix elements, $\langle E^{(0)}(N_e)|X_{m,0}|E^{(n)}(N_e - 1)\rangle$, to be truncated.

The density of f states below the Fermi level at $T = 0$, $\rho_f^<(\omega)$, is of order $1/N$ (this follows from (8.11) as n_f is restricted to the range $0 < n_f < 1$). To carry out the calculation to this order it is sufficient to use the wavefunction $|\phi^{(0)}\rangle$ calculated earlier in (8.44) which is exact to order $(1/N)^0$ for the ground state $|E^{(0)}(N_e)\rangle$. This is because the resolvent term in (8.59) contributes a term of order $1/N$. The state $X_{0,m}|\phi^{(0)}\rangle$ corresponds to a single hole in the Fermi sphere, with the f state empty, as illustrated in figure 8.21.

The only state to the same order in $1/N$ which is mixed with this state by the hybridization term in the Hamiltonian is $|(\epsilon_k^h, m)(\epsilon_{k'}^h, \epsilon_f, m')\rangle$ for $m \neq m'$ (the term $m = m'$ gives a contribution of higher order in $1/N$). This state and the states which give the next order contributions are indicated in figure 8.21. The matrix elements needed to evaluate (8.60)

PES

BIS

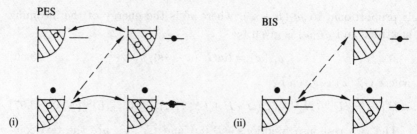

(i)   (ii)

Figure 8.21 A schematic representation of the basis states used in the evaluation of (i) the photoelectric spectrum (PES) and (ii) the bremsstrahlung isochromat spectrum (BIS). The notation used is as in figure 8.20.

to order $1/N$ are

$$\langle \epsilon_k^h, m | \omega - E^{(0)}(N_e) + H | \epsilon_{k'}^h, m' \rangle = (\omega - E^{(0)} - \epsilon_k)\delta_{k,k'}\delta_{m,m'}, \quad (8.61)$$

$$\langle (\epsilon_{k'}^h, \epsilon_f, m')(\epsilon_k^h, m) | \omega - E^{(0)}(N_e) + H | \epsilon_{k_1}^h, m_1 \rangle = V\delta_{k,k_1}\delta_{m,m_1} + \mathrm{O}(1/N), \quad (8.62)$$

$$\langle (\epsilon_{k'}^h, \epsilon_f, m')(\epsilon_k^h, m) | \omega - E^{(0)}(N_e) + H | (\epsilon_{k_2}^h, m_2)(\epsilon_{k_1}^h, \epsilon_f, m_1) \rangle =$$
$$(\omega - E^{(0)} - \epsilon_k^h - \epsilon_{k'}^h + \epsilon_f)\delta_{k',k_1}\delta_{k,k_2}\delta_{m',m_1}\delta_{m,m_2}. \quad (8.63)$$

The matrix element required in the evaluation of (8.59) is $\langle \epsilon_k^h, m | (\omega - E^{(0)}(N_e) + H)^{-1} | \epsilon_{k'}^h, m' \rangle$, which can be calculated by writing $\bar{H}$ in block form,

$$\bar{H} = \begin{bmatrix} \bar{H}_{1,1} & \bar{H}_{1,2} \\ \bar{H}_{2,1} & \bar{H}_{2,2} \end{bmatrix}, \quad (8.64)$$

where the blocks are generated from the subspaces spanned by $|\epsilon_k^h, m\rangle$ and $|\epsilon_k^h, m(\epsilon_{k'}^h, \epsilon_f, m')\rangle$, and then using

$$(\bar{H}_{1,1})^{-1} = (\bar{H}_{1,1} - \bar{H}_{1,2}\bar{H}_{2,2}^{-1}\bar{H}_{2,1})^{-1}, \quad (8.65)$$

which simplifies the calculation as $\bar{H}_{2,2}$ is diagonal so $\bar{H}_{2,2}^{-1}$ is trivially calculated. Using these results to calculate (8.59) we find

$$G_f^<(\omega) = \frac{A^2}{N} \sum_{k}^{occ} \frac{\alpha_k^2}{\omega - E^{(0)} - \epsilon_k - \Sigma^{(0)}(\omega - E^{(0)} - \epsilon_k - \epsilon_f)}, \quad (8.66)$$

where $\Sigma^{(0)}(\omega)$ is given by

$$\Sigma^{(0)}(\omega) = N \sum_{k}^{occ} \frac{|V_k|^2}{(\omega - \epsilon_k)}. \quad (8.67)$$

Using the result for α_k in (8.48) and $A^2 = (1 - n_f^{(0)})$, we find for a flat band, $\rho_{\mathrm{f}}^<(\omega)$ deduced from (8.66) is

$$\rho_{\mathrm{f}}^<(\omega) = \frac{(1 - n_{\mathrm{f}}^{(0)})N\Delta}{N\pi^2}$$

$$\mathrm{Im}\int_{-D}^0 \frac{1}{(\epsilon - k_{\mathrm{B}}T_{\mathrm{A}})^2} \frac{1}{(\omega - E^{(0)} - \epsilon - \Sigma^{(0)}(\omega - E^{(0)} - \epsilon - \epsilon_{\mathrm{f}}) - is)}\, d\epsilon.$$

(8.68)

The integrand has an isolated pole at $\epsilon = \omega$ and a branch cut for $\epsilon > \omega + k_{\mathrm{B}}T_{\mathrm{A}}$. For $-k_{\mathrm{B}}T_{\mathrm{A}} < \omega \leq 0$, only the isolated pole makes a contribution to the integral and gives

$$\rho_{\mathrm{f}}^<(\omega) = \frac{(1 - n_{\mathrm{f}}^{(0)})^2}{N}\frac{N\Delta}{\pi}\frac{1}{(\omega - k_{\mathrm{B}}T_{\mathrm{A}})^2},$$

(8.69)

on using equations (8.49), (7.16) and (7.20). As $\tilde{\Delta} = \pi n_{\mathrm{f}}^{(0)}k_{\mathrm{B}}T_{\mathrm{A}}/N = (1 - n_{\mathrm{f}}^{(0)})\Delta$, this can be seen to correspond to the tail below the Fermi level in the many-body quasi-particle resonance at $\omega = k_{\mathrm{B}}T_{\mathrm{A}}$ (evaluated to order $1/N$) and multiplied by the wavefunction renormalization factor $z = (1 - n_{\mathrm{f}}^{(0)})$. The continuum contribution appears for ω below $\omega = -k_{\mathrm{B}}T_{\mathrm{A}}$. This contribution has the f-spectral weight associated with the f level at $\omega \sim \epsilon_{\mathrm{f}}$ seen in earlier results for the density of states (see figure 8.6). It arises from the resonance in the integrand in (8.68) for $\epsilon \sim \omega - E^{(0)}$, giving a peak at $\omega \sim \epsilon_{\mathrm{f}}$ for $D \gg -\epsilon_{\mathrm{f}} \gg \Delta$, with a width of the order of $N\Delta$. This peak is not present in the leading order $(1/N)^0$ calculation of the spectral density which is simply a delta function quasi-particle peak above the Fermi level at $\omega = k_{\mathrm{B}}T_{\mathrm{A}}$. For the non-interacting Anderson model, which corresponds to the model with $N = 1$, there is only the lower peak at $\omega \sim \epsilon_{\mathrm{f}}$. A very severe test of the approximation used here is to compare the results for $N- = 1$ with the exact results. Gunnarsson & Schönhammer consider the case with ϵ_{f} well below the Fermi level and $N = 1$ and find results very close to the exact ones when all the states shown in the two rows of figure 8.21(i) are used. To this order the results are significantly better than those for the NCA, which still has vestiges of a Kondo peak at the Fermi level due to the neglect of vertex corrections. This result suggests that the 'intermediate states' method of Gunnarsson & Schönhammer to this order should give good results for $\rho_{\mathrm{f}}^>(\omega)$ for all values of N.

We can see within the intermediate states calculations, as in the zero band width calculations in appendix C, that the weight in $\rho_{\mathrm{f}}^<(\omega)$ associated with the Kondo resonance at the Fermi level arises because, after the f electron is removed from the initial state, there is a probability that a conduction electron will fall into the empty f level to give an f^1 final state. The total energy cost of this process is due to the hole created at the Fermi level, and any further particle–hole excitations. The energy involved is relatively small and so $\rho_f^<(\omega)$ is peaked at $\omega = 0$.

Calculation of the BIS part of the spectrum $\rho_f^>(\omega)$ from (8.57) proceeds along similar lines. The intermediate states used in this calculation are $|\epsilon_f, m\rangle$, $|\epsilon_k^p, m\rangle$ and $|(\epsilon_f, \epsilon_{k'}^h, m', m)(\epsilon_k^p m)\rangle$. The final result for $G_f^>(\omega)$ is

$$G_f^>(\omega) = \frac{A^2}{\omega - \epsilon_f + E^{(0)} - \Sigma^{(1)}(\omega)}, \qquad (8.70)$$

where

$$\Sigma^{(1)}(\omega) = \frac{\Delta}{\pi} \int_0^D \frac{d\epsilon}{\omega - \epsilon + E^{(0)} - \Sigma^{(0)}(\epsilon + \epsilon_f - \omega - E^{(0)})}, \qquad (8.71)$$

for a flat band density of states for the conduction electrons. The integrand in (8.71) has a pole at $\epsilon = \omega$. This pole contribution gives for the imaginary part of $\Sigma^{(1)}(\omega)$ for $0 \leq \omega < k_B T_A$,

$$\operatorname{Im} \Sigma^{(1)}(\omega - i0^+) = (1 - n_f^{(0)})\Delta. \qquad (8.72)$$

As the real part of $\Sigma^{(1)}(\omega)$ is of order $1/N$ it can be neglected in a calculation of $\rho_f^>(\omega)$ to order $1/N$. Substituting (8.72) into (8.70) and then evaluating for $\rho_f^>(\omega)$ gives to order $1/N$,

$$\rho_f^>(\omega) = \frac{(1 - n_f^{(0)})^2}{N} \frac{N\Delta}{\pi} \frac{1}{(\omega - k_B T_A)^2}, \qquad (8.73)$$

for $0 \leq \omega < k_B T_A$, so that this joins smoothly the result for $\rho_f^<(\omega)$ at $\omega = 0$. The total spectral weight in $\rho_f^>(\omega)$ is of the order $(1/N)^0$, as can be seen from (8.10), in contrast with that for $\rho_f^<(\omega)$ from (8.11) which is of order $(1/N)^1$. The missing order $(1/N)^0$ weight in $\rho_f^>(\omega)$ is associated with the pole in (8.70) which to order $(1/N)^0$, neglecting $\Sigma^1(\omega)$ which is of order $1/N$, gives a delta function contribution at $\omega = k_B T_A$,

$$\rho_f^>(\omega) = (1 - n_f^{(0)})\delta(\omega - k_B T_A), \qquad (8.74)$$

which we found earlier, equation (8.8). For $\omega > k_B T_A$ there is a continuum contribution of order $1/N$ arising from (8.71). We know from our earlier results for the quasi-particle density of states (7.69) that $\rho_f^>(\omega)$ has a resonance of width $n_f \pi k_B T_A / N$ at $\omega = k_B T_A$. The results, (8.73) and (8.74), are consistent with that for the quasi-particle resonance on expanding to first order in $1/N$ as $(1 - n_f^{(0)})\Delta$ is equal to $n_f \pi k_B T_A / N$ to leading order in $1/N$.

If $\Sigma^1(\omega)$ in (8.71) is retained in the denominator of (8.70), instead of making an expansion to order $1/N$, a non-singular form is found for $\rho_f^>(\omega)$ with a peak at $\omega \sim k_B T_A$. If this is used for $\rho_f^>(\omega)$ near

$\omega = 0$ there is some slight mismatch with $\rho_f^<(\omega)$ due to the different approximations which have been used for the two quantities.

In considering the BIS spectrum it is unrealistic to put $U = \infty$ as there is an observable level at $\omega \sim \epsilon_f + U$ corresponding to the addition of a further f electron to an occupied f state. This is the peak seen above the Fermi level in earlier calculations of the spectral density for $N = 2$ (see figures 5.7–5.10). This f^2 peak has been included in the calculations of Gunnarsson & Schönhammer (1985). The calculations we have presented are also unrealistic in that we have included only the f states arising from the lowest spin–orbit split f configurations. In the original calculations of Gunnarsson & Schönhammer the spin-orbit splitting was neglected so the degeneracy factor N was the orbital degeneracy of the f states so $N = 14$. Later calculations (Gunnarsson & Schönhammer, 1987) for Ce took account of the spin-orbit splitting into the ground state set $j = \frac{5}{2}$, $N = 6$, and the excited states $j = \frac{7}{2}$, $N = 8$. We show a comparison of the results of this type of calculation with experiment in the next chapter.

Other spectra calculated within the $1/N$ approach by Gunnarsson & Schönhammer are for core level X-ray photoemission spectroscopy (XPS) and X-ray absorption spectroscopy (XAS). To calculate the core level spectroscopy a term H_c is added to the Anderson model to describe the core level and the interaction of the core hole with the f electrons,

$$H_c = \epsilon_c c^\dagger c + U_{fc}(1 - c^\dagger c)\sum_m n_{f,m}, \qquad (8.75)$$

where ϵ_c is the core level, $c^\dagger$, c the corresponding creation and annihilation operators and U_{fc} (~ 10 eV) the matrix element of the Coulomb interaction between the f electrons and the core hole. This interaction has a strong effect on the f states, an f^2 state a few electron volts above the Fermi level can be pulled down several electron volts below the Fermi level when a core hole is created. As a result from an initial, primarily f^1, state there can be final states corresponding to f^0, f^1, and f^2, giving three peaks in the resulting spectrum. These can be observed in the spectra of Ce compounds as we shall see in the next chapter. Calculation of the spectrum involves the evaluation of response functions of the form (8.54) with the creation operators $X_{m,0}$ and $X_{0,m}$ replaced by $c^\dagger$ and c, and an intermediate states calculation using the $1/N$ truncation proceeds along similar lines to the PES and BIS cases.

9

Theory and Experiment

9.1 Introduction

In chapter 1 we introduced the s-d and the Anderson models as the basic models for magnetic impurities in simple metallic hosts. In the succeeding chapters we have outlined techniques for predicting the static and dynamic behaviour of these models over most of the possible parameter regimes. These techniques give results which are either exact or which are within well controlled approximations (for a detailed summary of these results see appendix K). We know from the comparisons between theory and experiment which we have made so far, that the physical picture which emerges is in broad agreement with the experimental observations. The parameter regime of primary interest is the Kondo regime where there is local moment behaviour at high temperatures or high magnetic fields with a Curie law susceptibility. This undergoes a broad transition or crossover at temperatures of the order T_K (in weak or zero fields) to Fermi liquid behaviour and a Pauli susceptibility at low temperatures. The $\ln T$ contributions calculated by Kondo explain the resistance minimum, and the Fermi liquid theory gives the power law behaviour of the resistance at very low temperatures. The narrow peak in the one electron density of states, the Kondo resonance, qualitatively explains the basic trends in the change in thermodynamic behaviour with temperature, such as the peak in the specific heat. One appealing feature of these theories is that there is only one relevant energy scale, $k_B T_K$, for the thermodynamic and low frequency behaviour, and hence only one parameter to be determined. Once this is fixed we have complete predictions for a range of physical observables, such as $\chi_{\text{imp}}(T)$, $C_{\text{imp}}(T)$, $M_{\text{imp}}(H)$, for all relevant temperatures and field strengths. However, these models are somewhat oversimplified for most physical situations. For rare earth impurities, for example, crystal field terms should be included, and other spin–orbit split multiplets as well as the ground state one in some cases should also be included. The effect of these terms can be calculated by most of the techniques we have described. Such terms do, however, introduce other relevant energy scales which complicates the comparison with experiment. Whether such terms play an important role depends on the energy scale being considered. On some energy scales these additional terms can be absorbed as a renormalization of the parameters of the simple models, giving a change in T_{K} which does not affect the basic physics.

In taking the theory a stage further, to make possible a more detailed comparison between theory and experiment, and also to be able to look at a greater range of experimental observations, we classify the experiments into two broad categories according to the energy scale involved. The first category is for experiments on a high energy scale, of the order of several electron volts, and includes experiments such as optical experiments and photoemission. The second category includes experiments on the thermodynamic scale, 1–500 K. This covers a diverse range of experiments which can be subdivided into measurement of (i) thermodynamic properties (susceptibility, specific heat), (ii) transport properties (resistivity, magnetoresistance, thermopower), and (iii) dynamic responses on this energy scale such as neutron scattering. Finally there are experiments which we have not considered so far which involve hyperfine coupling to the nuclei, such as nuclear magnetic resonance (NMR), nuclear orientation (NO), and Mössbauer experiments (ME). These are local probes and give detailed information about the charge and spin densities induced in the host in the neighbourhood of the impurity. The energy scale these experiments sample depends on the distance from the impurity; the asymptotic behaviour at large distances depends on the excitations close to the Fermi level while the near neighbour response depends on an average over the complete energy scale associated with the impurity. There have been several reviews over the years devoted primarily to experimental work on impurity systems; Daybell & Steyert, 1968; Wohlleben & Coles, 1973; Daybell, 1973; Rizzuto, 1974; Grüner, 1974, and Grüner & Zawadowski, 1978.

It seems logical to start with the highest energy scale, so we begin with the optical and photoemission experiments. This also gives us continuity as this was the last topic considered in chapter 8.

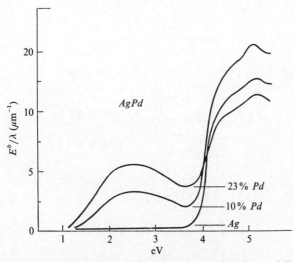

Figure 9.1 Optical interband absorption for dilute alloys of Pd in Ag as a function of the photon energy (Myers, Walldén & Karlsson, 1968). An extra absorption peak develops at $\omega \sim 2.6$ eV with increase in the concentration of Pd.

9.2 High Energy Spectroscopies

Optical experiments in the 1960s were one of the first types of measurement to give direct evidence in support of the virtual bound state picture of impurity states in a metallic host. To detect these levels large impurity concentrations were used, of the order of 10–20%. The positions of the virtual bound states do not, however, vary significantly with concentration, increasing the concentration mainly has the effect of increasing the weight of the impurity excitations in the total spectrum. The main experiments are absorption measurements in thin films, where the photons induce electronic interband transitions. The intensity of absorption depends on the densities of the initial and final states which are separated by ω, the frequency of the absorbed photon. Results of early experiments of this type by Myers, Walldén & Karlsson (1968) for different concentrations of Pd in Ag are shown in figure 9.1. The extra absorption due to the impurity in this case gives a broad peak at $\Delta E = 2.6$ eV. This was interpreted as due to 4d levels of Pd at $\epsilon_d = -2.6$ eV below the Fermi level in the form of a broad resonance of width $\Delta \sim 2.8$ eV. The impurity absorption is clearly seen in this case because it falls below the strong interband absorption in Ag which starts at 4 eV. The impurity absorption may be difficult to detect if it

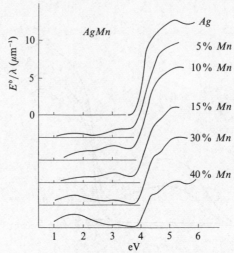

Figure 9.2 Optical interband absorption for dilute alloys of Mn in Ag as a function of the photon energy (Myers, Walldén & Karlsson, 1968). Two peaks develops with increase of Mn content at $\omega \sim 1.6$ eV and $\omega \sim 3.2$ eV.

falls within a region of strong conduction electron absorption of the host metal.

In the same series of experiments measurements were made on Mn impurities in Ag which has local moment behaviour. In this case impurity absorption peaks were seen at $\Delta E = 1.6$ eV and $\Delta E = 3.2$ eV as seen in figure 9.2. These peaks are from the joint density of states and it is not clear from these results alone whether these peaks are from states above or below the Fermi level but this can be resolved from additional information from photoemission data. On taking the photoemission data into account the second peak was found to be due to excitations from occupied d levels below the Fermi level, $\epsilon_d \sim -3.2$ eV, and the lower energy peak due to a vacant level at $\epsilon_d + U \sim 1.6$ eV. This would imply a value of $U \sim 4.8$ eV. The peak width Δ was estimated as 0.5 eV. As $U > \pi\Delta, |\epsilon_d|$ this is consistent within a simple Anderson model description (1.66) with the observation of local moment behaviour. This model, however, neglects the orbital degeneracy and is clearly not sufficient as a description of Mn which has a moment corresponding to 4–5 unpaired electrons. A model appropriate for Mn impurities will be considered in the next section.

Impurity states have been observed by optical absorption for Ni in Cu at much lower impurity concentrations, 0.5-4 atomic %, by Drew & Doezema (1972). An impurity induced absorption peak was found at

$\Delta E = 0.75$ eV with a width $\Delta = 0.27$ eV. This was interpreted as a partially filled virtual bound state resonance below the Fermi level at $\epsilon_d \sim -0.75$ eV. Ni in Cu does not have local moment behaviour. If we have a d states below the Fermi level with a single hole above the Fermi level, corresponding to a configuration $3d^9$, and yet there is no local moment behaviour this can be due to one of two reasons, either because U is less than $\max\{\pi\Delta, |\epsilon_d|\}$, or because T_K is large, say greater than 500 K, so that in the usual thermal energy range the moment is largely compensated. Larger T_K values occur as $\epsilon_d \to \epsilon_F$ in the approach to the mixed valence regime where the Schrieffer–Wolff transformation begins to break down. With a lower resonance width $\Delta \sim 0.27$ eV and $|\epsilon_d| = 0.75$ eV it seems rather unlikely that U is less than $\pi\Delta$ or $|\epsilon_d|$ for Ni. A more likely explanation for the absence of a Curie–Weiss term for Ni in Cu is that it is close to the mixed valence regime with a large T_K. If this is the case then an absorption peak associated with the d hole above the Fermi level should be seen at $\Delta E \sim \epsilon_d + U$. With $U \sim 5$ eV as in Mn this would correspond to $\Delta E = 4.25$ eV, which falls outside the range of the original experiments which were for $0.1 eV < \Delta E < 3 eV$. It would be interesting to investigate this system further to understand in more detail why there is no moment. In any realistic model of this situation the orbital degeneracy should be taken into account. An explicit model for an impurity with one hole in the d shell d^9, which is applicable to Ni, will be introduced in the next section.

In photoemission (PES) and bremsstrahlung isochromat spectroscopy (BIS) an electron is removed or added to the system. We saw in the last chapter that in the sudden approximation these spectra can be directly related to the spectral density of the one electron Green's function. As a consequence there have been more detailed comparisons between theory and experiment for these type of experiments, particularly recently for rare earth systems. If we take a rare earth system in a ground state configuration $4f^n$, then in f electron PES we are considering the transition $4f^n \to 4f^{n-1}$, to the final state $4f^{n-1}$, and in BIS, $4f^n \to 4f^{n+1}$, to the final state $4f^{n+1}$. The PES experiments devised to measure the 4f spectra must be at a photon energy where the 4f cross-section is much greater than for the other channels so that the f transitions predominate. If we adopt the simple zero configuration width ionic model (1.80) these excitation energies are given by

$$E_n - E_{n-1} = \epsilon_f + U(n-1), \quad E_{n+1} - E_n = \epsilon_f + Un, \qquad (9.1)$$

where the energy level ϵ_f is measured relative to the Fermi level. These

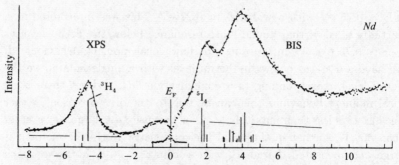

Figure 9.3 Photoemission and BIS spectra for Nd. The vertical bars indicate the positions and intensities of the 4f final states calculated within intermediate coupling (Lang, Baer & Cox, 1981).

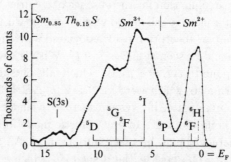

Figure 9.4 The measured photoemission spectrum for the mixed valent alloy $Sm_{0.85}Th_{0.15}S$ (Campagna & Wertheim, 1974). Two distinct peaks are seen associated with the transition $4f^6 \rightarrow 4f^5$ (Sm^{2+}, near the Fermi level) and $4f^5 \rightarrow 4f^4$ (Sm^{3+}, well below the Fermi level).

excitations should show up as peaks in the PES and BIS spectra respectively. In figure 9.3 experimental results on Nd for these two spectra are put together, the two peaks are clearly seen (the second peak in the BIS spectrum is from a higher energy multiplet). The energy difference between the peaks, $E_{n+1} - 2E_n + E_{n-1}$, is equal to U in the zero configuration width approximation. This gives an estimate of $U \sim 6$ eV which is typical for 4f elements. If the ground state is an admixture of the $4f^n$ and $4f^{n-1}$ configurations, the average f occupation n_f lies between n and $n - 1$, we then have a mixed or intermediate valence situation. In the model with $V = 0$ this occurs when the excitation $4f^n \rightarrow 4f^{n-1}$ is at the Fermi level, $E_n - E_{n-1} \sim \epsilon_F$. In this case there are three main peaks to be seen in the spectra of the order of U apart. As well as the peak at the Fermi level, $4f^n \leftrightarrow 4f^{n-1}$, and the BIS peak, $4f^n \rightarrow 4f^{n+1}$, there is a photoemission peak corresponding to the tran-

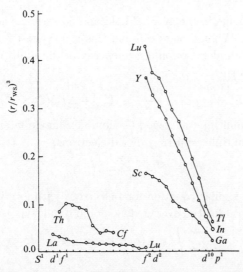

Figure 9.5 The ratio of the d and f shell volume to the Wigner–Seitz volume for transition elements in the 3d, 4d, 5d and 4f, 5f series (van der Marel & Sawatzky, 1988).

sition, $4f^{n-1} \rightarrow 4f^{n-2}$. The two lower peaks are seen clearly in the experimental PES results in figure 9.4 corresponding to the transitions $4f^{n-1} \rightarrow 4f^{n-2}$, $4f^n \rightarrow 4f^{n-1}$, for Sm in $Sm_{0.85}Th_{0.15}S$ which is a mixed valence alloy.

There is clear evidence from the profiles and subsidiary peaks in figures 9.2 and 9.3 of the multiplet structure which has been neglected in the derivation of (9.1), in making the zero configuration width approximation (1.80). The structure arises from the fact that in removing or adding an electron there is a probability that the system will be left in an excited state rather than the ground state in the final configuration $4f^{n-1}$ or $4f^{n+1}$. The splittings between these excited configurations can be estimated from atomic calculations using Russell–Saunders coupling, classifying the states in terms of $|n, L, S, \lambda\rangle$, the total orbital quantum number L, spin S, and seniority quantum number λ. As discussed briefly in section 1.4, the 3d and 4f electrons are quite localized, due to the orbital contribution to the effective potential in the equation which determines the spatial part of the wavefunction, and lie well within the Wigner–Seitz cell in solids. This is seen clearly for the 4f electrons in cerium in figure 1.4. It can be seen for the other elements in the lanthanide series in figure 9.5 where the ratio of the f and d volumes (calculated within Hartree–Fock theory) to the Wigner–Seitz cell is plotted for elements in the 4f, 5f, 3d, 4d and 5d series. They all show localization in the Wigner–Seitz cell and this is strongest for the 4f series. In Russell–Saunders coupling the energy of the multiplets can be expressed in the form,

$$E_n(L, S, \lambda) = n\epsilon_0 + \frac{n(n-1)}{2} U_{av} + \Delta E_n(L, S, \lambda), \qquad (9.2)$$

where U_{av} is the multiplet averaged Coulomb exchange interaction and $E_n(L, S, \lambda)$ the multiplet splitting. For d electrons,

$$U_{av}^{dd} = F_l^0 - \frac{2}{4l+1} J(l, l), \qquad (9.3)$$

with $l = 2$, and a similar equation for f electrons with $l = 3$, where F_l^0 is the Slater integral for the averaged Coulomb repulsion,

$$F_l^0 = \frac{1}{2l+1} \sum_{m < n} \int |\psi_{lm}(\mathbf{r}_1)|^2 \frac{1}{r_{12}} |\psi_{ln}(\mathbf{r}_2)|^2 \, d\mathbf{r}_1 d\mathbf{r}_2, \qquad (9.4)$$

and $J(l, l)$ is the average value of the Hund's rule exchange integral for electrons in the same shell,

$$J(l, l) = \frac{1}{2l+1} \sum_{m < n} \int \psi_{lm}^*(\mathbf{r}_1) \psi_{ln}^*(\mathbf{r}_2) \frac{1}{r_{12}} \psi_{lm}(\mathbf{r}_2) \psi_{ln}(\mathbf{r}_1) \, d\mathbf{r}_1 d\mathbf{r}_2.$$
$$(9.5)$$

Estimates of the multiplet splitting $\Delta E_n(L, S, \lambda)$ from spectra in the solid are of the same order as those obtained from Hartree–Fock calculations in the atom, and empirical estimates from atomic optical experiments. However, F_l^0 is very much reduced in a solid from its value in the 3d and 4f shells of an atom where it is of the order of 20 eV. The reduction is due to the screening effects of the conduction electrons in the solid. Herring (1966) gave a prescription for estimating the screened interaction for 3d electrons in the solid. Instead of taking the energy differences between the configurations d^{n+1}, d^{n-1}, and $2(d^n)$ in the atom, as in (9.1), he took the energy difference between the atomic configurations d^{n+1}, $d^{n-1}s^2$, and $2(d^n s)$. The s electron screens the d hole, and mimics the screening effects of the conduction electrons in the solid. Theoretical and experimental estimates for the peaks in the PES and BIS spectra, taking the screening into account in this way, are in good agreement (Herbst et al, 1978; Cox, Lang & Baer, 1981). Estimates of the multiplet splittings $\Delta E_n(L, S, \lambda)$ from atomic calculations and their relative weights appear to be quite consistent with the peaks and line profiles of the spectra in figures 9.1 and 9.2, and for most of the other rare earth elements (Cox et al, 1981).

The PES and BIS spectra for Ce and Ce intermetallic compounds however cannot be explained satisfactorily in this way. The interpretation of the spectra for these systems has been the subject of controversy. The peak at about 2 eV below the Fermi level seen in the PES of Ce has been interpreted as due to the transition $4f^1 \to 4f^0$. A transition $4f^1 \to 4f^2$ has been observed in the BIS spectrum around $\omega \sim 4$ eV giving $U \sim 6$ eV, in line with theoretical estimates. Subsidiary peaks, however, are seen near the Fermi level in both α and γ Ce and many Ce intermetallic compounds. These peaks do not fit in with the ionic scheme, at least for $V = 0$. A promotional or mixed valence model would give a $f^1 \to f^0$ peak at the Fermi level but would leave unexplained the peak at -2 eV.

The most extensive theoretical interpretation of the spectra of cerium compounds has been the work of Gunnarsson & Schönhammer (1983, 1987) based on the Anderson model. Use of the impurity model is based on the assumption that the intersite correlations between the f states are weak and can be neglected on the energy scale of the PES and BIS experiments. Evidence in support of this assumption will be discussed in the next chapter. For the moment we take it as a working hypothesis and see whether with this assumption we can explain the anomalous features in the spectra of the cerium compounds. To be able to make predictions within a well controlled approximation a number of simplifications have to be made. In the Gunnarsson & Schönhammer theory the multiplet splittings are neglected, or in some cases put into the calculation 'by hand' . The Coulomb interaction between the f and conduction electrons, which can be expressed in the form,

$$U_{\text{fc}} \sum_{k,k',\sigma,\sigma'} n_{\text{f},\sigma'} c_{k,\sigma}^\dagger c_{k',\sigma}, \tag{9.6}$$

is also not included explicitly. This term is unlikely to be small as the f electrons being rather localized contribute to screening the nucleus so a change in f occupation gives a significant change in the Coulomb interaction on the conduction electrons. The assumption in the use of the Anderson model is that this term can be taken into account via a renormalization of the parameters of the model such as ϵ_f and U_{ff}, which are to be regarded as effective parameters. We shall discuss this point later. The Gunnarsson & Schönhammer calculations were based on the $1/N$ expansion as outlined in section 8.4. Their later calculations do include a finite U_{ff} and take account of spin–orbit splitting of the states.

As the energy scale being considered in the PES and BIS experiments is of the order of several electron volts it is unreasonable to neglect the

ω dependence of the width function $\Delta(\omega)$ by taking it to be independent of ω as in the flat band case. In the earlier calculations of Gunnarsson & Schönhammer the semi-elliptical form (4.16) was used. In later calculations the ω dependence of $\Delta(\omega)$ was deduced from the observed spectra at a photon energy where the spectrum is primarily due to conduction electrons. An instrumental width was allowed for by convoluting the calculated spectra with either a Gaussian or Lorentzian. Finite lifetime effects outside the scope of the Anderson model were taken into account by a Lorentzian broadening Γ which in the BIS spectrum was taken to be of the form,

$$\Gamma(\omega) = \Gamma_0 + \Gamma_1(\omega - \epsilon_F)^2. \qquad (9.7)$$

The multiplet effects in the final f^2 peak were simulated by performing calculations with $U_i = U + \epsilon_i$, where ϵ_i is the shift of a particular multiplet in the f^2 final state. The spectrum corresponding to the different multiplets was then built up by combining the spectra for the different ϵ_is with relative weights w_i, where ϵ_i and w_i were taken from the atomic calculations within the intermediate coupling scheme of Lang et al (1981). Thus the overall weight of the f^2 peak was calculated but the relative positions of the multiplets and their relative weights were put in by hand. For further details of the fitting procedures in individual cases we refer the reader to the original papers.

In figure 9.6 we show a fitting of the Gunnarsson & Schönhammer theory to the experimental results for the BIS, PES and 3d XPS spectra for the compound $CeNi_2$. The spin–orbit splitting of the f levels is included. The inset shows the form of $\Delta(\omega)$ deduced experimentally and used in the PES calculations. What is seen clearly in the PES and BIS spectra are peaks in the neighbourhood of the Fermi level. These arise from a f^1 final state in each case. Excitations of this type can be clearly seen in the zero band width calculations in appendix C. The leading contribution to the peaks at ϵ_F is the Kondo resonance. This has significant weight in this compound due to the relatively large admixture of the f^0 configuration in the ground state, $n_f \sim 0.8$, $\Delta_{av} \sim 0.1$ eV. In the PES spectrum the f^0 peak, which might have been expected to be near $\epsilon_f \sim -1.6$ eV is split because the hybridization factor $\Delta(\omega)$ has a peak in this region. Some of the weight is lowered to give a peak at $\omega \sim -3.5$ eV and some of the weight is pushed to higher energies and contributes to the weight seen near ϵ_F. Other factors contribute to the weight near ϵ_F. In finite U calculations there is some weight due to the admixture of f^2 in the ground state, and there are also continuum

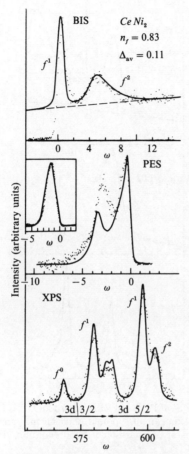

Figure 9.6 The 3d XPS, 4f PES and BIS spectra for $CeNi_2$ (Allen et al, 1986). The dots show the experimental points and the full curve the theoretical fit to the calculations of Gunnarsson & Schönhammer (1987). Spectral weight in the vicinity of the Fermi level is clearly seen in the PES and BIS spectra.

contributions. The f^1 peak on the BIS side is more prominant than in the PES as expected due to the large degeneracy factor. The f^2 peak seen in the BIS spectrum about 5 eV above the Fermi level has multiplet effects included as described earlier. The XPS spectrum has 3d 3/2 and 3d 5/2 components which partially overlap. Each component has f^0, f^1 and f^2 peaks, and multiplet effects are not included here, which might explain why the f^1 and f^2 peaks are narrower than the experimental ones. Other cerium compounds with features at the Fermi level which

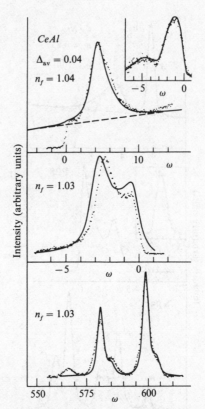

Figure 9.7 The 3d XPS, 4f PES and BIS spectra for $CeAl$ (Allen et al, 1986). The dots show the experimental points and the full curve the theoretical fit to the calculations of Gunnarsson & Schönhammer (1987).

have been interpreted in a similar way are $CeRu_2$ and $CeIr_2$ (Allen et al, 1986).

Results for the compound $CeAl$ shown in figure 9.7 in contrast have no distinctive Fermi level feature. The parameters extracted from the fit in this case are consistent with this picture as $n_f \sim 1.03$ and $\Delta_{av} \sim 0.04$, so there is very little f^0 weight in the ground state and a much smaller T_K. The Kondo peak being rather narrow, of the order of T_K/N, and only having a weight of order T_K/Δ, is not observable spectroscopically in this case.

An interesting general question is whether the parameters, ϵ_f and Δ_{av}, obtained from fitting the spectra (high energy scale) are consistent with the parameters deduced from fits to the thermodynamic data (low energy scale). Gunnarsson & Schönhammer (1987) calculated n_f and

χ_{imp} within the same large N method and found values consistent with the thermodynamic data. They found that there were two important factors in obtaining this agreement. One was the degeneracy factor N. The f^1 peak has a width of order $N\Delta$ rather Δ and so attempts to fit the data with a non-degenerate model can lead to estimates of Δ which are different by an order of magnitude. The other important factor in obtaining agreement is that the systems are predominantly in the Kondo regime so that the low energy scale is determined by Kondo temperature T_{K}. If they were in the mixed valence regime (as in the promotional model) the low energy scale would be of the order of Δ which would be too large to explain the thermodynamic properties.

The conclusion of Gunnarsson & Schönhammer that the same set of parameters can explain both the photoemission and thermodynamic data has been questioned by Patthey et al (1990). They fit the photoemission spectra using the results of NCA calculations for the single impurity model. The parameters which they find give good fits to the spectra predict values of γ and $\chi(0)$ which are about a factor of 10 larger than those observed experimentally (apart from $YbAl_3$, which is a mixed valent compound, where smaller values than those observed are predicted). We will see in the next section that this point is a difficult one to settle as the Kondo temperature T_{K}, setting the scale for the low energy thermodynamic behaviour, depends on the crystal field splittings and the coupling to the higher multiplets, so it is almost impossible to calculate T_{K} at all accurately from first principles. It may also be important to use the lattice model to predict the low temperature thermodynamics, but not so necessary in calculating the on-site f Green's function for the photoemission spectra.

One interaction which could affect the high and low energy scales differently is the Coulomb interaction between the localized d or f electrons. Terms of this type were omitted in the calculations of Gunnarsson & Schönhammer due to the difficulty in finding a reliable approximation to take them into account along with the hybridization, degeneracy and the Coulomb interaction U_{ff}. It is possible, however, to include these interactions within the numerical renormalization group scheme as described in section 5.6. Brito & Frota (1990) have calculated the XPS spectra for $N = 2$ in this way including a Coulomb interaction U_{dc} of the form (9.6) between the d core hole state and the conduction electrons. For $U_{\text{dc}} > U_{\text{ff}}$ they found a qualitative change in the spectrum and reduced values of the effective hybridization needed to explain the peak widths (the effective hybridization is that which would be required

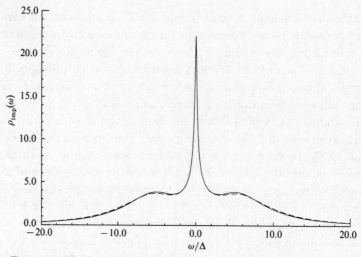

Figure 9.8 The f spectral density for the Anderson model with $U_{fc} = U_{ff} = 4\pi\Delta$, $\Delta = 0.01$ (dashed line) compared with that for the standard model with $U_{fc} = 0$, $U_{ff} = 2.4\pi\Delta'$, $\Delta' = 0.015 = 1.5\Delta$(solid line) (Costi & Hewson, 1992a) .

to give similar results using the model with $U_{dc} = 0$). They conclude that for this spectrum, contrary to the conclusions of Gunnarsson & Schönhammer, different effective parameters may be needed to explain the XPS and low energy data.

Costi & Hewson (1992a) have used the numerical renormalization group to investigate the effect of the f electron–conduction electron interaction, U_{fc} in (9.6), on the 4f spectral density and thermodynamic properties. For $0 < U_{fc} < U_{ff}$ in the parameter regime with a distinct Kondo peak they deduced effective parameters, ϵ'_f, U'_{ff} and Δ' for fitting with a model with $U_{fc} = 0$. The effective width Δ' was calculated from the density of states at the Fermi level, assuming the Friedel sum rule (for $U_{fc} = 0$), and ϵ'_f and U'_{ff} from the PES and BIS peak positions. The 4f spectrum and thermodynamic properties were recalculated with $U_{fc} = 0$, using the effective parameters, and then compared to the results of the original calculation. The thermodynamic properties as a function of temperature were very well reproduced by the calculation with the renormalized parameters. The 4f spectrum also agreed well, as can be seen in the comparison for the symmetric model shown in figure 9.8. The effective hybridization width Δ' obtained was significantly larger than Δ. The asymmetry of the spectral line due to the excitation of particle–hole pairs (Doniach & Sunjić, 1970), which can be clearly seen in results for $U_{ff} = 0$, $U_{fc} \neq 0$, is very much reduced for $U_{ff} > U_{fc}$ and is not observable in figure 9.8. They conclude that, except in extreme parameter regimes (e.g. $\Delta \to 0$), the high and low energy behaviour for U_{fc} in the range $0 < U_{fc} < U_{ff}$ can be described by the standard Anderson model ($U_{fc} = 0$) provided all the parameters, including the hybridization, are renormalized. If there is a discrepancy between the parameters needed to explain the photoemission and thermodynamic results (apart possibly from the XPS data) it does not appear to be due to the neglect of the U_{fc} interaction.

One further point of controversy is whether the single impurity Anderson model is sufficient to describe the temperature dependence of the peaks in the photoemission spectra near the Fermi level. Patthey et al (1990) fit the photoemission peaks just below the Fermi level in $CeSi_2$ to the high temperature collapse of the Kondo resonance as calculated within the NCA. This interpretation has been challenged by Joyce et al (1992) who see no effects beyond those which can be attributed to the Fermi factor and phonon broadening. Lawrence et al (1991) come to similar conclusions for the peaks seen just below the Fermi level in $YbCu_2Si_2$. If this is correct it could be due to the temperature dependence of the chemical potential which has to be taken into account in lattice models, as we shall see in the next chapter, but is absent in single impurity models. For the photoemission spectra for anomalous rare earth ions there are some very convincing fittings of the overall features to the theory, but questions as to the consistency of the low energy and high energy parameters, the temperature dependence, and the validity of the impurity model need further clarification.

9.3 Thermodynamic Measurements

The thermodynamic measurements of primary interest for magnetic impurity systems are measurements of the impurity contributions to the magnetic susceptibility, the specific heat and the induced magnetization in an applied field. The calculation of these quantities has been our main concern so far. The models on which these calculations have been based have neglected spin–orbit splittings and crystal field splittings which need to be taken into account in making predictions for specific systems for comparison with experimental results. There is a difference between the transition metal and rare earth impurities in the relative order of these effects. As the 4f electrons in rare earth ions lie much

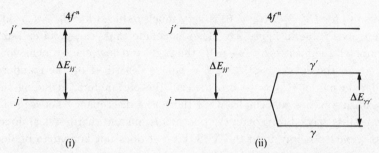

Figure 9.9 Lower multiplets associated with a rare earth ion in a configuration $4f^n$: (i) two multiplets with a spin–orbit splitting $\Delta E_{jj'}$ and (ii) with the lowest multiplet split by a crystal field with an excitation energy $\Delta E_{\gamma\gamma'}$.

closer to the nucleus than the 3d electrons in transition metal ions, and lie within the 5d shells, they are to a large extent shielded from the local environment. As a consequence the spin–orbit interactions are stronger than the crystal field interactions for rare earth impurities. For a rare earth atom the spin–orbit splittings are typically of the order of 0.1 eV, and the crystal field splittings of the order of 50 K. For transition metal impurities the crystal field terms are the most important, of the order of 0.2-1 eV. If the orbital degeneracy of the ground state multiplet is removed entirely (fully quenched) the spin–orbit interaction cannot cause further splitting. There can, however, be a g factor correction and g factor anisotropies in this case due to the mixing in of higher unquenched states. If the orbital degeneracy is not fully removed by the crystal fields there can be further splittings of the levels due to the spin–orbit interaction. Because of the differences between the transition metal and rare earth cases we consider them separately. We look at the rare earth case first.

Consider the case of a rare earth ion with a stable $4f^n$ configuration with a ground state Hund's rule multiplet $|n, L, S\rangle$. We assume that this is split by a spin–orbit coupling into multiplets, $|n, L, S, j\rangle$ and $|n, L, S, j'\rangle$ with energies E_j, $E_{j'} = E_j + \Delta E_{jj'}$ with $\Delta E_{jj'} > 0$ so that the j multiplet lies lowest (see figure 9.9(i)). The degeneracy factor will be denoted by N_j, where $N_j = 2j + 1$. The one-body hybridization V mixes in states of the $4f^{n+1}$ and $4f^{n-1}$ configurations. Not to get too complicated we take into account the lowest of the $4f^{n-1}$ multiplets only, which we take to be non-degenerate of energy E_0. This would be appropriate to the case of Ce with $n = 1$, $E_j - E_0 + \epsilon_F = \epsilon_f$, $j = \frac{5}{2}$ and $j' = \frac{7}{2}$. The Hamiltonian of the ionic model for this situation is given by

$$H = \sum_{m'',j''=j,j'} E_{j'',m''} X_{j'',m'';j'',m''} + \sum_{k,m'',j''=j,j'} \epsilon_{k,j''} c^\dagger_{k,j'',m''} c_{k,j'',m''} +$$
$$E_0 X_{0;0} + \sum_{k,m'',j''=j,j'} (V_k^{j''} X_{j'',m'';0} c_{k,j'',m''} + V_k^{j''*} c^\dagger_{k,j'',m''} X_{0;j'',m''}),$$

(9.8)

where we will take V_k^j to be independent of j for simplicity.

We can calculate the ground state variationally using the arguments of section 8.4. We use the ansatz,

$$|\phi\rangle = A\{1 + X_{j,m:0} \sum_{k<k_{\rm F}} \alpha_k c_{k,j,m} + X_{j',m':0} \sum_{k<k_{\rm F}} \alpha'_k c_{k,j',m'}\}|\Omega\rangle, \quad (9.9)$$

where $|\Omega\rangle$ is the Fermi sphere. If $\Delta E_{jj'} \to \infty$ then $\alpha'_k \to 0$ and this becomes equivalent to (8.44). In that case the energy gain in forming a singlet ground state was found to be $-k_{\rm B}T_{\rm A}$ with $T_{\rm A}$ given by (7.17). In the Coqblin–Schrieffer regime, $E_j - E_0 - \epsilon_{\rm F} < 0$, $|E_j - E_0 - \epsilon_{\rm F}| \gg N_j\Delta$, the gain is $-k_{\rm B}T_{\rm K}$, with $T_{\rm K}$ given by

$$k_{\rm B}T_{\rm K} \sim De^{-\pi|E_j - E_0 - \epsilon_{\rm F}|/N_j\Delta}, \quad (9.10)$$

for the simplest variational ansatz (8.44). We can ask how this result is modified by taking $\Delta E_{jj'}$ finite and including the higher order multiplet ($\alpha'_k \neq 0$). The equation for the ground state energy $E^{(0)}$ becomes

$$E^{(0)} = N_j|V|^2 \sum_{k<k_{\rm F}} \frac{1}{E^{(0)} - E_j + E_0 - \epsilon_k}$$
$$+N_{j'}|V|^2 \sum_{k<k_{\rm F}} \frac{1}{E^{(0)} - E_{j'} + E_0 - \epsilon_k},$$

(9.11)

where we have taken the matrix element V for the mixing with the conduction electrons to be the same for j and j'. Writing $E^{(0)} = E_j - E_0 - \epsilon_{\rm F} - k_{\rm B}T_{\rm A}$ leads to a generalized equation for $T_{\rm A}$,

$$k_{\rm B}T_{\rm A} = E_j - E_0 - \epsilon_{\rm F} - \frac{N_j\Delta}{\pi}\ln\left|\frac{k_{\rm B}T_{\rm A}}{D}\right| - \frac{N_{j'}\Delta}{\pi}\ln\left|\frac{k_{\rm B}T_{\rm A} + \Delta E_{jj'}}{D}\right|,$$

(9.12)

In the Coqblin–Schrieffer limit $T_{\rm A} \to T_{\rm K}$, $T_{\rm K}$ is given by

$$k_{\rm B}T_{\rm K} \sim D\left(\frac{D}{k_{\rm B}T_{\rm K} + \Delta E_{jj'}}\right)^{N_{j'}/N_j} e^{-\pi|E_j - E_0 - \epsilon_{\rm F}|/N_j\Delta}. \quad (9.13)$$

If $\Delta E_{jj'} \gg k_{\rm B}T_{\rm K}$ then $T_{\rm K}$ is given by

$$k_{\rm B}T_{\rm K} \sim D\left(\frac{D}{\Delta E_{jj'}}\right)^{N_{j'}/N_j} e^{-\pi|E_j - E_0 - \epsilon_{\rm F}|/N_j\Delta}. \quad (9.14)$$

For Ce with a lower lying $j = \frac{5}{2}$ state and an excited multiplet $j = \frac{7}{2}$,

$N_{j'}/N_j = 4/3$, and as $D/\Delta E_{jj'}$ can be typically of the order of 10, the enhancement of the prefactor due to the j' multiplet can be significant. Hence even though $\Delta E_{jj'} \gg k_B T$, and thermal fluctuations to the excited multiplet are not important, virtual fluctuations can lead to significant renormalizations of T_K, which is the parameter which determines the low energy scale. For $k_B T_K \gg \Delta E_{jj'}$, then from (9.13)

$$k_B T_K \sim D e^{-\pi |E_j - E_0 - \epsilon_F|/(N_j + N_{j'})\Delta}, \tag{9.15}$$

which is the Kondo temperature associated with a multiplet without spin–orbit splitting of degeneracy $N_j + N_{j'} = (2L+1)(2S+1)$.

The same arguments apply to the case of crystal field splittings (Hanzawa, Yamada & Yosida, 1985). If the lowest multiplet $|n, L, S, j\rangle$ is split into two multiplets, $|n, L, S, j, \gamma\rangle$ and $|n, L, S, j, \gamma'\rangle$ as in figure 9.9(ii), with degeneracies, N_γ and $N_{\gamma'}$, and energies, E_γ, $E_{\gamma'} = E_\gamma + \Delta E_{\gamma, \gamma'}$ ($\Delta E_{\gamma, \gamma'} > 0$, $N_\gamma + N_{\gamma'} = N_j$), we can consider the effect on the ground state energy by a variational ansatz of the form (9.9). The arguments are equivalent with $N_j \to N_\gamma$, $N_{j'} \to N_{\gamma'}$, $E_j \to E_\gamma$ and $E_{j'} \to E_{\gamma'}$. We conclude therefore that if $\Delta E_{\gamma\gamma'} \ll k_B T_K$ then the low temperature thermodynamics is governed by the energy scale $k_B T_K^{(j)}$ with $T_K^{(j)}$, the Kondo temperature associated with the unsplit multiplet, given by (9.15) as $N_\gamma + N_{\gamma'} = N_j$. If on the other hand $\Delta E_{\gamma\gamma'} \gg k_B T_K$, T_K will be that appropriate to the lowest crystal field multiplet with a degeneracy factor N_γ and an enhancement factor, $(D/\Delta E_{\gamma, \gamma'})^{N_{\gamma'}/N_\gamma}$. These results can be applied to Ce $n = 1$, $j = \frac{5}{2}$ split by a cubic crystal field into a Γ_7 doublet ($N_{\Gamma_7} = 2$) and a Γ_8 quartet ($N_{\Gamma_8} = 4$). Note that in assessing the relative importance of the crystal field splitting in the Coqblin–Schrieffer regime the relevant ratio is $T_K^{(j)}/\Delta E_{\gamma, \gamma'}$ and not $\Delta/\Delta E_{\gamma, \gamma'}$ (this is only a relevant ratio in the mixed valence regime).

We can come to similar conclusions using the poor man's scaling arguments. As the conduction band width is reduced we can take into account the renormalizations due to the spin–orbit splittings and crystal fields using generalizations of the arguments in sections 3.3 and 3.4. We have the advantage in the scaling derivation of being able to stop the scaling at a finite temperature T. We apply the arguments to the crystal field split case after having transformed to an equivalent Coqblin–Schrieffer model, the exchange part of which takes the form,

$$H_{ex} = J_\gamma \sum_{k,m,m'} X_{\gamma,m:\gamma,m'} c^\dagger_{k,\gamma,m'} c_{k',\gamma,m}$$
$$+ J_{\gamma'} \sum_{k,m,m'} X_{\gamma',m:\gamma',m'} c^\dagger_{k,\gamma',m'} c_{k',\gamma',m}. \tag{9.16}$$

If we derive this from a generalized Anderson model of the form (9.8) with hybridization matrix elements V independent of γ and $|E_\gamma - E_0 - \epsilon_F| \gg \Delta$, $|E_\gamma - E_{\gamma'}| \ll |E_\gamma - E_0 - \epsilon_F|$ then $J_\gamma = J_{\gamma'} = J$. The leading order perturbative scaling equation generated by reducing D, using the arguments given in section 3.3 and appendix D, is

$$\frac{dJ\rho_0}{dD} = (J\rho_0)^2 \left(\frac{N_\gamma}{D} + \frac{N_{\gamma'}}{D + \Delta E_{\gamma\gamma'}} \right). \qquad (9.17)$$

Integrating this equation using the condition $\tilde{J}\rho_0 \to \infty$ for $\tilde{D} = k_B T_K$ gives the equation for T_K,

$$k_B T_K \sim D \left(\frac{D}{k_B T_K + \Delta E_{\gamma\gamma'}} \right)^{N_{\gamma'}/N_\gamma} e^{-\pi|E_\gamma - E_0 - \epsilon_F|/N_\gamma \Delta}, \qquad (9.18)$$

which is the equivalent of (9.13), where we have used $J\rho_0 = \Delta/\pi(E_\gamma - E_0 - \epsilon_F)$. Terminating the scaling at $\tilde{D} \sim k_B T$ gives the effective coupling,

$$\tilde{J}\rho_0 = \frac{1}{N_\gamma \ln(T/T_K) + N_{\gamma'} \ln((k_B T + \Delta E_{\gamma\gamma'})/(k_B T_K + \Delta E_{\gamma\gamma'}))}, \qquad (9.19)$$

with T_K given by (9.18).

These results can be rewritten in terms of the Kondo temperature associated with the unsplit multiplet $T_K^{(j)}$. The equation for T_K then becomes

$$T_K = T_K^{(j)} \left(\frac{T_K^{(j)}}{T_K + \Delta E_{\gamma\gamma'}} \right)^{N_{\gamma'}/N_\gamma}. \qquad (9.20)$$

For $\Delta E_{\gamma\gamma'} \ll T_K^{(j)}$, $T_K \to T_K^{(j)}$ and for $\Delta E_{\gamma\gamma'} \gg T_K$,

$$T_K \to T_K^{(j)} \left(\frac{T_K^{(j)}}{\Delta E_{\gamma\gamma'}} \right)^{N_{\gamma'}/N_\gamma}, \qquad (9.21)$$

which is the Kondo temperature T_K^γ associated with the lower crystal field split state. The expression (9.19) for $\tilde{J}\rho_0$ can be written in the form,

$$\tilde{J}\rho_0 = \frac{1}{N_\gamma \ln(T/T_K^{(j)}) + N_{\gamma'} \ln((k_B T + \Delta E_{\gamma\gamma'})/k_B T_K^{(j)})}. \qquad (9.22)$$

This form corresponds to a universal scaling law for the thermodynamics in the absence of an applied field of the form $F(T/T_K^{(j)}, \Delta E_{\gamma\gamma'}/T_K^{(j)})$ for given N_γ and $N_{\gamma'}$.

Note that in deriving these expressions we have used the leading order scaling equations and the simplest possible form for the variational ansatz. If we go beyond leading order in the scaling, as in appendix D, or beyond the simplest variational ansatz, as in section 8.4, we will

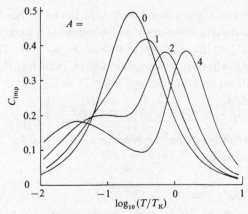

Figure 9.10 The specific heat of the $j = 3/2$ Coqblin–Schrieffer model as a function of temperature for various crystal field splittings $\Delta_{\gamma\gamma'}$, with $A = \Delta_{\gamma\gamma'}/T_{\rm K}^{(3/2)}$, of the quadruplet into two Kramers doublets (Desgranges & Rasul, 1987).

generate higher order corrections to the prefactors for $T_{\rm K}$ but they will not affect our general conclusions.

The scaling laws can be verified from exact Bethe ansatz results for the Coqblin–Schrieffer model in the presence of a crystal field (model (9.16) with $J_{\gamma'} = J_\gamma$). If the couplings J_γ are all equal the integrability of the linear dispersion model is not affected and the crystal field splittings are included in a similar way to the splittings due to an external field, through the boundary conditions on the thermodynamic equations (Schlottmann, 1984b; Kawakami & Okiji, 1986; Desgranges & Rasul, 1985). The crystal field has formally also to be applied to the conduction electrons but this has no real consequences in calculating the impurity response other than a small renormalization of the band width cut-off D'. Using arguments similar to those used in section 6.6 it can be verified that the thermodynamic quantities for $H = 0$ are universal functions of $T/T_{\rm K}^{(j)}$ and $\Delta E_{\gamma,\gamma'}/T_{\rm K}^{(j)}$. The specific heat coefficient can be expressed in the form,

$$\gamma_{\rm imp} = \frac{a}{T_{\rm K}^{(j)}} \left(\frac{\Delta E_{\gamma,\gamma'}}{k_{\rm B} T_{\rm K}^{(j)}} \right)^{N_{\gamma'}/N_\gamma} + f\left(\frac{\Delta E_{\gamma,\gamma'}}{k_{\rm B} T_{\rm K}^{(j)}} \right), \qquad (9.23)$$

where a is a constant and the leading term is that corresponding to the lowest crystal field state so $f(\Delta E_{\gamma,\gamma'}/k_{\rm B} T_{\rm K}^{(j)}) \to 0$ as $\Delta E_{\gamma,\gamma'} \to \infty$.

The behaviour of the full specific heat curve $C_{\rm imp}(T)$ for a $j = \frac{3}{2}$ state split by a crystal field into two doublets calculated from the Bethe ansatz solution is shown in figure 9.10. As the ratio A of the crystal field split-

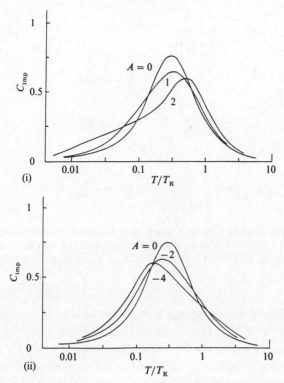

Figure 9.11 The specific heat of the $j = 5/7$ Coqblin–Schrieffer model as a function of temperature for various cubic crystal field splittings: (i) with the Γ_7 state lowest ($N_{\Gamma_7} = 2$), (ii) with the Γ_8 state lowest ($N_{\Gamma_8} = 4$)(Kawakami & Okiji, 1986).

ting $\Delta E_{\gamma,\gamma'}$ relative to $T_K^{(3/2)}$ increases from zero the peak associated with the quartet develops a shoulder on the low temperature side which develops into a Kondo peak associated with the lowest doublet for large A. The main peak moves to higher temperatures and is approximately described by a Schottky type term for large crystal field splitting,

$$C_{\text{imp}}^{\text{Sch}}(T) = k_B \left(\frac{\Delta E_{\gamma,\gamma'}}{k_B T_K^{(j)}} \right)^2 \frac{e^{\beta \Delta E_{\gamma,\gamma'}}}{(1 + e^{\beta \Delta E_{\gamma,\gamma'}})^2}. \tag{9.24}$$

For a cerium impurity with a ground state multiplet with $j = \frac{5}{2}$, which splits into Γ_7 and Γ_8 states in a cubic crystal field, the way the shoulder develops with increasing field depends upon which state lies lowest. Results of Kawakami & Okiji (1986) are shown in figure 9.11. When the Γ_7 lies lowest as in figure 9.11(i) the shoulder develops on the low energy side as in the case shown in figure 9.10. When the Γ_8 is the ground

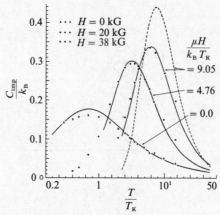

Figure 9.12 The specific heat as a function of temperature for various applied magnetic fields for the $S = 1/2$ s-d model fitted to experimental data (Bader et al, 1975) for Ce impurities in $LaAl_2$ (Rajan, Lowenstein & Andrei, 1982).

state the shoulder develops on the high energy side as is seen in figure 9.11(ii).

On the assumption that the Γ_7 doublet lies well below the Γ_8 quartet Bethe ansatz results for the specific heat in an applied field have been used to fit measurements on cerium impurities in $LaAl_2$. The g factor of the Γ_7 state is $10/7$ so the only free parameter in the fitting is the value of T_K. The results are well fitted with $T_K = 0.2\pm0.002$ K, as shown in figure 9.12, and as the crystal field splitting $\Delta E_{\Gamma_7\Gamma_8} \sim 150$ K the neglect of the higher crystal field term other than through T_K should be reasonable. In the applied field the magnetic field energy in the larger fields is greater than the binding energy $k_B T_K$ so as the magnetic field energies become comparable and greater than $k_B T_K$ the specific heat peak is significantly modified, becoming sharper and moving to higher temperatures. The change of the results with the varying field strength is well reproduced by the theory. The high field results begin to approximate to those of a free spin Schottky anomaly, $C_{imp}/k_B \sim x_0^2 \mathrm{sech}^2 x_0$ with $x_0 = \mu_B/k_B T$. There is a significant difference, however, which is due to the logarithmic Kondo corrections to the free spin theory.

Results for the impurity magnetization as a function of $\mu_B H/k_B T$ for the same system are shown in figure 9.13. The agreement of the Bethe ansatz results with experiment is reasonably good. The experimental results lie systematically above the theoretical curve and this difference could be due to a temperature independent van Vleck term due to a small

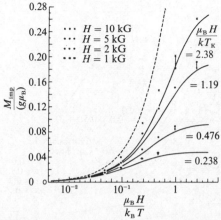

Figure 9.13 The magnetization in constant magnetic field as a function of temperature for the $S = 1/2$ s-d model fitted to experimental data (Felsch, Winzer & von Minnigerode, 1975) for Ce impurities in $LaAl_2$ (Rajan, Lowenstein & Andrei, 1982).

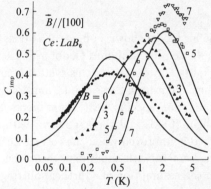

Figure 9.14 The specific heat as a function of temperature for various applied magnetic fields for the $j = 3/2$ Coqblin–Schrieffer model fitted to experimental data (Gruhl & Winzer, 1986) for Ce impurities in LaB_6 (Schlottmann, 1987a).

admixture of the Γ_8 state in the ground state. The van Vleck term has been estimated for each field strength, ignoring any modification due to the conduction electrons, and is indicated by the arrows in figure 9.13. The importance of the logarithmic corrections in explaining the deviations from the free spin results is evident in this case also.

For cerium impurities in LaB_6 the Γ_8 state lies lowest with $\Delta E_{\Gamma_8\Gamma_7} \sim$ 530 K. A fit to the specific heat in a magnetic field to the results of the

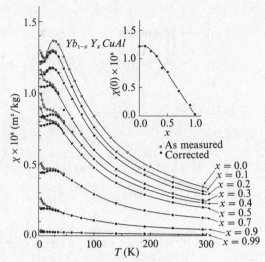

Figure 9.15 Experimental results for the susceptibility as a function of temperature for $Yb_{1-x}Y_xCuAl$ for various values of x. The inset shows the $T = 0$ susceptibility as a function of x (Mattens, 1980).

$j = \frac{3}{2}$ Coqblin–Schrieffer model shown in figure 9.14 does not reproduce the data so well in this case. The standard Coqblin–Schrieffer model in a magnetic field is not quite appropriate to this system in a magnetic field as the Zeeman splittings differ from those of a Γ_8 multiplet which depend on the field direction. The specific heat experiments are for a field applied along the four-fold axis and the agreement with experiment in the presence of the field is improved when the correct Zeeman splittings are used. A better agreement with experiment has been obtained with the zero field data on the assumption of a further splitting of the Γ_8 into two doublets.

There are some rare earth alloys of the form $R_xM_{1-x}C$ where a magnetic rare earth component R replaces a non-rare earth element M, which are either paramagnetic at low temperatures or only order at very low temperatures usually with a reduced magnetic moment. Their similarities with impurities systems have led to them being classified as *dense Kondo systems* or *Kondo lattices*. A good example is $Yb_xY_{1-x}CuAl$. In this case the susceptibility as a function of temperature for $0 < x < 1$, is shown in figure 9.15, where from the inset it can be seen that $\chi(0)$ scales linearly with x over a large concentration range while the position of the maximum is almost independent of x. Though the high temperature susceptibility corresponds to that of a Yb^{3+} ion with a g

factor $g = 8/7$, this system remains paramagnetic down to the lowest temperatures measured even for the stochiometric compound $x = 1$. In these systems the interactions between the magnetic moments seem to be very much reduced compared to normal magnetic rare earth compounds. Their behaviour at higher temperatures can often be quite well explained by neglecting all intersite interactions using the results of the impurity model.

To show how well the single impurity models can fit the observed behaviour for some of these systems in figure 9.16 we give a fit for three thermodynamic properties, $\chi_{\mathrm{imp}}(T)$, $C_{\mathrm{imp}}(T)$ and $M_{\mathrm{imp}}(H)$ (at $T = 0$) for the exact results of the $j = \frac{7}{2}$ Coqblin–Schrieffer model, which is appropriate for a Yb^{3+} ion without crystal field splittings ($k_{\mathrm{B}}T_{\mathrm{K}}^{(j)} \gg \Delta E_{\gamma,\gamma'}$). With $g = 8/7$ there is only one free parameter T_{K} which was obtained by fitting the susceptibility at zero temperature to $\chi_{\mathrm{imp}}(0)$. The theoretical curves give a remarkably good fit. The features depend sensitively on the degeneracy factor N ($= 2j + 1$), the model only gives a positive curvature for $M_{\mathrm{imp}}(H)$ and a peak in $\chi_{\mathrm{imp}}(T)$ for $N \geq 4$. Impurity fits have been made to the data for other Kondo lattice compounds such as $CeSn_3$ and $Ce_xLa_{1-x}Pb_3$ (see Schlottmann, 1989). A naive interpretation of these results in terms of individually compensated Kondo impurities is untenable (Nozières, 1985). We shall leave the discussion of this point to the next chapter where we consider heavy fermions which display even more marked low temperature anomalies than the Kondo lattices.

For Ce and Yb impurities we are dealing with one electron or one hole in the f shell and these systems can be described by the degenerate Anderson model (1.84) or the Coqblin–Schrieffer model (1.89). The predicted ground state for these models is a singlet state consistent with the experimental observations on these systems. For more stable rare earth ions such as Gd the Heisenberg exchange term for the Coulomb interaction with the conduction electrons, which has been omitted from the models we have considered and which is ferromagnetic in sign, is more important. In these systems the f levels lie too far below the Fermi level for the induced antiferromagnetic exchange due to the virtual hopping via the hybridization into the higher configurations to be important. As a consequence these impurities do not lead to Kondo anomalies. There are one or two other rare earth elements where the f levels in some compounds lie close to the Fermi level, such as Sm and Tm, giving low temperature anomalies. The Kondo effect has been observed in dilute alloys with Sm^{3+} ions as impurities. This ion has a magnetic ground

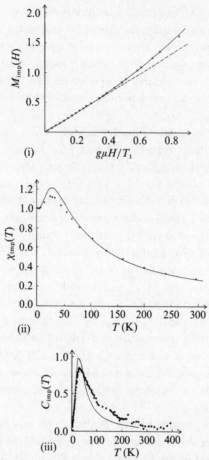

Figure 9.16 Comparison of experimental results for the compound $YbCuAl$ with exact results for the $j = 7/2$ Coqblin–Schrieffer model with $T_K = 66$ K, (i) the magnetic isotherm $M(H)$ as a function of field ($T = 4.2$ K), (ii) the susceptibility $\chi(T)$, and (iii) the specific heat $C(T)$ as a function of temperature. The experimental data is from Mattens (1980), the theoretical curve (i) is from Hewson, Rasul & Newns (1983) and curves (ii) and (iii) from Rajan (1983).

state configuration $^6H_{5/2}$ ($n = 5, L = 5, S = \frac{5}{2}, j = \frac{5}{2}$) which hybridizes with an excited non-magnetic configuration $^0F_3$, corresponding to the half full f shell. It can be described by an ionic model very similar to the degenerate ($U = \infty$) Anderson model or the Coqblin–Schrieffer model. There are also several intermediate valence compounds of Sm where the f levels, corresponding to the excitations $4f^n \rightarrow 4f^{n-1}$, lie at the Fermi level. We shall consider examples, such as SmS under pressure and SmB_6, in the next chapter.

The case of Tm, which has a tendency to mixed valence in the ionic states, Tm^{2+} and Tm^{3+}, is more complicated. This is because the 2+ ionic state corresponds to a configuration $4f^{13}$ which has a ground state, $L = 3, S = \frac{1}{2}, j = \frac{7}{2}$, and the 3+ state, corresponding to $4f^{12}$, has a ground state, $L = 5, S = 1, j = 6$, which also has a magnetic moment. This situation, as both the ground state multiplets are degenerate, cannot be modelled by a degenerate Anderson or Coqblin–Schrieffer model. If we set up an ionic or Hirst type of model for this system, the ionic part without the hybridization can be written in the form,

$$
H = \sum_{-6 \le M \le 6} E_{12} X_{6,M;6,M} + \sum_{-7/2 \le m' \le 7/2} E_{13} X_{\frac{7}{2},m';\frac{7}{2},m'}
$$
$$
+ \sum_{k,j,m} \epsilon_{k,j,m} c^{\dagger}_{k,j,m} c_{k,j,m},
$$

(9.25)

on expressing the conduction electron states in terms of partial waves which are eigenstates of total angular momentum about the impurity site corresponding to quantum numbers j, m. If we assume spherical symmetry about the impurity site the hybridization conserves orbital angular momentum and spin about the impurity site so the hybridization with an f state can only mix in conduction state partial waves with $j = \frac{5}{2}, \frac{7}{2}$. The total angular momentum of a individual f electron is not a good quantum number for a Russell–Saunders' multiplet so the hybridization term between the f states and the conduction electrons takes the form,

$$
H_{\text{hyb}} = \sum_k V_k^{5/2} \sum_{M,m,m'} \langle 6, M | \frac{7}{2} m', \frac{5}{2} m \rangle c^{\dagger}_{k,\frac{5}{2},m} X_{6,M:\frac{7}{2},m'}
$$
$$
+ \sum_k V_k^{7/2} \sum_{M,m'',m'} \langle 6, M | \frac{7}{2} m', \frac{7}{2} m'' \rangle c^{\dagger}_{k,\frac{7}{2},m''} X_{6,M:\frac{7}{2},m'} + \text{h.c.}
$$

(9.26)

where the summations are such that $-6 \le M \le 6$, $-5/2 \le m \le 5/2$ and $-7/2 \le m', m'' \le 7/2$. The terms in the angular brackets are the conventional Clebsch–Gordon coefficients. There are as yet no definitive results for the model obtained by combining (9.25) and (9.26). Diagonalization in the zero band width limit, equivalent to the calculation for the non-degenerate Anderson model as given in appendix C but involving many more states, gives a doublet ground state (Baliña & Aligia, 1990).

Other calculations for variants of this model, using different coupling schemes such that only the $j = \frac{7}{2}$ conduction states are coupled, predict a singlet ground state (Yafet, Varma & Jones, 1985; Read, Dharamvir, Rasul & Newns, 1986; Saso, 1989). There is a version of this model involving two magnetic configurations which has proved to be integrable (Schlottmann, 1985; Aligia, Balseiro & Proetto, 1986). The results show a partial spin compensation in the ground state. A general feature of the solutions of the models with two magnetic configurations is that new low energy scales play a role and give additional specific heat peaks.

The most well known of the anomalous Tm compounds is $TmSe$ which is a mixed or intermediate valence compound. It is unusual for this class of materials because it orders magnetically at low temperatures. It has been argued that this is a consequence of the two lowest configurations being magnetic (Varma, 1976). Dilute alloys with Tm impurities have strong similarities with the compound $TmSe$ and experimental evidence indicates the ground state may be a doublet rather than a singlet. We will look at the neutron scattering evidence for this later.

The theoretical problem of describing thulium impurities has similarities with the case of uranium impurities. There are uranium compounds with anomalous properties which are thought to be due to fluctuations between configurations $5f^2$ and $5f^3$, both of which have low lying magnetic multiplets. Some of the most interesting of the heavy fermion compounds (as we shall see in the next chapter) are uranium compounds. However, due to the complexity of the models in these cases, and the inconclusive results obtained so far, we do not feel we can profitably pursue the subject further here. We refer the interested reader to the papers cited, and the references they quote. We turn instead to look at the case of 3d transition metal impurities.

Models for transition metal impurities, including orbital degeneracy and crystal field splittings, have been considered by Okada & Yosida (1973) and Nozières & Blandin (1980). The simplest case to consider is that for Mn^{2+} because it corresponds to a half full d shell. Due to the Hund's rule coupling it has a spin $S = \frac{5}{2}$ and is an orbital singlet. Experimentally Mn impurities, like other transition metal magnetic impurities, have a $\chi_{\text{imp}}(T)$ which is well described by a Curie–Weiss law at high temperatures which remains finite as $T \to 0$ indicating a singlet ground state. We found in section 6.5 that the s-d model with a local exchange interaction, a $J_{\mathbf{k},\mathbf{k}'}$ independent of $\mathbf{k}, \mathbf{k}'$, for $S = \frac{5}{2}$ predicts a ground state with only partial spin compensation corresponding to $S = 2$. Such an exchange interaction is only appropriate for the scattering of s ($l = 0$) conduction electrons and is not useful for transition metals where the predominant scattering is of d ($l = 2$) conduction electrons. One way of setting up an appropriate model for transition metal impurities is to go back to the Anderson model and include explicitly the orbital degeneracy. The Hamiltonian is of the form,

$$H = \sum_{k,m} \epsilon_k c^\dagger_{k,m,\sigma} c_{k,m,\sigma} + \sum_{k,m} (V_k c^\dagger_{d,m,\sigma} c_{k,m,\sigma} + V_k^* c^\dagger_{k,m,\sigma} c_{d,m,\sigma}) + H_d,$$

(9.27)

where H_d describes the on-site interactions in the impurity d shell. For a spherically symmetric impurity potential hybridization with partial waves of angular momentum $l = 2$ only have to be taken into account and the m index is the z-component of angular momentum which runs from $m = -2$ to $m = 2$. This is similar to the situation for Ce and Yb impurities considered in section 1.9. There we hybridized only with partial wave states with $j = \frac{5}{2}$ (Ce) and with $j = \frac{7}{2}$ (Yb); the difference being that we had to work with j rather than l in that case due to the significant spin-orbit splitting within the the f states. Any deviation from spherical symmetry about the impurity site will result in some hybridization with states with $l \neq 2$ but this should be much weaker than the $l = 2$ contribution and it should be a reasonable approximation to neglect them.

For H_d we take the form,

$$H_d = \sum_{m,\sigma} \epsilon_d c^\dagger_{d,m,\sigma} c_{d,m,\sigma} + \frac{(U + J_H)}{2} \sum_{mm',\sigma\sigma'} c^\dagger_{d,m,\sigma} c^\dagger_{d,m',\sigma'} c_{d,m',\sigma'} c_{d,m,\sigma}$$

$$- \frac{J_H}{2} \sum_{mm',\sigma\sigma'} c^\dagger_{d,m,\sigma} c^\dagger_{d,m',\sigma'} c_{d,m,\sigma'} c_{d,m',\sigma} ,$$

(9.28)

which is characterized by two parameters, U the Coulomb interaction and a Hund's rule coupling J_H (< 0). The Hund's rule coupling aligns the spins so that the ground state multiplet has the maximum spin S. This is not the most general form for the Hamiltonian H_d, as there can be terms which are non-diagonal in m and σ, $c^\dagger_{m_1,\sigma_1} c^\dagger_{m_2,\sigma_2} c_{m_3,\sigma_3} c_{m_4,\sigma_4}$, such that $m_1 + m_2 = m_3 + m_4$ and $\sigma_1 + \sigma_2 = \sigma_3 + \sigma_4$. If these are included there are in total ($l + 1$) parameters (the Slater coefficients) required to specify H_d. If crystal field and spin–orbit interactions are ignored and we have rotational symmetry in both orbital and spin space the most general exchange type of Hamiltonian we can construct for a conduction electron interacting with an L, S multiplet has terms of the form,

$$(\mathbf{S} \cdot \mathbf{s})_{\sigma\sigma'}^{p} (\mathbf{L} \cdot \mathbf{l})_{mm'}^{q}, \qquad (9.29)$$

where $\mathbf{s}$ and $\mathbf{l}$ are the spin and orbital operators for the conduction electron. The index p can take values 0 or 1 (as $s = \frac{1}{2}$), and q the values 0 to $|2\mathrm{Min}(L,l)+1|$. This means in general that we have $2|2\mathrm{Min}(L,l)+1|$ independent exchange parameters. In the case of an orbital singlet $L = 0$, such as Mn^{2+}, we need only two terms. We will derive an explicit form for the effective interaction in the Kondo regime in this case based on the form (9.28) for H_{d}. In the presence of the Hund's rule term this calculation is still a little complicated even in the restricted case of an orbital singlet so we make a further approximation of taking the limit for strong Hund's rule coupling and only include virtual transitions from the ground state configuration $3d^n$ with $n = 2l + 1$ to the lowest multiplets of the configurations $3d^{(n+1)}$, $3d^{(n-1)}$ which are mixed in via the hybridization. This leads to an exchange interaction of the form,

$$H_{\mathrm{exc}} = 2J \sum_{k,k',m} \mathbf{S} \cdot c_{k,m,\sigma}^{\dagger} (\mathbf{s})_{\sigma\sigma'} c_{k',m,\sigma'}, \qquad (9.30)$$

for $-l \le m \le l$, plus a potential scattering term. The spin operator $\mathbf{S}$ is for a localized d spin with $S = (2l + 1)/2$. The exchange coupling J is given by

$$J = \frac{|V|^2}{(2l + 1)} \left(\frac{1}{(\epsilon_{\mathrm{F}} - \bar{\epsilon}_{\mathrm{d}})} + \frac{1}{(\bar{\epsilon}_{\mathrm{d}} + U + 2l|J_{\mathrm{H}}|)} \right), \qquad (9.31)$$

where $\bar{\epsilon}_{\mathrm{d}} = \epsilon_{\mathrm{d}} + 2l(U - |J_{\mathrm{H}}|)$. The model for $l \ne 0$ differs from the s-d model used in section 6.5 ($J_{\mathbf{k},\mathbf{k}'}$ independent of $\mathbf{k}$ and $\mathbf{k}'$) in that the spin S is coupled to $(2l + 1)$ conduction electron channels.

This model can also be generated from the general s-d model (1.64) with a non-local interaction $J_{\mathbf{k},\mathbf{k}'}$ (Schrieffer, 1967). In the case of spherical symmetry about the impurity site $J_{\mathbf{k},\mathbf{k}'}$ depends only of the angle between $\mathbf{k}$ and $\mathbf{k}'$ so we can expand it in terms of the Legendre polynomials $P_l(\cos\theta_{\mathbf{k}\mathbf{k}'})$ (similar to (5.5)),

$$J_{\mathbf{k},\mathbf{k}'} = \sum_{l'} J_{l'} P_{l'}(\cos\theta_{\mathbf{k}\mathbf{k}'}) = \sum_{l'} \frac{4\pi}{2l' + 1} J_{l'} \sum_{m'=-l'}^{l'} Y_{l'}^{m'}(\Omega_{\mathbf{k}}) Y_{l'}^{m'}(\Omega_{\mathbf{k}'}), \qquad (9.32)$$

where $Y_{l'}^{m'}(\Omega_{\mathbf{k}})$ is the spherical harmonic associated with the wavevector $\mathbf{k}$. If the conduction electron states are expressed in terms of partial waves, and only the scattering of conduction electrons of given angular momentum l, m is taken into account, then only the term with $l' = l$ and $m' = m$ is retained in the sum in (9.32). The Hamiltonian is then equivalent to (9.30).

Exact solutions have been generated for this model, (9.30), (known as the n channel Kondo model) via the Bethe ansatz. This is not a straightforward task because the use of linear dispersion in this case leads to a complete uncoupling of the spin and orbital channels and an unphysical solution. Andrei & Destri (1984) overcame this problem by adding a quadratic term to the conduction electron dispersion so $\epsilon(k) = k - k_F + (k - k_F)^2/\Lambda$. Counter-terms can be added to preserve integrability. What is missing in the linear dispersion case is the presence of higher spin complexes of the conduction electrons which can interact with the impurity. When non-linear terms are included these composites can be dynamically generated as the particles no longer move with the same velocity. Once the physically appropriate solution is generated the limit $\Lambda \rightarrow \infty$ can be taken. The counter-terms are irrelevant in this limit. Tsvelick & Wiegmann (1984), who also solved the model, adopted a different strategy. They used linear dispersion but for the Anderson model, (9.27) plus (9.28), for which these problems do not arise, with a special choice of parameters, $U = -J_H$ and $|J_H|$ large. The model has a ground state multiplet with $n_d = n$ if $\bar{\epsilon}_d = \epsilon_d - |J_H|(n - 1) < \epsilon_F < \epsilon_d + |J_H|n$ for $V = 0$. A Schrieffer–Wolff transformation to take into account virtual transitions to the excited configurations to second order in V, as we have just performed in a more general case, leads to a model with an exchange interaction of the form (9.31). Hence solutions corresponding to the model of interest, (9.30), can be obtained with this particular Anderson model in the appropriate parameter regime. With $S = n/2$, as is the case for a the half full shell of orbital angular momentum l with $n = 2l + 1$, the ground state is a singlet. This is in contrast to the s-d model with $S > \frac{1}{2}$ for s-wave scattering ($n = 1$) where a ground state with a spin $S - \frac{1}{2}$ was found in section 6.5. The ground state for (9.30) corresponds to the strong coupling fixed point $J \rightarrow \infty$ in which a conduction electron of spin $\frac{1}{2}$ in each of the channels is antiferromagnetically bound to the localized spin. As $S = n/2$, where n is the number of channels this leads to a fully compensated ground state.

The situation is basically the same as the spin $\frac{1}{2}$ Kondo problem for which the Fermi liquid theory applies. The phenomenological Fermi liquid theory for the low temperature and low field behaviour has been extended to this more general case by Nozières & Blandin (1980) to derive the χ_{imp}/γ_{imp} ratio. We give a version of their derivation based on the approach used in section 5.1 and 7.4. We assume that near the strong coupling fixed point the quasi-particles and their interactions can be described by an Anderson Hamiltonian of the form (9.28) but with renormalized parameters, $\tilde{V}$, $\tilde{U}$ and $\tilde{J}_H$. As mentioned earlier this is not the most general form with a local interaction, there can be off diagonal terms in m. Off diagonal terms do not appear in (9.30); this follows from general symmetry considerations and is not just due to the restrictive form (9.28) used in the derivation. It is assumed therefore that inclusion of such terms is not necessary. The interaction in (9.28) has to be normal ordered in terms of particles and holes so that it comes into play only for finite temperatures or finite fields. We make a Hartree–Fock or molecular field approximation, as in section 5.1, and calculate the specific heat coefficient, spin and charge susceptibility,

$$\gamma_{\mathrm{imp}} = \frac{\pi^2 k_B^2}{3} 2n\tilde{\rho}_{\mathrm{imp}}(\epsilon_F), \tag{9.33}$$

$$\chi_{\mathrm{imp}} = \frac{(g\mu_B)^2}{2} n\tilde{\rho}_{\mathrm{imp}}(\epsilon_F)(1 + (\tilde{U} - \tilde{J}_H(n-1))\tilde{\rho}_{\mathrm{imp}}(\epsilon_F)), \tag{9.34}$$

$$\chi_{\mathrm{c,imp}} = 2n\tilde{\rho}_{\mathrm{imp}}(\epsilon_F)(1 - (\tilde{U}(2n-1) + 3\tilde{J}_H(n-1))\tilde{\rho}_{\mathrm{imp}}(\epsilon_F)). \tag{9.35}$$

The Kondo limit $\chi_{\mathrm{c,imp}} \to 0$ gives us just one condition on the quasiparticle parameters,

$$(\tilde{U}(2n-1) + 3\tilde{J}_H(n-1))\tilde{\rho}_{\mathrm{imp}}(\epsilon_F) = 1. \tag{9.36}$$

Since in the scattering of conduction electrons in (9.28) the channel index m does not change there should be no change in the m channel occupation number $\delta n_{m,\sigma}$ if the chemical potential changes in one of the other channels $m' \neq m$. The response in the m channel to such a change is proportional to $(2\tilde{U} + 3\tilde{J}_H)\tilde{\rho}(\epsilon_F)$, which we shall equate to zero giving $\tilde{J}_H = -2\tilde{U}/3$. Substituting this into (9.35) and equating $\chi_{\mathrm{c,imp}}$ to zero enables us to calculate $\tilde{U}$ in the Kondo limit. With particle–hole symmetry the quasi-particle resonance must be at the Fermi level, and from the Friedel sum rule, which applies here, we find $\tilde{\rho}_{\mathrm{imp}}(\epsilon_F) = 1/\pi\tilde{\Delta}$. All the renormalized parameters can be then expressed in terms of the Kondo temperature T_L, which we define for general n via $\gamma_{\mathrm{imp}} = \pi^2 n k_B/6T_L$, giving

$$\tilde{U} = -\frac{3}{2}\tilde{J}_H = \pi\tilde{\Delta} = 4k_B T_L. \tag{9.37}$$

Note that for $n = 1$ $\tilde{J}_H$ does not vanish but it does not appear in the equations as it is always multiplied by the factor $(n-1)$. These agree with the earlier results (5.22) for $n = 1$. With these values for the renormalized parameters the expression for χ_{imp} becomes

$$\chi_{\mathrm{imp}} = \frac{(g\mu_B)^2 n(n+2)}{12 k_B T_L}. \tag{9.38}$$

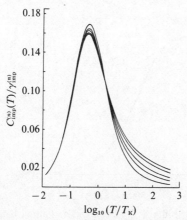

Figure 9.17 The specific heat $C_{\mathrm{imp}}^{(n)}(T)$ of the n channel Kondo model normalized to the linear coefficient of specific heat $\gamma_{\mathrm{imp}}^{(n)}$ for $n = 2S = 1, 2, 3, 5$ and 7 as a function of the logarithm of the temperature (diagram from Schlottmann (1989) based on calculations of Desgranges (1985)).

The Wilson or χ/γ ratio is then given by

$$R = \frac{\chi_{\mathrm{imp}}/\chi_{\mathrm{c}}}{\gamma_{\mathrm{imp}}/\gamma_{\mathrm{c}}} = \frac{4\pi^2 k_{\mathrm{B}}^2}{3(g\mu_{\mathrm{B}})^2} \frac{\chi_{\mathrm{imp}}}{\gamma_{\mathrm{imp}}} = \frac{2(n+2)}{3}. \tag{9.39}$$

This is the Nozières & Blandin result, which was first derived by Yoshimori (1976) using Ward identities under the restrictive assumption of scattering terms diagonal in m only, and more generally by Mihály & Zawadowski (1978). It has been verified by the exact Bethe ansatz results.

Numerical solutions of the integral equations that determine the thermodynamics from the Bethe ansatz solution have been obtained by Desgranges (1985). Results for the specific heat $C_{\mathrm{imp}}(T)$ and the susceptibility $\chi_{\mathrm{imp}}(T)$ as a function of temperature for various values of n are shown in figures 9.17 and 9.18. What is remarkable about these results is the very weak dependence with n over most of the temperature range. In the Fermi liquid regime this follows from the fact that the quasiparticle parameters (9.37) are independent of n. The weak dependence on n extends well beyond the Fermi liquid regime for the specific heat, and extends over the complete temperature range for the susceptibility. The susceptibility monotonically decreases with T presumably because the many-body resonance is always symmetrically placed at the Fermi level independent of n. The approach of the high temperature susceptibility to a Curie law for an isolated moment is very slow for large n and perturbation theory in the coupling is only valid for $T \gg e^n T_{\mathrm{K}}$. In the

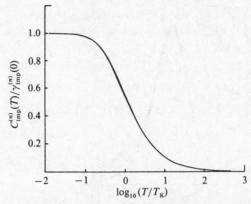

Figure 9.18 The susceptibility $\chi_{\mathrm{imp}}^{(n)}(T)$ of the n channel Kondo model normalized to the value to $T = 0$ for $n = 2S = 1, 2, 3, 5$ and 7 as a function of the logarithm of the temperature. The ratio $\chi_{\mathrm{imp}}^{(n)}(T)/\chi_{\mathrm{imp}}^{(n)}(0)$ is almost independent of n (diagram from Schlottmann (1989) based on calculations of Desgranges (1985)).

temperature range $T_{\mathrm{K}} \ll T \ll e^n T_{\mathrm{K}}$, $T\chi_{\mathrm{imp}}^{(n)}(T) \sim n\ln(T/T_{\mathrm{K}})$. These results are in contrast to those for the N-fold degenerate model for rare earth systems where the Kondo resonance becomes asymmetric with respect to the Fermi level and narrows with increasing N (see figure 7.14) and $\chi_{\mathrm{imp}}(T)$ has a maximum (for $N > 3$).

The results of the n channel Kondo model are in qualitative agreement with experimental observations, such as those of Hedgcock & Li (1970) on Mn impurities in Al where $\chi_{\mathrm{imp}}(T)$ decreases with increase of T from $T = 0$ according to $\chi_{\mathrm{imp}}(T) = A + BT^2$ ($B < 0$) as shown in figure 9.19. The inset shows the linear dependence of the coefficient B with Mn concentration. The same form of decrease is found for other 3d impurity systems such as Fe in Cu where Triplett & Phillips (1971) deduced $\chi_{\mathrm{imp}}(T) = \chi_{\mathrm{imp}}(0)[1 - 15(T/29)^2]$ for $T < 2$ K from measurements of the specific heat at low temperatures in a magnetic field and the thermodynamic relation (6.76). Above 2 K direct susceptibility measurements on this system at weak concentrations can be fitted to the Curie–Weiss form $\chi_{\mathrm{imp}}(T) = \chi_{\mathrm{imp}}(0)[29/(T + 29)]$ (Tholence & Tournier, 1970). Triplett & Phillips also measured the heat capacity of Cr impurities in Cu over a wide temperature range. Their results for $C_{\mathrm{imp}}(T)$ per mole Cr are shown in figure 9.20 fitted to the Hamann theory (1967) (equivalent to the summation of parquet diagrams). The entropy change ΔS was estimated to be in good agreement with $\Delta S = R\ln(2S + 1)$, that required for complete spin compensation, for $S = \frac{3}{2}$.

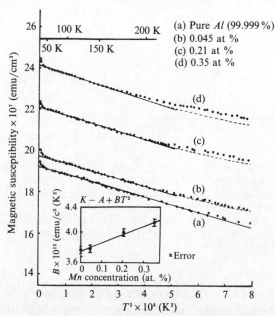

Figure 9.19 Experimental results of Hedgcock & Li (1970) for $\chi_{\text{imp}}(T)$ versus T^2 for Mn impurities in Al. The insert shows the linear dependence of the T^2 coefficient with Mn concentration.

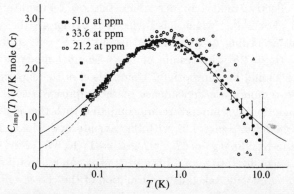

Figure 9.20 The specific heat for Cr in Cu. The solid surve is the fit to the Hamann theory scaled and shifted to fit the peak. The dashed line is a linear extrapolation to $T = 0$ (Triplett & Phillips, 1971).

Sacramento and Schlottmann (1990a) have recently made a detailed comparison of predictions for the n-channel Kondo model with a range of experimental results for Fe impurities in Cu. The n-channel model (9.30) is not strictly appropriate for Fe because unlike Mn^{2+} the ground

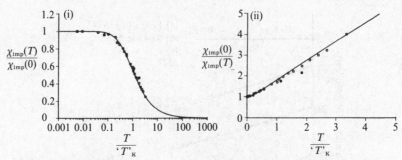

Figure 9.21 Results of the 4-channel Kondo model, (i) for the zero field susceptibility normalized to the value at $T = 0$ and (ii) the inverse susceptibility as a function of T compared with experimental data for dilute Fe in Cu (open squares, results of Steiner et al, 1973; filled squares, results from Tholence & Tournier, 1970). The figure is from Sacramento & Schlottmann (1990) and 'T_K' with their definition corresponds to 'T_K' $= 6T_L/\pi(n + 2) = 18$ K.

state multiplet of Fe^{2+} is not an orbital singlet and should be described by a more general model. The experimental results indicate, however, that the system has a ground state which is a spin and orbital singlet. If the orbital degeneracy is lifted on an energy scale much larger than the Kondo temperature then it would seem reasonable to use the n-channel model as an effective model for describing the low temperature spin compensation. For Fe^{2+}, which has 6 electrons (4 holes), the ground state multiplet according to Hund's rule has a spin $S = 2$, so for spin compensation ($n = 2S$) the results of the n-channel model with an effective $n = 4$ should be appropriate. Comparison of the experimental results for the temperature dependence of the impurity susceptibility ($g = 2$) as a function of temperature are compared with the predictions of the $n = 4$ model in figure 9.21 with Sacramento and Schlottmann's definition of the Kondo 'T_K' $= 6T_L/\pi(n + 2) = 18$ K. We noted earlier that $\chi_{imp}(T)$ normalized to $\chi_{imp}(0)$ is a very weak function of n so that this agreement cannot be taken as confirmation of the choice $n = 4$.

The same parameters, however, are used for a comparison with the specific heat data for the model with $n = 1, 2, 3, 4, 5$ in figure 9.22(i) and this comparison confirms the choice $n = 4$. The same parameters give good agreement with experimental results on the induced magnetization M_{imp} as a function of $g\mu_B H/k_B T$ for various magnetic field strengths as can be seen in figure 9.22(ii). As there is essentially only one unknown parameter T_L involved in the fitting the agreement between theory and experiment is excellent. Sacramento and Schlottmann (1990b) have also

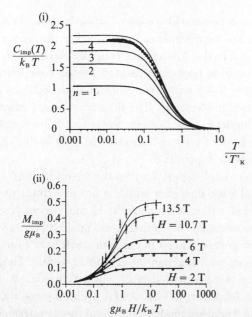

Figure 9.22 (i) $C_{\mathrm{imp}}(T)/k_BT$ as a function of T for the n-channel Kondo model for $n = 1, 2, 3, 4, 5$ compared with experimental data (filled squares) from Triplett & Phillips (1971) for dilute Fe in Cu, and (ii) $M_{\mathrm{imp}}(T)$ as a function of $g\mu_B H/k_B T$ for $S = 2$, $g = 2$, and $n = 4$ for the n-channel Kondo model for various magnetic field strengths (indicated in Tesla) compared with experimental data also for dilute Fe in Cu from Steiner, Hüfner & Zdrojewski (1974) and Frankel et al (1967). Both figures are from Sacramento and Schlottmann (1990) with their definition, 'T_K'= $6T_L/\pi(n+2) = 18$ K.

made a comparison of the experimental results for Cr in Cu with the n-channel model with $S = \frac{5}{2}$, $n = 5$, $g = 2$ and 'T_K' = $6T_L/\pi(n+2) =$ 1.5 K. This is based on the assumption that the Cr ion retains its 5d electrons and so according to Hund's rules has an orbital singlet like Mn^{2+}. Because of the lower value of the exchange coupling and T_L this makes it a less favourable case for comparison than for Fe in Cu due to the difficulty in eliminating the interimpurity interaction effects at low temperatures. Nevertheless Sacramento and Schlottmann have obtained a reasonable agreement with experiment for a range of properties with the single free parameter T_L.

The n channel Kondo model (9.30) has also been investigated for cases where $n \neq 2S$. If $S > n/2$ the results are essentially the same as those found in section 6.5 for the special case $n = 1$. The ground state has a spin $S - n/2$, corresponding to a strong coupling fixed point, and the spin is only partially compensated by a spin $\frac{1}{2}$ in each of the n channels. The correction terms are logarithmic as in (6.81). For the case $n > 2S$ some new physics is involved. The strong coupling fixed point is unstable as the impurity spin is over-compensated by the conduction electrons giving a net spin $n/2 - S$ which is antiferromagnetically coupled to the rest of the conduction electrons. The final stable fixed point is unusual for this type of model in that it corresponds to a finite coupling $J = J^*$ and results in a form of critical behaviour as $T, H \to 0$. This situation is discussed briefly in appendix J.

For a transition metal impurity without spin–orbit coupling or crystal fields in a ground state multiplet which is not an orbital singlet, $L \neq 0$, we have in general a Hamiltonian in the Kondo regime involving both orbital and spin exchange scattering built up from terms of the form (9.29). The most general model in this limit will have $(4|\min(L, l) + 1)$ exchange constants plus a potential scattering term. These exchange couplings can, in principle, be derived via a Schrieffer–Wolff transformation from the Anderson model. As $l + 1$ parameters are required to specify H_{d} in the Anderson model the general case is rather complicated. To see what is involved we look at the particular case of one localized d electron (or hole) taking the simplified form for H_{d} given in (9.28). The Schrieffer–Wolff transformation for this case has been derived by Okada & Yosida (1973). The exchange part of the Hamiltonian is

$$H = \sum_{km\sigma} \epsilon_k c^{\dagger}_{k,m,\sigma} c_{k,m,\sigma} + J_1 \sum_{kk'mm'\sigma\sigma'} c^{\dagger}_{k',m',\sigma'} c_{k,m,\sigma} c^{\dagger}_{\mathrm{d},m,\sigma} c_{\mathrm{d},m',\sigma'}$$

$$+ J_2 \sum_{kk'mm'\sigma\sigma'} (c^{\dagger}_{k',m',\sigma} c_{k,m,\sigma} c^{\dagger}_{\mathrm{d},m,\sigma'} c_{\mathrm{d},m',\sigma'} - c^{\dagger}_{k',m,\sigma} c_{k,m,\sigma} c^{\dagger}_{\mathrm{d},m',\sigma'} c_{\mathrm{d},m',\sigma'}),$$

$$(9.40)$$

with the constraint $\sum_{m,\sigma} c^{\dagger}_{\mathrm{d},m,\sigma} c_{\mathrm{d},m,\sigma} = 1$. The first exchange term (J_1) exchanges both orbital and spin states and the second and third terms (J_2) exchange spin and orbital states respectively. The exchange couplings are given by

$$J_1 = -|V|^2 \frac{(U + J_{\mathrm{H}})(\epsilon_{\mathrm{d}} + U + J_{\mathrm{H}}) - J_{\mathrm{H}}^2}{\epsilon_{\mathrm{d}}(\epsilon_{\mathrm{d}} + U)(\epsilon_{\mathrm{d}} + U + 2J_{\mathrm{H}})}, \qquad (9.41)$$

and

$$J_2 = -|V|^2 \frac{J_{\mathrm{H}}}{(\epsilon_{\mathrm{d}} + U)(\epsilon_{\mathrm{d}} + U + 2J_{\mathrm{H}})}. \qquad (9.42)$$

Okada & Yosida looked at the case $|J_{\mathrm{H}}| \ll U$ and neglected J_2 compared to J_1. Using the variational ansatz in which any number of particle–hole pairs are taken into account within logarithmic accuracy they found a ground state with a binding energy,

$$\Delta E^{(0)} = -De^{-1/2(2l+1)\rho_0 J_1},\tag{9.43}$$

corresponding to a spin and orbital singlet. We could have deduced (9.43) from results on the Coqblin–Schrieffer model as the models in this limit are formally identical with (m,σ) replacing m. The total degeneracy factor is then $N = 2(2l + 1)$ so (9.43) corresponds to (7.24). We can anticipate therefore if the next leading order logarithmic corrections are taken into account there will be a prefactor $(\rho_0 J_1)^{1/2(2l+1)}$ in (9.43) as in (D.18). We can also deduce the χ/γ ratio in this limit using $N = 2(2l + 1)$ in (7.47).

For $J_H = 0$ this result can easily be generalized to the case of n d electrons. The model in this limit has only one non-zero exchange coupling $J_1(J_H = 0)$ corresponding to the first term of (9.40) restricted to the subspace $\sum_{m,\sigma} c^\dagger_{d,m,\sigma} c_{d,m,\sigma} = n$. The binding energy $\Delta E^{(0)}$ has the same form as (9.43) but multiplied by a factor n. The limit $J_H = 0$ is not, however, the physically relevant situation. We found earlier in the more physically appropriate situation of strong Hund's rule coupling that for a half full shell $n = 2l + 1$ the exponent in the exponential is much reduced and hence predicts a much lower binding energy.

There is no general proof that the ground state is an orbital and spin singlet for all fillings of the d shell. The singlet state seems to be favoured because of the energy gain through hybridization with higher degenerate states. This was clearly seen in the calculation for the N-fold degenerate Anderson model ($U = \infty$) in section 8.4. Experimentally all transition metal impurities which show a Kondo effect are found to have a finite susceptibility as $T \to 0$, indicating to a singlet ground state. However, it is essential to include the crystal field splittings in a realistic model for transition metal impurities, except in the case of a half full d shell (orbitally non-degenerate). We consider what modifications can occur when we take these terms into account.

Due to the reduction in symmetry the effect of crystal field interactions is to reduce the orbital degeneracy of the impurity levels and cause further splittings. For an orbital singlet there is no degeneracy to remove so the effects are relatively minor causing some small differences in the exchange couplings in the various channels, which are now classified by the crystal field index γ rather than the orbital angular momentum quantum number m. If we introduce a cubic crystal field we need to express the $l = 2$ partial waves of the conduction electrons, which are the ones which predominately hybridize with the d states of the impurity, in terms of states which transform according to the cubic symmetry group. The states are classified by the irreducible representations, $\gamma = \Gamma_3, \Gamma_5$; Γ_5 is two-fold degenerate and the Γ_3 is three-fold degenerate. The differences in exchange couplings, J_{Γ_3} and J_{Γ_5}, for an orbital singlet arise only due to the differences in the hybridization matrix elements V_{Γ_3} and V_{Γ_5}. These differences should in most cases be small. We then expect to scale to the strong coupling fixed point in each of the channels so that any small initial difference in the couplings becomes irrelevent at low temperatures. Our conclusions therefore about the ground state and low temperature behaviour for the orbital singlet case remain unchanged.

For $L \neq 0$ further splittings of the $|L, S\rangle$ multiplet will be induced. There are two cases where the ground state multiplet in the crystal field is left with only spin degeneracy. These correspond to $3d^3$ and $3d^8$. The ground state Hund's rule multiplet for $3d^3$ is ^4F ($L = 3, S = \frac{3}{2}$) and for $3d^8$ is ^3F. These both split into states whose total wavefunctions transform as Γ_2 ($N_{\Gamma_2} = 1$), Γ_5 ($N_{\Gamma_5} = 3$) and Γ_4 ($N_{\Gamma_4} = 3$) with the non-degenerate state Γ_2 lying lowest. In terms of the individual d states for $3d^3$ we can think of this as a half full shell of the lower lying Γ_5 d states, and for $3d^8$ a half full shell of the lower lying hole Γ_3 states. The corresponding exchange Hamiltonian in the Kondo limit must be of the form (9.30). For large crystal field splittings the exchange interactions with the Γ_5 and Γ_3 conduction electron states will be quite different due to the level splittings. The exchange interaction with the lower lying levels will be greater so that $|J_{\Gamma_5}| \gg |J_{\Gamma_3}|$ for $3d^3$ and $|J_{\Gamma_3}| \gg |J_{\Gamma_5}|$ for $3d^8$. We might then expect to scale to the fixed points $J_{\Gamma_5} \to \infty$, $J_{\Gamma_3} \to 0$ for $3d^3$ and $J_{\Gamma_5} \to 0$, $J_{\Gamma_3} \to \infty$ for $3d^8$. We would then have a fully compensated ground state as in the case of an orbital singlet where the number of channels n is equal to $2S$. This situation should be applicable to V (d^2) or Co (d^8) impurities in a strong cubic crystal field.

If we return to the case of a single d electron, $3d^1$, and impose a cubic crystal field the d states split into Γ_3 and Γ_5 which hybridize with the conduction states of the same symmetry (for a spherically symmetric potential), with the Γ_5 state lowest. The exchange Hamiltonian in the Kondo limit can be derived by a Schrieffer–Wolff transformation, similar to the derivation of (9.30) but more complicated in that the d states are no longer all degenerate. This calculation has been done by Cragg, Lloyd & Nozières (1980). Their result, as one would expect, now has exchange terms between the crystal field channels as well as spin exchange. The result is rather long so we do not reproduce it here but refer the interested reader to the original paper. The possibility of a singlet ground state for such a system has been analysed in general terms by Nozières & Blandin (1980). Combining a Γ_5 d state with either a Γ_5 or Γ_3 conduction state gives states with a spatial part of the wavefunction which transforms as

$$\Gamma_5 \otimes \Gamma_5 = \Gamma_1 + \Gamma_3 + \Gamma_4 + \Gamma_5, \quad \Gamma_3 \otimes \Gamma_5 = \Gamma_4 + \Gamma_5, \quad (9.44)$$

where $N_{\Gamma_1} = 1$, $N_{\Gamma_4} = 3$, with the spin part as a singlet or triplet. A similar analysis can be made of the one hole case, $3d^9$, where the Γ_3 state lies lowest. The combined states in this case transform as

$$\Gamma_3 \otimes \Gamma_3 = \Gamma_1 + \Gamma_2 + \Gamma_3, \quad \Gamma_3 \otimes \Gamma_5 = \Gamma_4 + \Gamma_5, \quad (9.45)$$

where $N_{\Gamma_2} = 1$. We might guess that in each case the singlet ground state is favoured but there is no explicit calculation to verify this. An attempt to specify the form of the Fermi liquid Hamiltonian for the quasi-particles would be difficult. As symmetry considerations are not sufficient to reduce the number of parameters to one or two, as in the other cases we have considered, the χ/γ ratio is not expected to be universal here.

To summarise this section on transition metal impurities: the simple s-d model with $J_{\mathbf{k},\mathbf{k}'}$ independent of $\mathbf{k}$, $\mathbf{k}'$, is inappropriate for all cases. The more general s-d model for $S = \frac{5}{2}$ with d wave scattering is equivalent to the n-channel Kondo model (9.30) for $n = 2S = 5$ and is applicable to Mn^{2+} impurities. Solutions have been obtained for the thermodynamics which are in general agreement with experiment in the case of Cr in Cu where detailed comparison has been made. The cases of $3d^3$ and $3d^8$ which are orbital singlets in strong cubic crystal field can be described by the same theory. For impurity ions in other configurations exchange scattering in the orbital channel must be taken into account. In these cases there are some general observations and speculations, based on insights gained through scaling, renormalization group and Fermi liquid theory, but little in the way of exact results. The predictions of the $n = 2S = 4$ channel Kondo model applied to Fe in Cu of Sacramento and Schlottmann (1990a) are, however, in excellent agreement with the experimental results.

9.4 Transport Properties

It was the marked effect of magnetic impurities on the resistivity of simple metals that provoked the early interest in these systems. The important experimental observation was the correlation in the occurrence

and size of a Curie–Weiss term in the susceptibility with the occurrence
and magnitude of the resistance minimum (Sarachik et al, 1964). This
stimulated the work of Kondo leading to his discovery of $\ln T$ terms due
to the spin flip scattering of conduction electrons with a local moment,
and the consequent explanation of the resistance minimum. The anomalous increase in the scattering of conduction electrons on lowering the
temperature affects other transport properties as well as the resistivity.
They are all enhanced at low temperatures due to the density of low
energy excitations in the Kondo resonance which always lies at or very
close to the Fermi level. The conductivity $\sigma(T)$, thermopower $S(T)$, and
the Lorentz ratio $L(T)$ $(= \kappa(T)/\sigma(T)T$, where $\kappa(T)$ is the thermal conductivity) can all be related to moments $K^{(n)}$ of the transport lifetime
$\tau_{\text{tr}}(\omega)$ weighted with the derivative of the Fermi function,

$$\sigma_{\text{imp}}(T) = \frac{e^2 k_F^3}{3\pi^2 m} K^{(0)}, \qquad S_{\text{imp}}(T) = -\frac{1}{eT} \frac{K^{(1)}}{K^{(0)}}, \tag{9.46}$$

and

$$L_{\text{imp}}(T) = \frac{1}{e^2 T} \left(\frac{K^{(2)}}{K^{(0)}} - \frac{(K^{(1)})^2}{(K^{(0)})^2} \right), \tag{9.47}$$

where

$$K^{(n)} = \int_{-\infty}^{\infty} \omega^n \tau_{\text{tr}}(\omega) \left(-\frac{\partial f}{\partial \omega} \right) d\omega. \tag{9.48}$$

The derivative of the Fermi function appearing in (9.48) means that
these quantities reflect the form of $\tau_{\text{tr}}(\omega)$ near the Fermi level. We
found earlier that for the Anderson model, with predominant scattering
of conduction electrons in a single angular momentum channel (d, $l = 2$
or f, $l = 3$) $\tau_{\text{tr}}^{-1}(\omega)$ is proportional to the density of states $\rho_d(\omega)$ or
$\rho_f(\omega)$ near the Fermi level. For $T = 0$, the resistivity depends on the
density of states at the Fermi level, $\rho_{d(f)}(0)$, which is determined by the
Friedel sum rule and the condition for charge neutrality. From (5.39)
and (5.82), and using the Friedel sum rule, (5.50) or (7.80), we have for
the resistance at $T = 0$,

$$R_{\text{imp}}(0) \propto \sin^2 \left(\frac{\pi n_{d(f)}}{N_{\text{sc}}} \right), \tag{9.49}$$

where N_{sc} is the number of scattering channels, which is equal to $2(2l+1)$
if we neglect multiplet splittings, crystal fields and spin–orbit interactions. This corresponds to (2.16) and (2.20) with a single phase shift for
the resonant states, $\eta_{2(3)}(\epsilon_F) = \pi n_{d(f)}/N_{\text{sc}}$. The form of (9.49) reflects
the filling of a single resonance in the density of states near the Fermi
level whose position is determined by the requirement that it must ac-

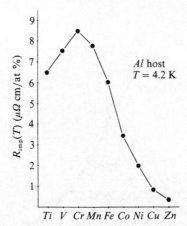

Figure 9.23 Low temperature impurity resistivity for 3d transition elements in Al (from Grüner, 1974).

commodate the appropriate number of d or f electrons contributing to overall charge neutrality.

There is some support for this relation from experimental results on Al based alloys with different transition metal impurities as shown in figure 9.23 ($N_{sc} = 10$). The experimental curve, however, has a distinct asymmetry and the maximum in $R_{imp}(0)$ occurs at Cr rather than Mn. The theoretical prediction (9.49), however, neglects the s and p phase shifts, including these non-resonant terms in (2.20) can account for the observed asymmetry (Grüner, 1974). In general Hund's rule terms, crystal fields and spin–orbit interactions split the resonance and, though the Friedel sum rule will still be valid, the phase shifts will no longer all be the same and the relation (9.49) will not apply. If these splittings are large then there may only be one resonance in the immediate vicinity of the Fermi level in which case (9.49) will again apply, but with a reduced degeneracy factor N_{sc}. For example for the $U = \infty$ N-fold degenerate Anderson model (large spin–orbit splitting) in the Kondo regime, where only the phase shifts associated the conduction electrons of total angular momentum j are retained ($2j + 1 = N$), (9.49) will be valid with $N_{sc} = N$.

The temperature dependence of the impurity contribution to the resistivity for systems which have only one relevant energy scale T_K at low temperatures should correspond to a universal form, $R_{imp}(T) = F(T/T_K)$, the parameters of the particular system combining into the single parameter T_K which can vary widely due to the exponential de-

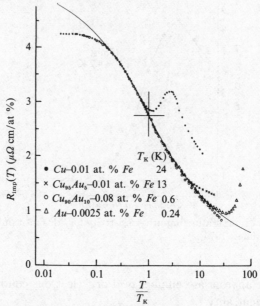

Figure 9.24 Temperature dependence of the impurity resistivity of Fe in Cu_xAu_{1-x} alloys normalized to a Kondo temperature T_K. The full line is a fit to the Hamann result (3.12) with $S = 0.77$ (Loram, Whall & Ford, 1970).

pendence on J. It is difficult to probe a particular system over the full temperature range from $T \ll T_K$ to $T_K \gg T$. For systems with low T_K it is difficult to eliminate the effects of impurity–impurity interactions, particularly for 3d systems where impurity interactions can be observed even for very low concentrations. For those systems with higher T_K there can be difficulties in the range $T \gg T_K$ in subtracting out the phonon contribution, which increases rapidly with T, and arriving at an accurate estimate of the impurity contribution. For this reason Loram, Whall & Ford (1970) explored systems of Fe impurities in Cu, Au and $CuAu$ alloys with values of T_K in the range 0.24 K$< T_K <$2.4 K. In figure 9.24 results for $R_{\mathrm{imp}}(T)/c_{\mathrm{imp}}$ for four different compositions are shown plotted against $\log(T/T_K)$. There is a deviation above $5T_K$ for the Fe in Cu case due to an extra phonon contribution caused by impurities. Apart from this the results give clear evidence in support of a universal curve. If a reasonable correction is made for the phonon upturn for the Fe in Cu then the results in this case also agree with the universal curve at higher temperatures. The full line in figure 9.24 is the Hamann result

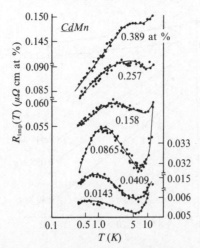

Figure 9.25 The effect of magnetic ordering on the resistivity at higher concentration for Mn impurities in Cd due to impurity–impurity interactions giving a low temperature maximum (Hedgcock & Rizzuto, 1967).

(3.12) with $S = 0.77$. This gives a good fit for $T \geq T_\mathrm{K}$ but clearly fails at low temperatures $T \ll T_\mathrm{K}$.

There were difficulties in the early experimental work in extracting the temperature dependence of the single impurity contribution to $R_\mathrm{imp}(T)$ for $T \ll T_\mathrm{K}$ due to interaction effects which at higher concentrations can lead to spin glass behaviour and magnetic ordering. Interaction effects leading to magnetic order and a suppression of the low energy spin scattering can result in a resistance maximum as well as a minimum as can be seen in the results of Hedgcock & Rizzuto (1967) for Mn in Cd at low temperatures shown in figure 9.25.

A simple T^2 behaviour (3.17) in the Kondo regime was first reported by Caplin & Rizzuto (1967) for Mn in Al with $\theta_R = 530 \pm 30$ K, which is significantly smaller than estimates of the hybridization width Δ so that the results could not be explained by a single particle theory (see figure 9.26).

This was first interpreted within a local RPA spin fluctuation theory. The RPA theory is not valid in the Kondo regime, however, and so this was not a satisfactory explanation. The T^2 form at very low concentrations was confirmed by Star & Nieuwenhuys (1969) for Fe in Cu as shown in figure 9.27 and interpreted in terms of a non-interacting resonant level model (see Grüner, 1974). The exact T^2 form was derived in the Fermi liquid theories of Nozières (1974) and Yamada (1975) for

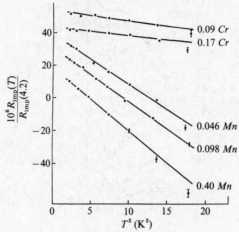

Figure 9.26 Relative change in the impurity resistance for Mn and Cr impurities in Al (in parts per million) versus T^2. The vertical position of each curve is arbitrary (Caplin & Rizzuto, 1968).

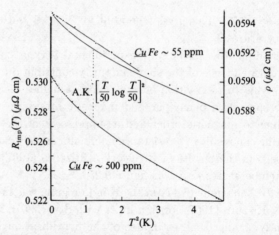

Figure 9.27 The resistivity of two Fe in Cu alloys showing that a T^2 dependence gives a better fit to the low concentration data than an earlier predicted form, $(T/T_K)^2 \ln^2(T/T_K)$ by Appelbaum & Kondo (1969) (from Star & Nieuwenhuys, 1969).

the non-degenerate s-d and Anderson models as discussed in chapter 5. The T^2 dependence is in accord with the results of Loram et al shown in figure 9.24. The more general Fermi liquid result for $R_{\text{imp}}(T)$ for the n-channel Kondo model with $n = 2S$ (Yoshimori,1976; Mihály &

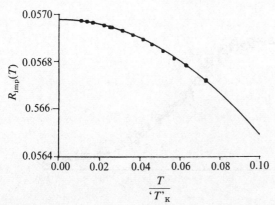

Figure 9.28 A fit of the low temperature resistivity for dilute Fe in Cu (data from Star et al, 1972) to a T^2 law by Sacramento and Schlottmann (1990). If the coefficient c_2 corresponding to the n-channel Kondo model from (9.50) for $n = 4$ is used the fit gives 'T_K' $= T_L/\pi = 23$ K (see text).

Zawadowski, 1978) is

$$R_{\text{imp}}(T) = R_{\text{imp}}(0)(1 - c_2 T^2), \quad \text{with} \quad c_2 = \left(\frac{\pi^2}{12T_L}\right)^2 (4n+5). \quad (9.50)$$

which agrees with (5.105) for $n = 1$. The Fermi liquid form for the impurity resistivity has been fitted to the low temperature resistivity results for Fe in Cu by Sacramento and Schlottmann (1990). They used the same parameters for the earlier fit to experimental results on this system shown in figures 9.21 and 9.22. The results shown in figure 9.28 are clearly well fitted by the T^2 law. Unfortunately the coefficient c_2 they quote and use for the $n = 4$ model is incorrect and does not correspond to that given in (9.50). If the correct coefficient as given in (9.50) is used the fit corresponds to a slightly larger value for 'T_K' $= 6T_L/\pi(n + 2) = 23$ K than the value 'T_K' $= T_L/\pi = 18$ K used in the earlier fits.

The resistivity as a function of temperature over the full temperature range is one of the most difficult quantities to calculate as it requires the density of states over the low frequency range at all temperatures. One way of calculating this quantity for the higher degenerate models is via the NCA. Results of calculations of Bickers, Cox & Wilkins (1985) are shown for the resistivity for the $N = 6$ model are shown in figure 9.29 for various values of the f occupation number at $T = 0$, $n_f(0)$, in the range $0.71 \leq n_f(0) \leq 0.97$ including in some cases the higher spin multiplet $j' = \frac{7}{2}$. To a good approximation the results give almost universal behaviour as a function of T/T_0, where T_0 is the position of

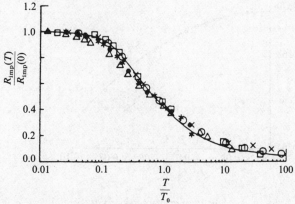

Figure 9.29 NCA calculations for the resistivity ratio $R_{\mathrm{imp}}(T)/R_{\mathrm{imp}}(0)$ as a function of $\log(T/T_0)$. The solid curve is a fit to the data for Ce impurities in LaB_6 (Winzer, 1975) with $T_0 = 1.6$ K (from Bickers et al, 1985). Key: circle $n_{\mathrm{f}}(0) = 0.97$, square $n_{\mathrm{f}}(0) = 0.92$, filled circle $n_{\mathrm{f}}(0) = 0.81$, asterisk $n_{\mathrm{f}}(0) = 0.71$, all with $N = 6$, $N' = 0$ and, cross $n_{\mathrm{f}}(0) = 0.96$, triangle $n_{\mathrm{f}}(0) = 0.86$ both with $N = 6$ $N' = 8$.

the peak in the Kondo resonance in the spectral density $\rho_{\mathrm{f}}(\omega)$. In the Fermi liquid regime the results correspond to (9.50) with $c_2 = 5/T_0^2$. The c_2 coefficient has been calculated for the same model by Houghton et al (1987) to order $1/N$ who find a value $c_2 = (1 - 8/3N)\pi^2/T_0^2$ which gives $c_2 = 5.6/T_0^2$ for $N = 6$, in approximate agreement with the NCA result.

Interimpurity interactions are much less of a problem for rare earth than for transition level impurities and it is consequently easier to obtain results over a wide temperature range. The solid curve in figure 9.29 corresponds to experimental results for dilute Ce in LaB_6 (Winzer, 1975) over several decades and agrees well with the theoretical predictions based on the NCA.

For the $N = 2$ model the methods based on large N approximations, such as the NCA, are no longer adequate. The exact T^2 coefficient for low temperatures is known from Fermi liquid theory, equation (5.105) which is equivalent to (9.50) with $n = N = 2$, and the Hamann result (3.12) is valid for $T > T_{\mathrm{K}}$. The numerical renormalization group method, as described in section 5.6, has recently been extended to the calculation of the temperature dependence of the dynamic response functions (Costi & Hewson, 1992b). The resistivity in the crossover regime can be deduced from the results for the spectral density $\rho_{\mathrm{d}}(\omega, T)$ using (5.83), and the results of these calculations in the Kondo regime are shown in figure

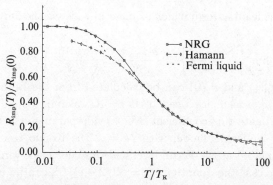

Figure 9.30 The resistivity $R_{\mathrm{imp}}(T)$ for the non-degenerate Anderson model calculated using a numerical renormalization group approach over the temperature range $0 \le T \le 100T_{\mathrm{L}}$ for $U/\pi\Delta = 6$, $\epsilon_{\mathrm{d}} = -5\Delta$. The curve marked Hamann is a fit to (3.12) with $S = 1/2$ and $T_{\mathrm{K}} = 1.2T_{\mathrm{L}}$, and the dashed curve is a fit to the T^2 form (5.105) (Costi & Hewson, 1992b).

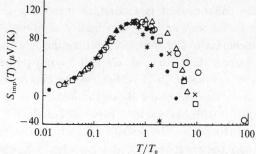

Figure 9.31 The thermopower calculated within the NCA for Ce impurities (Bickers et al, 1985).

9.30. Similar results in the crossover region have been generated from Monte Carlo calculations (we briefly describe this approach in the next section) coupled with a maximum entropy method for improving the continuation of the imaginary time results to real frequencies (Jarrell, Gubernatis, Silver & Sivia, 1991).

Results for the thermopower using the NCA have been calculated by Bickers et al for $N = 6$ and the same range of values of $n_{\mathrm{f}}(0)$ and these results are shown in figure 9.31. The extra power of ω in the calculation of this quantity means that it is more sensitive to the higher energy excitations at higher temperatures so that the results for $T > T_0$ for this range of values of $n_{\mathrm{f}}(0)$ deviate from universality.

Another approach to the calculation of the thermopower, at least the coefficient of the leading term which is linear in T, is via the Fermi liquid relation (E.12),

$$\lim_{T \to 0} \left\{ \frac{S_{\text{imp}}(T)}{T} \right\} = \frac{2\gamma_{\text{imp}}\pi}{eN} \cot \left(\frac{\pi n_{\text{f}}(0)}{N} \right). \qquad (9.51)$$

The values of γ_{imp} and $n_{\text{f}}(0)$ can be calculated from the Bethe ansatz results. This approach has been used by Kawakami, Usuki & Okiji (1987) to investigate this coefficient over a wide parameter range. For the symmetric Anderson model with $N = 2$ the Kondo resonance is symmetrically placed at the Fermi level and this coefficient vanishes. In the asymmetric case the quantity $\delta E = (U/2 + \epsilon_{\text{f}})/\Delta$ can be used as a measure of the degree of asymmetry for $N = 2$. The sign of the linear coefficient (9.51) then reflects the sign of δE, the sign of the gradient of the density of states at $\omega = 0$ depending on whether the resonance is below or above the Fermi level. For a given asymmetry factor δE the magnitude of the coefficient of the thermopower is exponentially enhanced ($\propto 1/T_{\text{K}}$) in the Kondo regime for large U. In the mixed valence regime the enhancement is modest as increasing U does not appreciably narrow the resonance in the density of states. There is virtually no enhancement in the empty orbital regime.

In the N-fold degenerate models for Ce and Yb impurities the sign of the thermopower coefficient (9.51) differs in the two cases. For Ce impurities the Kondo resonance lies above the Fermi level and is only fractionally occupied, $n_{\text{f}}(0)/N$, for $0 < n_{\text{f}}(0) < 1$ and $N = 6$ (without crystal field splittings), so the density of states at the Fermi level is steeply rising and the coefficient is positive. For Yb, which is the particle–hole image of the Ce case, the Kondo resonance lies below the Fermi level with a fraction of $n_{\text{f}}^{\text{h}}(0)/N$ holes above, where $0 < n_{\text{f}}^{\text{h}}(0) < 1$ and $N = 8$, so the coefficient is then negative.

These arguments neglect the contribution from the non-resonant scattering channels. These should be included and are especially important when the Kondo resonance is symmetrically placed at the Fermi level, as the contribution to (9.51) from the resonant term alone vanishes. These non-resonant phase shifts have been taken into account by Zlatić & Rivier (1974) and we outline their calculation. The general expression for the linear in T term in the thermopower is

$$S = -\frac{\pi^2}{3} \frac{k_{\text{B}}^2 T}{e} \frac{\partial}{\partial \omega} \ln \tau_{\text{tr}}(\omega)|_{\omega = \epsilon_{\text{F}}}, \qquad (9.52)$$

which follows from (9.46) on using the Sommmerfeld expansion (1.6). We then express τ_{tr} in terms of the phase shifts using (2.20), keeping only the $l = 0, 1$ and 2 phase shifts and neglecting any spin–orbit interactions.

If the ω dependence of all but the $l = 2$ phase shift is neglected (for 3d systems) this gives

$$S = -\frac{\pi^2 k_B^2 T}{3e} \left[\frac{3\sin(2\eta_2) - 2\sin(2(\eta_1 - \eta_2))}{\sin^2(\eta_0 - \eta_1) + 2\sin^2(\eta_1 - \eta_2) + 3\sin^2(\eta_2)} \right] \frac{\partial \eta_2}{\partial \omega}\bigg|_{\omega = \epsilon_F}. \tag{9.53}$$

We then use (5.48) for the $l = 2$ phase shift $\eta_2(\omega)$ and differentiate with respect to ω. For a symmetric Kondo resonance at the Fermi level, $\epsilon_d + \Sigma^R(\epsilon_F) = \epsilon_F$, as would be the case for Mn,

$$\left(\frac{\partial \eta_2}{\partial \omega} \right)_{\omega = \epsilon_F} = \frac{1}{z\Delta} = \frac{1}{\tilde{\Delta}}. \tag{9.54}$$

As $\tilde{\Delta} \sim T_K$ this coefficient is enhanced in the Kondo regime, it can be estimated from other measurements such as the specific heat or the T^2 term in the resistivity. If the non-resonant phase shifts are neglected, $\eta_0 = \eta_1 = 0$, the term in square brackets in (9.53) vanishes as $\eta_2(\epsilon_F) = \pi/2$ from (5.48). Hence the non-resonant terms are essential to explain the large linear in T term in the measurements for Mn in Al (Cooper, Vučić & Babić, 1974). For Mn with an atomic configuration $(4s^2, 3p^6, 3d^5)$ relative to Al with a configuration $(3p)$ (common core $(1s^2, 2s^2, 2p^6, 3s^2)$) we might take $\eta_0 = \pi$, $\eta_1 = -\pi/6$, and $\eta_2(\epsilon_F) = \pi/2$, on the assumption that the screening cloud is atomic-like and each angular momentum shell is screened separately (the Friedel sum rule only requires that the *total* excess charge $\Delta Ze = 12e$ is screened). In this way Zlatić & Rivier explained the decrease in the measured thermopower of Boato & Vig (1967) in the sequence Mn, Fe, Co and Ni in Al.

The magnetic field dependence of the resistivity, the magnetoresistance, can be deduced at $T = 0$ from the Friedel sum rule. Applying the sum rule to the individual z-components of angular momentum m for the N-fold degenerate Coqblin–Schrieffer model gives

$$\frac{R_{\text{imp}}(0)}{R_{\text{imp}}(H)} = \frac{\sin^2(\pi/N)}{N} \sum_m \frac{1}{\sin^2(\pi n_m(H))}, \tag{9.55}$$

where $n_m(H)$ is the occupation number of the m channel in the presence of a magnetic field H. The occupation numbers can be evaluated from the $T = 0$ Bethe ansatz equations for this model. For $N = 2$, using $M_{\text{imp}} = g\mu_B(n_\uparrow(H) - n_\downarrow(H))/2$, and $n_\uparrow(H) + n_\downarrow(H) = 1$, (9.55) can be expressed in terms of the impurity magnetization,

$$R_{\text{imp}}(H) = R_{\text{imp}}(0)\cos^2\left(\frac{\pi M_{\text{imp}}}{g\mu_B} \right). \tag{9.56}$$

This result is compared with experimental data for Ce impurities in $LaAl_2$ in figure 9.32 using the same parameters as the fit for the specific

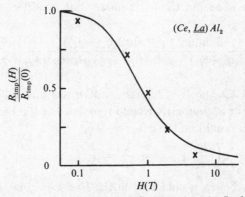

Figure 9.32 Magnetoresistance of Ce impurities in $LaAl_2$. The theoretical curve (Andrei, 1982) is based on equation (9.56) using the same data as in the fits for the specific heat and magnetization for the same system (figures 9.12 and 9.13). The experimental points are from Felsch & Winzer (1973).

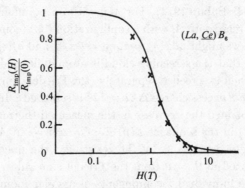

Figure 9.33 Magnetoresistance of Ce impurities in LaB_6 normalized to the zero field value. The theoretical curve was calculated by Schlottmann (1987b) based on equation (9.55) using the same parameters as in the fit to the specific heat (figure 9.14). The experimental points are from Samwer & Winzer (1976).

heat (figure 9.12) and the magnetization (figure 9.13). Experimental results for Ce impurities in LaB_6 in figure 9.33 are compared with predictions based on (9.55) for the $N = 4$ model using the same parameters as in figure 9.14.

The low temperature behaviour of the magnetoresistance has been deduced by Kawakami & Okiji (1986) for the $N = 2$ symmetric Anderson model in the Kondo regime using the low frequency and low temperature form for the self-energy,

$$\Sigma_\sigma(\omega,H) = \frac{U}{2} - \omega\left(\frac{1}{z}-1\right) - \frac{i}{2z^2\Delta}[\omega^2+(\pi k_B T)^2]+\Sigma_\sigma(H)+..., \quad (9.57)$$

which can be deduced using (5.102) in limit with $R \to 2$ and $z \to 4k_B T_L/\pi\Delta$. The term $\Sigma_\sigma(H)$ can be deduced from the Bethe ansatz equations in an arbitrary field at $T = 0$. The temperature dependence of the resistivity is reduced in the presence of a large magnetic field as the spin flip scattering and spin compensated ground state is suppressed. The low temperature thermal conductivity and Lorentz ratio can be determined in a similar way.

From the comparison with theory undertaken so far the evidence is that the main features of the impurity induced transport properties as a function of temperature and magnetic field can be satisfactorily explained. There is clearly scope, however, for more detailed comparison of experimental results with the more accurate theories.

9.5 Neutron Scattering

One of the most direct ways of gaining information about the magnetic excitations of a system over the thermal energy range is by slow neutron scattering. This technique has been applied to dilute magnetic alloys to investigate the behaviour of single impurity ions in their local environment. Examples of systems studied are Mn in Cu (Murani & Tholence, 1977) and Fe in Au (Scheuer, Loewenhaupt & Schmatz, 1977). As with other types of spectral measurements there are difficulties working with low enough concentrations to avoid the effects of interimpurity interactions. There are, however, a number of concentrated alloys and compounds of Ce and Yb whose neutron scattering response seems to reflect single ion behaviour. These are systems which are mainly paramagnetic and appear to have weak intersite interactions. Several of them, such as $YbCuAl$ and $CeSn_3$ have been mentioned already in previous sections.

The differential scattering cross-section for a given energy transfer (ω) and wavevector transfer (Q) is given by

$$\frac{d^2\sigma}{d\Omega d\omega} = N_{ion}\left(\frac{\gamma e^2}{mc^2}\right)^2 \frac{k}{k'}S(Q,\omega), \quad (9.58)$$

where N_{ion} is the number of magnetic ions, $\gamma e^2/mc^2$ the coupling constant, k and k' the incoming and out-going wavevectors. The scattering function $S(Q,\omega)$ is related to the imaginary part of the Q and ω dependent magnetic susceptibility, $\chi(Q, \omega, T)$, and takes the form,

$$S(Q, \omega) = \frac{2}{\pi} \frac{F^2(Q)}{(1 - e^{-\omega/k_B T})} \frac{\mathrm{Im}\chi(0, \omega, T)}{(g\mu_B)^2}, \tag{9.59}$$

for mutually independent well defined magnetic moments with a form factor $F(Q)$. The static susceptibility can be deduced from $\mathrm{Im}\chi(\omega, T)$ using the Kramers–Kronig relation,

$$\chi(T) = \frac{1}{\pi} \int_{-\infty}^{\infty} \frac{\mathrm{Im}\chi(\omega, T)}{\omega} \, d\omega. \tag{9.60}$$

If we introduce a spectral function $f(\omega, T)$, which we define by the relation $\mathrm{Im}\chi(\omega, T) = \pi\omega\chi(T)f(\omega, T)$, then $f(\omega, T)$ is an even function of ω, and from (9.60) its integral over all frequencies is unity.

If we describe the interaction of the impurity spin with its metallic environment by the s-d model (1.64), and calculate its relaxation due to conduction electron scattering to second order in J then we obtain the Korringa result (Korringa, 1950),

$$f(\omega, T) = \frac{1}{\pi} \frac{\Gamma(T)}{\omega^2 + \Gamma(T)^2}, \tag{9.61}$$

corresponding to a quasi-elastic line with a line width $\Gamma(T) \propto T$ (half line width to be precise, and so in much neutron scattering literature this is multiplied by an extra factor of $1/2$). The corresponding form for $\mathrm{Im}\chi(\omega, T)$ has a peak at $\omega = \Gamma(T)$ and falls off as $1/\omega$ for large ω. When third order coupling terms are taken into account there are $\ln T$ corrections and $\Gamma(T)$ has the form,

$$\Gamma(T) = 4\pi(2J\rho_0)^2 k_B T[1 - 4(2J\rho_0)\ln(T) + ...], \tag{9.62}$$

for $S = \frac{1}{2}$, where the leading term is the Korringa term (Walker, 1970). This can clearly not apply as $T \to 0$. If we use the Korringa or Shiba relation (E.10),

$$\lim_{\omega \to 0} \frac{\mathrm{Im}\chi_{\mathrm{imp}}(\omega, T)}{\pi\omega} = \frac{N\chi_{\mathrm{imp}}^2}{(g\mu_B)^2}, \tag{9.63}$$

for $N = 2$, then $f(0,0) = 2\chi_{\mathrm{imp}}/(g\mu_B)^2$. In the Kondo regime $\chi_{\mathrm{imp}} \propto 1/T_K$ which suggests that for $T \to 0$, $\Gamma(0) \propto T_K$, assuming the form (9.61) is applicable in this limit. Results of calculations of $\mathrm{Im}\chi_{\mathrm{imp}}(\omega, T)$ for the non-degenerate Anderson model via the numerical renormalization group approach, shown in figure 9.34, are consistent with this conclusion. The limit $\omega \to 0$, $T \to 0$, is in good agreement with the relation (9.63) and the width of the peak centred at the origin is of the order T_K.

The observed quasi-elastic line widths as a function of temperature for Fe in Au and Mn in Cu over the range 30 K$<T<$300 K are shown in figure 9.35. The temperature range is above the Kondo temperature

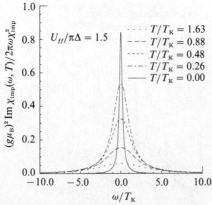

Figure 9.34 $(g\mu_{\mathrm{B}})^2\mathrm{Im}\chi_{\mathrm{imp}}(\omega, T)/2\pi\omega\chi_{\mathrm{imp}}^2$ as a function of ω for the non-degenerate Anderson model at different temperatures calculated within a numerical renormalization group approach (Costi & Hewson, 1991).

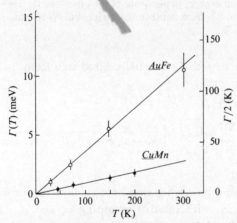

Figure 9.35 Temperature dependence of the quasi-elastic line width $\Gamma(T)$ for single magnetic moments in dilute alloy of Fe in Au and Mn in Cu (Loewenhaupt & Schmatz, 1977).

($T_{\mathrm{K}} \sim 1$ K for Fe in Au and $T_{\mathrm{K}} \sim 0.01$ K for Mn in Cu) and the results are in accord with Korringa relaxation $\Gamma(T) \propto T$. The Fe in Cu system has a higher Kondo temperature with $\theta_x \sim 29$ K, and the line width $\Gamma(T)$ as a function of T, deduced from both neutron scattering and NMR experiments (see figure 9.36), shows a clear deviation from a Korringa law with a finite line width as $T \to 0$ of order $k_{\mathrm{B}}T_{\mathrm{K}}$.

For rare earth impurities crystal field splittings in the thermal energy range should give inelastic lines in the neutron spectrum. If inelastic

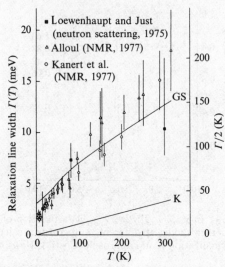

Figure 9.36 Temperature dependence of the quasi-elastic line width $\Gamma(T)$ for Fe impurities in Cu from neutron scattering and NMR data (Loewenhaupt & Schmatz, 1977).

lines are observed, $f(\omega, T)$ is usually fitted to a form,

$$f(\omega, T) = \frac{a_0 \Gamma_0(T)}{\omega^2 + \Gamma_0^2(T)} +$$

$$\sum_i a_i \Gamma_i(T) \left\{ \frac{1}{(\omega - \omega_i)^2 + \Gamma_i^2(T)} + \frac{1}{(\omega + \omega_i)^2 + \Gamma_i^2(T)} \right\},$$

$$(9.64)$$

corresponding to a broadened quasi-elastic line plus broadened inelastic lines at $\omega = \pm\omega_i$, with relative weighting a_0 and a_i. In normal rare earth ions the low lying crystal field excitations can be deduced from the positions of the inelastic lines.

There are Kondo lattice systems, such as $YbCuAl$ and $CeSn_3$, where there is evidence from neutron scattering of inelastic lines which cannot be satisfactorily interpreted as due to crystal field excitations. In fitting the data for $YbCuAl$ at 5 K with the form Murani, Mattens, de Boer & Lander (1985) found inelastic peaks at $\omega_1 \sim 10$ meV, $\omega_2 \sim 23$ meV. As the temperature was raised to 30 K (which is approximately the temperature at which $\chi(T)$ has a peak) this inelastic structure disappears. The width of the quasi-elastic component $\Gamma_0(T)$, as can be seen in figure 9.37, is a non-monotonic function of T. Experiments on $CeSn_{3-x}In_x$ in the range $0 < x < 1$ also reveal a broad inelastic peak, whose position

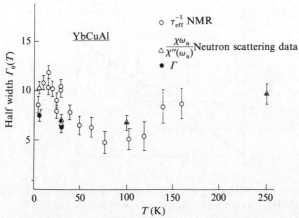

Figure 9.37 Temperature dependence of the quasi-elastic line width $\Gamma_0(T)$ for $YbCuAl$ as deduced from NMR (open circles, Laughlin et al, 1979) and neutron scattering data (triangles and filled circles, Murani et al, 1985).

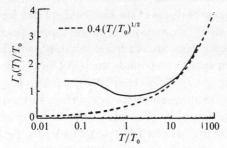

Figure 9.38 The inelastic line width $\Gamma_0(T)$ as a function of temperature in the Kondo regime calculated using the NCA (Bickers et al, 1987).

ω_i scales inversely with $\chi(0)$ and γ as the composition is varied (Murani, 1987).

The NCA calculations of $\chi_{\text{imp}}(\omega, T)$ give an inelastic hump in $f(\omega, T)$ as we saw earlier in figure 8.10. The peak in $f(\omega, T)$ seems to behave almost precisely like the peak in the f spectral density. For large N, $\omega_i \sim T_K$ whereas for $N = 2$ we see only a quasi-elastic peak as in figure 9.34 $\omega_i = 0$ (compare this to the change of $\tilde{\rho}_f(\omega)$ with N seen in figure 7.14). From the NCA calculations Bickers et al, (1987) deduced a temperature dependent relaxation time $\Gamma(T)$ from the position of the peak in $\text{Im}(\omega, T)$ which is non-monotonic in T (see figure 9.38) and has clear similarities with $\Gamma_0(T)$ shown in figure 9.37. At higher temperatures the calculated $\Gamma_0(T)$ behaves approximately as $T^{1/2}$. This form of temperature dependence has been observed at higher temperatures in $CeAl_3$

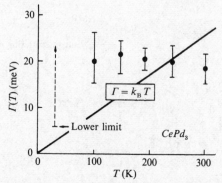

Figure 9.39 Temperature dependence of the quasi-electric line width for $CePd_3$ (Holland-Moritz, Wohlleben & Loewenhaupt, 1982).

and several other Kondo lattice compounds (see Murani, 1987). These results suggest that a single ion response based on the degenerate models could explain the main features of the observed spectra for $YbCuAl$ and $CeSn_3$. More detailed comparisons between theory (including crystal field terms) and experiment are needed to establish whether this is so.

Rare earth mixed valent compounds, such as $CePd_3$ (Holland-Moritz, Loewenhaupt, Schmatz & Wohlleben, 1976), have a very broad almost temperature independent quasi-elastic scattering. In these cases charge fluctuations are playing more of a role in the impurity spin relaxation resulting in broader line widths $\Gamma(T) \sim 20$ meV, see figure 9.39. One exception is the mixed valent compound $TmSe$. This compound, as we discussed in the last section, orders antiferromagnetically at $T_N \sim 3.5$ K, and has valence fluctuations between $4f^{12}$ and $4f^{13}$, the lowest multiplets of both of these configurations being magnetic. Between 120 K and 300 K the observed neutron spectrum can be described by a broad quasi-elastic peak with a line width $\Gamma_0(T)$, which only slightly increases with temperature from 6 meV to 7 meV, similar to other mixed valence materials. Below this temperature the quasi-elastic line width decreases as the temperature is lowered down to the ordering temperature 3.5 K; at 5 K the line width is 0.5 meV. At the same time an inelastic line appears, $\omega_i \sim 6$ meV at 60 K, which narrows and increases as the temperature is lowered until at 5 K, $\omega_i \sim 10$ meV, as seen in figure 9.40 (Loewenhaupt & Holland-Moritz, 1978). At low temperatures the elastic and quasi-elastic lines are well separated. With the on-set of magnetic order the quasi-elastic line disappears and an inelastic line at $\omega_i \sim 1$ meV, which has been attributed to magnon excitations, develops. Apart from the

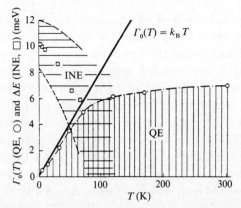

Figure 9.40 The temperature dependence of the quasi-elastic line width (circles) and the position of the inelastic line (squares) in $TmSe$ as a function of temperature (Holland-Moritz et al, 1982).

magnetic order the dilute systems, $Tm_{1.05}La_{0.95}Se$, $Tm_{0.05}Y_{0.95}Se$, behave in a similar way which suggests that the main features of the spectra can be explained in a single impurity framework (Holland-Moritz, 1983).

Several calculations of the dynamic susceptibility have been made using single ion Tm models but there are as yet no definitive results. Schlottmann (1984c) predicts a magnetic ground state with quasi-elastic and inelastic components in the spectrum. The line width of the quasi-elastic peak as a function of temperature looks very much like the observed form. The theory does not seem to work so well, however, for the inelastic peak.

The theories for the rare earth systems predict, at least qualitatively, the features found experimentally. Whether the single ion theories can explain in detail the temperature dependence of the position of the inelastic humps and the line widths is not clear at this stage.

9.6 Local Measurements

The introduction of an impurity into a metal causes a local charge and spin disturbance to the host conduction electrons. This can be detected experimentally via the hyperfine coupling to the host nuclei. The states of the nuclei, both at the impurity and host sites, can be probed by a variety of techniques such as nuclear magnetic resonance (NMR), Mössbauer measurements (ME) and nuclear orientation (NO). The extra charge introduced by the impurity, relative to that of the host metal, is the primary disturbance and though screened out by the conduction electrons on a scale of an atomic spacing, as discussed in chapter 1, induces longer range Friedel oscillations due to the sharpness of the Fermi surface (see equation (1.34)). In the Kondo regime there are spin–spin correlations induced between the localized spin and the conduction electrons which, for $T < T_K$, magnetically screen the local moment. The coherence length ξ for this magnetic screening must of the order of $\xi \sim v_F/k_B T_K$, where v_F is the Fermi velocity of the conduction electrons. We look first of all at what theoretical estimates can be made of these local screening effects and then briefly survey the experimental methods of probing them.

The local charge and spin densities can be deduced from the conduction electron Green's function $G_{k\sigma,k'\sigma}(\omega)$. We will calculate this for the non-degenerate Anderson model (1.66). From the equation of motion of this Green's function we found earlier, equation (5.38),

$$G_{k\sigma,k'\sigma}(\omega) = \frac{\delta_{k,k'}}{(\omega - \epsilon_k)} + \frac{V_k}{(\omega - \epsilon_k)} G_{d\sigma,d\sigma}(\omega) \frac{V_{k'}^*}{(\omega - \epsilon_{k'})}. \qquad (9.65)$$

Taking the inverse Fourier transform of this with respect to the radial vector $\mathbf{r}$, as measured from the impurity site, gives

$$\sum_{k,k'} G_{k\sigma,k'\sigma}(\omega) e^{i(\mathbf{k}-\mathbf{k}')\cdot\mathbf{r}} = \sum_k \frac{1}{(\omega - \epsilon_{k,\sigma})} + g_\sigma^2(\omega,r) G_{d\sigma,d\sigma}(\omega), \qquad (9.66)$$

where

$$g_\sigma(\omega,r) = \sum_k \frac{V_k e^{i\mathbf{k}\cdot\mathbf{r}}}{(\omega - \epsilon_{k,\sigma})}. \qquad (9.67)$$

The change in conduction electron density at $\mathbf{r}$ due to the presence of the impurity, $\Delta\rho_c(\mathbf{r},T)$ $(= \rho_c(\mathbf{r},T) - \rho_c^{(0)}(\mathbf{r},T))$ deduced from (9.66) is given by

$$\Delta\rho_c(\mathbf{r},T) = -\frac{\mathrm{Im}}{\pi} \sum_\sigma \int_{-\infty}^\infty f(\omega) g_\sigma^2(\omega^+,r) G_{d\sigma,d\sigma}(\omega^+) \, d\omega. \qquad (9.68)$$

Evaluation of this integral requires a specific form for V_k and ϵ_k. To proceed further we make the assumption that V_k is independent of $\mathbf{k}$ and take plane waves for the conduction electron states so that the integral in (9.67) can be performed analytically. This gives

$$g(\omega^+,\mathbf{r}) = -\frac{\pi\rho_0 V}{k_F r} e^{ir(2m(\omega+\epsilon_F))^{1/2}}, \qquad (9.69)$$

where $\epsilon_F = k_F^2/2m$ and ρ_0 is the density of states at the Fermi level, $\rho_0 = mk_F/2\pi^2$ (Ishii, 1976).

To find the induced charge density for all values of $\mathbf{r}$ requires a knowledge of $G_{d\sigma,d\sigma}(\omega^+)$ for all frequencies ω. The asymptotic form for large r, however, can be deduced by an integration by parts and only requires a knowledge of $G_{d\sigma,d\sigma}(\omega^+)$ at the Fermi level $\omega = 0$. Using (9.69) and integrating by parts twice for $T = 0$ gives

$$\Delta\rho_c(\mathbf{r}, 0) = \frac{\Delta}{4\pi r^3}\left[\sum_\sigma \mathrm{Re}\{e^{2ik_F r}G_{d\sigma,d\sigma}(0^+)\}\right.$$

$$\left. -\frac{1}{2k_F r}\left(\mathrm{Im}\left\{e^{2ik_F r}(G_{d\sigma,d\sigma}(0^+) + 2\epsilon_F\frac{dG_{d\sigma,d\sigma}(\omega^+)}{d\omega})\right\}_{\omega=0} + \mathrm{O}(r^{-2})\right)\right],$$

(9.70)

where $\Delta = \pi\rho_0|V|^2$. Using equation (5.44) continued for real frequencies for $G_{d\sigma,d\sigma}(\omega^+)$, and evaluating for $\omega = 0$, gives

$$G_{d\sigma,d\sigma}(0^+) = \frac{1}{-E_{d,\sigma} + i\Delta} = \frac{\sin\eta_\sigma(\epsilon_F)}{\pi\Delta}e^{i\eta_\sigma(\epsilon_F)}, \qquad (9.71)$$

where $\eta_\sigma(\epsilon_F)$ is the phase shift at the Fermi level for spin σ,

$$\eta_\sigma(\epsilon_F) = \frac{\pi}{2} - \tan^{-1}\left(\frac{E_{d,\sigma}}{\Delta}\right), \qquad (9.72)$$

and $E_{d,\sigma} = \epsilon_{d,\sigma} + \Sigma^R(0)$, measuring $\epsilon_{d,\sigma}$ relative to the Fermi level ϵ_F. Substituting (9.71) into the leading term in (9.70) gives the Friedel form (1.34) we derived earlier for purely potential scattering,

$$\Delta\rho_c(\mathbf{r}, 0) \sim -\frac{1}{2\pi^2 r^3}\sin\eta(\epsilon_F)\cos(2k_F r + \eta(\epsilon_F)), \qquad (9.73)$$

but now including the effects of interactions. This result only applies to s wave scattering because we have taken $V_\mathbf{k}$ to be independent of $\mathbf{k}$. It can be written in terms of the occupation number of the impurity level n_d via the Friedel sum rule, $\eta_\sigma(\epsilon_F) = \pi n_d/2$. With $n_d = \Delta Z$, so that the excess impurity charge, ΔZe, is fully screened, this asymptotic form depends only on the charge to be screened and not on the details of the impurity potential and local interactions. To find any evidence of the Kondo effect we must look at the next term in (9.70). This, on evaluation using (5.44), gives

$$\frac{\sin\eta(\epsilon_F)}{2\pi^2 k_F r^4}\left\{\frac{\epsilon_F}{\tilde{\Delta}}\sin\eta(\epsilon_F)\sin(2k_F r + 2\eta(\epsilon_F)) + \frac{1}{2}\sin(2k_F r + \eta(\epsilon_F))\right\},$$

(9.74)

where the effective resonance width $\tilde{\Delta}$ is as defined earlier, $\tilde{\Delta} = z\Delta$, and z is the wavefunction renormalization factor $z = 1/(1 - \partial\Sigma/\partial\omega)_{\omega=0}$. The first term has the many-body enhancement factor $\Delta/\tilde{\Delta}$ relative to the non-interacting model. In the Kondo regime with $\tilde{\Delta} = 4k_B T_K/\pi w$ (5.22) this enhancement factor can be large for $\Delta \gg k_B T_K$. The second term in (9.74) arises from the ω dependence in $g(\omega, \mathbf{r})$ and has no explicit many-body factors. For $k_F r \sim \epsilon_F/\tilde{\Delta}$ the first term in (9.74) becomes comparable with the leading term (9.73). Hence we find in the Kondo regime that the Friedel result only applies for $r > \xi$ with $\xi = v_F \pi w/8 k_B T_K$. In the non-interacting case the Friedel result is valid for $r > r_0 \sim \epsilon_F/k_F \Delta$. Taking typical values of $\epsilon_F/\Delta \sim 10$, r_0 is of the order of a few lattice spacings. In the Kondo regime ξ is typically 10–100 times larger.

For a transition metal impurity we must take the $\mathbf{k}$ dependence of $V_{\mathbf{k}}$ into account. For a spherically symmetric impurity potential $V_{\mathbf{k}}$ can be expanded in terms of spherical harmonics $Y_l^m(\Omega_{\mathbf{k}})$ associated with the angular momentum quantum numbers l, m, $V_{\mathbf{k}} = \sum_{l,m} V_l Y_l^m(\Omega_{\mathbf{k}})$. Retaining only the l wave scattering the results (9.73), (9.74), for $\Delta\rho_c(\mathbf{r}, 0)$ becomes generalized to

$$\Delta\rho_c(\mathbf{r}, 0) \sim -\frac{(-1)^l(2l+1)}{2\pi^2 r^3} \sin\eta_l(\epsilon_F)$$
$$\left\{ \cos(2k_F r + \eta_l(\epsilon_F)) - \frac{1}{k_F r}\left(\frac{\epsilon_F}{\tilde{\Delta}}\sin\eta_l(\epsilon_F)\sin(2k_F r + \eta_l(\epsilon_F)) \right.\right.$$
$$\left.\left. + \left(l(l+1) + \frac{1}{2}\right)\sin(2k_F r + \eta_l(\epsilon_F)) + (r^{-2})\right)\right\},$$

$$(9.75)$$

where $\eta_l(\epsilon_F) = \pi n_d/2(2l+1)$ from the Friedel sum rule (Šokčević, Zlatić & Horvatić, 1989). This form is only applicable if we can neglect Hund's rule terms, crystal fields and spin orbit interactions.

Evaluation of (9.68) for the near and pre-asymptotic region $r < \xi$ requires an expression for $G_{d\sigma,d\sigma}(\omega^+)$ over the full frequency range and not just in the vicinity of the Fermi level. Šokčević et al, (1989) have performed calculations in this region for $l = 0$ using the form of $G_{d\sigma,d\sigma}(\omega^+)$ deduced from perturbation theory in U. One set of their results is shown in figure 9.41 and illustrates the suppression of the charge oscillations as the value of U is increased and the resonance in the density of states at the Fermi level narrows (see figure 5.7). The period and the phase of the oscillations seem to be almost unaffected by the value of U. We look at the way these effects have been probed experimentally in the discussion of experimental techniques at the end of this section. We now look at the local magnetic effects caused by the impurity.

In a magnetic field there will be an extra contribution to the magnetization of the conduction electrons due to the presence of the impurity which we shall denote by $\Delta M_c(\mathbf{r}, T)$. This can be calculated from the

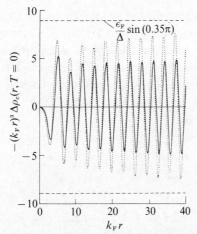

Figure 9.41 A plot of $-(k_F r)^3 \Delta \rho_c(r, 0)$ as a function of $k_F r$ for $n_d = 0.7$ and $U/\pi\Delta = 2.5$ (solid line) and $U/\pi\Delta = 0$ (dotted line). The dashed line is the envelope of the Friedel result (9.73) (from Šokčević et al, 1989).

difference in induced charge densities for the spin up and spin down electrons,

$$\Delta M_c(\mathbf{r}, T) = -\frac{g\mu_B}{2\pi} \mathrm{Im} \sum_\sigma \sigma \int_{-\infty}^{\infty} f(\omega) g_\sigma^2(\omega^+, r) G_{d\sigma, d\sigma}(\omega^+) \, d\omega, \quad (9.76)$$

where $\sigma = 1(\uparrow)$ or $-1(\downarrow)$. For simplicity we neglect the field dependence of $g_\sigma(\omega^+, \mathbf{r})$ and concentrate on the field dependence due to $G_{d\sigma, d\sigma}(\omega^+)$. This should give the dominant contribution to $\Delta M_c(\mathbf{r}, T)$, and it is relatively straightforward to take any correction arising from $g_\sigma(\omega^+, \mathbf{r})$ into account. With this assumption we can derive an expression for the asymptotic form of $\Delta M_c(\mathbf{r}, T)$ for the s wave model, which is basically the same as (9.70) but with an extra factor σ within the σ-summation. To evaluate this expression we need the field dependence of $G_{d\sigma, d\sigma}(\omega^+)$ and its derivative at $\omega = 0$. Using (9.71) with $\eta_\sigma(\epsilon_F) = \pi n_{d,\sigma}$. we deduce

$$G_{d\sigma, d\sigma}(0^+) = \frac{e^{i\pi n_d/2}}{\Delta} \sin(\pi n_d/2) - \frac{\pi \chi_{\mathrm{imp}} H}{\Delta(g\mu_B)} e^{i\pi n_d} + \mathrm{O}(H^2), \quad (9.77)$$

to first order in the magnetic field H at $T = 0$. For the derivative evaluated to the same order we find

$$\frac{dG_{d\sigma, d\sigma}(\omega^+)}{d\omega} = \frac{\sin^2(\pi n_d/2) e^{i\pi n_d}}{\tilde{\Delta}\Delta} - \frac{2\pi i \sigma}{\tilde{\Delta}\Delta} \sin(\pi n_d/2) e^{3i\pi n_d/2}$$
$$+ \frac{\sigma H}{\Delta^2} \sin^2(\pi n_d/2) e^{i\pi n_d} \left(\frac{\partial^2 \Sigma(\omega^+)}{\partial H \partial \omega} \right)_{\omega = H = 0}. \quad (9.78)$$

The derivative $\partial^2 \Sigma / \partial H \partial \omega$ is difficult to calculate in the general case.

Ishii (1976) has shown that for the symmetric model this term is an odd function of ω and hence vanishes also for $\omega = 0$. Using equation (5.80), and the fact that Σ is an odd function of H, we can relate this derivative to $\partial \chi_c^* / \partial H$ and hence deduce that it vanishes in the Kondo limit where $\chi_c^* \to 0$. Restricting discussion to the case $n_d = 1$ where we are justified in neglecting this term we obtain the result of Ishii (1976)

$$
\Delta M_c(\mathrm{r}, 0) = \frac{\chi_{\mathrm{imp}} H}{4\pi r^3 (g\mu_B)} \left[\cos(2k_F r) \right.
$$
$$
\left. - \frac{1}{2k_F r} \left(\sin(2k_F r) + \frac{4\epsilon_F}{\tilde{\Delta}} \cos(2k_F r) + \mathrm{O}(r^{-2}) \right) \right],
$$

(9.79)

for $\Delta M_c(r, 0)$ to first order in H. This again only applies beyond the screening cloud $r \gg \xi$. In the leading term the Kondo effect is only manifest through the susceptibility χ_{imp}. For the symmetric model we can consider the change in (9.79) on increasing U. For $U > \pi\Delta$, χ_{imp} increases markedly and the region over which this asymptotic result applies, $r \gg \xi$, gets pushed to larger values of r as ξ increases (ξ increases exponentially with U, $\xi \sim e^{U/\pi\Delta}$ for U large).

Calculation of $\Delta M_c(\mathrm{r}, T)$ in the near and pre-asymptotic regime in this case is more difficult than in the charge case because it requires a knowledge of $G_{d\sigma, d\sigma}(\omega^+)$ not only at all frequencies but also as a function of the magnetic field strength H. The form of $\Delta M_c(\mathrm{r}, T)$ in this regime will not be universal but will depend on details of the band structure and the form of $V_{\mathbf{k}}$. As far as the author is aware there are no reported calculations in this regime in the literature to date.

Calculation of $\Delta M_c(\mathrm{r}, T)$ even in the regime $r < \xi$ will not give any direct information on how the conduction electrons screen the local moment in the Kondo regime. This is best examined by calculating the correlation function between the localized spin and the conduction electrons at a distance r from the impurity, $\langle \mathbf{S} \cdot \mathbf{S}_c(\mathbf{r}) \rangle$. The best way of calculating this quantity is using the Monte Carlo method, an approach we have not considered so far. There are several ways of setting up Monte Carlo calculations but a particularly useful way for the Anderson model is via the discrete Hubbard–Stratonovich transformation (Hirsch, 1983). The Hubbard–Stratonovich transformation is based on the identity,

$$
e^{\mathbf{A}^2/2} = \sqrt{2\pi} \int_{-\infty}^{\infty} e^{-x^2/2 - x\mathbf{A}} \, dx,
$$

(9.80)

where $\mathbf{A}$ is an operator. It is a way of linearizing an interaction term $\mathbf{A}^2$ into a single-body term coupled to an auxiliary field x. This identity can be applied to the interaction term $U n_{\mathrm{d},\uparrow} n_{\mathrm{d},\downarrow}$ in the Anderson model by re-expressing it in the form $-U(n_{\mathrm{d},\uparrow} - n_{\mathrm{d},\downarrow})^2/2$, the terms linear in $n_{\mathrm{d},\sigma}$ $(n_{\mathrm{d},\sigma}^2 = n_{\mathrm{d},\sigma})$ can be cancelled by shifting the d level ϵ_{d} to $\epsilon_{\mathrm{d}} + U/2$. In the discrete case the x-field is replaced by an Ising variable σ' which takes only two values $\sigma' = \pm 1$. The identity in this case becomes

$$e^{\alpha(n_{\mathrm{d},\uparrow} - n_{\mathrm{d},\downarrow})^2} = \sum_{\sigma'} e^{K(n_{\mathrm{d},\uparrow} - n_{\mathrm{d},\downarrow})\sigma'}, \qquad (9.81)$$

where $\cosh K = e^{\alpha}$.

To apply these identities the partition function Z is written in the form,

$$Z = \operatorname{Tr} e^{-\beta H} = \operatorname{Tr} e^{-\Delta\tau H} e^{-\Delta\tau H} \ldots e^{-\Delta\tau H}, \qquad (9.82)$$

where $\Delta\tau = \beta/L$ and the right hand side of (9.82) contains L factors. The individual terms are then factorized approximately,

$$e^{-\Delta\tau H} \approx e^{-\Delta\tau H_0} e^{-\Delta\tau H_{\mathrm{I}}}, \qquad (9.83)$$

where $H = H_0 + H_{\mathrm{I}}$ and H_{I} is the interaction term. The error is of the order of $(\Delta\tau)^2$. The transformation (9.81) is applied to $e^{-\Delta\tau H_{\mathrm{I}}}$ in each 'time slice'. The errors in the approximation are controlled by making L large ($\Delta\tau$ small). In principle the limit $L \to \infty$ should be taken so that the system becomes equivalent to a non-interacting model in the presence of a 'time' dependent Ising field $\sigma'(\tau)$. In practice L is taken to be finite but as large as possible. The final result for the partition function is in the form of a path integral with a sum over all possible distributions of the auxiliary Ising fields $\sigma'(\tau) = \pm 1$ for each time slice τ_n where $\tau_n = n\Delta\tau$ ($\sigma'(\tau)$ is like $V(\tau)$ in figure 3.2).

This approach is a particularly convenient one for calculating correlation functions because before the sum over the auxiliary fields is taken the model is a non-interacting one and two particle correlation functions can be factorized into single body terms and related to products of single particle Green's functions. The correlations due to the interactions arise from averaging these products over all possible configurations of the auxiliary Ising fields. The Monte Carlo algorithm samples the configuration space so that averages are made over a representative set which gives the dominant contributions at a particular temperature T. In practice only the d Green's function need be calculated for each configuration. For details of the method of selection of the configurations and procedures for updating the Green's function we refer the reader to the work of Blankenbecler, Scalapino & Sugar (1981), and Hirsch & Fye (1986).

Gubernatis, Hirsch & Scalapino (1987) have calculated the quantity

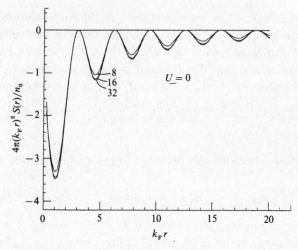

Figure 9.42 The spin–spin correlation $4\pi(k_{\mathrm{F}}r)^2 S(r)/n_0$ $(n_0 = k_{\mathrm{F}}^3/3\pi^2)$ as a function of $k_{\mathrm{F}}r$ for $U = 0$, $\Delta = 0.5$ and $1/k_{\mathrm{B}}T = 8, 16, 32$ (Gubernatis et al, 1987).

$S(\mathbf{r})$ defined by

$$S(\mathbf{r}) = 4\langle S_z S_{\mathrm{c},z}(\mathbf{r}) \rangle = 2\langle n_{\mathrm{d},\uparrow} n_{\mathrm{c},\uparrow}(\mathbf{r}) - n_{\mathrm{d},\downarrow} n_{\mathrm{c},\uparrow}(\mathbf{r}) \rangle, \qquad (9.84)$$

which from rotational symmetry is related to $\langle \mathbf{S} \cdot \mathbf{S}_{\mathrm{c}}(\mathbf{r}) \rangle$ by $\langle \mathbf{S} \cdot \mathbf{S}_{\mathrm{c}}(\mathbf{r}) \rangle = 3S(\mathbf{r})/4$. The asymptotic form of this correlation function for $r \gg \xi$ has been calculated by Ishii (1978) using an approach similar to the derivation of (9.79). The leading term is given by

$$S(\mathbf{r}) \sim -\frac{R\epsilon_{\mathrm{F}}\sin^2(\eta(\epsilon_{\mathrm{F}}))(1 + \cos(2k_{\mathrm{F}}r + 2\eta(\epsilon_{\mathrm{F}})))}{\pi^3 \tilde{\Delta} k_{\mathrm{F}} r^4}, \qquad (9.85)$$

where R is the Wilson χ/γ ratio. The correction term is difficult to calculate in the general case. Ishii, however, has derived the leading term plus corrections for the symmetric model $\eta(\epsilon_{\mathrm{F}}) = \pi/2$,

$$S(\mathbf{r}) \sim -\frac{2R\epsilon_{\mathrm{F}}}{\pi^3 \tilde{\Delta} k_{\mathrm{F}} r^4} \left[\sin^2(k_{\mathrm{F}}r) \left(1 - \frac{4\epsilon_{\mathrm{F}} R}{k_{\mathrm{F}} r \tilde{\Delta}}\right) + \frac{4}{k_{\mathrm{F}} r} \sin(2k_{\mathrm{F}}r) \right] \qquad (9.86)$$

which is valid for $r \gg 4\epsilon_{\mathrm{f}} R/k_{\mathrm{F}}\tilde{\Delta}$, equivalent to $r \gg \xi$.

Results of calculations of Gubernatis et al (1987) for the symmetric model in the regime $r \ll \xi$, with $\Delta = 0.5$, at various temperatures are shown in figures 9.42 and 9.43. The antiferromagnetic correlations in the results for $S(\mathbf{r})$ for $U = 0$ and small r simply reflect Fermi hole correlations due to the exclusion principle resulting in a depletion in $n_{\mathrm{c},\uparrow}$ when $n_{\mathrm{d},\uparrow}$ is occupied. In the case $U = 4$ shown in figure 9.43 there is an enhanced antiferromagnetic correlation due to the interaction. More surprisingly there are also ferromagnetic correlations. The enhanced

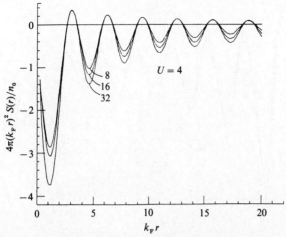

Figure 9.43 The spin–spin correlation $4\pi(k_F r)^2 S(r)/n_0$ $(n_0 = k_F^3/3\pi^2)$, as a function of $k_F r$ for $U = 4$, $\Delta = 0.5$ and $1/k_B T = 8, 16, 32$ (Gubernatis et al, 1987).

antiferromagnetic correlations disappear as the temperature is raised above the Kondo temperature reflecting the dispersion of the screening cloud. The ferromagnetic correlations on the other hand show very little temperature dependence. As in the charge case, figure 9.41, the period and phase of the oscillations are practically unchanged by the interaction.

These results can be used to investigate the screening of the impurity moment. The impurity susceptibility can be expressed in terms of spin–spin correlations using (3.20),

$$\chi_{\text{imp}}(T) = \frac{(g\mu_B)^2\{\langle(S_z + S_{c,z}^{\text{tot}})^2\rangle - \langle(S_{c,z}^{\text{tot}})^2\rangle_0\}}{k_B T}. \qquad (9.87)$$

From equation (5.38), or (9.65), the total induced polarization of the conduction electrons in the presence of a magnetic field, $\langle S_{c,z}^{\text{tot}}\rangle$, can be estimated. The extra contribution due to the presence of the impurity depends on the factor $\sum_k |V_k|^2/(\omega - \epsilon_k)^2$ and can be shown to lead to a negligible contribution in the flat wide band limit $D \to \infty$ (this is known as the Clogston–Anderson compensation theorem (Clogston & Anderson, 1961). In this limit the contribution to (9.87) from $\langle(S_{c,z}^{\text{tot}})^2\rangle - \langle(S_{c,z}^{\text{tot}})^2\rangle_0$ can be neglected so it can be rewritten in the form,

$$\chi_{\text{imp}}(T) = \frac{(g\mu_B)^2}{k_B T}\{\langle S_z^2\rangle + 2\int \langle S_z S_{c,z}(\mathbf{r})\rangle \, d^3 r\}. \qquad (9.88)$$

For χ_{imp} to be finite as $T \to 0$ the correlation term, $\langle(S_z S_{c,z}(\mathbf{r})\rangle$, inte-

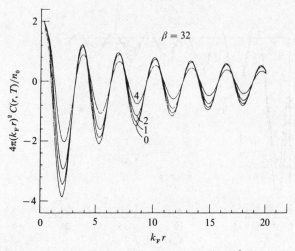

Figure 9.44 The charge–charge correlation $4\pi(k_F r)^2 C(r,T)/n_0$ as a function of $k_F r$ for $U = 0, 1, 2$ and 4, $\Delta = 0.5$ and $1/k_B T = 32$ (Gubernatis et al, 1987).

grated over the whole of space must cancel out the local moment term $\langle S_z^2 \rangle$. Integrating the asymptotic contribution (9.86) over the region $r > \xi$ accounts for about 20% of the screening. The main contribution, $\sim 80\%$ comes from the region within the coherence length, $r < \xi$. This was confirmed by Gubernatis et al (1987) on integrating their results.

Monte Carlo results for the charge-charge correlation $C(\mathbf{r}, T)$, where $C(\mathbf{r}, T) = \langle n_d n_c(\mathbf{r}) \rangle$, shown in figure 9.44, are similar to those for the induced charge $\Delta \rho_c(\mathbf{r})$ in figure 9.42. They clearly show a suppression of the charge oscillations within the coherence regime $r < \xi$. We now turn to look in more detail at some of the ways in which these effects can be probed experimentally by the interaction with the impurity or host nuclei.

There are several types of magnetic interactions between the nuclear moment and the surrounding electrons. There is a Fermi contact term (s electrons only), a dipole–dipole interaction with the electronic spin and a term arising from the moment induced by the orbital motion of the electron. If these interactions are weak, and the relaxation time of the electronic states is short compared to the time for a nuclear Larmor precession, these interactions can be treated as a time dependent molecular fields $\mathbf{H}^L(t)$. The thermal average of the local field $\langle \mathbf{H}^L \rangle$ is a static field which shifts the frequency required for a resonant excitation between the nuclear Zeeman levels. This shift as a function of temperature is one of the important pieces of information obtained from Mössbauer experiments and NMR.

There is no direct contribution to the Fermi contact term from electrons in a d shell. The main coupling to the spin S of electrons in an unfilled 3d shell is via the polarization this induces in the core s states in the presence of a magnetic field. This is also the case for 4f electrons but they give a smaller contribution as the 4f shell is more localized than the 3d. For an orbital singlet such as Mn^{2+}, or an orbitally quenched crystal field state, $\langle \mathbf{H}^L \rangle$ at the impurity site will have a contribution proportional to $\langle S_z \rangle$ and in weak magnetic field H this will be proportional to $\chi_{\mathrm{imp}}(T)H$. The coupling of the electronic spin S the conduction electron spin σ at the nucleus, and the nuclear spin I can be expressed as an exchange interaction,

$$H = A_s \mathbf{S} \cdot \mathbf{I} + A_\sigma \sigma \cdot \mathbf{I}, \qquad (9.89)$$

where A_s, A_σ are the respective couplings, so that the hyperfine field is given by $\langle \mathbf{H}^L \rangle = -(A_s \langle \mathbf{S} \rangle + A_\sigma \langle \sigma \rangle)/g_N \mu_N$, where μ_N is the nuclear Bohr magneton and g_N the nuclear g factor. The $\sigma \cdot \mathbf{I}$ term can be written out explicitly as in (1.64). This contribution can be usually be neglected due to the enhanced susceptibility associated with the localized spin S. Due to the mixing in of higher multiplets in the ground state there can also be a temperature independent orbital contribution (van Vleck term). On the host site a distance r from the impurity the spin contribution to the hyperfine field, $\langle \mathbf{H}_s^L \rangle$ will be proportional to $\Delta M_c(r, T)$. The quantity which is usually experimentally determined is the ratio, $K = \langle H^L \rangle / H$, which is independent of H for weak fields.

As well as giving information on the static fields at the nucleus via ΔK, NMR measurements also give information about the dynamics of these fields via the longitudinal and transverse relaxation times of the nuclear moment, T_1 and T_2. These relaxation times in the standard NMR theory (Narath, 1973) can be related to the Fourier transforms of the fluctuations in the local fields,

$$\frac{1}{T_1} = \frac{(g_N \mu_N)^2}{2} \int_{-\infty}^{\infty} \langle (H_x^L(t)H_x^L(0) + H_y^L(t)H_y^L(0)) \rangle e^{i\omega_0 t} \, dt, \qquad (9.90)$$

and

$$\frac{1}{T_2} = \frac{1}{2T_1} + \frac{(g_N \mu_N)^2}{2} \int_{-\infty}^{\infty} \langle \delta H_z^L(t) \delta H_z^L(0) \rangle \, dt, \qquad (9.91)$$

where $\delta H_z^L = H_z^L - \langle H_z^L \rangle$ and ω_0 is the nuclear Larmor frequency. Corresponding to the different contributions to the local field there are also different contributions to the relaxation times. For a nucleus at the impurity site local spin–spin contribution to (9.90) and (9.91) can be related to the imaginary part of the dynamic susceptibility via the fluctuation dissipation theorem,

$$\frac{2\mathrm{Im}\chi_{s,\alpha,\beta}(\omega)}{(1 - e^{-\beta\omega})} = (g\mu_\mathrm{B})^2 \int_{-\infty}^{\infty} \langle S_\alpha(t)S_\beta(0)\rangle e^{i\omega t}\, dt, \qquad (9.92)$$

where $\alpha, \beta = x, y$ or z. As this only has to be evaluated at low frequencies in the temperature range $\omega = \omega_0 \ll k_\mathrm{B}T$ it can be simplified to

$$\mathrm{Im}\chi_{s,\alpha,\beta}(\omega) = \frac{\omega(g\mu_\mathrm{B})^2}{2k_\mathrm{B}T} \int_{-\infty}^{\infty} \langle S_\alpha(t)S_\beta(0)\rangle e^{i\omega t}\, dt. \qquad (9.93)$$

In weak fields spin contribution to the resonance shift and the relaxation time can be related to real and imaginary parts of the dynamic local spin susceptibility,

$$K_s = \frac{1}{g\mu_\mathrm{B}} A_s \mathrm{Re}[\chi_{s,z,z}(0,0)], \qquad (9.94)$$

and

$$\frac{1}{T_{s,1}} = \frac{k_\mathrm{B}T}{(g\mu_\mathrm{B})^2}(g_\mathrm{N}\mu_\mathrm{N})^2 A_s^2 \lim_{\omega\to 0}\mathrm{Im}\left[\frac{\chi_{s,x,x}(\omega) + \chi_{s,y,y}(\omega)}{\omega}\right], \qquad (9.95)$$

where A_s is the coupling of the spin moment to the impurity in (9.89). Eliminating A_s between (9.94) and (9.95) gives the relation

$$K_s^2 T_{s,1}T = \frac{[\mathrm{Re}[\chi_{s,z,z}(0,0)]]^2}{k_\mathrm{B}\lim_{\omega\to 0}\mathrm{Im}\left[[\chi_{s,x,x}(\omega) + \chi_{s,y,y}(\omega)]/\omega\right]}, \qquad (9.96)$$

and using the Korringa–Shiba relation (E.1) for $T \to 0$, $H \to 0$,

$$K_s^2 T_{s,1}T \to \frac{(g\mu_\mathrm{B})^2}{(g_\mathrm{N}\mu_\mathrm{N})^2}\frac{1}{4\pi k_\mathrm{B}}, \qquad (9.97)$$

which is the classical Korringa constant for a free electron gas. There are similar independent equations to (9.94), (9.95) and (9.96) for the orbital contribution if crystal field and spin–orbit interactions are neglected. The orbital hyperfine coupling is of opposite sign to the spin term so that the overall sign (negative for spin, positive for orbital) is a useful guide as to which term gives the dominant effect.

Theoretical calculations of $1/T_1T$ as a function of temperature ($H \to 0$) for the non-degenerate Anderson model, scaled using (9.97) to correspond to 1 as $T \to 0$, have been obtained using a numerical renormalization group method. Results for the symmetric model are shown in figure 9.45. Similar results have been obtained by using Monte Carlo calculations supplemented by use of the maximum entropy principle (Jarrell, Gubernatis & Silver, 1991).

We look first of all at the experimental evidence from NMR of the shift in hyperfine fields due to the impurity induced spin polarization

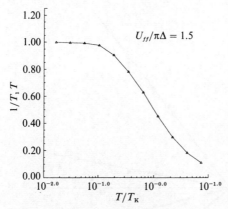

Figure 9.45 A plot of $1/T_1T$ as a function of T as calculated within a numerical renormalization group approach (Costi & Hewson, 1991).

$\Delta M_\mathrm{c}(r, T)$ in the host metal. A well studied example is the resonance at the Cu nuclei in dilute alloys of Fe in Cu. What is seen in this type of experiment is an inhomogeneously broadened line about the un-shifted line of pure Cu, plus satellites arising from near neighbour Cu nuclei of the Fe impurities. The broadened line arises from the different hyperfine fields $\langle H_\mathrm{L}(r)\rangle$ at Cu nuclei beyond a distance r_0 (the 'wipe out' radius). For $r > r_0$ the shifts, $\langle H_\mathrm{L}(r) - H_\mathrm{L}(\infty)\rangle$, of the lines from the position of that for pure Cu, arising from $\Delta M_\mathrm{c}(r, T)$, are so small and the nuclei so numerous that they merge to give an inhomogeneous line broadening. The response of individual Cu nuclei which are near neighbours of an impurity site can be seen as satellites to the main line. The effort in the early work in this field was concentrated on measuring the temperature and field dependence of the broadening $\Delta H_\mathrm{L}(H, T)$ of the main line (which was the only one seen at the time) as a way of gain-ing information about the conduction electron polarization cloud about the impurity (see Heeger, 1969; Potts & Welsh, 1972). The measured values of the ratio $\Delta H_\mathrm{L}(H, T)/H$ were found to be proportional to the local susceptibility for Fe in Cu ($\propto 1/(T + 29)$) for $T > 20$ K but to deviate significantly at low temperatures and high fields. It is not easy to calculate the line broadening $\Delta H_\mathrm{L}(H, T)$ theoretically as it requires $\Delta M_\mathrm{c}(r, T)$ to be estimated for $r > r_0$, which includes the pre-asymptotic region for typical values of r_0, for a single impurity and then averaged over a random distribution of impurities. As a result these are difficult experiments to interpret. Early interpretations based on perturbation

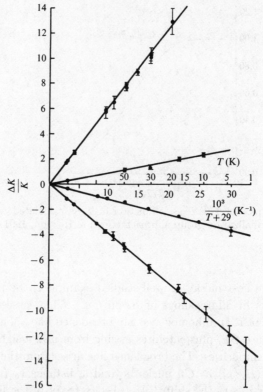

Figure 9.46 The relative shifts in hyperfine field $\Delta K/K$ as a function of $1/(T+29)$ for four ^{63}Cu NMR satellites for dilute Fe in Cu (after Boyce & Slichter, 1974).

theory which were used to estimate the exchange coupling J for the s-d model are unlikely to give reliable estimates of this quantity.

Observation of the satellite lines associated with the near neighbour shells of Cu atoms show shift ratios, $\Delta K_s(r) = (\langle H_L(r) - H_L(\infty)\rangle)/H$, which for weak fields scale as the impurity susceptibility over the full temperature range. Early results of Boyce & Slichter (1974), shown in figure 9.46, show lines whose positions all scale to a good approximation with the Curie–Weiss form $\chi_{imp} \propto 1/(T+29)$. With more measurements in the low temperature range Alloul (1975) was able to get a more precise fit in the range $T \ll T_K$ finding $\Delta K_s(T) \propto (1 - (T/\theta_K)^2)$ with $\theta_K = 17$, plus a small temperature independent contribution which was attributed to an orbital contribution ΔK_{orb}. Though $\chi_{imp}(T)$ has a similar T^2 dependence the coefficient deduced by Triplett & Phillips (1971), and quoted earlier, is about a factor of 4 larger. The behaviour

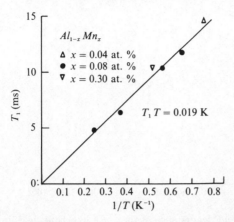

Figure 9.47 The NMR longitudinal relaxation time T_1 plotted as a function of $1/T$ for ^{55}Mn impurities in Al for three different compositions (from Narath & Weaver, 1969).

of the satellite lines is similar in the case of Mn in Cu but differences have been observed for Cr in Cu where the temperature dependence differs from the Curie–Weiss form (1.62). This has been investigated in more detail by Abbas, Aton & Slichter (1982) who conclude that this difference can be satisfactorily explained on the basis of an ionic model which includes the effects of crystal fields and spin–orbit coupling on the ground state L–S multiplet.

The fact that the different satellite lines to a good approximation scale in the same way with temperature led to the postulate that $\Delta K_s(r, T) \propto \chi_{imp}(T)F(r)$, where $F(r)$ is independent of T. This can be understood on the basis of equation (9.68) for sites close to the impurity where the ω dependence of $g_\sigma(\omega)^2$ is not so important and this factor can approximately be factorized out of the ω integration giving $\Delta M_c(r, T) \propto \chi_{imp}(T)$. It is certainly not true for the asymptotic regime $r > \xi$, where the oscillations of $g_\sigma(\omega)^2$ in the integrand means that the dominant contribution comes from the ω range near the Fermi level. Explicit calculation of the T dependence of the induced charge density $\Delta\rho_c(r, T)$ shows no such factorization (Šokčević et al, 1989).

The temperature dependencies of the hyperfine fields deduced from NMR measurements at the *impurity* nuclei are also found to correspond to Curie–Weiss laws over a wide temperature range. To give a specific example, the temperature dependence of the hyperfine field shift for ^{51}V in Au can be fitted over a wide temperature range by the Curie–Weiss form $\Delta K \propto 1/(T + 290)$ (Narath, 1969). The low temperature

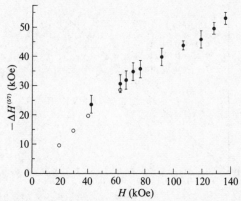

Figure 9.48 The field dependence of the ^{57}Fe hyperfine field for Fe in Cu at low temperatures ($T \ll T_K$, open circles— data of Kitchens et al (1965), full circles — Frankel et al (1967) (taken from Narath, 1972).

relaxation rate, $1/T_1 \propto T$, corresponding to (9.97), has been observed in systems with widely differing Kondo temperatures. Results for T_1 as a function of temperature for ^{55}Mn in Al are shown in figure 9.47. For more detailed information on these and other NMR results we refer the reader to the reviews of Narath (1972, 1973) and Alloul (1977). There is scope for a more detailed analysis of the experimental results in this field as a consequence of more recent theoretical developments.

Mössbauer measurements on suitable isotopes of the impurity nuclei give similar information about the hyperfine field shifts when the relaxation time for the electronic spin is very much faster than the time for the nuclear Larmor precession. This type of measurement is suitable for systems with a low T_K because the signal becomes weaker at higher temperatures. The temperature dependence of the hyperfine field at the impurity site, for example, for Fe in Cu has been deduced from measurements on the isotope ^{57}Fe (Kitchens, Steyart & Taylor, 1965; Frankel, Blum, Schwartz & Kim, 1967). The results can be described by a Curie–Weiss law with $\theta = 29$ K over a wide temperature range. In the low temperature range a T^2 dependence is observed in agreement with the NMR results on the Cu sites (Steiner, Zdrojewska, Gumprecht & Hüfner, 1973; Alloul, 1977). The hyperfine field as a function of the applied field H is shown in figure 9.48. The Kondo temperature is too high in this case to see any saturation of the local moment with the applied field.

Nuclear orientation (NO) is another technique applicable at low temperatures as again the signal becomes weaker at higher temperatures,

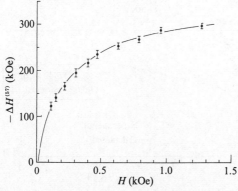

Figure 9.49 Field dependence of the ^{54}Mn hyperfine field for Mn in Ag from NO measurements at 0.0013 K (from Flouquet, 1970).

and can also be used for very dilute systems. In these experiments the state of a nucleus $|I, M\rangle$ in a static magnetic field can be deduced from the angular distribution of particles or gamma rays emitted. This is restricted to very low temperatures by the requirement that the Zeeman splitting, $E_M - E_{M-1} < k_B T$; if $k_B T \gg E_M - E_{M-1}$ there is no useful information because the distribution is isotropic as all the M states contribute equally. For gamma emission the states $|I, M\rangle$ and $|I, -M\rangle$ give the same distribution so the sign of the local field cannot be deduced. As systems with very low T_K can be measured with this technique saturation of the hyperfine field can be observed as can be seen in the results for Mn in Ag shown in figure 9.49. So far there has been no quantitative interpretation of these results on the basis of the n channel Kondo model ($n = 5$).

For very small values of T_K there is the possibility that T_K can become comparable or smaller than the coupling of the nucleus with the impurity spin. If this interaction has the form given in (9.89) ($A_\sigma = 0$) then for $k_B T_K \gg A_s$ this interaction can be taken into account as a hyperfine field contributing to $\langle \mathbf{H}^L \rangle$. For $A_s \gg k_B T_K$, the strong hyperfine coupling case, it is appropriate to start with the eigenstates of (9.89) in the absence of the conduction electron–impurity spin coupling. These states can be classified by the states of total angular momentum $\mathbf{F} = \mathbf{S} + \mathbf{I}$. The residual coupling to the conduction electrons in this limit can then be expressed as an exchange interaction of the form (1.64) with $\mathbf{F}$ replacing $\mathbf{S}$, the impurity electronic spin. For a singlet state $F = 0$ the Kondo effect can disappear, and also when $\mathbf{I}$ and $\mathbf{S}$ are antiferromagnetically coupled with $S < I$ as in this case the exchange

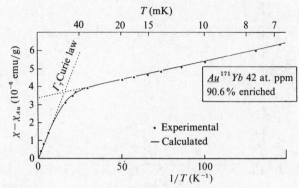

Figure 9.50 Susceptibility of ^{171}Yb impurities in Au versus $1/T$ (Frossati et al, 1976).

interaction between $\mathbf{F}$ and the conduction electrons becomes ferromagnetic. There is evidence that Yb impurities in Au are in the regime $A_s \gg k_B T_K$. There is an isotope of Yb, ^{171}Yb with $2A_s = 127$ mK and a Γ_7 crystal field ground state (effective spin of $\frac{1}{2}$) which satisifies the condition for an electronic-nuclear singlet $F = 0$. The crossover to the ground state electronic-nuclear singlet has been observed in the susceptibility measurements of Frossati, Mignot, Thoulouze & Tournier (1976) shown in figure 9.50. The susceptibility obeys a Curie law in the higher temperature part of the range and makes a crossover to a much weaker Curie law at a temperature of the order of the singlet-triplet separation. The weaker Curie law in the lower part of the range was attributed to contributions from other impurities. Resistivity measurements show similar anomalies with a resistance peak at $T \sim 90$ mK as the Kondo scattering disappears at temperatures below the singlet-triplet separation (see Flouquet, 1978, for further details and a general review of NO measurements).

The charge distribution induced by the impurity $\Delta\rho_c(\mathbf{r})$ can be detected experimentally via the coupling of the associated electric field gradient with the quadrupolar moment of the host nuclei. The electric field gradient $q(\mathbf{r})$ is a functional of $\Delta\rho_c(\mathbf{r})$ which at large distances from the impurity approximates to a linear relation $q(\mathbf{r}) \sim (8\pi/3)\mu\Delta\rho_c(\mathbf{r})$, where μ is an enhancement factor due to core polarization. The coupling of this electric field gradient to the quadrupolar moments of the host nuclei leads to a splitting of the nuclear levels. In nuclear quadrupolar resonance (NQR) the condition for resonant excitations between these states is determined and the splittings due to the electric field gradient are measured.

Information about the charge densities can also be deduced from NMR if the nucleus has a quadrupolar moment. The quadrupolar interaction modifies the Zeeman levels so that the energy differences between consecutive states ($\Delta m = 1$) are no longer independent of m and the Zeeman lines no longer coincide. To first order in the electric field gradient, this causes side bands which are shifted by an amount proportional to the gradient, for $2I + 1$ odd only the central line ($1/2 \rightarrow -1/2$) is unaffected. Averaging over a distribution of fields leads to inhomogeneous line broadening and a loss of intensity of the central line. For host nuclei close to the impurity the strength of the electric field gradient is such that second order terms have to be taken into account. To this order there is also a shift of the central line. Hence the second order effect can also result in a loss of intensity of the central line; the nuclei with a large gradient close to an impurity have their central lines displaced and do not contribute intensity to the central line of the pure host. From an analysis of these spectra information on the induced charge density $\Delta\rho_c(\mathbf{r})$ can be obtained. This is not a straightforward task and for a review of these techniques and their interpretation we refer the reader to the review of Grüner & Minier (1977). The interpretation of the data from such experiments for magnetic impurities supports a picture of reduced charge oscillations in the vicinity of the impurity with an observable temperature dependence in the range $0 < T < T_K$ due to the narrowed resonance $\tilde{\Delta}$ in the spectral density at the Fermi level; in fact it was partly due to these experiments that such a picture was advocated as being necessary to explain the experimental results (Grüner, 1974; Grüner & Zawadowski, 1978).

9.7 The Possibility of First Principles Calculations?

The model Hamiltonian approach has proved to be an effective way of gaining insight into the behaviour of magnetic impurities; the reasons for the occurrence of high temperature moments and then their disappearance at low temperatures resulting in an enhanced impurity susceptibility and specific heat. This insight was gained by working with simplified models, throwing away all the interactions that were thought not to have a direct bearing on the central questions. Then, to make a comparison with the experimental results, some of the complexities arising in real systems due to crystal fields, excited multiplets, and Coulomb interactions with the conduction electrons, have been added. Some of these parameters, such as the crystal field splittings, can be estimated by direct measurements such as neutron scattering; some can be absorbed as renormalizations of existing parameters and others have to be determined by fitting the results to the specific heat or susceptibility as a function of temperature. Clearly if there are more than one or two parameters which are not known a priori but have to be obtained by fitting to the experimental results the predictive power of the theory is diminished, and the theory may not be put to a real test. The question, therefore, naturally arises as to whether there might be some reliable first principles way of calculating the parameters needed.

There have been some pioneering attempts at calculations of the parameters of the Anderson model from first principles. One might speculate as to whether some complete first principles calculation might be possible, as in the calculation of the ground state properties of many metals using the density functional approach without any arbitrary parameters. The subtleties involved in the low energy correlations due to the Kondo effect would seem to rule this out, at least in the short term, so the first principles calculation of the model Hamiltonian seems to be a more realistic proposition.

The calculations of Herbst, Watson & Wilkins (1978) of the local Coulomb interaction U and the effective one electron excitation energies ϵ_f for the 4f and 5f series are early calculations of this type. They are based on the idea originally put forward and used by Herring (1966) (as discussed earlier in section 9.2) that in the solid each atom remains locally neutral, so that if a d or f electron is removed the charge of the hole remaining is locally, and almost instantaneously, screened out by the conduction electrons. With this assumption the effective interaction U, which takes account of the screening and relaxation of the other electrons in the atom and solid, can be estimated from calculations for a single atom. Herring estimated U for 3d systems from atomic data to be of the order 2 eV. Calculations based on the Herring approach by Cox, Coulthard & Lloyd (1974) give values of U which monotonically increase from 1.35 eV for Sc to 3.3 eV for Ni. Estimates of U by Herbst, Watson & Wilkins for the 4f series varied between 5–7 eV (except for Gd with $U \sim 12$ eV, reflecting the extra stability of the half full shell) and are in good agreement with estimates deduced from photoemission and BIS spectra.

More recent estimates of the Anderson model parameters have been based on band calculations using the local density approximation (LDA). As these are for the solid they do not involve the assumption of complete on-site screening. It might be questioned whether such calculations, which for strongly correlated systems often give qualitatively incorrect predictions for the magnetic nature of the ground state are suitable for this purpose. However, on the energy scale U associated with charge displacements the LDA calculations are expected to be reliable and give good results for bulk properties such as lattice constant, bulk moduli and phonon spectra. It is only on the energy scale associated with the much smaller energy differences between different magnetic states that the results are not always satisfactory. Different methods have been devised for deducing the parameters U, $\epsilon_{d(f)}$ and the hybridization matrix elements $V_{\mathbf{k}}$ or the derived quantity $\Delta(\omega)$. The methods for estimating U are all based on some form of constrained LDA calculation (Dederichs, Blügel, Zeller & Akai, 1984) in which the local d or f charge density is constrained to have a definite value n in calculating the total energy $E_{\text{tot}}(n)$ by adding an extra Lagrange multiplier term to the local density functional. On calculating the energies for different values of n the effective U is deduced from $U = \partial^2 E_{\text{tot}}(n)/\partial n^2$. As in each calculation the other electrons adjust to the particular value of n chosen, this generates an effective U which includes many-body renormalizations. To allow for the fact that the hybridization terms should not be doubly counted, as they are explicitly taken into account in the final model Hamiltonian, the hybridization terms to the particular 3d or 4f orbitals being considered are switched off in the calculations of the constrained energy $E_{\text{tot}}(n)$ (see, for instance, McMahan, Martin & Satpathy, 1988; Gunnarsson, Andersen, Jepson & Zaanan, 1988). Anisimov and Gunnarsson (1991) have studied the cases of metallic Fe and Ce using this type of calculation, and find distinct differences in the two cases. Very efficient on-site screening was found in the Ce case, almost entirely by the 5d electrons in the Wigner–Seitz sphere, with results for U in agreement with calculations for Herbst, Watson & Wilkins. For Fe on the other hand the on-site screening was just over 50%. In this case the screening is due to 4s and 4p electrons which are much less efficient because they are located largely outside the core region due to the requirement that they are orthogonal to the s and p core states. The calculation gave $U \sim 6$ eV, rather larger than the estimates of Cox, Coulthard & Lloyd who assume perfect on-site screening. This estimate is also much larger than the estimates from experiment, $U \sim 1$–2 eV, so there still appears to be a discrepancy to resolve for metallic 3d systems.

The ordinary self-consistent LDA calculations have been used to calculate the hybridization matrix elements $V_\mathbf{k}$. In the procedure used by Monnier, Degiorgi & Koelling (1986), for example, a Slater–Koster linear combination of atomic orbitals fit is made to the LDA band structure. The results are then used to calculate the extended states which do not involve the 4f orbitals. The hybridization matrix elements for the 4f crystal field states are calculated using the parameters derived from the Slater–Koster fit. An alternative approach used by Gunnarsson et al (1988) uses the local density approximation to calculate the G_{ff} (or G_{dd}) Green's function which is identified with the $U = 0$ Green's function for the Anderson model, and $\Delta_f(\omega)$ is then deduced from

$$\Delta_f(\omega) = \pi |V_f(\omega)|^2 = -\mathrm{Im}[G_{ff}(\omega - is)]^{-1}. \qquad (9.98)$$

Yet another variant which has been used for extracting parameters for strongly correlated lattice models is given by Hybertson, Schlüter & Christensen (1989), where the one-electron parameters are obtained under the assumption that the ordinary LDA bands are interpreted as the Hartree–Fock (mean field) solution of the model Hamiltonian. The Coulomb interaction terms are obtained from constrained LDA calculations. These calculations are made consistent to generate a complete set of one electron and two particle interaction terms for the model being considered.

These prescriptions all generate effective interactions which take account of the relaxation of all the other electrons in the solid which are not explicitly included in the model. These calculations represent the most serious attempts so far, when combined with the NCA or the intermediate states approximation, at first principles many-body predictions of physical properties ranging from photoemission to the magnetic susceptibility.

10

Strongly Correlated Fermions

10.1 Introduction

Our main concern so far has been with theories of a single magnetic impurity in a simple host metal. These theories have been developed over a period of more than thirty years and most of the models put forward for explaining these systems are now well understood. The concepts for understanding their behaviour in terms of scaling trajectories, fixed points, spin compensation, quasi-particles, Fermi liquid theory and Kondo resonances have been developed and there are exact solutions for the thermodynamic behaviour of many of the models as well as good approximations for their dynamic response (for the reader who has skipped the details of chapters 4–8 the important single impurity results are summarized in appendix K). As we saw in the previous chapter there is still much to be done in the way of detailed predictions for specific systems to compare with experiment. There is also the possibility of new types of experiments to probe the theory further (for example, the possibility of probing the Kondo resonance by resonant tunnelling has recently been discussed by Hershfield, Davies and Wilkins, 1991). These may throw up new puzzles to be answered. The general belief, however, is that if these occur they will be relatively minor and will not require a fundamental reworking of the theory. With these provisos it may be claimed the single impurity problem has been 'solved'.

The solution of any good problem is likely to lead to questions and further problems, just as challenging and just as interesting, in its wake. The Kondo problem is no exception. In the last chapter we compared the predictions for the thermodynamics of impurity models to one or two of the 'concentrated Kondo' systems or 'Kondo lattices' finding remarkably good agreement with experiment. There are problems in explaining this agreement. In forming a singlet ground state the conduction electrons compensate or screen the impurity moment and this screening takes place over a scale in real space of the order $\xi \sim v_F/k_B T_K$. It is possible in alloys with concentrations of rare earth impurities, such that the distance apart of the impurities $r_{imp-imp}$ is very much greater than ξ, for the interaction between impurities to be small enough to be neglected except at very low temperatures. However, for concentrated alloys such that $r_{imp-imp} \leq \xi$, and in the pure compounds where $r_{imp-imp} \sim a \ll \xi$, where a is the lattice spacing, the conduction electron screening clouds overlap and interimpurity interactions must be important. As pointed out by Nozières (1985) compensation of a localized moment in an impurity-like way would be impossible as there would not be enough occupied conduction electron states in the region of the Fermi level for this to be possible. The physics of concentrated systems must be different. The agreement with the impurity model predictions is only for some aspects of their behaviour. The transport properties of the 'Kondo' compounds behave quite differently from the predictions of the impurity model at low temperatures, usually increasing as the temperature is reduced from room temperature until a maximum is reached at low temperatures and then decreasing rapidly to give a very small residual resistivity at $T \to 0$. We will look at results for specific systems later.

The anomalous behaviour of the Kondo lattices must be due to the strong on-site Coulomb interaction U_{ff}, as for the impurity. This suppresses fluctuations of the f charge but does not in all circumstances lead to a localization of the f electrons. In 'normal' rare earth compounds the f electrons are localized and their interactions with the conduction electrons can be described by the s-f model, a Heisenberg exchange interaction between the f moments and the conduction electrons with a ferromagnetic coupling. In the most stable of the normal systems the f levels lie too far below the Fermi level for the indirect exchange by virtual hopping into the conduction band to be important, and the direct exchange term of the Coulomb interaction between the f and conduction electrons, which is ferromagnetic, dominates. This interaction then leads to an intersite exchange interaction between the localized f moments mediated by the conduction electrons, the RKKY interaction, which may be antiferromagnet, ferromagnetic, or both because it extends beyond nearest neighbour pairs (Ruderman & Kittel, 1954). The competing interactions induce a variety of types of magnetic order; this is a characteristic feature of the normal rare earth compounds (for an extensive survey of these systems see Coqblin, 1977).

In anomalous rare earth compounds the f levels lie nearer the Fermi level. In the case of Kondo lattices virtual excitations into the conduction band lead to an antiferromagnetic exchange interaction with the conduction electrons. Though the f electrons lie below the Fermi level the antiferromagnetic exchange interaction modifies the states in the region of the Fermi level leading to anomalous low temperature behaviour. These anomalies are particularly marked in the case of the so-called 'heavy fermions', a category which includes actinide as well as rare earth compounds. There are other examples of anomalous systems, the 'mixed valent' or 'intermediate valence' systems, where the f levels lie at the Fermi level giving a narrow f band.

In the next two sections we describe the behaviour of these anomalous systems starting first with the mixed valent and Kondo lattices, and then moving on to the more dramatic cases of heavy fermion behaviour. This latter includes unconventional superconductivity and metamagnetic behaviour. We then describe some of the theoretical ideas that have been put forward to explain this behaviour. Some of the techniques which were developed for the impurity models can be extended to these systems. There are, as yet, no exact solutions for the lattice models and their complexity would seem to rule out the possibility of finding such solutions. Some of the approximate techniques used successfully for the impurity problems, such as perturbation theory and the $1/N$ expansion, can be applied here. There are some limitations to their use in the lattice case which we will discuss later in section 10.5. The theoretical ideas developed so far do give some insight into aspects of heavy fermion behaviour. However, there is much to be explained and these ideas are only a beginning.

10.2 Anomalous Rare Earth Compounds

The first experiments to reveal delocalized 4f states were pressure experiments on the compound SmS. At normal pressures SmS is a black semi-conductor with Sm^{2+} ions in a ground state multiplet $^6F_0$. At a pressure of 6.5 kbar it undergoes a first order phase transition to a gold metallic phase with a decrease in volume of 13%. With pressure the 5p6s empty band above the Fermi level broadens, due to the increased

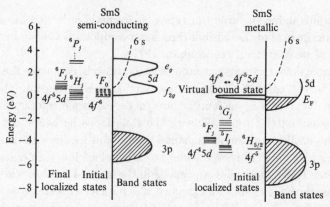

Figure 10.1 Energy level diagram of semi-conducting and metallic SmS, (i) and (ii), as deduced from optical reflectivity data (from Batlogg, Kaldis, Schlegel & Wachter, 1976).

overlap, until the bottom of the band falls below the $4f^6-4f^5$ excitation (see figure 10.1) and some of the f electrons enter this conduction band. The experimental evidence from lattice structure, plasmon frequency and Mössbauer measurements all indicate that less than one electron per Sm site enters the 5p6s band so that on average Sm is in the ionic state $Sm^{\nu+}$ with $\nu \sim 2.8$. This does not necessarily imply itinerancy of the f electrons, there are many mixed valence compounds with ions in two different stable valence states, these are usually ordered but the system could also be a random alloy of Sm^{2+} and Sm^{3+}. The isomer shifts for the nuclei of the two types of ions are different. These shifts can be detected in Mössbauer measurements and for a static array of the two types of ions two distinct lines are seen. Measurements on gold SmS, however, give a single line intermediate between that for the two types of ions. This implies that the ionic states have a lifetime much less than the observation time of a Mössbauer experiment, less than 10^{-9} s: hence the term *intermediate valence*. To indicate the difference of this situation from that of a static array of Sm^{2+} and Sm^{3+} ions a distinction was often made in the early literature on this subject between *mixed valence* (static array) and *intermediate valence* (itinerant 4f electrons). In recent years this distinction has generally been abandoned and the terms *mixed valence, intermediate valence* and *valence fluctuator* are all used to describe this situation.

Gold SmS is paramagnetic down to low temperatures, has a specific heat coefficient $\gamma \sim 145$ mJ/mole K^2 which is about 15 times greater than normal rare earth metals, a high temperature susceptibility with a Curie–Weiss law and a magnetic moment approximately intermediate between that for Sm^{2+} ($^7F_0$, zero moment) and Sm^{3+} ($^6H_{5/2}$, $g = 2/7$, $j = 5/2$). This intermediate state can be also induced by alloying, for example with Th as in $Sm_{1-x}Th_xS$ which is in the gold phase for $x > 0.26$. Photoemission experiments on this material shown in the last chapter, figure 9.4, show clearly excitations $f^6 \rightarrow f^5$ of the Sm^{2+} state near the Fermi level and lower excitations corresponding to $f^5 \rightarrow f^4$ in Sm^{3+} well below the Fermi level.

This intermediate, or mixed valence, state having been identified, it rapidly became apparent that it is not so uncommon; it exists in many compounds at normal pressures and temperatures. The rare earth ions involved are those which are known to exist in more than one valence state such as Sm, Ce, Yb, and to a lesser extent Eu and Tm. Examples are SmB_6, $CePd_3$, $CeSn_3$, $YbInAu$ and $EuCu_2Si_2$. Their main characteristic is that they remain paramagnetic down to very low temperatures, in contrast to most rare earth local moment materials, but with high magnetic susceptibilities and specific heats. Their neutron scattering spectrum usually consists of a broad almost temperature independent quasi-elastic peak, as was first observed in the case of $CePd_3$ (see figure 9.36, and for a review of neutron scattering for these materials see Holland-Moritz et al, 1982). Their low temperature electrical properties vary, most of them being definitely metallic down to low temperatures but there are systems which appear to be insulators at $T = 0$, such as SmB_6 where the resistivity increases dramatically at low temperatures (see figure 10.2). Point contact spectroscopy measurements support the conjecture that SmB_6 is an insulator and indicate a small gap in the low energy excitation spectrum (Frankowski & Wachter, 1982). The compound $TmSe$, which has been identified as intermediate valent, is exceptional in that it orders antiferromagnetically below 3.2 K. This difference could be due to the fact that the lowest multiplets of Tm^{2+} and Tm^{3+} both carry a moment, as noted in the last chapter.

For further details of these systems and a more comprehensive reference list we refer the reader to the early reviews, Jayaraman, Dernier & Longinotti (1975, experimental), Varma (1976, theoretical) and the proceedings of recent international conferences devoted wholly or in part to these systems, Köln (1984), Bangalore (1987) and Santa Fé (1989).

There is direct evidence from de Haas–van Alphen measurements that the metallic state is a Fermi liquid in the mixed valence compounds which are metals at low temperatures. The first measurement to show this was for the compound $CeSn_3$ (Johanson, Crabtree, Edelstein &

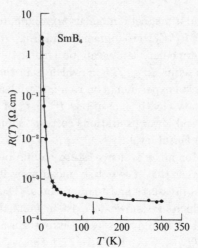

Figure 10.2 Temperature of the resistivity in SmB_6 (Allen, Batlogg & Wachter, 1979).

McMasters, 1981). Fermi surfaces can be deduced from these measurements. The enhanced effective masses observed, $m^*/m \sim 3.3 - 6.0$, indicate relatively flat states corresponding to f-like states at the Fermi level. Enhancements of this order of magnitude can be explained by band calculations based on the density functional theory using the local density approximation (LDA) (see section 1.1).

The Anderson impurity model has a mixed valence regime as $\epsilon_f^* \to \epsilon_F$ and the excitation $f^1 \to f^0$ approaches the Fermi level. There is also mixed valency when the excitation $f^2 \to f^1$ approaches the Fermi level as indicated in figure 1.9. In the first situation for $U \to \infty$ the susceptibility behaves as $\chi \sim 1/k_B T_A$ (equation (7.23)) with $k_B T_A \to \Delta$ as $\epsilon_f^* \to \epsilon_F$. For $\Delta \sim 0.1$ eV the susceptibility in this regime has only a weak temperature dependence in the usual thermal range. This model would also predict a quasi-elastic line of width $\sim k_B T_A$ with $k_B T_A \to \Delta$, showing very little temperature dependence in the same parameter regime, and, with $k_B T_A \gg \Delta E_{\gamma\gamma'}$, where $\Delta E_{\gamma\gamma'}$ is the ground to excited state crystal field splitting, there would be no evidence of the crystal field levels. This is qualitatively in line with the experimental results for mixed valence compounds. One might expect similar behaviour for a lattice model with Δ being replaced by W_{ff}, the effective f band width.

The intermediate valence materials are unusual in revealing itinerant 4f electron behaviour. The weak temperature dependence in the low temperature electronic properties, the lack of magnetic order and purely

quasi-elastic neutron scattering spectrum ($TmSe$ excepted), indicates
that there are no particularly strong many-body renormalizations in this
regime resulting from the strong intra-f Coulomb interactions. There
can be quite significant phonon anomalies however. The size of the
rare earth ion depends on the number of f electrons; when an f electron
is removed from the 4f shell the 5s and 5p shells contract due to the
reduced screening of the nucleus by the f electrons and the ion contracts
by about 15%. This leads to strong coupling to the lattice and can result
in softening of the lattice modes in some systems (Mook et al, 1978)
and anomalous elastic constants (Boppart, Treindel, Wachter & Roth,
1980). The coupling of the f states to the ionic volume also affects the
overlap of the electronic states. When f electrons are squeezed out and
enter a conduction band the lattice contracts and there is an increased
overlap of the electronic states which enhances the hybridization and
band widths. This can result in a positive feedback on the occupation
of the f states. The wider conduction bands can accommodate more of
the displaced f electrons. The Coulomb interaction U_{fc}, which is usually
omitted explicitly from most models, can act in a similar way. The
effective attractive potential induced in creating a f hole acting on the
conduction electrons draws down some of the conduction electron levels,
as in the virtual bound state problem discussed in section 1.3, allowing
the extra conduction electron to be accommodated. To take these effects
into account in a model Hamiltonian description the couplings to the
lattice and conduction electrons should be included explicitly, or taken
into account implicitly by allowing the hybridization and band width
parameters to depend on the occupation of the f states.

For lattice models, as opposed to impurity models, the chemical potential has to be calculated self-consistently. The chemical potential
change gives a negative feedback to any change in f occupation in the
mixed valence state. Due to the much lower density of states for the
conduction electrons than for the f electrons at the Fermi level there is
not much room in the conduction band to accommodate any electrons
removed from the f states. If one attempts to make a large change in the
occupation of the f states (ignoring volume effects) resulting, say, from
an increase of temperature, there will be a shift in the chemical potential
almost restoring the f occupation to its previous value; the net effect is
that the chemical potential is tied to the f levels, and only a small change
of $n_f(T)$ with T is possible. Experimental observations show that there
are compounds which have relatively big valence changes as a function of
temperature. This behaviour clearly cannot be explained on the basis of

a model Hamiltonian without taking the positive feedback mechanisms such as coupling to the lattice into account. Though the behaviour of the impurity model in the mixed valence regime is qualitatively similar to the behaviour of many mixed valence compounds the results of the impurity model (sections 4.8 and 6.8) are not directly applicable as they take no account of either the chemical potential or volume changes.

Systems which have been identified as Kondo lattices display more marked anomalies in their electronic properties at low temperatures, indicating stronger many-body correlations due to the Coulomb interaction. There is a preponderance of cerium compounds in this class such as $CeAl_2$, $CeIn_3$, $CeSi_2$ and $CeSb$. Other compounds identified as Kondo lattices are $YbCuAl$ and $NpSn_3$. Some of them magnetically order, such as $CeAl_2$ which orders antiferromagnetically at $T_N \sim 3.9$ K, whereas others remain paramagnetic down to the lowest temperatures. The specific heat coefficient γ tends to be slightly larger than for the intermediate valence compounds, $YbCuAl$ is typical with $\gamma \sim 260$ mJ/mole K^2. The distinguishing feature of this class is that the number of f electrons is very close to being integral corresponding, for example, to Ce^{3+} or Yb^{3+}. The quasi-elastic peak seen in the neutron scattering displays more temperature dependence than for the mixed valence compounds, the line width $\Gamma(T) \sim T^{1/2}$ as noted earlier (see figures 9.34 and 9.35), and it is possible in some cases to identify crystal field excitations. The anomalies in the resistivity at higher temperatures have been interpreted as due to the Kondo effect. The Kondo lattice materials are close to integral valence so it is possible to compare their behaviour with results for the impurity models in the integral valence regime (s-d and Coqblin–Schrieffer models), as there will be no significant chemical potential changes. There is generally good agreement in the higher temperature range and in many cases the agreement extends down to low temperatures ($T < T_K$) for the thermodynamic behaviour. We have seen a good example already in the case of $YbCuAl$, figure 9.16, where the experimental results are well described by theoretical predictions based on the $N = 8$ Coqblin–Schrieffer model with only one free parameter T_K.

Another remarkable case is the alloy $Ce_{1-x}La_xPb_3$ for $0 < x < 1$ where the excess susceptibility $\chi(T)$ and specific heat $C(T)$ relative to $LaPb_3$ scale precisely with x as can be seen in the results of Lin et al (1987), shown in figures 10.3(i) and (ii). In figure 10.4 the results for the specific heat per mole of Ce are well fitted to the exact results of the $S = \frac{1}{2}$ s-d model over the whole temperature range with $T_L = 3.3$ K,

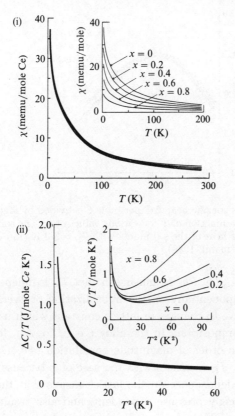

Figure 10.3 (i) Magnetic susceptibility per mole Ce, and (ii) excess specific heat ΔC per mole Ce divided by T, for $Ce_{1-x}La_xPb_3$ as a function of T for various values of x. The insets show the total values per mole (not per mole Ce) for the same samples (Lin et al 1987; figure from Schlottmann, 1989).

independent of x. The change in the low temperature resistivity $R(T)$ with x for this alloy is shown in figures 10.5(i) and (ii). It illustrates the difference between the lattice case $x = 0$ where the scattering is coherent so the resistivity becomes very small as $T \to 0$ so $R(T)$ develops a maximum as well as a minimum, and the relatively dilute case $x = 0.8$ where the scattering is mostly incoherent. One can roughly identify a temperature T_{coh} (coherence temperature) below which the single impurity response is modified and the effects of coherence begin to dominate. In the incoherent case the maximum disappears and there is a large residual resistivity as $T \to 0$. There are intermediate situations between these limits where the maximum slowly disappears on increasing x.

322 *Strongly Correlated Fermions*

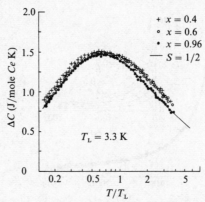

Figure 10.4 Excess specific heat ΔC per mole Ce divided by $\log(T/T_{\rm L})$ for $Ce_{1-x}La_xPb_3$ as a function of T for various values of x. The solid curve is a fit to the specific heat of the s-d model with $T_{\rm L} = 3.3$ K (Lin et al 1987; figure from Schlottmann, 1989).

The case of $Ce_{1-x}La_xPb_3$ is unusual in that $T_{\rm K}$ is independent of x. As $T_{\rm K}$ depends exponentially on the hybridization or overlap integral V it is very sensitive to changes in lattice constant which usually occur with changes of composition. In the alloy $Ce_{1-x}La_xAl_2$, for example, $T_{\rm K}$ decreases by an order of magnitude on dilution. For $Ce_{1-x}Y_xAl_2$, on the other hand, $T_{\rm K}$ increases with increase of x because the Y ions are smaller. This dependence on the lattice means that the electronic properties are strongly pressure dependent, and also results in strong coupling to phonons. These lattice effects are the subject of a review by Thalmeier & Lüthi (1991), and are also dealt with at some length in the heavy fermion review of Fulde, Keller & Zwicknagl (1988).

The dependence of $T_{\rm K}$ on the volume is the basis of the 'Kondo collapse' theory of the $\gamma \to \alpha$ transition in Ce (Allen & Martin, 1982). Early explanations of the $\gamma \to \alpha$ transition, a first order isostructural transition with a volume decrease of 15%, were based on the assumption that the f levels move up to the Fermi level giving a mixed valent α state (the promotional model). As both states have a valence close to Ce^{3+} and are Kondo lattice systems this explanation cannot be correct. In the Kondo collapse theory the volume decrease results in a large increase in $T_{\rm K}$ providing the electronic stabilization energy of the new phase.

There is no precise distinction between intermediate valence and the Kondo lattice regimes. There are alloys such as $Ce(In_{1-x}Sn_x)_3$ in which it is possible to study the continuous changeover from mixed valent $CeSn_3$ to the Kondo lattice $CeIn_3$ by varying x over the range $0 < x < 1$ (Lawrence, 1979).

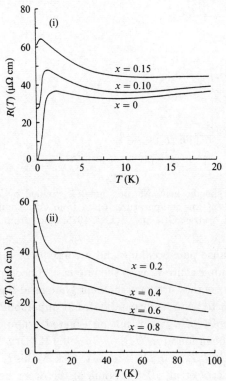

Figure 10.5. The resistivity difference $\Delta R(T)$ of $Ce_{1-x}La_xPb_3$ minus that of $LaPb_3$ for the concentrations indicated, (i) shows the on-set of coherence for small La concentrations and, (ii) the loss of the maximum and impurity-like behaviour for smaller concentrations of Ce (Lin et al 1987; figure from Schlottmann, 1989).

10.3 Heavy Fermions

The term *heavy fermions* has been coined to describe Kondo lattices which have specific heat coefficients greater than 400 mJ/moleK². This class includes $CeAl_3$ ($\gamma \sim 1620$ (mJ/moleK²)), $CeCu_2Si_2$ ($\gamma \sim 1100$), $CeCu_6$ ($\gamma \sim 1600$). It also includes actinide compounds, mainly compounds of uranium such as UPt_3 ($\gamma \sim 420$), U_2Zn_{17} ($\gamma \sim 400$), UBe_{13} ($\gamma \sim 1100$). The lower cut-off of 400 mJ/moleK² is somewhat arbitrary as these systems are part of a continuum extending down through Kondo lattices to mixed valence systems. They are the ones, however, with the most dramatic low temperature anomalies. The most surprising discovery was that of Steglich et al (1979) of superconductivity in the compound $CeCu_2Si_2$ below 0.7 K. Magnetic impurities in normal

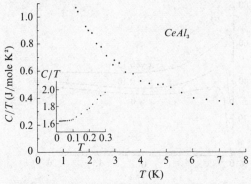

Figure 10.6. The specific heat ΔC per mole Ce divided by T for $CeAl_3$ as a function of T at low temperatures (data from Ott, Rudiger, Fisk & Smith, 1984a, and Andres, Graebner & Ott, 1975; figure from Ott, 1987).

superconductors cause pair-breaking and a consequent rapid reduction in the transition temperature T_c with the impurity concentration. Heavy fermions, therefore, as they are local moment systems at higher temperatures, would appear to be unlikely candidates for superconductivity. The superconductivity of $CeCu_2Si_2$ is not an isolated example. Superconductivity has also been found in UBe_{13}, $T_c \sim 0.9$ K (Ott, Rodiger, Fisk & Smith, 1983), UPt_3, $T_c \sim 0.5$ K (Stewart, Fisk, Willis & Smith 1984), and URu_2Si_2 (Palstra et al, 1985). Some of the other heavy fermions order magnetically at low temperatures. Examples are U_2Zn_{17}, which orders antiferromagnetically below 10 K with a moment $0.8\mu_B$ (much less than that associated with the Curie–Weiss law at higher temperatures where the moment is of the order 2.2-3μ_B), and UCd_{11} which orders antiferromagnetically at 5 K. There are also transitions to antiferromagnetic order at higher temperatures in the superconductors URu_2Si_2 ($T_N = 17$ K) with evidence for some antiferromagnetic order in the superconducting state in URu_2Si_2. The compound $CeCu_6$, on the other hand, is paramagnetic down to the lowest temperatures measured and the same was thought to be the case for $CeAl_3$ but there is recent evidence of antiferromagnetic order with tiny magnetic moments of the order $10^{-2}\mu_B$.

Though these compounds have been put together in a single class, the common feature being the very large low temperature specific heat coefficient, they show great diversity. Even the behaviour of the specific heat at low temperatures is not the same for all the compounds. None of them fit the usual metallic form, $C(T) = \gamma T + \beta T^3$, with temperature independent coefficients γ and β, as can be seen in the plots of $C(T)/T$ versus T^2 for $Ce_{1-x}La_xPb_3$ in figure 10.3(ii), and $C(T)/T$ versus T for

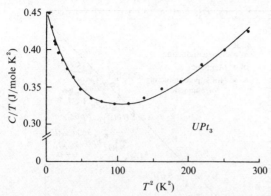

Figure 10.7. $C(T)/T$ as a function of T^2 for UPt_3 between 1.5 and 17 K. The solid curve is a fits to equation (10.1) (Stewart et al, 1984).

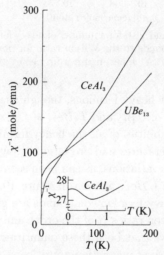

Figure 10.8. The inverse susceptibility $1/\chi$ versus T for $CeAl_3$ and UBe_{13}. The Curie–Weiss form above 100 K corresponds to local moment behaviour of the f electrons. The inset shows saturation indicating itinerant f electron behaviour at low T (Fisk, Ott, Rice & Smith, 1986).

$CeAl_3$ in figure 10.6. The inset in figure 10.6 shows a constant γ can only really be defined for $CeAl_3$ in the temperature range $T < 0.1$ K. The specific heat of UPt_3 has been satisfactorily fitted with the addition of a $T^3\ln T$ term,

$$C(T) = \gamma T + \beta T^3 + \delta T^3\ln T, \tag{10.1}$$

as shown in figure 10.7. There is no evidence of a $T^3\ln T$ term in the

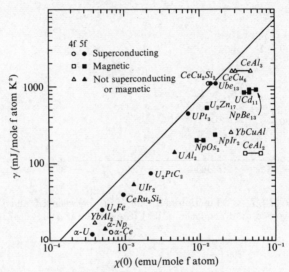

Figure 10.9. A plot of γ and $\chi(0)$ for a number of heavy fermion compounds. The straight line corresponds to the Wilson ratio for non-interacting electrons $R = 1$ (data from B.A. Jones, figure from Lee et al, 1986).

specific heat of the other heavy fermions, though such a term has been seen in the specific heat of UAl_2 and $TiBe_2$.

The magnetic susceptibilities of all the heavy fermions obey a Curie–Weiss law at high temperatures and are large at low temperatures but also display considerable variations, as can been seen in the plots of $1/\chi$ versus T for $CeAl_3$ and UBe_{13} as seen in figure 10.8. Plots of γ and $\chi(0)$ for a number of heavy fermion compounds are given in figure 10.9. They correspond to Wilson ratios R of the order unity (the straight line in figure 10.9); this shows that both these quantities are enhanced in a similar way, as in the case of magnetic impurities in the Kondo regime, the common cause presumably being the f spin fluctuations.

The very high value of the specific heat jump ΔC at the superconducting transition temperature, which is proportional to the density of states at the Fermi level, gives evidence that that the heavy mass electrons are directly involved in the superconductivity. There are systems such as the ternary-boride superconductors where there is competition between magnetism and superconductivity but these involve essentially different sets of electrons. The fact that the magnetic and superconducting behaviour concerns the same set of electrons for heavy fermions is the feature that has attracted a wide interest in these systems. The ratio of the specific heat jump ΔC to γT_c is one of the universal ratios

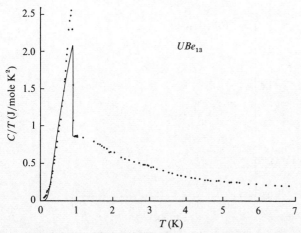

Figure 10.10 $C(T)/T$ versus T for UBe_{13} between 0.15 and 7 K. The solid curve a fit using BCS theory for the superconducting state (from Ott et al, 1984b).

of the weak coupling BCS theory of superconductivity, $\Delta C/\gamma T_c \sim 1.43$. It is of interest therefore to estimate this quantity for the heavy fermion superconductors. There are problems, however, in obtaining a precise estimate as γ effectively varies with T in this temperature range and the specific heat jump cannot be estimated very accurately due to uncertainties associated with the breadth of the transition region. For UBe_{13} estimates are of the order of 50% above the BCS weak coupling prediction and more in line with with the strong coupling theory (see figure 10.10).

There are many observations, such as measurement of NMR relaxation rates and ultrasonic attenuation, which indicate that the superconducting gap is anisotropic in most of the superconductors, and goes to zero at points on the Fermi surface. Other evidence comes from the specific heat measurements in the superconducting phase which fit a power law at low temperatures rather than the exponential form corresponding to a singlet BCS state. In UBe_{13}, for example, $C_s(T)$ has been fitted by Ott et al (1984) to a T^3 power law. Analogies have been drawn with the superfluid state of ^3He, where spin fluctuations lead to p wave pairing and an anisotropic gap. The p wave pairing is favoured as this keeps the ^3He atoms further apart than in a singlet state, which reduces the effect of the strong local Coulomb repulsion. Similar considerations could lead to higher momentum pairing in the heavy fermion superconductors but the situation is complicated by the spin–orbit interactions which mean

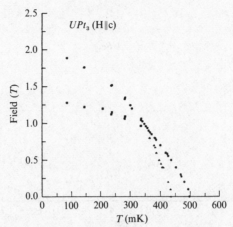

Figure 10.11 Phase diagram for UPt_3 with the magnetic field parallel to the c axis as deduced by Adenwalla et al (1990). The different phase boundaries are indicated by (i) triangles (T_{c^*}), (ii) circles (H_{c2}), and (iii) squares (H_{FL}).

that the pairing states cannot be classified in terms of the orbital angular momentum of the pairs. Group theory has been used to classify the possible symmetries which may be different in the different compounds. There can also be states of different symmetry in the same compound. In zero field in UPt_3 has two nearly degenerate but distinct transitions at T_c, T_{c^*} at 420 and 540 mK. These transitions move closer together in the presence of an applied field and merge on increasing the field strength to the value 9.5 kG for the field parallel to the c axis (it is lower when parallel to the basal plane), at what appears to be a tetracritical point. The phase diagram in the presence of a magnetic field, as shown in figure 10.11, has three distinct phases.

The similarities in the behaviour of the heavy fermion superconductivity with the superfluid behaviour of ^3He raise the question as to whether the origin of the superconductivity is the standard BCS mechanism, pairing due virtual phonon exchange, or whether it could be due to electronic interactions alone.

The electrical resistivities of several of the heavy fermion compounds as a function of temperature are shown in figure 10.12. Broadly speaking they are the same as the Kondo lattice compounds, increasing as the temperature is lowered from room temperature to a maximum followed by a rapid decrease at low temperatures, but again there are exceptions. The resistivity of UPt_3 decreases monotonically with decrease of temperature from room temperature and looks more like that of a normal

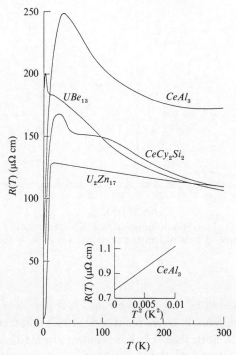

Figure 10.12 Temperature dependence of the electrical resistivity $R(T)$ for $CeAl_3$, UBe_{13}, $CeCu_2Si_2$ and U_2Zn_{11}. The inset shows the T^2 behaviour for $CeAl_3$ for $T < 0.1$ K (from Fisk et al, 1986).

metal. In the non-superconductors at low T, such as $CeAl_3$, the resistivity is proportional to T^2 at very low temperatures. This behaviour can be satisfactorily explained within a Fermi liquid theory. This interpretation is supported by de Haas–van Alphen observations of long lived quasi-particle excitations at the Fermi surface. The most extensive experimental studies using this technique have been on UPt_3 where most parts of the Fermi surface have been mapped out (Taillefer et al, 1987). The results have been compared with local density approximation (LDA) calculations. These band calculations have been successful in accounting for the shape and dimensions of the various sheets of the Fermi surface. They cannot, however, satisfactorily predict the values of the observed effective masses m^*. These range from 15 to 150 times the free electron mass m; they are an order of magnitude higher than the calculated band masses m_B and such that $12 < m^*/m_B < 26$. The predicted value of γ differs from the observed value by a similar factor. The agreement with the Fermi surface calculations is only satisfactory if the

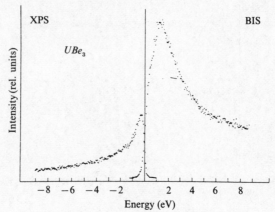

Figure 10.13 The combined photoemission and BIS spectrum of UBe_{13} with the intensities adjusted to give continuity at the Fermi level (Wuilloud et al, 1984).

5f electrons are treated as valence electrons. It is not clear that this is the case for the compounds CeB_6 (Joss et al, 1987) and $CeCu_6$ (Chapman et al, 1990). In both these cases, in contrast to UPt_3, the effective masses deduced from the de Haas–van Alphen measurements are field dependent as are also those deduced from specific heat measurements (Bredl, 1987; Amato et al, 1987).

The U heavy fermion compounds have photoemission and BIS spectra quite unlike those of the Ce Kondo lattice compounds considered in the last chapter (see figures 9.6 and 9.7). The spectrum for UBe_{13} is shown in figure 10.13, with the intensities of the photoemission and BIS spectra matched to give continuity at the Fermi level. The composite spectrum consists of a single broad peak with a maximum at 1.3 eV. Similar spectra have been observed for UAl_2 and UPt_3 (Allen et al, 1986), and also in the dilute compound $U_xY_{1-x}Al_2$ with $x = 0.1, 0.02$ (Kang et al, 1987). As there are no well separated peaks the results suggest a modest value for U_{ff} and an interpretation in terms of a normal 5f band structure. Band calculations give 5f widths, however, which are too small to explain the extent of the spectra, but also far too big to explain the enhanced specific heat values. It is possible that the spectra can still be explained on the basis of an Anderson model, despite the marked difference with the Ce compounds, because the excitations involved are from $5f^2$ and $5f^3$ configurations which have many more excited states than those for the Ce $4f^0$ and $4f^1$ configurations. Also it could be that

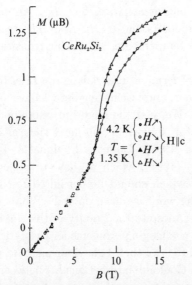

Figure 10.14 The field dependence of the magnetization in $CeRu_2Si_2$ for a field direction along the c axis. The very rapid increase near 8 Tesla is interpreted as a metamagnetic transition which becomes more abrupt at lower temperatures (Haen et al, 1987).

Ce and U systems behave quite differently in the mixed valence regime giving rather different spectra (we discuss this point later).

Neutron scattering measurements have revealed some weak form of antiferromagnetic order in the compounds UPt_3 and URu_2Si_2 associated with remarkably small magnetic moments. In UPt_3, for example, strong magnetic correlations are found on an energy scale of 5 meV with weak ferromagnetic alignment in the basal planes and antiferromagnetic alignment between planes (Aeppli et al, 1988). A modified form of this magnetic order is found at 0.3 meV with moments of the order of $0.02\mu_B$. This state appears to have a finite correlation length and does not show up as a second order transition in thermodynamic measurements. It can be stabilized by substitution of 5% Pd for Pt or Th for U with an enhanced moment of about $0.6\mu_B$. Similar weak antiferromagnetism is observed in URu_2Si_2 with moments of $0.03\mu_B$ and also reported in $CeAl_3$, previously thought to be a stable paramagnet. In this case, however, evidence for the on-set of this ordering can be seen in the specific heat measurements.

A further anomaly found in several heavy fermion compounds is a metamagnetic transition, a very rapid change in the induced magnetization in a magnetic field at a critical field H_c. An example is shown in figure 10.14 for the compound $CeRu_2Si_2$. The transition becomes sharper the lower the temperature. A similar transition has been observed in UPt_3.

This survey of heavy fermion materials is too brief to convey the diversity of their behaviour, and fascinating details have had to be omitted through lack of space. We refer the reader for more details to the comprehensive experimental reviews of this field such as Stewart (1984), Fisk et al (1986), Ott (1987), Taillefer, Flouquet & Lonzarich (1991) and Grewe & Steglich (1991), which we have drawn upon extensively in this summary and which should be consulted for a comprehensive list of references to the work we have quoted. What we hope to have conveyed to the reader, despite the brevity of the survey, is that these systems present many challenging problems for theoretical investigation. The complexity and variety of their behaviour must be a consequence of the strong short range Coulomb interactions U_{ff} between the f electrons, inducing strong local correlations. The concepts and techniques which have been developed to understand the role of this interaction in impurity systems provide a starting point for tackling these problems. Some success has already been achieved in this enterprise in understanding the very large effective masses for the quasi-particle excitations seen in the de Haas–van Alphen experiments. Some of this work will be introduced in the next two sections. The subtle questions concerning the interplay of the magnetic and superconducting states, and the origins of these effects, remain to be answered and are topics of current debate.

10.4 Fermi Liquid Theory and Renormalized Bands

Though band calculations based on the local density approximation have been successful in giving substantial agreement with Fermi surface measurements on a number of the heavy fermion compounds there is an order of magnitude discrepancy between the predicted and experimental values of the effective masses of the quasi-particles; the experimental values being much higher. To account for this discrepancy it would seem appropriate, therefore, to generalize the models we have used for the strong correlation impurity problems and set up similar models for a periodic lattice. The natural generalization of the Anderson impurity model (1.66) to a lattice is

$$H = \sum_{i,\sigma} \epsilon_{\mathrm{f}} c^{\dagger}_{\mathrm{f},i,\sigma} c_{\mathrm{f},i,\sigma} + U \sum_{i} n_{\mathrm{f},i,\uparrow} n_{i,\downarrow} + \sum_{\mathbf{k},\sigma} \epsilon_{\mathbf{k}} c^{\dagger}_{\mathbf{k},\sigma} c_{\mathbf{k},\sigma}$$
$$\sum_{i,\mathbf{k},\sigma} (V_{\mathbf{k}} e^{i\mathbf{k}\cdot\mathbf{R}_i} c^{\dagger}_{\mathrm{f},i,\sigma} c_{\mathbf{k},\sigma} + V^{*}_{\mathbf{k}} e^{-i\mathbf{k}\cdot\mathbf{R}_i} c^{\dagger}_{\mathbf{k},\sigma} c_{\mathrm{f},i,\sigma}),$$

$$\tag{10.2}$$

which is known as the *periodic Anderson model*. The creation and anni-hilation operators for the f electrons, $c^{\dagger}_{\mathrm{f},i,\sigma}$ and $c_{\mathrm{f},i,\sigma}$, now carry a site index i and there is a Coulomb interaction at each site which has lo-calized f electrons. The other states are described by delocalized Bloch states which have creation and annihilation operators, $c^{\dagger}_{\mathbf{k},\sigma}$ and $c_{\mathbf{k},\sigma}$. The hybridization $V_{\mathbf{k}}$ between the conduction and f electrons is multi-plied by a phase factor $e^{i\mathbf{k}\cdot\mathbf{R}_i}$, where $\mathbf{R}_i$ is the position vector of the site i. To gain some insight into the behaviour of the excitations for this system in the region near the Fermi level we generalize some of the ar-guments we developed in section 5.2 following the treatment of Edwards (1988). The one electron Green's functions for the conduction electrons and f electrons, as a function of ω and $\mathbf{k}$, can be written in the form,

$$\begin{bmatrix} \omega - \epsilon_{\mathrm{f}} - \Sigma_{\sigma}(\omega,\mathbf{k}) & -V_{\mathbf{k}} \\ -V_{\mathbf{k}} & \omega - \epsilon_{\mathbf{k}} \end{bmatrix} \begin{bmatrix} G^{\mathrm{ff}}_{\mathbf{k},\sigma} & G^{cf}_{\mathbf{k},\sigma} \\ G^{fc}_{\mathbf{k},\sigma} & G^{cc}_{\mathbf{k},\sigma} \end{bmatrix} = I, \tag{10.3}$$

where $\Sigma_{\sigma}(\omega,\mathbf{k})$ is the proper self-energy. This equation follows from a diagrammatic expansion in powers of the Coulomb interaction U and the hybridization $V_{\mathbf{k}}$. There is only one proper self-energy $\Sigma_{\sigma}(\omega,\mathbf{k})$ as the two-body interaction is solely between the f states. This self-energy is zero for $U = 0$, in which case the f electron Green's function has the form,

$$G^{\mathrm{ff}}_{\mathbf{k},\sigma}(\omega) = \frac{1}{\omega - \epsilon_{\mathrm{F}} - \epsilon_{\mathrm{f},\mathbf{k}} - |V_{\mathbf{k}}|^2/(\omega - \epsilon_{\mathbf{k}})}. \tag{10.4}$$

Assuming $\Sigma(\omega,\mathbf{k})$ can be expanded in a Taylor's series about a point on the Fermi surface, $\omega = \epsilon_{\mathrm{F}}$, $\mathbf{k} = \mathbf{k}_{\mathrm{F}}$, we have

$$\Sigma(\omega,\mathbf{k}) = \Sigma^{\mathrm{R}}(\epsilon_{\mathrm{F}},\mathbf{k}_{\mathrm{F}})$$
$$+ (\mathbf{k} - \mathbf{k}_{\mathrm{F}}) \cdot \nabla \Sigma^{\mathrm{R}}(\epsilon_{\mathrm{F}},\mathbf{k})_{k=k_{\mathrm{F}}} + (\omega - \epsilon_{\mathrm{F}}) \left(\frac{\partial \Sigma^{\mathrm{R}}(\omega,\mathbf{k}_{\mathrm{F}})}{\partial \omega} \right)_{\omega=\epsilon_{\mathrm{F}}} + ...,$$

$$\tag{10.5}$$

using the Luttinger result that $\Sigma^{\mathrm{I}}(\omega,\mathbf{k}_{\mathrm{F}}) \sim (\omega - \epsilon_{\mathrm{F}})^2$ in this region. If we retain terms in (10.5) to first order in $\omega - \epsilon_{\mathrm{F}}$ and $\mathbf{k} - \mathbf{k}_{\mathrm{F}}$, we can define a quasi-particle Green's function $\tilde{G}^{\mathrm{ff}}_{\mathbf{k},\sigma}(\omega)$,

$$\tilde{G}^{\mathrm{ff}}_{\mathbf{k},\sigma}(\omega) = \frac{1}{\omega - \epsilon_{\mathrm{F}} - \tilde{\epsilon}_{f,\mathbf{k}} - |\tilde{V}_{\mathbf{k}}|^2/(\omega - \epsilon_{\mathbf{k}})}, \qquad (10.6)$$

which is such that $G^{\mathrm{ff}}_{\mathbf{k},\sigma}(\omega) = z_{\mathbf{k}_{\mathrm{F}}} \tilde{G}^{\mathrm{ff}}_{\mathbf{k},\sigma}(\omega)$ near the Fermi surface, where

$$\tilde{\epsilon}_{f,\mathbf{k}} = z_{\mathbf{k}_{\mathrm{F}}}(\epsilon_f - \epsilon_{\mathrm{F}} + \Sigma(\epsilon_{\mathrm{F}}, k_{\mathrm{F}}) + (\mathbf{k} - \mathbf{k}_{\mathrm{F}}) \cdot \nabla\Sigma^{\mathrm{R}}(\epsilon_{\mathrm{F}}, k_{\mathrm{F}})) \qquad (10.7)$$

and

$$|\tilde{V}_{\mathbf{k}}|^2 = z_{\mathbf{k}_{\mathrm{F}}} |V_{\mathbf{k}}|^2, \quad \text{and} \quad z_{\mathbf{k}_{\mathrm{F}}} = \left(1 - \left(\frac{\partial\Sigma^{\mathrm{R}}(\omega, \mathbf{k}_{\mathrm{F}})}{\partial\omega}\right)_{\omega=\epsilon_{\mathrm{F}}}\right)^{-1}$$

$$(10.8)$$

The quasi-particle Green's function for the conduction electrons has the form,

$$\tilde{G}^{cc}_{\mathbf{k},\sigma}(\omega) = \frac{1}{\omega - \epsilon_{\mathbf{k}} - |\tilde{V}_{\mathbf{k}}|^2/(\omega - \tilde{\epsilon}_{f,\mathbf{k}} - \epsilon_{\mathrm{F}})}, \qquad (10.9)$$

and $G^{cc}_{\mathbf{k},\sigma}(\omega) = \tilde{G}^{cc}_{\mathbf{k},\sigma}(\omega)$ in this region near the Fermi surface.

The quasi-particle energies correspond to the poles of (10.6) and (10.9) at energies $\omega = \tilde{\epsilon}^+_{\mathbf{k}}, \tilde{\epsilon}^-_{\mathbf{k}}$, and are the solutions of the quadratic equation,

$$(\omega - \epsilon_{\mathrm{F}} - \tilde{\epsilon}_{f,\mathbf{k}})(\omega - \epsilon_{\mathbf{k}}) - |\tilde{V}_{\mathbf{k}}|^2 = 0, \qquad (10.10)$$

and hence,

$$\tilde{\epsilon}^\pm_{\mathbf{k}} = \frac{\epsilon_{\mathrm{F}} + \tilde{\epsilon}_{f,\mathbf{k}} + \epsilon_{\mathbf{k}} \pm \sqrt{(\epsilon_{\mathrm{F}} + \tilde{\epsilon}_{f,\mathbf{k}} - \epsilon_{\mathbf{k}})^2 + 4|\tilde{V}_{\mathbf{k}}|^2}}{2}. \qquad (10.11)$$

These excitations are of the same form as the one electron bands in the non-interacting case $U = 0$, except that that parameters $\epsilon_{f,\mathbf{k}}$ and $V_{\mathbf{k}}$ are modified, hence they can be described as renormalized band excitations.

There is a one-to-one correspondence between the renormalized particles and the original electrons so the total number of particles n_{tot} can be calculated either from the sum of one electron Green's functions, $G^{cc}_{\mathbf{k},\sigma}(\omega) + G^{\mathrm{ff}}_{\mathbf{k},\sigma}(\omega)$, or the quasi-particle Green's functions, $\tilde{G}^{cc}_{\mathbf{k},\sigma}(\omega) + \tilde{G}^{\mathrm{ff}}_{\mathbf{k},\sigma}(\omega)$ as the factor $z_{\mathbf{k}_{\mathrm{F}}}$ cancels out. At $T = 0$

$$n_{\mathrm{tot}} = -\frac{1}{\pi} \int_{-\infty}^{\epsilon_{\mathrm{F}}} \sum_{\mathbf{k}} \mathrm{Im}[G^{cc}_{\mathbf{k},\sigma}(\omega^+) + G^{\mathrm{ff}}_{\mathbf{k},\sigma}(\omega^+)] \, d\omega, \qquad (10.12)$$

using the same steps as in the proof of the Friedel sum rule, (5.45)–(5.47), this is equal to

$$-\frac{1}{\pi} \int_{-\infty}^{\epsilon_{\mathrm{F}}} \sum_{\mathbf{k}} \mathrm{Im}\left\{ \frac{\partial}{\partial\omega} \ln[(\omega^+ - \epsilon_{\mathbf{k}})(\omega^+ - \epsilon_f - \Sigma(\omega, \mathbf{k})) - |V_{\mathbf{k}}|^2] \right.$$

$$\left. + \frac{\partial\Sigma(\omega^+, \mathbf{k})}{\partial\omega} G^{\mathrm{ff}}_{\mathbf{k},\sigma}(\omega^+) \right\} d\omega, \qquad (10.13)$$

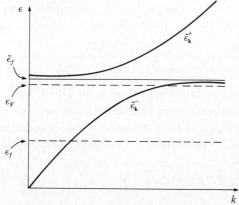

Figure 10.15 A schematic representation of the renormalized quasi-particle bands in the vicinity of the Fermi level.

and the last term in the integrand vanishes on using the Luttinger result (v) (in section 5.2). The integration over ω can then be evaluated and the result written in the form,

$$n_{\text{tot}} = \frac{2}{(2\pi)^3} \int_{-\infty}^{\infty} \theta[\mathbf{k}_F - \mathbf{k}] \, d\mathbf{k}, \qquad (10.14)$$

where $\Sigma^I(\epsilon_f, \mathbf{k}_F) = 0$. Hence we get Luttinger's theorem that $n_{\text{tot}} = 2/(2\pi)^3 \times$ volume of the Fermi surface, which is the same result as for non-interacting electrons.

In the limit $U \to \infty$ the effective f band $\tilde{\epsilon}_{f,\mathbf{k}}$ must lie above the Fermi level ϵ_F so that the total occupation of the f level per site, n_f, is such that $0 < n_f \leq 1$. This is the same reason the quasi-particle level $\tilde{\epsilon}_f$ lies at or above the Fermi level in the impurity case. If $V_\mathbf{k}$ and $\tilde{\epsilon}_{f,\mathbf{k}}$ are only weakly k dependent the quasi-particle excitation spectrum has a gap at $\omega \sim \epsilon_F + \tilde{\epsilon}_f$ between the upper $(\tilde{\epsilon}_\mathbf{k}^+)$ and lower bands $(\tilde{\epsilon}_\mathbf{k}^-)$. The form of the bands in this case is shown schematically in figure 10.15. If the ground state is metallic the Fermi level must lie in the lower band so that $\tilde{\epsilon}_{\mathbf{k}_F}^- = \epsilon_F$, which defines the Fermi surface, and the curves flatten to increase the volume to accommodate the extra f electron. Using (10.11) this equation becomes,

$$\epsilon_{\mathbf{k}_F} = \epsilon_F + \frac{|\tilde{V}_{\mathbf{k}_F}|^2}{\tilde{\epsilon}_{f,\mathbf{k}_F}}. \qquad (10.15)$$

It is only in the neighbourhood of the Fermi surface that the quasi-particle excitations have a long lifetime and are well defined. If we take $\epsilon_\mathbf{k}$ to be isotropic, $\epsilon_\mathbf{k} = \mathbf{k}^2/2m$, and neglect the k dependence of $V_{\mathbf{k}}$ we can define an effective mass for the quasi-particles via $\tilde{\epsilon}_{\mathbf{k}}^{-} = \epsilon_F + k_F(k - k_F)/m^*$ to lowest order in $(k - k_F)$. Using (10.11) we find

$$\frac{m^*}{m} = \frac{1 + |\tilde{V}_{k_F}|^2/\tilde{\epsilon}_{f,k_F}^2}{\left[1 + m|\tilde{V}_{k_F}|^2/k_F\tilde{\epsilon}_{f,k_F}^2 (\partial\Sigma/\partial k)_{k_F}\right]}. \tag{10.16}$$

We can compare these results with those from band theory. The effective f level ϵ_f^{eff} in any self-consistent band calculation must be well above the Fermi level if the f occupation is to be less than or equal to unity as the f band can hold up to 14 electrons. The effective f level can be identified as

$$\epsilon_f^{\text{eff}} = \epsilon_f + \Sigma(\epsilon_F, k_F), \tag{10.17}$$

following Fulde (1988). As the factor $z_{\mathbf{k}_F}$ cancels in equation (10.15) it can be rewritten in the form,

$$\epsilon_{k_F} = \epsilon_F + \frac{|V_{k_F}|^2}{\epsilon_f^{\text{eff}}}, \tag{10.18}$$

which is exactly the same Fermi surface as that generated by a band calculation with an effective f level ϵ_f^{eff}. The effective masses, however, can be quite different. We find an effective mass m_B for the band calculation by expanding $\epsilon_{\mathbf{k}}^{-}(U = 0)$ about $\mathbf{k}_F$,

$$\frac{m_B}{m} = 1 + \frac{|V_{k_F}|^2}{(\epsilon_f^{\text{eff}})^2}. \tag{10.19}$$

If for simplicity we take the ω dependence of $\Sigma^R(\omega, \mathbf{k})$ to be dominant and neglect the $\mathbf{k}$ dependence, then m_B and m^* are related by

$$\frac{m^*}{m} = 1 + \frac{1}{z}\left(\frac{m_B}{m} - 1\right), \tag{10.20}$$

and for $z \ll 1$, $m^*/m_B \sim 1/z$. Hence it is possible to explain the apparent paradox that the band calculations can predict the correct Fermi surface and yet give effective masses for the quasi-particle excitations which are an order of magnitude too small.

It should be noted that the number of f electrons must be calculated from $G_{\mathbf{k},\sigma}^{\text{ff}}(\omega)$ and is not equal to the number calculated from $\tilde{G}_{\mathbf{k},\sigma}^{\text{ff}}(\omega)$. This was strictly speaking also the case for the impurity problem but for a flat wide conduction band, where $n_{\text{imp}} = n_f$ (or n_d), the difference between them was insignificant. This may not always be so in the lattice case.

The total density of quasi-particle states is given by

$$\tilde{\rho}(\omega) = \sum_{\mathbf{k}}[\delta(\omega - \tilde{\epsilon}_{\mathbf{k}}^{+}) + \delta(\omega - \tilde{\epsilon}_{\mathbf{k}}^{-})]. \tag{10.21}$$

When the Fermi level lies in the flattened portion of the quasi-particle band $\tilde{\rho}(\omega)$ will be enhanced giving enhanced values of the specific heat coefficient γ and the susceptibility χ, as we can see by using the Fermi liquid results of section 5.3,

$$\gamma = \frac{2\pi k_{\rm B}^2}{3} \tilde{\rho}(\epsilon_{\rm F}), \qquad (10.22)$$

and for the susceptibility,

$$\chi = \frac{(g\mu_{\rm B})^2}{2} \alpha' \tilde{\rho}(\epsilon_{\rm F}), \qquad (10.23)$$

where α' is a many-body enhancement factor in the energy shift of the quasi-particles at the Fermi surface with magnetic field, $\tilde{\epsilon}(H) = \epsilon_{\rm F} - \alpha'\sigma h$, where $h = g\mu_{\rm B}H/2$. Using equations (10.11) and (10.21) it can be shown (Yamada & Yosida, 1985) that

$$\tilde{\rho}(\epsilon_{\rm F}) = \rho_{\rm c}^{(0)}(\epsilon_{\rm F}) + \sum_{\bf k} \left(1 - \frac{\partial \Sigma(\omega,{\bf k})}{\partial \omega}\right)_{\omega=\epsilon_{\rm F}} \rho_{{\bf k},f}^{(0)}(\epsilon_{\rm F}), \qquad (10.24)$$

where $\rho_{\rm c}^{(0)}$ is the density of states of the conduction states and $\rho_{{\bf k},f}^{(0)}$ the spectral density of $G_{{\bf k},\sigma}^{\rm ff}(\omega)$ for $U = 0$. If the $\bf k$ dependence of $\Sigma(\omega,{\bf k})$ is neglected then this becomes

$$\tilde{\rho}(\epsilon_{\rm F}) = \rho_{\rm c}^{(0)}(\epsilon_{\rm F}) + \frac{\rho_f^{(0)}(\epsilon_{\rm F})}{z}, \qquad (10.25)$$

as in the impurity case, where $\rho_f^{(0)}$ is the total density of the f states for $U = 0$.

We can find the form of the conduction electron Green's function in the neighbourhood of the Fermi level. Using the expansion $\tilde{\epsilon}_k^- = \epsilon_{\rm F} + k_{\rm F}(k - k_{\rm F})/m^*$ near the Fermi surface, neglecting for simplicity any anisotropy and $\bf k$ dependence of $V_{\bf k}$ and $\tilde{\epsilon}_{f,{\bf k}}$, the conduction electron Green's function corresponding to (10.9) can be written approximately as

$$G_{{\bf k},\sigma}^{cc}(\omega) \approx \frac{m/m^*}{\omega - \epsilon_{\rm F} - k_{\rm F}(k - k_{\rm F})/m^*}. \qquad (10.26)$$

This is quite different from the single impurity case where any modifications are of the order of $1/N_{\rm s}$ ($N_{\rm s}$, number of sites). The reduced weight for a strong correlation, $m^*/m \gg 1$, is due to the fact that the many-body effects flatten ϵ_k^- giving it more f character. The quasi-particle f propagator, on the other hand, from (10.6) becomes

$$\tilde{G}_{{\bf k},\sigma}^{\rm ff}(\omega) \approx \frac{1 - m/m^*}{\omega - \epsilon_{\rm F} - k_{\rm F}(k - k_{\rm F})/m^*}. \qquad (10.27)$$

The corresponding f electron Green's function has an additional weight factor z, which for strong correlation is of the order m/m^*. Hence from (10.26) and (10.27) the total weight of the renormalized bands at the

Fermi level is small (of order m/m^*) for strong correlation, like the weight of the quasi-particle resonance in the impurity case.

Yamada & Yosida (1986) have also extended their perturbation theory of sections 5.3 and 5.4 to the periodic Anderson model (10.2). In the lattice case the antisymmetrized vertex part for parallel spins $\Gamma_{\uparrow\uparrow}(\mathbf{k_1}, \mathbf{k_2}; \mathbf{k_3}, \mathbf{k_4})$ does not vanish due to momentum conservation. As a consequence the universal Fermi liquid results obtained for the impurity case do not hold for $U \neq 0$. With an attractive interaction between parallel spin quasi-particles the Wilson ratio can, in principle, exceed the value 2, the value obtained in the impurity case in the Kondo limit. Yamada and Yosida use the Kubo formula (2.23) and Ward identities to derive exact expressions for the T^2 coefficient in the resistivity. They show that if the $\mathbf{k}$ dependencies of $\Gamma_{\uparrow\uparrow}$ and $\Gamma_{\downarrow\uparrow}$ are neglected, and U is taken to be large so that the charge susceptibility of the f electrons vanishes, then the resistivity T^2 coefficient is enhanced and proportional to γ^2 as in the impurity case.

The similarities in the behaviour of some of the heavy fermion compounds with that of liquid ^3He (for example, UPt_3 has a $T^3\ln T$ term in the specific heat and is an odd parity superconductor) have led to similar Fermi liquid theories being proposed (Pethick et al, 1986). The mass enhancements in the one component Landau theories of this type arise from strong ferromagnetic fluctuations. These theories, however, predict enhanced values for the χ/γ or Wilson ratio, which experimentally are found to be of the order unity (see figure 10.9). Generalizations of these theories to two band models, with antiferromagnetic rather than ferromagnetic fluctuations, may be able to give predictions more in accord with the experimental results (Sanchez-Castro and Bedell, 1991).

10.5 Mean Field Theory

The slave boson approach for impurity models in the mean field approximation (section 7.5) gave quasi-particle parameters in terms of states of bare parameters of the Anderson model. For the N-fold degenerate model these results were shown to be exact as $N \to \infty$. At first sight it might seem quite straightforward to generalize these results to the lattice model (10.2) for $U \to \infty$. In the impurity case, however, a classification of terms in $1/N$ was possible because of the basis used for the conduction electrons. There was a unique site, the impurity site, for a partial wave expansion of the conduction states in terms of total angular momentum (j, m). It was then only necessary to retain explicitly the partial waves with the same set of quantum numbers as the localized f states, as these are the ones with which they predominantly hybridize. The single sum over the angular momentum variable m made it possible to classify the contributions in a $1/N$ expansion. In the lattice case there is no unique site for such an expansion so the original description of the conduction electron states in terms of the vector $\mathbf{k}$ and spin σ has to be retained. The hybridization matrix elements between an f state with z-component of angular momentum m at site i with conduction electron $\mathbf{k}, \sigma$ then depend on all three variables, $V_m^i(\mathbf{k}, \sigma)$, and we lose the feature of a single sum over the variable m at each vertex, and the $1/N$ classification. A mean field approximation can still be made but we lose a useful tool for making systematic approximations. A way to circumvent this problem is simply to make the obvious mathematical generalization of the $U = \infty$ model (7.2) to a lattice,

$$H = \sum_{i,m} E_1 X^i_{1m,1m} + \sum_i E_0 X^i_{0,0} + \sum_{\mathbf{k},m} \epsilon_{\mathbf{k}} c^\dagger_{\mathbf{k},m} c_{\mathbf{k},m}$$
$$\sum_{i,\mathbf{k},m} (V_{\mathbf{k}} e^{i\mathbf{k}\cdot\mathbf{R}_i} X^i_{1m,0} c_{\mathbf{k},m} + V^*_{\mathbf{k}} e^{-i\mathbf{k}\cdot\mathbf{R}_i} c^\dagger_{\mathbf{k},m} X^i_{0,1m}),$$
(10.28)

where i is a site index and the index m has $2j+1 = N$ values. The model is unphysical except for the case $m \to \sigma$, $N = 2$ when it is equivalent to the $U \to \infty$ limit of (10.2). We can now classify terms by $1/N$ as in the impurity case and use this as a basis for making approximations. The hope is that the extrapolation of the large N results to $N = 2$ will provide a reasonably good description of the basic physics, as it did in the impurity case.

Introducing slave bosons b_i, $b_i^\dagger$, and fermion operators f_i, $f_i^\dagger$, at each site i,

$$X^i_{1m,0} = b_i f^\dagger_{i,m}, \quad \text{and} \quad X^i_{0,1m} = b_i^\dagger f_{i,m},$$
(10.29)

with the constraint,

$$b_i^\dagger b_i + \sum_m f^\dagger_{i,m} f_{i,m} = \sum_m X^i_{1m,1m} + X^i_{0,0} = I.$$
(10.30)

Then introducing a constraint field λ_i as in (7.90) we obtain a modified Hamiltonian,

$$H = \sum_{i,m} \epsilon_f f^\dagger_{i,m} f_{i,m} + \sum_{\mathbf{k},m} \epsilon_{\mathbf{k}} c^\dagger_{\mathbf{k},m} c_{\mathbf{k},m} + \sum_i \lambda_i \Big(\sum_m f^\dagger_{i,m} f_{i,m} + b_i^\dagger b_i - 1 \Big)$$
$$+ \sum_{i,\mathbf{k},m} (V_{\mathbf{k}} e^{i\mathbf{k}\cdot\mathbf{R}_i} b_i f^\dagger_{i,m} c_{\mathbf{k},m} + V^*_{\mathbf{k}} e^{-i\mathbf{k}\cdot\mathbf{R}_i} c^\dagger_{\mathbf{k},m} b_i^\dagger f_{i,m}).$$
(10.31)

In the mean field approximation the b_i, $b_i^\dagger$ operators are replaced by their expectation values, $\langle b_i \rangle = \langle b_i^\dagger \rangle = \sqrt{z}$ so that the Hamiltonian is converted into a one-body Hamiltonian,

$$H_{\mathrm{MF}} = \sum_{\mathbf{k},m} \epsilon_{\mathbf{k}} c_{\mathbf{k},m}^\dagger c_{\mathbf{k},m} + \tilde{\epsilon}_{\mathrm{f}} \sum_{i,m} f_{i,m}^\dagger f_{i,m}$$
$$+ \sum_{\mathbf{k},i,m} (\tilde{V}_{\mathbf{k}} e^{i\mathbf{k}\cdot\mathbf{R}_i} f_{i,m}^\dagger c_{\mathbf{k},m} + \tilde{V}_{\mathbf{k}}^* e^{-i\mathbf{k}\cdot\mathbf{R}_i} c_{\mathbf{k},m}^\dagger f_{i,m}) + N_{\mathrm{s}}(\tilde{\epsilon}_{\mathrm{f}} - \epsilon_{\mathrm{f}})(z-1),$$

$$(10.32)$$

which on diagonalization can be written in the form,

$$H_{\mathrm{MF}} = \sum_{\mathbf{k},m} (\tilde{\epsilon}_{\mathbf{k}}^+ c_{\mathbf{k},+,m}^\dagger c_{\mathbf{k},+,m} + \tilde{\epsilon}_{\mathbf{k}}^- c_{\mathbf{k},-,m}^\dagger c_{\mathbf{k},-,m}), \qquad (10.33)$$

where

$$|\tilde{V}_{\mathbf{k}}|^2 = z|V_{\mathbf{k}}|^2, \quad \tilde{\epsilon}_{\mathrm{f}} = \epsilon_{\mathrm{f}} + i\bar{\lambda}, \quad z = 1 - n_{\mathrm{f}}, \qquad (10.34)$$

and $n_{\mathrm{f}} = \sum_m \langle n_{\mathrm{f},i,m} \rangle$.

As with the impurity problem the free energy can be minimized with respect to $\bar{\lambda}$ and z. At $T = 0$ minimization with respect to z gives

$$\sum_{\mathbf{k},m} \tilde{V}_{\mathbf{k}} \langle f_{\mathbf{k},m}^\dagger c_{\mathbf{k},m} \rangle = z(\epsilon_{\mathrm{f}} - \tilde{\epsilon}_{\mathrm{f}}), \qquad (10.35)$$

and minimization with respect to $\bar{\lambda}$ gives the averaged constraint condition, $z = (1 - n_{\mathrm{f}})$. The chemical potential μ ($= \epsilon_{\mathrm{F}}$ at $T = 0$) must also be determined self-consistently to give the correct number of electrons n_{tot},

$$\sum_{\mathbf{k},m} \langle c_{\mathbf{k},m}^\dagger c_{\mathbf{k},m} \rangle + n_{\mathrm{f}} = n_{\mathrm{tot}}. \qquad (10.36)$$

The expectation values, n_{f}, $\langle f_{\mathbf{k},m}^\dagger c_{\mathbf{k},m} \rangle$ and $\langle c_{\mathbf{k},m}^\dagger c_{\mathbf{k},m} \rangle$, can be calculated from the one-electron Green's functions, $G_{\mathrm{MF}}^{\mathrm{ff}}(\mathbf{k},\omega)$, $G_{\mathrm{MF}}^{fc}(\mathbf{k},\omega)$, $G_{\mathrm{MF}}^{cc}(\mathbf{k},\omega)$, from the mean field Hamiltonian (10.32) in the standard way. The closed set of equations derived from the equations of motion for these are

$$\begin{bmatrix} \omega - \tilde{\epsilon}_{\mathrm{f}} - \epsilon_{\mathrm{F}} & -\tilde{V}_{\mathbf{k}} \\ -\tilde{V}_{\mathbf{k}} & \omega - \epsilon_{\mathbf{k}} \end{bmatrix} \begin{bmatrix} G_{\mathrm{MF}}^{\mathrm{ff}}(\mathbf{k},\omega,\sigma) & G_{\mathrm{MF}}^{cf}(\mathbf{k},\omega,\sigma) \\ G_{\mathrm{MF}}^{fc}(\mathbf{k},\omega,\sigma) & G_{\mathrm{MF}}^{cc}(\mathbf{k},\omega,\sigma) \end{bmatrix} = I. \quad (10.37)$$

Equations (10.35), (10.36) and the averaged constraint condition then give sufficient equations to determine $\tilde{\epsilon}_{\mathrm{f}}$, z and μ.

To simplify the discussion we follow the treatment of Newns & Read (1987) and take a flat band density of states (1.55). The self-consistent equations can then be written in the form,

$$n_{\text{tot}} = n_{\text{f}} + N(\mu + D')\rho_0, \tag{10.38}$$

$$n_{\text{f}} = \rho_0 N |\tilde{V}|^2 \left[\frac{1}{\tilde{\epsilon}_{\text{f}}} - \frac{1}{\tilde{\epsilon}_{\text{f}} + D' + \mu} \right], \tag{10.39}$$

$$\epsilon_{\text{f}} - \tilde{\epsilon}_{\text{f}} = \rho_0 N |V|^2 \ln[(\tilde{\epsilon}_{\text{f}} - \epsilon_{\text{F}})(\tilde{\epsilon}_{\text{f}} + D')], \tag{10.40}$$

where the modified lower band cut-off $-D'$ is given by

$$D' = \frac{1}{2}\{-\tilde{\epsilon}_{\text{f}} + D + [(\tilde{\epsilon}_{\text{f}} + D)^2 + 4|\tilde{V}|^2]^{1/2}\}, \tag{10.41}$$

and $\rho_0 = 1/2D$ if the density of states is normalized to 1. For $\tilde{V}_{\mathbf{k}}$ independent of $\mathbf{k}$, as we have assumed in (10.38) to (10.41), there must be a gap in the excitation spectrum above the Fermi level at $\tilde{\epsilon}_{\text{f}}$.

We first of all consider the metallic case where the Fermi level lies below the gap in the lower band with $n_{\text{tot}} \sim N/2$. It is possible to simplify (10.41) in this limit,

$$D' \approx D + \frac{|V|^2}{(\tilde{\epsilon}_{\text{f}} + D)}, \tag{10.42}$$

so that (10.39) can be written in the form,

$$n_{\text{f}} = \frac{N\Delta/\pi k_{\text{B}} T^*}{(1 + N\Delta/\pi k_{\text{B}} T^*)}, \tag{10.43}$$

where $\Delta = \pi |V|^2 \rho_0$, and T^* is given by

$$k_{\text{B}} T^* = \epsilon_{\text{f}} - \epsilon_{\text{F}} - \frac{N\Delta}{\pi} \ln \left(\frac{k_{\text{B}} T^*}{D + \epsilon_{\text{F}}} \right). \tag{10.44}$$

These equations look very similar to (7.20) and (7.17) for the impurity case ($T_{\text{A}} \rightarrow T^*$, $\epsilon_{\text{F}} = 0$) and the same is true for the results for the susceptibility and specific heat coefficients,

$$\chi = \frac{(g\mu_{\text{B}})^2 j(j+1)}{3} \left[\rho_0 N + \frac{n_{\text{f}}}{k_{\text{B}} T^*} \right], \tag{10.45}$$

and

$$\gamma = \frac{\pi^2 k_{\text{B}}^2}{3} \left[\rho_0 N + \frac{n_{\text{f}}}{k_{\text{B}} T^*} \right], \tag{10.46}$$

for the f contribution.

If the f states are well below the Fermi level, $\epsilon_{\text{f}} - \epsilon_{\text{F}} \ll 0$, then

$$k_{\text{B}} T^* \sim (D + \epsilon_{\text{F}}) e^{-\pi |\epsilon_{\text{F}} - \epsilon_{\text{f}}|/N\Delta}, \tag{10.47}$$

so that in this approximation $T^* \sim T_{\text{K}}$, where T_{K} is the Kondo temperature for the equivalent impurity model.

The charge susceptibility $\chi_{\text{c}} = dn_{\text{f}}/d\mu$ is the same as that in the impurity model and is given by (7.26). The derivative of n_f with respect to ϵ_f, however, is given by

$$\frac{dn_f}{d\epsilon_f} = \frac{\pi n_f^2 (1 - n_f)}{N\Delta(1 + n_f^2(1 - n_f)^2 \pi/2N^2 \Delta \rho_0)}. \tag{10.48}$$

This derivative is equal to the charge susceptibility in the impurity case but differs for the lattice due to the dependence of μ on ϵ_f. The expression (10.48) illustrates the negative feedback of the chemical potential. If ϵ_f is changed by $\delta\epsilon_f$ a large number of f electrons are displaced due to the high density of f states. As there is not room to accommodate them in the conduction band, which has a much lower density of states, the chemical potential must change by an amount $\delta\epsilon_F$ so that the difference $\delta\epsilon_f - \delta\epsilon_F$ is very small and the original situation is almost restored. This is the mechanism by which the effective f level is tied to the Fermi level ϵ_F giving a strong negative feedback tending to suppress any change in n_f. This leaves out coupling to the lattice and the Coulomb interaction between the f and conduction electrons, which as we discussed earlier, can give positive feedback terms.

The 'χ/γ' ratio at this level of approximation is universal and is equal to unity,

$$R = \frac{\pi^2 k_B^2}{j(j+1)(g\mu_B)^2} \frac{\chi}{\gamma} = 1, \tag{10.49}$$

reflecting the fact that, in the mean field theory, the quasi-particles, though strongly renormalized, are non-interacting.

The mean field calculation may not be very accurate. We know for the impurity case for $N = 2$ that some of the predictions for physical quantities, such as the 'χ/γ' ratio can be in error by a factor of 2. We are considering here many-body renormalizations from a degeneracy temperature T_D $(= \epsilon_F/k_B)$ to T^*, which due to the exponential dependence on the bare parameters can be very large indeed. If such a simple theory can make predictions of the right order of magnitude then it is rather remarkable.

The mean field theory illustrates how it is possible for the lattice model to give predictions for thermodynamic quantities which are rather similar to those obtained for the impurity models; this may go some way to explaining the good agreement of the predictions for the impurity models with some of the experimental results for Kondo lattices. The paramagnetic singlet ground state generated in this approximation is a fully coherent one built up from 'renormalized' Bloch states. It is quite different from an impurity ground state with the f moments screened solely, and independently, by the conduction electrons.

There are important differences between the mean field theory for the lattice and impurity at finite temperatures. For the lattice model the chemical potential has to be calculated self-consistently and, due to the negative feedback, $n_f(T)$ increases only slowly with temperature. As a consequence the collapse of the quasi-particle bands in the mean field solution as $n_f(T) \to 1$, $z \to 0$, occurs only at very high temperatures $T \gg T^*$, so the equations can be solved in the crossover region $T \sim T^*$, in contrast the the impurity equations which breakdown for $T \sim T_K$. The solutions at higher temperatures, however, in most situations closely resemble those of the impurity; the susceptibility, for instance, has a crossover from Pauli to Curie–Weiss behaviour with a peak at low T for $N > 4$ (Evans, Chung and Gehring, 1989).

The results can also be applied to systems which have an insulating ground state. For example if we take the case $N = 2$, $n_{tot} = 2$; the lower quasi-particle band is full and ϵ_F falls within the 'hybridization' gap giving an insulating ground state. We can apply these ideas to a situation analogous to SmS under pressure if we work in terms of holes rather than electrons. We consider a full hole s band containing two holes per site with empty hole excitations at ϵ_f^h ($f_h^0 \to f_h^1$ to mimic $f^6 \to f^5$) and $\epsilon_f^h + U$ ($f_h^1 \to f_h^2$ to mimic $f^5 \to f^4$) above the band as in black SmS. If under pressure the hole state at ϵ_f^h moves into the band as in gold SmS we can model this with (10.31) for $N = 2$ and $U \to \infty$. Filling all the available holes below the gap gives two electrons per site so that the Fermi level and the renormalized f level $\tilde{\epsilon}_f^h$ falls within the gap and we still have an insulator at $T = 0$. Without hybridization with ϵ_f^h in the lower band this would be metallic with $\epsilon_F = \epsilon_f^h$. For $V \neq 0$ ($U \to \infty$) the f level is renormalized to $\tilde{\epsilon}_f^h$ and a gap induced. The gap, though dependent on the hybridization V, is not a simple one-electron gap proportional to the hybridization but is renormalized due to many-body interactions. A straightforward manipulation of equations (10.38)–(10.41) for this situation gives a gap separation,

$$E_{gap} = 2De^{-(\epsilon_f - \epsilon_F)\pi/N\Delta}, \tag{10.50}$$

which goes to zero for $V \to 0$.

Mean field equations have also been formulated for models including U atoms. In this case two auxiliary Bose fields have to be introduced to restrict the possible configurations to $5f^2$ or $5f^3$. In contrast to the cerium case the charge fluctuations are suppressed relative to the spin

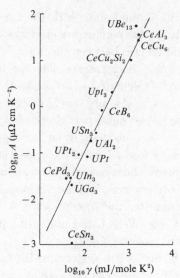

Figure 10.16 A log–log plot of the T^2 coefficient A in the resistivity versus the specific heat coefficient γ for a number of Kondo lattice and heavy fermion compounds (Kadowaki & Woods, 1986).

fluctuations, with large enhancements, strongly renormalized bands and a low energy scale T^* throughout the range $2 < n_f < 3$, including the intermediate valence regime $n_f \sim 2.5$ (Rasul & Harrington, 1987; Evans & Gehring, 1989).

For a fuller discussion of the mean field theory and further references we refer the reader to the review article of Newns & Read (1987).

There have been a number of calculations beyond mean field theory for (10.2) of the effects of second order or Gaussian fluctuations in the Bose and λ fields, along the lines of the impurity calculation given in section 8.3. Millis & Lee (1987), using the same formulation as in section 8.3, showed explicitly the cancellations to order $1/N$ in the physical quantities arising from the phase fluctuations of the Bose field. Using the alternative functional integral approach in the radial gauge, which circumvents the divergence problems, Auerbach & Levin (1986a) and Rasul & Desgranges (1986) found $T^3 \ln T$ contributions to the specific heat from single boson scattering, which is similar to the origin of such terms in ^3He through paramagnon scattering. The coefficient δ in equation (10.1) was found to be enhanced and to scale as γ^3. Calculations of the resistivity give a T^2 coefficient A which scales as γ^2. There is some empirical support for these relations. Auerbach & Levin (1986b)

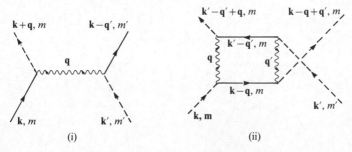

Figure 10.17 (i) A quasi-particle scattering diagram of order $1/N$ involving a single boson exchange, (ii) a spin exchange diagram of order $1/N^2$.

have analysed data on UPt_3 under pressure and find the changes in χ, A, and δ to be compatible with the scaling law $\propto \gamma^n$, with $n = 1, 2, 3$, respectively. Evidence in support of the relation $A \propto \gamma^2$ for a number of heavy fermion compounds is shown in figure 10.16.

The question naturally arises as to whether there might be attractive terms in the interaction between quasi-particles which might lead to superconductivity. The quasi-particle scattering diagrams of the type shown in figure 10.17(i) which involve a single boson exchange, which have a clear parallel with phonon exchange, have been considered by Lavagna, Millis & Lee (1987). The equation for the two particle irreducible vertex for repeated scattering of electrons with momenta $\mathbf{k}, \mathbf{k}'$ on the Fermi surface depends only on the angle θ between them and can be projected into distinct angular momentum channels. Lavagna et al found repulsion for $l = 0, 1$ but weak attraction in the d wave channel $l = 2$. The single boson exchange diagram, which is of order $1/N$, corresponds to a density fluctuation exchange. Houghton, Read & Won (1988) included the effects of spin fluctuation exchange which, though of order $1/N^2$, might to be expected to be physically more important as it is the exchange of spin fluctuations that the believed to induce p wave superfluidity in ^3He. These diagrams, as shown in figure 10.17(ii), involve two boson exchange. On taking these into account Houghton et al found attraction in the p channel, repulsion in the s channel and weak attraction in the d channel. These calculations suggest that the pairing and superconductivity of the heavy fermion metals might be explained by exchange of purely electron spin excitations.

The possible effects of quasi-particle interactions in inducing magnetic transitions has also been investigated. The RKKY interaction, which arises from perturbation theory to fourth order in the hybridization, and which leads to magnetic ordering in normal rare earth compounds,

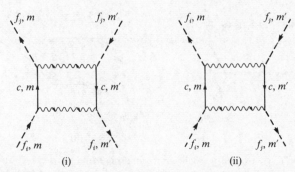

Figure 10.18 Fourth order perturbation diagrams for the effective interaction between f electrons at sites i and j.

corresponds to a diagram of the type 10.18(i) for the effective interaction between f electrons at sites i and j. There are other diagrams of the same order in this formulation such as shown in figure 10.18(ii), but which should be very small after the imposition of the constraint in the limit $n_f = 1$, because if the f states at i and j are initially occupied it implies double occupation of one of the levels in the intermediate state. Diagrams of this type, to order $1/N^2$, also occur as corrections to mean field theory on using the relevant mean field propagators. Ladder diagrams of the type 10.18(i) for repeated particle–hole scattering of f electrons with $m' = -m$, $\mathbf{k}' = \mathbf{k} - \mathbf{q}$, have been summed approximately by Doniach (1987) by taking into account only the dominant high frequency part of the Bose propagator. If this ladder sum is $\chi_{ff}(\omega, \mathbf{q})$ then it is useful to express it in the form,

$$\chi_{ff}(\omega, \mathbf{q}) = \frac{\bar{\chi}_{ff}(\omega, \mathbf{q})}{[1 - \bar{\chi}_{ff}(\omega, \mathbf{q})I(\omega, \mathbf{q})]}. \tag{10.51}$$

The results of Doniach's calculation are in this form with $\bar{\chi}_{ff}(\omega, \mathbf{q})$ as the mean field particle–hole propagator and $I(\omega, \mathbf{q}) = V^4/\epsilon_f^2 \sum_{\omega'} \bar{\chi}_{cc}(\omega', \mathbf{q})$ independent of ω, where $\bar{\chi}_{cc}(\omega, \mathbf{q})$ is the mean field particle–hole propagator for the conduction electrons. The possibility of an instability for a ferromagnetic ground state ($\mathbf{q} = 0$), or a spin density wave ($\mathbf{q} \neq 0$), can be investigated by seeing whether the denominator in (10.51) can vanish for $\omega = 0$. As $\sum_{\omega'} \bar{\chi}_{cc}(\omega', 0) \sim \rho_0$, $\bar{\chi}_{ff}(0,0) \sim 1/T^*$, the condition for a ferromagnetic state becomes essentially that the energy gain through ordering, which is of the order of J^2/N^2 with $J = NV^2/\epsilon_f$, be greater than that due to singlet formation $k_B T^*$, which was the criterion found in earlier work on this problem (for example, Jullien, Fields & Doniach, 1977). In practice heavy fermion systems do not seem to regard these as mutually exclusive and appear to settle for compromise, an ordered state but with moments reduced by magnetic screening. Evans (1991) has extended Doniach's calculation to include low energy contributions to $I(\mathbf{q})$. The low energy contribution is shown to be important in the Fermi liquid state where it can drive a magnetic transition at a temperature T_N if $T_N \ll T^*$. The transition arises due to the temperature dependence of the effective intersite interaction. This fits in well with the analysis of experimental data for U_2Zn_{17} by Broholm et al (1987), who used the form (10.51) and concluded that a temperature dependent $I(\mathbf{q})$ induces the transition in this compound.

Slave boson calculations have been used to investigate theoretically a number of other aspects of heavy fermion behaviour such as metamagnetic transitions (Evans, 1992), de Haas–van Alphen measurements (Rasul, 1989; Wasserman, Springford & Hewson, 1989), NMR relaxation and Knight shifts (Evans & Coqblin, 1991).

10.6 Further Theoretical Approaches

As well as the slave boson type of calculations considered in the last section, there are many other many-body techniques which have been applied to the heavy fermion models. One of the foremost of these is the variational Gutzwiller approach (Gutzwiller, 1965). This based on a relatively simple ansatz for the ground state wavefunction,

$$|\Psi_G\rangle = \prod_i (1 - (1 - g)n_{f,i,\uparrow}n_{f,i,\downarrow})|\Phi_0\{n_{\mathbf{k},\sigma}\}\rangle, \qquad (10.52)$$

where $|\Phi_0\rangle$ is the uncorrelated ground state $n_{\mathbf{k},\sigma} = 1$, $|\mathbf{k}| < k_F$ and 0 otherwise, and g is a variational parameter to take account of correlations in the occupation of the local f orbitals. Despite the simplicity of (10.52) the exact evaluation of $\langle\Psi_G|H|\Psi_G\rangle$ is very difficult. To make the calculations analytically tractable a further approximation, known as the Gutzwiller approximation, can be made in which the matrix elements are evaluated by calculating the statistical weights of the spin configurations which involves the neglect of spatial correlations. This approach applied to the periodic Anderson model ($U = \infty$), like the slave boson technique, gives a renormalized hybridization, $\tilde{V}_{\mathbf{k},m,\sigma} = q_{m,\sigma}V_{\mathbf{k},m,\sigma}$ (without spin–orbit coupling) where $q_{m,\sigma}$ is given by

$$q_{m,\sigma}^2 = \frac{1 - n_f}{1 - n_{f,m,\sigma}}, \qquad (10.53)$$

where $n_f = \sum_{m,\sigma} n_{f,m,\sigma}$ and $n_{f,m,\sigma}$ is the occupation number in the (m,σ) f orbital. The extra factor in the denominator compared with

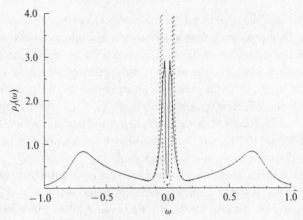

Figure 10.19 The f electron spectral density for the symmetric periodic Anderson model calculated in self-consistent second order perturbation theory in U. The full curve is the result in the local approximation, and the dashed curve the result with the full **k** dependence obtained by including summations up to the third neighbour shell (Schweitzer & Czycholl, 1990).

the slave boson form means that this goes over to the correct unrenormalized result $q = 1$ for a non-degenerate spinless model ($N = 1$). This extra factor being spin dependent affects the criterion for a magnetic ground state, this theory tends to favour a magnetically ordered ground state in contrast to the slave boson type of approximation where the possibility of magnetic transitions appears to be suppressed. There is a generalized slave boson approach of Kotliar & Ruckenstein (1982) for the finite U model which in mean field theory is equivalent to the Gutzwiller approximation. For a more accurate evaluation of $\langle \Psi_{\mathrm{G}} | H | \Psi_{\mathrm{G}} \rangle$ using the Gutzwiller wavefunction without making the Gutzwiller approximation, Monte Carlo techniques have been used. For further details of the Gutzwiller technique we refer the reader to the review of Vollhardt (1984), and for applications to the periodic Anderson model to Shiba & Fazekas (1990), and references therein. Another variational approach to the periodic Anderson model, which has also been generalized to finite temperatures, is that of Brandow (1986, 1988), leading to a similar picture of renormalized bands.

The weak coupling perturbation expansion in U about the unpolarized Hartree–Fock state, as described in section 5.5, has also been exploited for lattice models. Calculations have been made of the self-energy based on the second order U^2 diagram, figure 5.4, but with the propagators made fully self-consistent so that the Luttinger Fermi liquid theorems are

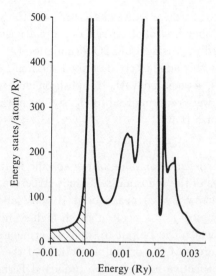

Figure 10.20 The quasi-particle density of states for $CeSn_3$ deduced from renormalized band structure calculations of Strange & Newns (1986).

satisfied. These calculations have exploited the idea of a $1/d$ expansion where d is the dimensionality. By appropriately scaling the variables the limit $d \to \infty$ can be taken, which gives a self-energy $\Sigma(\omega, \mathbf{k})$ which is on-site and a function of ω only. This turns out to be a good approximation to the fully $\mathbf{k}$ dependent three dimensional self-energy, and the corrections can be calculated by working in real space from the on-site limit, taking into account near neighbour shells. For $d = 3$ this gives rapidly convergent results. Shown in figure 10.19 is the f spectral density of states for the symmetric periodic Anderson model calculated in this way together with the results of the $d \to \infty$ calculation. Results can be obtained for the resistivity as a function of temperature for different choices of parameters which show the same range of behaviour as that observed in the heavy fermion compounds (Schweitzer & Czycholl, 1992). Metzner & Vollhardt (1989) have shown that $d \to \infty$ limit simplifies the evaluation of $\langle \Psi_G | H | \Psi_G \rangle$, giving contributions in this limit which are entirely on-site and equivalent to the Gutzwiller approximation.

There have been attempts to go beyond model calculations and calculate the quasi-particle excitations for specific systems from a band approach, modified to incorporate many-body renormalizations. One of the first schemes devised for carrying out a renormalized band calculations was that of Strange & Newns (1986) for the mixed valence compound $CeSn_3$. They first of all carried out a relativistic band calculation in the standard way using a LMTO formulation (Andersen, 1975). It was assumed that this adequately described the non-f electrons. Parameters C_3, $\phi_3(-)$, associated with the position and width of the f states, respectively, were renormalized to $\tilde{C}_3$, $\tilde{\phi}_3(-)$, in the spirit of the mean field theory, such that

$$[\tilde{\phi}_3(-)/\phi_3(-)]^2 = 1 - n_{\mathrm{f}}, \qquad (10.54)$$

and to obtain the experimental value of the specific heat coefficient γ. The density of states of the renormalized bands structure, shown in figure 10.20, has a narrow $j = \frac{5}{2}$ peak about 10 meV above the Fermi level, and an unoccupied $j = \frac{7}{2}$ peak at much higher energy. The features of the theory which can be put to some experimental test are the predictions for the areas of the orbits on sections of the Fermi surface and the associated effective masses. The predicted Fermi surface was similar to that obtained from the unrenormalized calculations but the effective masses were quite different and in much better agreement with the experimental results. Most of the effective masses are larger in the renormalized calculation but the mass for one of the orbits was lighter, yet also in better agreement with experiment.

Calculations, similar in spirit but differing in detail, have been performed by Sticht, d'Ambrumenil & Kübler (1986) for the Kondo lattice $CeCu_2Si_2$ based on an approach devised by Razafimandimby, Fulde & Keller (1984). In these calculations the phase shifts of the f states near ϵ_{F} for Ce at site i are taken to be of the form,

$$\eta_\tau^i(\omega) = \eta(\epsilon_{\mathrm{F}}) + \alpha(\omega - \epsilon_{\mathrm{F}}) + \sum_{j,\tau',\omega'} \Phi_{\tau,\tau'}^{i-j} \delta n_{\tau'}^j(\omega'), \qquad (10.55)$$

where i is a site index and τ a pseudo-spin index for the crystal field ground state (a Γ_7 doublet), with α as a parameter of the order $1/k_{\mathrm{B}}T^{'*}$. This form is analogous to that assumed by Nozières (1974) in deriving the Fermi liquid behaviour for the Kondo problem, the last term on the right hand side takes account of the quasi-particle interactions, with Φ as a molecular field, due to local departures $\delta n_{\tau'}^j$ from equilibrium in the occupation number $n_{\tau'}$ at site j. For states of other symmetries it was assumed that $\eta_l^i(\omega) = \eta_l^i(\epsilon_{\mathrm{F}})$. These phase shifts, defined for non-overlapping spheres around each ion, were then used as input for a Korringa–Kohn–Rostoker (KKR) band calculation for states near the Fermi level. To perform these calculations an expression for $\eta_\tau(\omega)$ is required which is valid over a larger frequency range than (10.55) so in

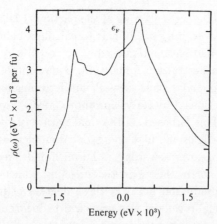

Figure 10.21 The quasi-particle density of states in the immediate vicinity of the Fermi level for $CeCu_2Si_2$ as deduced from renormalized band calculations of Sticht et al, (1986).

practice $\eta_\tau(\omega)$ with the resonance form,

$$\eta_\tau(\omega) = \cot^{-1}\left(\frac{\tilde{\epsilon}_f - \omega}{\tilde{\Delta}}\right), \qquad (10.56)$$

was used, for the Γ_7 states with $\tilde{\epsilon}_f = \epsilon_F$. This form reduces to (10.55) close to the Fermi level with $\alpha^{-1} = \tilde{\Delta}$ and $\eta_\tau(\epsilon_F) = \cot^{-1}(\tilde{\epsilon}_f - \omega)/\tilde{\Delta})$, for $T = 0$ when the quasi-particle molecular field term vanishes. The phase shifts associated with the excited crystal field state were neglected as the excitation energy $\Delta E_{\gamma\gamma'}$ in $CeCu_2Si_2$ is of the order 140 K while the heavy fermion degeneracy temperature T^* is only of the order 13 K, satisfying the condition $\Delta E_{\gamma\gamma'} \gg k_B T^*$. The results were found to be very sensitive to the assumed form for the wavefunction of the Γ_7 state. The calculated quasi-particle density of states in the immediate neighbourhood of the Fermi level is shown in figure 10.21. The value of α^{-1}, deduced by fitting the density of quasi-particle states $\tilde{\rho}(\epsilon_F)$ to the observed γ using (10.22), was found to be 22 K, satisfying the requirement $\alpha^{-1} \sim k_B T^*$. In this calculation the predictions of the theory can be compared with the observed form for the specific heat ratio $C(T)/T$ versus T (Bredl et al, 1984). The results for this ratio give a maximum, as in the experiment, but with a very much smaller peak than that observed and at a lower temperature. The form of the specific heat itself, a less sensitive quantity than the ratio, was found to be in better agreement with experiment. The important general point that emerges with these calculations is that the renormalized bands, though associated with an overall energy scale $k_B T^*$, can have fine features within this scale from the coherent scattering in the particular lattice structure. These features, however, may be difficult to calculate being sensitive to the approximations and parameters used.

To explain the rather puzzling magnetic behaviour of some of the heavy fermion compounds, such as weak antiferromagnetic order and metamagnetic transitions, Miyake & Kuramoto (1990) have put forward a 'duality' model. This is based on the dual nature of the excitations, low energy quasi-particles and high energy local moments. In the renormalization group crossover towards the Fermi liquid fixed point there should be a regime where the magnetic screening is incomplete giving rise to a residual interaction between these degrees of freedom. The phenomenological duality model was proposed to incorporate these interactions, and then applied both to metamagnetic and weak magnetic behaviour. Edwards (1991) has microscopically deduced such a model in the strongly polarized situation in a large magnetic field, and also applied it to the metamagnetic transition. In this, and earlier work (Edwards, 1988), Edwards discusses the possibility that the f electron is localized at this transition. The consequences of this theory can be discussed in terms of the simple quasi-particle band picture illustrated in figure 10.15 for a non-degenerate f level ($N = 2$). In an applied field the effective f level $\tilde{\epsilon}_{f,\uparrow}$ for the up spins moves down towards the chemical potential μ, flattening the lower quasi-particle band so as to accommodate the extra up spin f charge. The down spin level $\tilde{\epsilon}_{f,\downarrow}$ moves up so reducing the down spin Fermi surface. At the metamagnetic transition $\tilde{\epsilon}_{f,\uparrow}$ passes through the Fermi level, which now lies in the upper quasi-particle band. The up spin Fermi surface after the transition then closely resembles that of the conduction electrons in the absence of itinerant f electrons. Changes in the Fermi surface on passing through the metamagnetic transition have been observed in $CeRu_2Si_2$ and $CeAl_2$ (Lonzarich, 1988).

There have been at least two attempts to calculate theoretically the superconducting transition temperature T_c in particular compounds, based on certain assumptions about the nature of the electron pairing. Assuming s wave superconductivity due to coupling with phonons through the dependence of the characteristic temperature T^* on volume (as in the Kondo collapse theory of Allen & Martin, 1982) Razafimandimby et al (1984) estimate T_c for $CeCu_2Si_2$. Due to the uncertainties in the parameters involved such a calculation cannot be precise but they estimate that this particular mechanism can account for the order of magnitude of the observed transition temperature $T_c \sim 0.5$ K. Estimates of T_c for UPt_3 based on a purely electronic pairing mechanism due to spin fluctuations have been made by Norman (1987). In this calculation it is the $\mathbf{q}$ dependence of $\chi(\mathbf{q})$ that leads to an attractive interaction and the estimates of T_c lie in the range 0.1–.3 K ($T_c \sim 0.5$ K). The solution of the Eliashberg equations in the superconducting phase gives a gap which has a polar structure with a line of nodes on the Fermi surface. Calculations of this type are difficult but are needed as the eventual aim is to be able to predict specific features of the superconducting state to discriminate amongst the different possible mechanisms for pairing.

Though Ce and U heavy fermion compounds have been put together in a single class there are important differences which should not be over looked. Apart from obvious differences in the photoemission and BIS spectra, there are distinct differences in the magnetic response. For example UBe_{13} has a specific heat almost independent of field whereas $CeCu_2Si_2$ has a strongly field dependent specific heat. Their neutron scattering spectra are also quite different, $CeCu_2Si_2$ has a peak about 1 meV ($\sim T^*$) whereas the peak in UBe_{13} is much larger at 15 meV. This has led Cox (1987b, 1991) to put forward a rather different model of U in which the lowest crystal field multiplet of an assumed tetravalent state is a Γ_3 with no net moment. This hypothesis explains the lack of magnetic response but leaves the problem of explaining the large mass enhancements and the low energy scale T^* of the heavy fermion response. To explain this Cox assumes that the predominant coupling of the $5f^2$ Γ_3 state is with the $5f^1$ Γ_7 state, on an impurity model this coupling is through the hybridization with Γ_8 conduction electrons. A canonical transformation to eliminate the virtual excitations to the excited states leads to a model with a pseudo spin $\frac{1}{2}$ electric quadrupole moment coupled to the conduction electrons, equivalent to the n-channel Kondo model (9.30) with $S = \frac{1}{2}$, $n = 2$. As $n > 2S$ this corresponds to the over-compensated case discussed in appendix J which has a non-Fermi liquid behaviour at low temperatures due to a non-trivial $J = J^*$ fixed point (a possibility pointed out Nozières and Blandin, 1980). Critical behaviour is found for the over-compensated model for χ_{imp} and C_{imp}/T for low T for $n > 2$, and also as a function of field for $T = 0$. For the case $n = 2$, however, the critical exponents vanish to be replaced by a logarithmic dependence (the exact solution of the $S = \frac{1}{2}$, $n = 2$ model has been extensively studied by Sacramento and Schlottmann, 1989). The resistivity is believed to have the form $\Delta R_{\text{imp}}(T) \propto (T/T_K)^\nu$ for $T \ll T_K$ with an exponent ν near to unity in contrast to the Fermi liquid case $\nu = 2$. The compound $Y_{0.8}U_{0.2}Pd_3$, which has a low temperature resistivity of this form with $\nu = 1.12$, has been cited, together with evidence from specific heat and susceptibility measurements as an example of this type of behaviour (Seaman et al, 1991; Cox, 1991). This interpretation has been challenged by Andraka and Tsvelick (1991) who interpret the anomalous behaviour of this compound to a $T = 0$ phase transition, due to some unknown order parameter, rather than a single impurity effect. The low T resistivities of most heavy fermions, however, have a T^2 dependence at sufficiently low temperatures.

The more general question posed by this work is whether we should consider the heavy fermions as a single group or whether their similarities mask important underlying differences. The possibility of non-Fermi liquid behaviour also raises the question as to whether there are any close similarities with the normal state of the high T_c superconductors, whose anomalous behaviour we shall briefly look at in the next section.

In this brief survey we cannot expect to cover all aspects of recent activity. We refer readers interested in the theoretical work on transport properties to the resistivity calculations of Coleman (1987) and Cox & Grewe (1988), and for calculations on the Hall effect, which has an anomalous peak at $T \sim T^*$, to Coleman, Anderson & Ramakrishnan (1985) and Fert & Levy (1987), and references therein.

10.7 The High T_c Superconductors

Since the discovery of high temperature superconductivity in the compounds $La_{2-x}Ba_xCuO_4$, with $T_c \sim 35$ K, $YBa_2Cu_3O_7$, $T_c \sim 90$ K (Bednorz & Müller, 1986; Wu et al, 1987), and related compounds, there has been intense activity, both experimental and theoretical, to understand the behaviour of this class of materials. Many mechanisms, conventional and unconventional, have been proposed to explain the superconductivity. So far no theory has gained general acceptance, nor has any experiment provided specific enough clues to indicate a particular pairing mechanism to the exclusion of all others. The normal state behaviour of these ceramic materials is so markedly different from that of conventional metals it too has become the focus of much of the recent experimental and theoretical interest. One almost universal feature of the normal state of these materials is a linear in T dependence of the electrical resistivity with temperature (see figure 10.22).

The common feature of all these high T_c materials is the CuO_2 planes. The materials are very anisotropic and the superconductivity is virtually two dimensional. The superconductivity is s wave and, from the

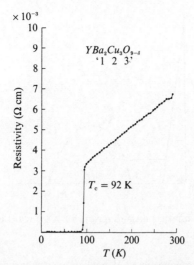

Figure 10.22 The resistivity of a single-phase $YBa_2Cu_3O_7$ sample as a function of temperature (Bednorz & Müller, 1988).

evidence of flux quantization, due to electron pairs as in the BCS theory. Estimates of the superconducting energy gap $E_{sc,gap}$ to the transition temperature, $E_{sc,gap}/k_BT_c$ of the order of 6 approximately a factor of 2 larger than the weak coupling BCS value 3.5. They are type II superconductors with relatively low critical fields H_{c1} (the on-set of flux penetration) and very high upper critical field H_{c2} (the on-set of the completely normal state). The very high upper critical field is associated with the small coherence lengths, $\xi_{sc}^{a,b} \sim 15$ Å in plane, $\xi_{sc}^c \sim 2$ Å interplane for $YBa_2Cu_3O_7$, as $H_{c2} \sim 1/\xi_{sc}^2$. These unusual features, the very high T_c values and the lack of a significant isomer effect on T_c, have been interpreted by many as indicating a purely electronic pairing mechanism.

The undoped material, La_2CuO_4 is a three dimensional antiferromagnetic insulator ($T_N \sim 300$ K) but with a very small interplane exchange coupling. The antiferromagnetic long range order rapidly disappears with hole doping on increasing x and the Néel temperature goes to zero, there then appears to be a spin glass phase and the superconductivity sets in for $x \sim 0.05$; $T_c(x)$ develops a plateau at $T_c(x) \sim 40$ K and drops sharply to zero for $x > 0.2$. The phase diagram for this material is sketched in figure 10.23. The interplay of antiferromagnetism and superconductivity has some similarities with the heavy fermion materials where it has also been suggested that the superconductivity is due to

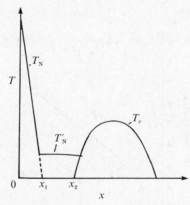

Figure 10.23 A schematic phase diagram for a typical high T_c material as a function of doping.

some novel pairing mechanism. Whether antiferromagnetic correlations in the high T_c materials play any role or not in inducing the interelectron attraction leading to pairing is not clear. Purely electronic pairing mechanisms involving magnetism have been proposed by Kampf & Schrieffer (1990) and Trugman (1990) amongst others.

A more general question is what role the antiferromagnetic correlations play in the normal state. Very soon after the discovery of the high T_c materials Anderson (1987) put forward the conjecture that these correlations in a two dimensional Heisenberg model would lead to a novel spin liquid or resonating bond state (RVB), one with strongly locally correlated singlet pairs, which on doping would naturally lead to superconductivity. Monte Carlo and other calculations, however, have established beyond reasonable doubt that the ground state of the two dimensional Heisenberg model has long range antiferromagnetic order. Doping with a low density of holes, however, could favour an RVB state. Such a state above T_c would not behave as a Fermi liquid; this would be remarkable and an unprecedented situation as in all cases of superconductivity or superfluidity in fermion systems, known so far, arise as instabilities in a Fermi liquid state.

These conjectures bring to the fore the question as to the precise nature of the observed normal state. Levin, Kim, Lu & Si (1991) have put forward the proposition that despite the fact that the behaviour of the normal state does not correspond to 'canonical' Fermi liquid behaviour, $\chi(T) \propto$ constant, $C(T) \propto \gamma T + \delta T^3 \ln T$, $R(T) \propto T^2$, $1/T_1 \propto T$, it could correspond to the 'non-canonical' Fermi liquid behaviour such as that observed in heavy fermion systems in the crossover region $T \sim T_{\text{coh}}$, cor-

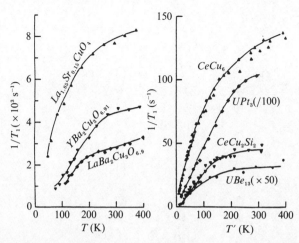

Figure 10.24 A comparison of Cu NMR relaxation $1/T_1$ in cuprates with NMR relaxation at non-f sites for heavy fermions. The transformed temperature scale T' for the heavy fermion case is explained in the text (from Levin et al, 1991).

responding to temperatures in the range 1–10 K. To bring out the analogy between these systems Levin et al plot results for the heavy fermion systems on a transformed temperature scale T', $T' = 100T(\gamma/\gamma_{CeCu_6})$, and then compare with the results for the high T_c materials on a normal temperature scale. An example is shown in figure 10.24, where the Cu NMR rates $1/T_1$ in the high T_c materials are compared with the relaxation rates at non-f sites for several heavy fermion compounds. This interpretation requires that $T_{coh} < T_c$ if the canonical Fermi liquid behaviour is not to be observed. When compared in this way the similarities between the two sets of results are often quite striking. Though there is a temperature range where $R(T)$ for heavy fermions is linear in T it is usually small in extent so it would be very difficult to explain results like those for the alloy $Bi_{2+x}Sr_{2-y}CuO_{6+x}$ which has a $T_c = 10$ K, and a resistivity $R_{ab}(T)$ in the CuO_2 plane which is linear with T over a temperature range of 700 K. However, this analogy cannot be pushed too far as the CuO_2 planes in the high T_c materials are two dimensional whereas heavy fermion materials are all basically three dimensional.

As a result of an analysis of a range of experimental data Varma et al (1989) have a quite different conjecture that the self-energy of the low energy excitations has the form,

$$\Sigma(\omega) \sim g^2\rho_0^2 \left\{ \omega\ln\left(\frac{x}{\omega_c}\right) - i\frac{\pi x}{2} \right\}, \qquad (10.57)$$

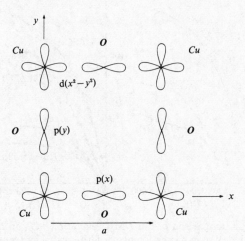

Figure 10.25 A schematic representation of the d($x^2 - y^2$) orbitals at the Cu sites and the p(x) and p(y) orbitals at the oxygen sites in the CuO_2 planes.

where $x = \max(|\omega|, T)$, ω_c is an ultraviolet cutoff and g a coupling constant. As $\Sigma^I(\omega) \to 0$ but only as $|\omega|$ at $T = 0$, and not as ω^2 as in the conventional Fermi liquid theory, and the wavefunction renormalization factor $z_{\mathbf{k}}$ goes to zero at the Fermi level as $1/\ln(\omega_c/|k - k_F|)$, the term 'marginal Fermi liquid' has been use to describe it.

We cannot hope to survey the vast number of papers which have appeared on this subject since the discovery of these materials. We suggest the interested reader consult one or two of the recent reviews or books: Bednorz & Müller (1988); Friedel (1989); Pickett, 1989; Los Alamos Symposium 1989, Batlogg, 1991; from which further references can be found.

The main reason for including some discussion of these materials at the end of a book concerned primarily with magnetic impurities and heavy fermions is because at a formal level the models for all these systems look remarkably similar. The d-p model which has been put forward to describe the low energy electronic excitations in the CuO_2 planes is composed of d hole states on the Cu sites (one hole per Cu in the undoped material) in d($x^2 - y^2$) orbitals which are strongly hybridized with pσ obitals on the neighbouring oxygen sites (see figure 10.25). As the undoped material is a magnetic insulator the holes on the Cu sites must be localized by a reasonably strong on-site Coulomb interaction

U_{dd}. The Hamiltonian describing these basic elements is

$$H_{\rm dp} = \sum_{i,\sigma} \epsilon_{\rm d} d^\dagger_{i,\sigma} d_{i,\sigma} + \sum_{l,\sigma} \epsilon_{\rm p} p^\dagger_{l,\sigma} p_{l,\sigma} + U_{dd} \sum_i n_{d,i\uparrow} n_{d,i\downarrow}$$

$$+ \sum_{\langle i,l\rangle,\sigma} V_{il}(d^\dagger_{i,\sigma} p_{l,\sigma} + p^\dagger_{l,\sigma} d_{i,\sigma}),$$

$$(10.58)$$

where i and l refer to Cu and O sites respectively, $d^\dagger_{i,\sigma}$ creates a hole in a d$(x^2 - y^2)$ Cu orbital of spin σ, and $p^\dagger_{l,\sigma}$ a hole in an O p(x) or p(y) orbital (the one pointing to the nearest neighbour Cu site) of spin σ (see figure 10.25). For $U_{dd} = 0$ this corresponds to hybridized bonding and antibonding bands,

$$\epsilon^\pm(\mathbf{k}) = \frac{\epsilon_{\rm p} + \epsilon_{\rm d}}{2} + \left[\left(\frac{\epsilon_{\rm p} - \epsilon_{\rm d}}{2}\right)^2 + 4V^2(\sin^2(k_x a/2) + \sin^2(k_y a/2))\right]^{1/2},$$

$$(10.59)$$

where a is the lattice spacing in the 2d CuO_2 plane. If direct oxygen–oxygen overlap $t_{\rm pp}$ is included in (10.58) the model is then a special case of the periodic Anderson model. Other interactions which should be included are a $U_{\rm pd}$ between the oxygen p holes and the copper d holes, and an on-site $U_{\rm pp}$ Coulomb interaction between the p holes.

There are a number of features which distinguish this situation from that for the heavy fermions, (i) the two dimensionality, (ii) the specific $\mathbf{k}$ dependence of $V(\mathbf{k})$ due to the tight-binding form of (10.58) and (iii) the relative magnitudes of the hybridization V compared with the direct oxygen–oxygen overlap $t_{\rm pp}$ (estimates give $V \sim 1.3$ eV and $t_{\rm pp} \sim 0.65$ eV).

If this model is an adequate description of the CuO_2 planes in the high T_c materials then the questions that arise are: (i) is there a region in the parameter space where the model is metallic and yet not a Fermi liquid? (ii) is there any possibility of an electron–electron attraction that would lead to pairing? In the undoped case with $\epsilon_{\rm p} - \epsilon_{\rm d} > 0$, treating the hybridization perturbatively as in section 1.7, an effective Heisenberg Hamiltonian can be generated between localized d holes on the Cu sites with an antiferromagnetic coupling $J_{dd} \sim V^4(U + 2\epsilon_{\rm p} - 2\epsilon_{\rm d})/U(\epsilon_{\rm p} - \epsilon_{\rm d})^3$, mediated by virtual transitions to the oxygen sites. This explains the antiferromagnetic ordering in the undoped materials. It is not clear what role this interaction plays in the normal and superconducting states of the doped systems.

It has been questioned whether it is necessary to use the two band (or three band as there is a non-bonding p band) d-p model (10.58) to understand the high T_c materials as the d and p states are so strongly hybridized. The argument is that in this strongly coupled situation the two band model can be reduced to an effective one band model which contains the essential physics. The effective one band model is the Hubbard model,

$$H = \sum_{\langle i,j \rangle, \sigma} t_{ij} c^\dagger_{i,\sigma} c_{j,\sigma} + U \sum_i n_{i\uparrow} n_{i\downarrow}, \qquad (10.60)$$

where $c^\dagger_{i,\sigma}$ creates as electron at site i of spin σ in a hybridized d-p orbital, t_{ij} is the intersite hopping matrix element and U the effective on-site Coulomb interaction (Hubbard, 1963). For $\epsilon_p - \epsilon_d > U$ the Hubbard model (10.60) can be derived from the d-p model by eliminating to lowest order the virtual excitations of the d holes to the oxygen sites using the method given in section 1.7. Zhang & Rice (1988) have shown that it can also be derived for $\epsilon_p - \epsilon_d < U$ with strong hybridization. In this case a d-hole and a hole in a combination of p orbitals around a Cu site form a strongly coupled singlet. The effective Hubbard model then describes this singlet pair hopping from Cu site to Cu site, with a strong Coulomb repulsion arising from the d electrons for two singlets on the same 'site'.

In one dimension the Hubbard model has been solved by the Bethe ansatz (Lieb & Wu, 1968). The densities of states of the pure charge and spin excitations (corresponding to (6.39) and figure 6.7 for the Anderson model), the holons and spinons, have been calculated from this solution by Kawakami & Okiji (1990). For $U \neq 0$ the low energy excitations of this model do not correspond to a Fermi liquid. The wavefunction renormalization factor $z(\mathbf{k}_F)$ vanishes, going to zero as a power law, $z(\mathbf{k}) \sim |\mathbf{k} - \mathbf{k}_F|^\alpha$, $0 < \alpha < 1$, as $\mathbf{k} \to \mathbf{k}_F$. For strong correlation near half-filling it can be shown that $\alpha \to 1/8$ (Kawakami & Yang, 1991, and references therein). This form of behaviour was first encountered in another one dimensional fermion model, the spinless Luttinger model (Luttinger, 1963; Mattis & Leib, 1965). Haldane (1981) has argued that this type of behaviour is typical of a general class of one dimensional interacting Fermi systems which he termed 'Luttinger liquids' in analogy with Fermi liquids. These systems have similarities with the marginal Fermi liquids and with Anderson's picture of the normal state of the high T_c materials (Anderson & Ren, 1990).

In three dimensions and $U \to \infty$ the ground state of the Hubbard model with one hole in an otherwise filled band was shown by Nagaoka (1966) to be ferromagnetic for most types of lattice. It has not been proved rigorously that this applies to a finite density of holes but calculations based on more and more general ansätze make it seem highly likely (von der Linden & Edwards, 1991, and references therein).

When U is large it is possible to perform a canonical transformation, or equivalently take into account the virtual excitations to doubly occupied states using the method of section 1.7, to derive an effective Hamiltonian for the low energy excitations keeping terms to order t_{ij}^2/U. If spin flip exchange hopping terms involving three sites can be neglected (they are likely to be less important for a low density of holes), we arrive at the t-J model,

$$H_{t-J} = \sum_{\langle i,j \rangle, \sigma} t_{i,j} c_{i,\sigma}^\dagger (1 - n_{i,-\sigma}) c_{j,\sigma} (1 - n_{j,-\sigma}) + J \sum_{\langle i,j \rangle} [\mathbf{S}_i \cdot \mathbf{S}_j - \frac{1}{4}] n_i n_j,$$

(10.61)

with an antiferromagnetic coupling $J = 2t^2/U$ for nearest neighbour hopping. For this model J is usually taken as an independent parameter. The $(1 - n_{i,\sigma})$ factors in the kinetic energy term are due to the fact that as U is large ($\to \infty$) the electrons can only move if there are unoccupied sites available. For one electron per site the electrons are localized and the model describes a Heisenberg antiferromagnet. When holes are introduced the electrons can move and both spin and charge excitations are possible. The interplay of the charge and spin degrees of freedom in this case is not well understood, particularly in two dimensions. The model poses the question of the interplay of these excitations in the simplest form relevant to the high T_c systems, and so has become the focus of much of the current work in the field.

These models, from the impurity Anderson model (1.66), the periodic Anderson model (10.2), the d-p model (10.58), the Hubbard model (10.60) to the t-J model (10.61), describe a rich diversity of physical behaviour associated with strongly correlated systems, metallic and insulating, Fermi liquids, paramagnetism, ferromagnetism, antiferromagnetism and more exotic types of magnetic order. Whether superconductivity (for $U > 0$) should be included in this list is at the time of writing an open question.

Appendix A

Scattering Theory

We have quoted several results from scattering theory in chapters 1 and 2. Here we provide brief details of their derivation. We consider the scattering of Bloch waves by the impurity potential using the Hamiltonian (1.11). If $|\psi\rangle$ is an eigenstate of this Hamiltonian H $(= H_0 + V)$ associated with the continuous spectrum corresponding to an energy ϵ then

$$(\epsilon - H_0)|\psi\rangle = V|\psi\rangle. \tag{A.1}$$

This equation can be rewritten by formally regarding it as an inhomogeneous equation, and then solving using the conduction electron Green's function $G_0(\epsilon)$ (see (1.15)). For solutions $|\psi^+\rangle$ corresponding to an outgoing scattered wave the equation takes the form,

$$|\psi^+\rangle = |\phi\rangle + G_0^+(\epsilon)V|\psi^+\rangle, \tag{A.2}$$

where $|\phi\rangle$ is any solution of the homogeneous equation $(\epsilon - H_0)|\phi\rangle = 0$. Use of the advanced Green's function $G^-(\epsilon)$ gives an incoming wave $|\psi^-\rangle$. In the r-representation (A.2) becomes

$$\psi^+(\mathbf{r}) = \phi(\mathbf{r}) + \int G_0^+(\mathbf{r}, \mathbf{r}'; \epsilon)V(\mathbf{r}')\psi^+(\mathbf{r}')\, d\mathbf{r}', \tag{A.3}$$

which is the Lippmann–Schwinger equation of scattering theory.

To obtain the scattered wave $\psi_{\mathbf{k}}^+(\mathbf{r})$ due to an incoming Bloch wave $\phi_{\mathbf{k}}(\mathbf{r})$ of energy $\epsilon_{\mathbf{k}}$ we need the corresponding Green's function,

$$G_0^+(\mathbf{r}, \mathbf{r}'; \epsilon) = \sum_{\mathbf{k}} \frac{\langle \mathbf{r}|\mathbf{k}\rangle\langle \mathbf{k}|\mathbf{r}'\rangle}{(\epsilon + is - \epsilon_{\mathbf{k}})} = \sum_{\mathbf{k}} \frac{\phi_{\mathbf{k}}(\mathbf{r})\phi_{\mathbf{k}}^*(\mathbf{r}')}{(\epsilon + is - \epsilon_{\mathbf{k}})}, \tag{A.4}$$

for $\epsilon = \epsilon_{\mathbf{k}}$ where $\langle \mathbf{r}|\mathbf{k}\rangle = \phi_{\mathbf{k}}(\mathbf{r})$. Using (A.4) equation (A.3) can be written as

$$\psi_{\mathbf{k}}^+(\mathbf{r}) = \phi_{\mathbf{k}}(\mathbf{r}) + \sum_{\mathbf{k}'} \frac{T_{\mathbf{k}'\mathbf{k}}\phi_{\mathbf{k}'}(\mathbf{r}')}{(\epsilon_{\mathbf{k}} - \epsilon_{\mathbf{k}'} + is)}, \tag{A.5}$$

where

$$T_{\mathbf{k}'\mathbf{k}} = \int \phi_{\mathbf{k}'}^*(\mathbf{r}')V(\mathbf{r}')\psi_{\mathbf{k}}^+(\mathbf{r}')\, d\mathbf{r}'. \tag{A.6}$$

Iterating (A.6) gives

$$T_{\mathbf{k}'\mathbf{k}} = V_{\mathbf{k}'\mathbf{k}} + \sum_{\mathbf{k}''} V_{\mathbf{k}'\mathbf{k}''} \frac{1}{(\epsilon_{\mathbf{k}} - \epsilon_{\mathbf{k}''} + is)} V_{\mathbf{k}''\mathbf{k}} + \dots \tag{A.7}$$

and establishes the connection with the T matrix defined by (1.22),

$T_{\mathbf{k'k}} = \langle \mathbf{k'}|T(\epsilon_\mathbf{k}^+)|\mathbf{k}\rangle$. For plane waves $\phi_\mathbf{k}(\mathbf{r}) = e^{i\mathbf{k}\cdot\mathbf{r}}/V^{1/2}$ subject to periodic boundary conditions for a box of volume V,

$$G_0^+(\mathbf{r},\mathbf{r}':\epsilon_\mathbf{k}) = \int \frac{e^{i\mathbf{k}\cdot(\mathbf{r}-\mathbf{r}')}}{(\epsilon_\mathbf{k}-\epsilon_{\mathbf{k'}}+is)}\frac{d\mathbf{k'}}{(2\pi)^3}, \tag{A.8}$$

where $\epsilon_\mathbf{k} = \hbar^2\mathbf{k}^2/2m$. Integrating over the angles in (A.8) gives

$$G_0^+(\mathbf{r},\mathbf{r}':\epsilon_\mathbf{k}) = \frac{-mi}{2\pi^2\hbar^2|\mathbf{r}-\mathbf{r}'|}\int_{-\infty}^{+\infty}\frac{e^{ik'|\mathbf{r}-\mathbf{r}'|}}{(k^2-k'^2+is)}\,dk'. \tag{A.9}$$

The final integration over k' can be evaluated by contour integration using a standard semi-circular contour in the upper half k' plane. The sole contribution is from the pole at $k' = k + i0$ giving

$$G_0^+(\mathbf{r},\mathbf{r}':\epsilon_\mathbf{k}) = -\frac{m}{2\pi\hbar^2}\frac{e^{ik|\mathbf{r}-\mathbf{r}'|}}{|\mathbf{r}-\mathbf{r}'|}. \tag{A.10}$$

On substituting into the integral equation (A.3) and using the asymptotic form for large r,

$$\frac{e^{ik|\mathbf{r}-\mathbf{r}'|}}{|\mathbf{r}-\mathbf{r}'|} \sim \frac{e^{ikr}}{r}e^{-i\mathbf{k'}\cdot\mathbf{r}'}, \tag{A.11}$$

where $\mathbf{k'} = k\mathbf{r}/r$, then the asymptotic form for the scattered wave function is

$$\psi_\mathbf{k}(\mathbf{r}) = e^{i\mathbf{k}\cdot\mathbf{r}} + \frac{f(\mathbf{k'},\mathbf{k})}{r}e^{ikr} \tag{A.12}$$

with the scattering amplitude $f(\mathbf{k'},\mathbf{k})$ (which has the dimensions of length) given by

$$f(\mathbf{k'},\mathbf{k}) = -\frac{m}{2\pi\hbar^2}T_{\mathbf{k'k}}. \tag{A.13}$$

The differential scattering cross-section $d\sigma/d\Omega$, where $d\Omega$ is an element of solid angle, is given by

$$\frac{d\sigma}{d\Omega} = |f(\mathbf{k'},\mathbf{k})|^2 = \frac{m^2}{4\pi^2\hbar^4}|T_{\mathbf{k'k}}|^2 \tag{A.14}$$

and the total scattering cross-section $\sigma(k)$ is given by integration over Ω,

$$\sigma(k) = \frac{m^2}{4\pi^2\hbar^4}\int|T_{\mathbf{k'k}}|^2\,d\Omega. \tag{A.15}$$

The transition rate for the scattering from state $\mathbf{k}$ to state $\mathbf{k'}$, $W_{\mathbf{kk'}}$ is given by the cross-section $\sigma(k)$ multiplied by the velocity of approach $\hbar k/m$. Using the relation,

$$\sum_\mathbf{k}|T_{\mathbf{k'k}}|^2\delta(\epsilon_\mathbf{k}-\epsilon_{\mathbf{k'}}) = \frac{mk}{(2\pi)^3\hbar^2}\int|T_{\mathbf{k'k}}|^2\,d\Omega, \tag{A.16}$$

the transition rate can be expressed in the form,

$$W_{\mathbf{k'k}} = \frac{2\pi}{\hbar}|T_{\mathbf{k'k}}|^2\delta(\epsilon_\mathbf{k}-\epsilon_{\mathbf{k'}}). \tag{A.17}$$

This corresponds to the lowest order perturbational result in V, the 'Golden Rule', with T replacing V. The expression for the cross-section $\sigma(k)$ can be cast into a simpler form by the use of an identity for $T(\epsilon^+) - T(\epsilon^-)$ derived from (1.22),

$$T(\epsilon^+) - T(\epsilon^-) = T(\epsilon^-)(G_0^+(\epsilon) - G_0^-(\epsilon))T(\epsilon^+), \qquad (A.18)$$

where use has been made of the operator identity,

$$(I - A)^{-1} - (I - B)^{-1} = (I - B)^{-1}(A - B)(I - A)^{-1}. \qquad (A.19)$$

On taking the diagonal elements of (A.18) with respect to $|\mathbf{k}\rangle$,

$$\langle \mathbf{k}|T(\epsilon^+)|\mathbf{k}\rangle - \langle \mathbf{k}|T(\epsilon^-)|\mathbf{k}\rangle =$$
$$\sum_{\mathbf{k}'} \langle \mathbf{k}|T(\epsilon^-)|\mathbf{k}'\rangle \langle \mathbf{k}'|(G_0^+(\epsilon) - G_0^-(\epsilon))|\mathbf{k}'\rangle \langle \mathbf{k}'|T(\epsilon^+)|\mathbf{k}\rangle.$$

$$(A.20)$$

Using (1.16) and (1.17) this becomes

$$\langle \mathbf{k}|\operatorname{Im} T(\epsilon^+)|\mathbf{k}\rangle = -\pi \sum_{\mathbf{k}'} |\langle \mathbf{k}'|T(\epsilon^+)|\mathbf{k}\rangle|^2 \delta(\epsilon - \epsilon_{\mathbf{k}'}). \qquad (A.21)$$

This can be related to the right hand side of (A.15) using (A.16) to put it in the form,

$$\sigma(k) = -\frac{2m}{k\hbar^2} \operatorname{Im} T_{\mathbf{kk}}, \qquad (A.22)$$

which, from (A.13), can be expressed in terms of the imaginary part of the forward scattering amplitude $f(\mathbf{k}, \mathbf{k})$. This is the well known optical theorem of scattering theory.

For a spherically symmetric scattering potential $T_{\mathbf{k}'\mathbf{k}}$ depends only on the angle θ between $\mathbf{k}$ and $\mathbf{k}'$, and can be expanded in terms of Legendre polynomials $P_l(\cos\theta)$ corresponding to partial waves of angular momentum l,

$$T_{\mathbf{k}'\mathbf{k}} = \sum_{l=0}^{\infty} (2l + 1)t_l(k)e^{i\eta_l(k)} P_l(\cos\theta), \qquad (A.23)$$

where $t_l(k)$ is real and $\eta_l(k)$ is the phase shift of the lth partial wave. The $(2l + 1)$ factor is due to the independence of the scattering on the z-component of the angular momentum m. We can use this expression in (A.18) to find the amplitude $t_l(k)$ of the lth partial wave. On performing the $\mathbf{k}'$ integration, and using the orthogonality relation,

$$\int P_l(\cos\theta)P_{l'}(\cos\theta)\,d\Omega = \frac{4\pi\delta_{ll'}}{(2l + 1)}, \qquad (A.24)$$

we find

$$t_l(k) = -\frac{2\pi\hbar^2}{mk}\sin\eta_l(k). \qquad (A.25)$$

Substituting (A.22) and (A.24) into (A.21), we find an expression for the total scattering cross-section in terms of phase shifts,

$$\sigma(k) = \frac{4\pi}{k^2} \sum_{l=0}^{\infty} (2l+1) \sin^2 \eta_l(k). \qquad (A.26)$$

The maximum contribution to $\sigma(k)$ in the angular momentum channel l is $4\pi(2l+1)/k^2$, which is known as the unitarity limit (it can be deduced from the unitarity of the S matrix, which is related to the T matrix by $S = 1 - 2\pi i \delta(\epsilon - H_0)T$).

To calculate the asymptotic form for the wave function $\psi_{\mathbf{k}}(\mathbf{r})$ for large r we need the expansion of the plane wave in terms of partial waves,

$$e^{i\mathbf{k}\cdot\mathbf{r}} = \sum_l (2l+1) \, i^l P_l(\cos\theta) j_l(kr), \qquad (A.27)$$

where $j_l(kr)$ is a spherical Bessel function of order l. Substituting this into equation (A.12) and using the asymptotic form,

$$j_l(kr) \to \frac{1}{kr} \sin\left(kr - \frac{l\pi}{2}\right), \qquad (A.28)$$

together with equations (A.13), (A.23) and (A.25), gives

$$\psi_{\mathbf{k}}(\mathbf{r}) \sim \frac{1}{kr} \sum_l (2l+1) i^l e^{i\eta_l(k)} \sin\left(kr + \eta_l(k) - \frac{l\pi}{2}\right) P_l(\cos\theta), \quad (A.29)$$

which when substituted into equation (1.32) gives (1.33).

Appendix B

Linear Response Theory and Conductivity Formulae

When a system is probed by a time dependent field it is often sufficient for comparison with experiment to calculate the response to first order in the applied field. Let us consider a system described by a time independent Hamiltonian H with a time dependent perturbation $V(t) = BF(t)$, where B is an operator which describes to the coupling of the system to the field and $F(t)$ is a time dependent scalar. The change in an observable A due to the field to first order in $V(t)$ can be expressed in the form,

$$\Delta A(t) = \int_{-\infty}^{t} \phi_{AB}(t - t')F(t')\,dt' , \qquad (B.1)$$

where the upper limit can be extended to $+\infty$ if the definition of $\phi_{AB}(t-t')$ is extended so that it is zero for $t' > t$.

We can derive an expression for the linear response function $\phi_{AB}(t-t')$ using equations for the density matrix $\rho(t)$ for the system in the presence of the field following the arguments of Kubo (1957). The expectation value of A at time t is given by $\overline{A(t)} = \text{Tr}\,(\rho(t)A)$, where $\rho(t)$ satisfies the equation of motion,

$$\frac{d\rho(t)}{dt} = \frac{1}{i\hbar}[H + V(t), \rho(t)]_- , \qquad (B.2)$$

subject to $\rho(-\infty) = \rho_{\text{eq}}$ for $V(-\infty) = 0$, where ρ_{eq} is the equilibrium density matrix, $\rho_{\text{eq}} = \exp(-\beta(H - \mu N_0))/Z$ and Z is the partition function, $Z = \text{Tr}\,(\exp(-\beta(H - \mu N_0)))$.

Let $\Delta\rho(t) = \rho(t) - \rho_{\text{eq}}$, then equation (B.2) for $\Delta\rho(t)$ to first order in V becomes

$$\frac{d\rho(t)}{dt} = \frac{1}{i\hbar}[H + V(t), \Delta\rho(t)]_- + \frac{F(t)}{i\hbar}[B, \rho_{\text{eq}}]_- . \qquad (B.3)$$

Transforming to $\Delta\rho_I = \exp(itH/\hbar)\Delta\rho(t)\exp(-itH/\hbar)$ and solving to first order in B,

$$\Delta\rho(t) = \frac{1}{i\hbar}\int e^{-i(t-t')H/\hbar}[B, \rho_{\text{eq}}]_- e^{i(t-t')H/\hbar}F(t')\,dt' , \qquad (B.4)$$

and

$$\overline{\Delta A(t)} = \text{Tr}\,(\Delta\rho(t)A) = \frac{1}{i\hbar}\text{Tr}\int_{-\infty}^{t}[B, \rho_{\text{eq}}]_- A(t - t')F(t')\,dt' , \qquad (B.5)$$

using the cyclic invariance of the trace to express A in the Heisenberg representation (2.25) (and also assuming N_0 commutes with A,B and H).

Hence, on comparing with (B.1), we find

$$\phi_{AB}(t) = -\frac{i}{\hbar}\theta(t)\langle[A(t),B(0)]_-\rangle,\qquad(B.6)$$

where $\langle\ldots\rangle$ indicates the thermal average (2.24). The response function $\phi_{AB}(t)$ is a form of double time thermal Green's function $\langle\langle A(t) : B(0)\rangle\rangle$ defined in (5.28).

The response function can be expressed in an alternative form using the identity,

$$[B, e^{-\beta H}]_- = e^{-\beta H}\int_0^\beta e^{\lambda H}[H,B]_- e^{-\lambda H}\,d\lambda,\qquad(B.7)$$

(which can be verified by differentiation with respect to β), in (B.5) and the Heisenberg equation of motion for $B(t)$,

$$\frac{dB(t)}{dt} = \frac{1}{i\hbar}[B(t),H]_-,\qquad(B.8)$$

to give

$$\phi_{AB}(t) = -\int_0^\beta\left\langle\frac{dB(0)}{dt}A(t+i\lambda\hbar)\right\rangle\,d\lambda.\qquad(B.9)$$

The Fourier transform $\phi(\omega)$, which is the response to a field of frequency ω, can be defined by

$$\phi(\omega) = \int_{-\infty}^\infty \phi_{AB}(t)e^{-(i\omega+s)t}\,dt,\qquad(B.10)$$

for $s\to+0$.

To calculate the conductivity $\sigma(\omega,T)$ we need to calculate the current j_ν ($\nu = x,y,z$), to first order in an applied field $\mathbf{E}(t)$. This corresponds to $V(t) = e\sum_i \mathbf{r}_i\cdot\mathbf{E}(t)$, where $\mathbf{r}_i$ is the position vector of conduction electron i. Using (B.9) and (B.10), and $j_{\nu'} = -e\sum_i dx_{\nu'i}/dt$, gives

$$\sigma_{\nu'\nu}(\omega,T) = \int_0^\infty\int_0^\beta\langle j_{\nu'}(0)j_\nu(t+i\lambda\hbar)\rangle d\lambda\,dt.\qquad(B.11)$$

For an isotropic system $\sigma_{\nu'\nu} = 0$ for $\nu\neq\nu'$ and, averaging over the directions x,y and z, gives

$$\sigma(\omega,T) = \frac{1}{3}\int_0^\infty\int_0^\beta\langle\mathbf{j}(0)\cdot\mathbf{j}(t+i\lambda\hbar)\rangle e^{-(i\omega+s)t}\,d\lambda dt\qquad(B.12)$$

corresponding to (2.23).

To derive the form (2.26) for a system with one-body potential scattering, we evaluate the trace in the thermal average using Slater determinants constructed from the one-body eigenstates $|\alpha\rangle$ of the Hamiltonian

(2.1) corresponding to eigenvalues ϵ_α. On performing the t and λ integration for $\omega = 0$ we obtain

$$\sigma = -\frac{e^2 \hbar}{3m^2} \sum_{\substack{\alpha \alpha' \\ \nu = x,y,z}} \frac{f(\epsilon_\alpha) - f(\epsilon_{\alpha'})}{(\epsilon_\alpha - \epsilon_{\alpha'})} \left[\frac{-i(\epsilon_\alpha - \epsilon_{\alpha'}) + \hbar s}{(\epsilon_\alpha - \epsilon_{\alpha'})^2 + s^2 \hbar^2} \right] \langle \alpha | p_\nu | \alpha' \rangle \langle \alpha' | p_\nu | \alpha \rangle,$$

(B.13)

where $f(\epsilon_\alpha)$ is the Fermi function. The first term within the square brackets vanishes as it is odd under the interchange $\alpha \leftrightarrow \alpha'$.

In the limit $s \to 0$,

$$\frac{f(\epsilon_\alpha) - f(\epsilon_{\alpha'})}{(\epsilon_\alpha - \epsilon_{\alpha'})} \left[\frac{\hbar s}{(\epsilon_\alpha - \epsilon_{\alpha'})^2 + s^2 \hbar^2} \right] \to \pi \frac{\partial f(\epsilon_\alpha)}{\partial \epsilon_\alpha} \delta(\epsilon_\alpha - \epsilon_{\alpha'}) \quad \text{(B.14)}$$

using (1.17). In this limit

$$\sigma = -\frac{\pi \hbar e^2}{3m^2} \sum_{\substack{\alpha \alpha' \\ \nu = x,y,z}} \frac{\partial f(\epsilon_\alpha)}{\partial \epsilon_\alpha} \delta(\epsilon_\alpha - \epsilon_{\alpha'}) \langle \alpha | p_\nu | \alpha' \rangle \langle \alpha' | p_\nu | \alpha \rangle, \quad \text{(B.15)}$$

or

$$\sigma = \frac{\hbar e^2}{3m^2} \sum_{\substack{\alpha \\ \nu = x,y,z}} \frac{\partial f(\epsilon_\alpha)}{\partial \epsilon_\alpha} \delta(\epsilon - \epsilon_\alpha) \text{Im} \langle \alpha | p_\nu G^+(\epsilon) p_\nu | \alpha \rangle, \quad \text{(B.16)}$$

using (1.16) and (1.17). Hence we obtain,

$$\sigma = -\frac{\hbar e^2}{3m^2 \pi} \sum_{\nu = x,y,z} \frac{\partial f(\epsilon)}{\partial \epsilon} \text{Tr} \left\{ \text{Im} \, G^+(\epsilon) p_\nu \text{Im} \, G^+(\epsilon) p_\nu \right\} d\epsilon, \quad \text{(B.17)}$$

and, on expanding $\text{Im} \, G^+(\epsilon)$ in terms of $G^+(\epsilon)$ and $G^-(\epsilon)$, gives (2.26).

Appendix C

The Zero Band Width Anderson Model

Here we look at a very simple model in which the conduction electrons are described by a single state at the Fermi level (zero band width limit). The simplicity of the model is such that we can calculate the many-body states by diagonalization of matrices no greater than 3×3. Nevertheless the results do give some insight as to why a singlet ground state occurs in the strong correlation limit, and why states with small weight are seen in the vicinity of the Fermi level when the impurity d or f electron is added or removed. The model is the non-degenerate Anderson model with a single conduction band state at the Fermi level ϵ_F,

$$
H = \sum_\sigma \epsilon_F c_{1,\sigma}^\dagger c_{1,\sigma} + \sum_\sigma \epsilon_d c_{d,\sigma}^\dagger c_{d,\sigma}
$$
$$
+ V \sum_\sigma (c_{1,\sigma}^\dagger c_{d,\sigma} + c_{d,\sigma}^\dagger c_{1,\sigma}) + U n_{d,\uparrow} n_{d,\downarrow}.
$$

$$(C.1)$$

This corresponds to the first step in the renormalization group calculations for the Anderson model (see sections 4.7 and 4.8).

It is sufficient for our purpose to consider states with only one or two electrons. The one electron eigenstates

$$
c_{\pm,\sigma}^\dagger |0\rangle = \alpha_\pm c_{d,\sigma}^\dagger |0\rangle + \beta_\pm c_{1,\sigma}^\dagger |0\rangle, \qquad (C.2)
$$

correspond to energies

$$
E_\pm = \frac{\epsilon_d + \epsilon_F \pm \left[(\epsilon_d - \epsilon_F)^2 + 4V^2 \right]^{1/2}}{2}. \qquad (C.3)
$$

For an impurity state well below the 'Fermi level' we have $\epsilon_F - \epsilon_d \gg V^2$. In this limit (C.3) simplifies to

$$
E_+ = \epsilon_d - \frac{V^2}{\epsilon_F - \epsilon_d}, \qquad E_- = \epsilon_F + \frac{V^2}{\epsilon_F - \epsilon_d}, \qquad (C.4)
$$

to leading order in V^2, with

$$
c_{+,\sigma}^\dagger = \alpha \left(c_{d,\sigma}^\dagger - \frac{V}{(\epsilon_F - \epsilon_d)} c_{1,\sigma}^\dagger \right), \qquad (C.5)
$$

$$
c_{-,\sigma}^\dagger = \alpha \left(c_{1,\sigma}^\dagger + \frac{V}{(\epsilon_F - \epsilon_d)} c_{d,\sigma}^\dagger \right), \qquad (C.6)
$$

with

$$
\alpha = 1 - \frac{V^2}{2(\epsilon_F - \epsilon_d)^2}. \qquad (C.7)
$$

The two electron states can be classified as singlets or triplets. In the triplet case the spatial part of the wavefunction must be antisymmetric and the U interaction plays no role, so these states can be built up as in the non-interacting case from products of the one electron states. The triplet states are $c_{+,\uparrow}^\dagger c_{-,\uparrow}^\dagger |0\rangle, (c_{+,\uparrow}^\dagger c_{-,\downarrow}^\dagger + c_{+,\downarrow}^\dagger c_{-,\uparrow}^\dagger)|0\rangle/\sqrt{2}$, and $c_{+,\downarrow}^\dagger c_{-,\downarrow}^\dagger |0\rangle$, corresponding to a total z-component of spin $M_S = 1, 0$ and -1. The total energy of the triplet states is $E_+ + E_-$ equal to $\epsilon_d + \epsilon_F$. There are three possible singlet states built up from the states, $c_{1,\uparrow}^\dagger c_{1,\downarrow}^\dagger |0\rangle$, $(c_{1,\uparrow}^\dagger c_{d,\downarrow}^\dagger - c_{1,\downarrow}^\dagger c_{d,\uparrow}^\dagger)|0\rangle/\sqrt{2}$, and $c_{d,\uparrow}^\dagger c_{d,\downarrow}^\dagger |0\rangle$, with energies $E_{1,2,3}$ corresponding to the solutions of the equation,

$$\begin{vmatrix} E - 2\epsilon_F & -\sqrt{2}V & 0 \\ -\sqrt{2}V & E - \epsilon_d - \epsilon_F & -\sqrt{2}V \\ 0 & -\sqrt{2}V & E - 2\epsilon_d - U \end{vmatrix}. \qquad (C.8)$$

In the limit $U \to \infty$, one of the singlet states is pushed to large energies (the state in which the impurity level is predominantly doubly occupied). The two remaining states in this limit have energies,

$$E_{d1,2} = \frac{\epsilon_d + 3\epsilon_F \pm \left[(\epsilon_d - \epsilon_F)^2 + 8V^2\right]^{1/2}}{2}. \qquad (C.9)$$

For $\epsilon_F - \epsilon_d \gg V^2$, the leading order terms are

$$E_1 = \epsilon_d + \epsilon_F - \frac{2V^2}{\epsilon_F - \epsilon_d}, \quad E_2 = 2\epsilon_F + \frac{2V^2}{\epsilon_F - \epsilon_d}, \qquad (C.10)$$

and the corresponding states are

$$|\Psi_{1s}\rangle = \alpha' \left\{ \frac{(c_{d,\uparrow}^\dagger c_{1,\downarrow}^\dagger - c_{d,\downarrow}^\dagger c_{1,\uparrow}^\dagger)}{\sqrt{2}} - \frac{\sqrt{2}V}{(\epsilon_F - \epsilon_d)} c_{1,\uparrow}^\dagger c_{1,\downarrow}^\dagger \right\} |0\rangle, \qquad (C.11)$$

$$|\Psi_{2s}\rangle = \alpha' \left\{ \frac{V}{(\epsilon_F - \epsilon_d)} (c_{d,\uparrow}^\dagger c_{1,\downarrow}^\dagger - c_{d,\downarrow}^\dagger c_{d\uparrow}^\dagger) + c_{1,\uparrow}^\dagger c_{1,\downarrow}^\dagger \right\} |0\rangle, \qquad (C.12)$$

with α' given by

$$\alpha' = 1 - \frac{V^2}{(\epsilon_F - \epsilon_d)^2}. \qquad (C.13)$$

In this large U limit we find a singlet ground state with an energy reduction of $2V^2/(\epsilon_F - \epsilon_d)$ due to the hybridization. This is equivalent to the energy gain in the s-d model of $2J$ with J corresponding to the Schrieffer–Wolff transformation (1.73) in the limit $U \to \infty$.

We can see why spectral weight is found in the d electron Green's function $G_{d\sigma,d\sigma}(\omega)$ near the Fermi level of the order of $1 - n_d$, where n_d is the occupation of the impurity d level. We could calculate this Green's function exactly in this case but it is not necessary just to demonstrate this. Consider an excitation in which a d,↑ electron is removed from the ground state $|\Psi_{1s}\rangle$ leaving a final state $c_{+,\downarrow}^{\dagger}|0\rangle$, corresponding to an excitation energy $E_{d,int} - E_{d,fin} = \epsilon_F - V^2/(\epsilon_F - \epsilon_d)$ just below the Fermi level. For $\epsilon_F - \epsilon_d \gg V^2$, this is not the most likely final state; this must be the state $c_{-,\downarrow}^{\dagger}|0\rangle$ corresponding to an excitation energy $E_{d,int} - E_{d,fin} = \epsilon_d - 3V^2/(\epsilon_F - \epsilon_d)$ well below the Fermi level. Nevertheless there is a probability of a transition to a final state $c_{+,\downarrow}^{\dagger}|0\rangle$, in which the d state occupation is close to unity. The probability for this final state will depend on the matrix element,

$$(\langle 0|c_{+,\downarrow})c_{d,\uparrow}(|\Psi_{1s}\rangle) = -\frac{\alpha\alpha'V}{\sqrt{2}(\epsilon_F - \epsilon_d)}. \tag{C.14}$$

Hence the probability for this transition to leading order in V^2 is proportional to $V^2/2(\epsilon_F - \epsilon_d)^2$. The expectation value of n_d in the ground state from (C.11) is given by

$$n_d = 1 - \frac{2V^2}{(\epsilon_F - \epsilon_d)^2}, \tag{C.15}$$

and so the weight of this transition is equal to $(1 - n_d)/4$. This weight would be seen in photoemission experiments (see sections 8.4 and 9.2). Most of the weight will be in the transition near the original d level at $\epsilon_d - 2V^2/(\epsilon_F - \epsilon_d)$. Using equation (C.14) with $c_{-,\downarrow}$ instead of $c_{+,\downarrow}$ this weight can be shown to be equal to $(3n_d - 1)/4$. Similarly there is an excitation at $2V^2/(\epsilon_F - \epsilon_d)$ above the Fermi level on adding a d, $\uparrow$ to the ground state $|\Psi_{1s}\rangle$ to give the state $c_{d,\uparrow}^{\dagger}c_{1,\uparrow}^{\dagger}c_{1,\downarrow}^{\dagger}|0\rangle$, which is an eigenstate of energy $2\epsilon_F + \epsilon_d$. This transition has weight which can be calculated from the matrix element,

$$(\langle 0|c_{1,\downarrow}c_{1,\uparrow}c_{d,\uparrow})c_{d,\uparrow}^{\dagger}(|\Psi_{1s}\rangle) = -\frac{\sqrt{2}\alpha'V}{(\epsilon_F - \epsilon_d)}, \tag{C.16}$$

equal to $1 - n_d$. This peak would be seen in the BIS spectrum (see sections 8.4 and 9.2).

These excitations and their weights in the spectral density of the one electron Green's function for the d electron are shown in figure C.1(i). For the degenerate models discussed in chapters 7 and 8, the relative weights of the excitations just above and just below the Fermi level are in the ratio, N to 1 for large N, where N is the degeneracy factor.

If we had worked with the symmetric model $\epsilon_F - \epsilon_d = U/2$, then $n_d = 1$ and for large U the excitations near the Fermi level are at $\epsilon_F \pm 3J/2$, with $J = 4V^2/U$. These excitations have equal weight, $18V^2/U^2$ for large U. As U is kept finite in this case the excitation to the state which is predominantly occupied by two d electrons is seen at $(U+5J)/2$ above the Fermi level (see figure C.1(ii)).

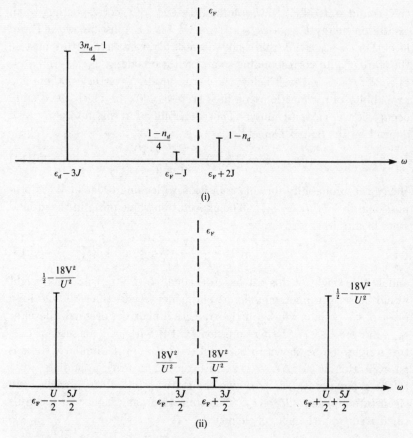

Figure C.1 A schematic representation of the excitations from the 2 electron ground state ($\epsilon_d \ll \epsilon_F$) on adding or removing a d electron with the weights in the spectral density of the one electron d Green's function indicated. Case (i) is for $U \to \infty$ with $J = V^2/(\epsilon_F - \epsilon_d)$. The total weight of the peaks is $1 - n_d/2$. Case (ii) is for the symmetric model with $\epsilon_F - \epsilon_d = U/2$ and $J = 4V^2/U$ with $V \ll U$. The total weight of the peaks in this case is equal to 1.

In the renormalization group calculations further sites are added to the chain to build up the band of conduction states. The binding of the singlet is increased from $2J$ to $k_B T_K$ as the band of conduction states is built up. The spectral weight at the Fermi level develops into the Kondo resonance, which has a small overall weight for $n_d \sim 1$.

Appendix D

Scaling Equations for the Coqblin–Schrieffer Model

In chapter 3, section 3.3, we derived scaling equations for the s-d model ($S = \frac{1}{2}$) perturbatively to second order in the coupling J using Anderson's 'poor man's approach'. Here we extend those calculations to third order in J for the more general N-fold degenerate model, the Coqblin–Schrieffer model (1.90), following the approach of Read & Hewson (1984). The earlier results correspond to the case $N = 2$. We follow the same approach as used in section 3.3 and reduce the band width for the conduction electrons from D to $D - |\delta D|$ but with a more general form for H_{12},

$$H_{12} = J \sum_{\substack{q,k' \\ m,m'}} \left(X_{m,m'} c_{q,m'}^{\dagger} c_{k',m} - \frac{1}{N} X_{m,m} c_{q,m'}^{\dagger} c_{k',m'} \right), \qquad (D.1)$$

where the energy of the conduction state $\mathbf{q}$, ϵ_q, lies within one of the band edges to be eliminated.

Second order diagrams of the type shown in figure 3.5(i), with a particle in an intermediate state at the band edge, are shown in figure D.1(i) and (ii).

The open circle represents scattering from the first term in (D.1), and the full circle from the second diagonal term. The result from figure D.1(i), corresponding to figure 3.5(i), is

$$J^2 \sum_{\substack{k,q \\ m,m'}} X_{m,m'} c_{k',m'}^{\dagger} c_{q,m} (E - H_{22})^{-1} \sum_{\substack{k,q' \\ m'',m'''}} X_{m'',m'''} c_{q',m'''}^{\dagger} c_{k,m''}, \qquad (D.2)$$

which to second order in J becomes

$$J^2 \rho_0 |\delta D| \sum_{\substack{k,k' \\ m,m'}} X_{m,m} c_{k',m'}^{\dagger} c_{k,m'} (E - D + \epsilon_k - H_0)^{-1}, \qquad (D.3)$$

and for $E, \epsilon_k \ll D$ is approximately

$$-J^2 \rho_0 \frac{|\delta D|}{D} \sum_{\substack{k,k' \\ m,m'}} X_{m,m} c_{k',m'}^{\dagger} c_{k,m'}. \qquad (D.4)$$

Similarly diagram D.1(ii) leads to an off diagonal interaction

$$J^2 \rho_0 \frac{|\delta D|}{ND} \sum_{\substack{k,k' \\ m,m'}} X_{m,m'} c_{k',m'}^{\dagger} c_{k,m}, \qquad (D.5)$$

within the same approximation. There are two more diagrams of the

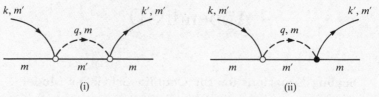

Figure D.1 Second order scattering diagrams which have a particle in an intermediate state at a band edge (represented by a dashed line).

type in figure 3.5(i), with a particle in the intermediate state, one as in D.1(i) with the open circle vertices replaced by the full circles and the other corresponding to D.1(ii) but with the two types of vertices interchanged. There are also four intermediate hole diagrams corresponding to figure 3.5(ii). The contributions from the two sets of diagrams which involve the full circle vertex cancel, leaving contributions from D.1(i), given by (D.4), and the counterpart hole diagram. The total contribution is

$$J^2 N \rho_0 \frac{|\delta D|}{D} \sum_{\substack{k,k' \\ m,m'}} \left(X_{m,m'} c^\dagger_{k',m'} c_{k,m} - \frac{1}{N} X_{m,m} c^\dagger_{k',m'} c_{k,m'} \right), \quad (D.6)$$

which has the same form as the original interaction and corresponds to increasing the coupling from J to $J + \delta J$, where

$$\delta J = N \rho_0 J^2 \frac{|\delta D|}{D}. \quad (D.7)$$

To calculate the scaling equations to higher order in perturbation theory it is necessary to take into account the shift in ground state energy to linear order in E for the terms we have just considered. For a calculation to third order we shall need to calculate this shift to second order for reasons that will become clear later. We need the ground state expectation value of terms such as (D.3), which we have to treat rather more carefully than earlier because we now have to integrate over ϵ_k and we can no longer restrict our consideration to the case $|\epsilon_k| \ll D$. The expectation value of (D.3) is

$$J^2 \rho_0 \frac{|\delta D|}{D} \sum_m X_{m,m} \int_{-D+|\delta D|}^0 \frac{\rho_0}{E - D + \epsilon} \, d\epsilon, \quad (D.8)$$

which we evaluate to order $|\delta D|$, assuming a constant density of states and then expanding in powers of E/D to first order. If all the contributions are treated in this way the total result to order E is

$$-\left(N - \frac{1}{N} \right) J^2 \rho_0^2 \frac{|\delta D|}{D} E \sum_m X_{m,m}. \quad (D.9)$$

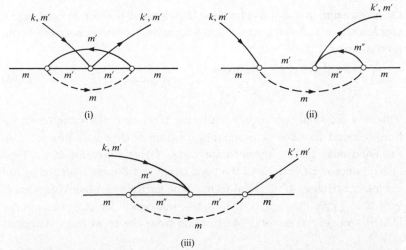

Figure D.2 Third order scattering diagrams with a particle in an intermediate state at a band edge (dashed line) which lead to a modified single particle scattering vertex.

This factor as we shall see later is essentially a wavefunction renormalization term due to the elimination of states.

Explicit terms to order J^3 arise from the expansion of factors such as $(E - H_{22})^{-1}$ in (D.2) in powers of J. If we expand this factor to order J then we obtain a third order contribution,

$$J^2 \sum_{\substack{q,k' \\ m,m'}} X_{m,m'} c^\dagger_{k',m'} c_{q,m} (E - H_0)^{-1} H_1 (E - H_0)^{-1}$$

$$\times \sum_{\substack{q',k'' \\ m,''m'''}} X_{m'',m'''} c^\dagger_{q',m'''} c_{k,m''},$$

$$\text{(D.10)}$$

where H_0 and H_I are the non-interacting and the interacting part of H_{22} respectively. There are six ways of contracting this term to give a modified single particle scattering vertex ; three ways are shown in figure D.2(i), (ii) and (iii), and the remaining three correspond to replacing the central circular vertex by a full circle vertex.

We retain only those terms in the expansion of $(E - H_0)^{-1}$ in which a scattered particle is in the range $-D + |\delta D| < \epsilon < D - |\delta D|$, otherwise it would give a contribution of order $1/D^2$ which can be neglected. For example the diagram in figure D.2(i) leads to a contribution,

$$J^3 \sum_{\substack{k,k' \\ m,m'}} X_{m,m} c^\dagger_{k',m'} c_{k,m'} \int_{D-|\delta D|}^0 \frac{\rho_0^2 |\delta D|}{(E - D + \epsilon - \epsilon_k)} \frac{1}{(E - D + \epsilon - \epsilon'_k)} \, d\epsilon.$$

$$\text{(D.11)}$$

Appendix D

On integrating over ϵ and expanding in powers of $1/D$ for $E, \epsilon_k, \epsilon'_k \ll D$, this leads to an effective interaction for particles or holes near the Fermi level of order $1/D$,

$$J^3 \rho_0^2 \frac{|\delta D|}{2D} \sum_{\substack{k,k' \\ m,m}} X_{m,m} c^\dagger_{k',m'} c_{k,m'}. \qquad (D.12)$$

This is a diagonal (potential) scattering term, and the same contribution is found for the corresponding diagram with a hole promoted to the band edge in the intermediate state. Other diagrams of the type D.2(i), where one or more of the open circle vertices are replaced by full circles, contribute off diagonal terms. The particle and hole diagrams of type D.2(i) give identical contributions, but cancel for diagrams of type D.2(ii) and (iii). The total contribution from the third order diagrams is

$$-J^3 \rho_0^2 \frac{|\delta D|}{ND} \sum_{\substack{k,k' \\ m,m'}} \left(X_{m,m'} c^\dagger_{k',m'} c_{k,m} - \frac{1}{N} X_{m,m} c^\dagger_{k',m'} c_{k,m'} \right). \qquad (D.13)$$

The effective Hamiltonian we have obtained, $\tilde{H}_{\text{eff}}(E)$ is E dependent if we include the E dependence to linear order arising from the second order terms, and has the form $\tilde{H}_{\text{eff}}(0) - SE$, where S is given by (D.9) and E is measured relative to the ground state energy of the conduction electrons. Substituting this into the secular equation, $|\tilde{H}_{\text{eff}}(E) - IE| = 0$, leads to

$$|\tilde{H}_{\text{eff}}(0) - (I + S)E| = 0. \qquad (D.14)$$

This corresponds to an E independent effective Hamiltonian H_{eff} given by $(I + S)^{-1/2} \tilde{H}_{\text{eff}}(0)(I + S)^{-1/2}$, where to lowest order we can approximate $(I + S)^{-1/2}$ by $(I - S/2)$. When the second and third order terms are all taken into account the new effective coupling due to the elimination of scattering into the band edges is

$$J \left\{ 1 - (N - \frac{1}{N})J^2 \rho_0^2 \frac{|\delta D|}{D} \right\} \left\{ (1 + NJ\rho_0 \frac{|\delta D|}{D} - J^2 \rho_0^2 \frac{|\delta D|}{ND} \right\}, \qquad (D.15)$$

where the first term in brackets arises from the wavefunction renormalization factor S, and the second and third terms from (D.6) and (D.13). Hence, to order J^3 the increase in coupling δJ is

$$\delta J = NJ^2 \rho_0 \frac{|\delta D|}{D} - NJ^3 \rho_0^2 \frac{|\delta D|}{D}, \qquad (D.16)$$

and gives the scaling equation of the form (3.44),

$$\frac{d(N\rho_0 J)}{d \ln D} = -(N\rho_0 J)^2 + \frac{1}{N}(N\rho_0 J)^3 + O(J^4). \qquad (D.17)$$

On inverting and integrating as in section 3.3, we obtain T_K as a scaling invariant given by

$$k_B T_K \sim D(N\rho_0 J)^{1/N} e^{-1/N\rho_0 J}. \qquad \text{(D.18)}$$

Appendix E

Further Fermi Liquid Relations

There are two further exact relations that have been proved within Fermi liquid theory. The first is known as the *Korringa relation*; it relates $\mathrm{Im}(\chi_{\mathrm{imp}}(\omega)/\omega)$ to the impurity susceptibility χ_{imp} and was proved by Shiba (1975) for the Anderson model without spin–orbit coupling, and generalized by Yoshimori & Zawadowski (1982). The second relation is between the thermopower $S_{\mathrm{imp}}(T)$ and the specific heat coefficient γ_{imp} and was proved by Houghton, Read & Won (1987) and Kawakami, Usuki & Okiji (1987) for the N-fold degenerate Anderson model (7.70).

We start first with the Korringa relation which for the non-degenerate Anderson model is

$$\lim_{\omega \to 0} \left\{ \frac{\mathrm{Im}\, \chi_{\mathrm{imp}}(\omega + is)}{2\pi\omega} \right\} = \frac{\chi_{\mathrm{imp}}^2}{(g\mu_{\mathrm{B}})^2}, \qquad (\mathrm{E.1})$$

for $T = 0$, and $s \to 0$. The dynamic susceptibility for real frequencies, $\chi_{\mathrm{imp}}(\omega + is)$ can be deduced from $\chi_{\mathrm{imp}}(i\omega_n)$, the Fourier coefficient of the imaginary time response which is defined by

$$\chi_{\mathrm{imp}}(i\omega_n) = -\frac{(g\mu_{\mathrm{B}})^2}{4} \int_0^\beta e^{i\omega_n\tau} \sum_{\sigma,\sigma'} \sigma\sigma' \langle T_\tau n_{d,\sigma}(\tau) n_{d,\sigma'}(0) \rangle \, d\tau, \quad (\mathrm{E.2})$$

where $\omega_n = 2\pi n/\beta$ and n is an integer. This quantity can be calculated using the perturbation theory of section 5.3 for finite U, and $\chi_{\mathrm{imp}}(\omega + is)$ deduced by the continuation $i\omega_n \to \omega + is$. We will prove (E.1) using the arguments of sections 5.3 and 5.4 for the symmetric Anderson model first of all, and then indicate how it can be generalized later. For $T \to 0$, $\omega_n \to \omega$, where ω is a continuous variable, we need to calculate the term in $\chi_{\mathrm{imp}}(i\omega)$ which is linear in $i\omega$. The complete set of diagrams for $\langle T_\tau n_{d,\sigma}(\tau) n_{d,\sigma'}(0) \rangle$ is represented in figure E.1(i).

To see how to carry out the calculation let us consider the lowest order diagram, as shown in figure E.1(ii), and differentiate with respect to $i\omega$ to pick out the $i\omega$ coefficient,

$$-\frac{\delta_{\sigma,\sigma'}}{2\pi} \int_{-\infty}^{\infty} \frac{dG_\sigma^0(i\omega + i\omega_1)}{di\omega} G_\sigma^0(i\omega_1) \, d\omega_1. \qquad (\mathrm{E.3})$$

This integral was evaluated earlier in equations (5.91) and (5.92) for the symmetric Anderson model and gives

$$\frac{i\delta_{\sigma,\sigma'}}{\pi\Delta^2} \mathrm{sgn}\,\omega. \qquad (\mathrm{E.4})$$

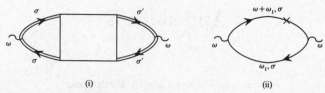

Figure E.1 (i) Diagrams of the type that contribute to $\chi_{\text{imp}}(\omega)$. The double lines indicate a full Green's function. (ii) The lowest order diagram which contributes to $\text{Im}(\chi_{\text{imp}}(\omega)/\omega)$ as $\omega \to 0$. The cross indicates the replacement of the propagator $G_\sigma^0(\omega)$ by $2\delta(\omega)/\Delta$.

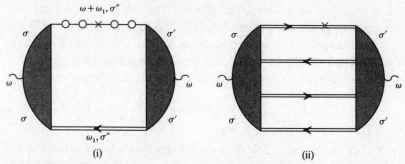

Figure E.2 (i) Diagrams that give a finite contribution to $\text{Im}(\chi_{\text{imp}}(\omega)/\omega)$ as $\omega \to 0$, and (ii) a diagram of the type which has more than one intermediate particle–hole pair and gives zero contribution.

In generalizing this procedure to higher order diagrams we find that only diagrams with two intermediate lines contribute, as these carry the same frequency in the limit $\omega \to 0$. This is similar to the evaluation of $\partial^2 \Sigma_\sigma(i\omega)/\partial(i\omega)^2$ in section 5.4, except that in this case we have one less derivative and one less line is involved.

Consequently the finite contributions to $\partial \chi_{\text{imp}}(i\omega)/\partial(i\omega)$ come from diagram (i) in figure E.2, while diagrams of type (ii) in figure E.2 give zero. The contribution from figure E.2(i) is

$$\frac{1}{\pi \Delta} \int \sum_{\sigma''} \delta(\omega + \omega_1) G_{\sigma''}(i\omega_1) F(\omega, \omega_1) \, d\omega_1, \qquad (\text{E.5})$$

where

$$F(\omega, \omega_1) = \frac{\Lambda_{\sigma\sigma''}(i\omega, i\omega + i\omega_1, i\omega_1)\Lambda_{\sigma'\sigma''}(-i\omega, -i\omega - i\omega_1, -i\omega_1)}{(1 - G_{\sigma''}^{(0)}(i\omega + i\omega_1)\Sigma_{\sigma''}(i\omega + i\omega_1))^2}. \qquad (\text{E.7})$$

The Λs are three point vertex functions and are related to the four point antisymmetrized vertex function, $\Gamma_{\sigma_1\sigma_2\sigma_3\sigma_4}(i\omega_1, i\omega_2, i\omega_3, i\omega_4)$, defined in

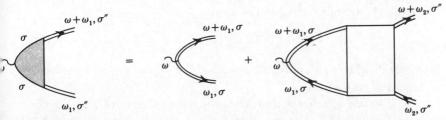

Figure E.3 The diagrammatic representation of equation (E.6) for the three point vertex function.

section 5.3 by

$$\Lambda_{\sigma\sigma''}(i\omega, i\omega + i\omega_1, i\omega_1) = \delta_{\sigma,\sigma''} +$$
$$\frac{1}{2\pi} \int G_\sigma(i\omega + i\omega_2) G_\sigma(i\omega_2) \Gamma_{\sigma\sigma''\sigma''\sigma}(i\omega + i\omega_2, i\omega + i\omega_1, i\omega_1, i\omega_2) \, d\omega_2,$$

(E.6)

as indicated diagrammatically in figure E.3.

The total contribution to $\partial\chi_{\text{imp}}(i\omega)/\partial(i\omega)$ as $\omega \to 0$ is

$$\frac{(g\mu_{\text{B}})^2}{4\pi\Delta^2} i \operatorname{sgn}\omega \sum_{\sigma''} \Big(\sum_\sigma \sigma\Lambda_{\sigma\sigma''}(0,0,0)\Big)^2,$$

(E.8)

which can be re-expressed as

$$\frac{(g\mu_{\text{B}})^2}{2\pi\Delta^2} i \operatorname{sgn}\omega \left(1 + \frac{\partial\Sigma_\uparrow(0)}{\partial\mu_\uparrow} - \frac{\partial\Sigma_\uparrow(0)}{\partial\mu_\downarrow}\right)^2,$$

(E.9)

on using (5.69) for each spin component. Using (5.79) and (5.57) with $\tilde{\epsilon}_d = 0$ we can express the right-hand side in terms of χ_{imp} and this then gives

$$\frac{(g\mu_{\text{B}})^2}{2\pi} i \operatorname{sgn}\omega \, \chi_{\text{imp}}^2.$$

(E.10)

The relation (E.1) follows on continuing to real frequencies $i\omega \to \omega + is$.

Though we have taken the case of a symmetric model, which meant that we could use the results of section 5.4, the result is true more generally. The proof in the general case follows along similar lines except that we cannot assume particle–hole symmetry. The result (E.4) becomes $i\delta_{\sigma,\sigma'}\pi\rho_d(0)\operatorname{sgn}\omega$. There are also other contributions to (E.3) arising from the first term in (5.70), which vanish in the symmetric case. These terms, however, do not contribute to the imaginary part of $\chi_{\text{imp}}(\omega)$ and hence can be ignored in deriving (E.1).

The Korringa relation can be generalized to the N-fold degenerate model (7.70), where it takes the form,

$$\lim_{\omega \to 0} \left\{ \frac{\operatorname{Im} \chi_{\mathrm{imp}}(\omega + is)}{3\pi\omega} \right\} = \frac{\chi_{\mathrm{imp}}^2}{(g\mu_{\mathrm{B}})^2 j(j+1)N}, \tag{E.11}$$

where $N = 2j + 1$. This can be proved by a similar argument using the results of section 7.4.

The second relation is that for the linear coefficient of T in the thermopower $S_{\mathrm{imp}}(T)$. For the N-fold degenerate Anderson model (7.70) the relation is

$$\lim_{T \to 0} \left\{ \frac{S_{\mathrm{imp}}(T)}{\gamma_{\mathrm{imp}} T} \right\} = \frac{2\pi}{eN} \cot \left(\frac{\pi n_f}{N} \right). \tag{E.12}$$

The thermopower is given by the ratio of the moments $K^{(1)}$ and $K^{(0)}$ of the transport lifetime, $\tau(\epsilon, T)$,

$$S_{\mathrm{imp}}(T) = -\frac{1}{eT} \frac{K^{(1)}}{K^{(0)}}, \tag{E.13}$$

where

$$K^{(n)} = \int \left(-\frac{\partial f}{\partial \epsilon} \right) \tau_{\mathrm{tr}}(\epsilon, T) \epsilon^{(n)} \, d\epsilon. \tag{E.14}$$

For the scattering by the impurity in a single angular momentum channel $1/\tau_{\mathrm{tr}}(\epsilon, T)$ is proportional to the T matrix of the conduction electron scattering, and hence via (5.83) to the impurity density of states $\rho_{\mathrm{imp}}(\epsilon, T)$. The constant of proportionality can be assumed to be independent of ϵ and hence cancels out in the evaluation of (E.13). Using the Sommerfeld expansion (1.6) for $K^{(1)}$ we find

$$K^{(1)} = \int \left(-\frac{\partial f}{\partial \epsilon} \right) \frac{\epsilon}{\rho_{\mathrm{imp}}(\epsilon)} \, d\epsilon. \tag{E.15}$$

$$= -\frac{\pi^2}{3} \frac{(k_{\mathrm{B}} T)^2}{(\rho_{\mathrm{imp}}(0))^2} \left(\frac{\partial \rho_{\mathrm{imp}}}{\partial \epsilon} \right)_{\epsilon=0} + \mathrm{O}(T^4), \tag{E.16}$$

so $S_{\mathrm{imp}}(T)$ is proportional to the derivative of the impurity density of states at the Fermi level to lowest order in T.

Using (5.49) for $\rho_{\mathrm{imp}}(\epsilon)$ with $\Sigma^{\mathrm{I}}(0) \propto \epsilon^2$ as $\epsilon \to 0$, this becomes

$$-\frac{\pi^2}{3} \frac{(k_{\mathrm{B}} T)^2}{(\rho_{\mathrm{imp}}(0))^2} \frac{2}{\pi} \frac{\Delta(\epsilon_f + \Sigma(0))}{[(\epsilon_f + \Sigma(0))^2 + \Delta^2]^2} \left(1 - \frac{\partial \Sigma}{\partial \epsilon} \right)_{\epsilon=0} \tag{E.17}$$

to order T^2.

For $K^{(0)}$ we have

$$K^{(0)} = \frac{1}{\rho_{\mathrm{imp}}(0)} + \mathrm{O}(T^2). \tag{E.18}$$

Using the Friedel sum rule, (5.47) or (7.79), which applies at $T = 0$, and the expression for the impurity density of states at the Fermi level,

(5.49) with $\Sigma^{\mathrm{I}}(0) = 0$, this gives for $S_{\mathrm{imp}}(T)$,

$$S_{\mathrm{imp}}(T) = \frac{2\pi^3 k_{\mathrm{B}}^2 T}{3eN} \cot\left(\frac{\pi n_f}{N}\right) N\rho_{\mathrm{imp}}(0) \left(1 - \frac{\partial\Sigma}{\partial\epsilon}\right)_{\epsilon=0} + \mathrm{O}(T^3).$$

(E.19)

The derivative of $\partial\Sigma/\partial\epsilon$ evaluated at $\epsilon = 0$ can be expressed in terms of γ_{imp} using (5.56) for $N = 2$, (7.74) for general N, and the result (E.12) follows.

Appendix F

The Algebraic Bethe Ansatz

There is an elegant algebraic method for solving equations of the type, (6.17) and (6.100), developed in the late 70s, known as the algebraic Bethe ansatz. The method is based on the transfer matrix approach to certain two dimensional classical statistical mechanical problems (Baxter, 1982), and the quantum inverse scattering approach to one dimensional quantum mechanical problems (Sklyanin & Faddeev, 1978; Takhatajan, 1985). Here we give the basic steps that lead to the diagonalization of problems (6.17) and (6.100). For a fuller explanation we refer the reader to the articles quoted (see also the review article of Thacker, 1981), and the papers cited therein.

We define operators,

$$\mathbf{R}_{i,j}(u) = a(u)\mathbf{I} + b(u)\mathbf{P}_{i,j}, \tag{F.1}$$

such that $a(u) + b(u) = 1$, where $\mathbf{P}_{i,j}$ permutes the spin states $S = \frac{1}{2}$ for at sites $i, j = 1, 2, \ldots N$, in the product spin space V. The $\mathbf{R}$ and $\mathbf{S}$ operators defined in (6.10) and (6.95) are special cases of (F.1) with $a(u) = u/(u + i/2)$, $\mathbf{R}_j = e^{-iJ/2}\mathbf{R}_{0,j}(\alpha_0 - \alpha_j)$ with $\alpha_0 = -1/2\tan J$, $\alpha_j = 0$, and $\mathbf{S}_{i,j} = \mathbf{R}_{i,j}(\alpha_i - \alpha_j)$ with $\alpha_i = \alpha(k_i)$, where $\alpha(k)$ is given in equation (6.92). An additional site is introduced with an auxiliary spin space C. A matrix $\mathbf{T}_c(u)$, dependent on the set of parameters $\{\alpha_i\}$, is introduced, as a generalization of the matrix $\mathbf{T}_1$ of section 6.2, and is defined by

$$\mathbf{T}_c(u) = \mathbf{R}_{c,N}(u - \alpha_N)\mathbf{R}_{c,N-1}(u - \alpha_{N-1})\ldots\mathbf{R}_{c,1}(u - \alpha_1), \tag{F.2}$$

which operates on the product spin space $C \otimes V$. Using the states of C as a basis $\mathbf{T}_c(u)$ can be expressed in the form,

$$\mathbf{T}_c(u) = \begin{bmatrix} \mathbf{A}(u) & \mathbf{B}(u) \\ \mathbf{C}(u) & \mathbf{D}(u) \end{bmatrix}, \tag{F.3}$$

where the operators, $\mathbf{A}(u)$, $\mathbf{B}(u)$, $\mathbf{C}(u)$ and $\mathbf{D}(u)$ operate on the spin space V and depend on the parameter set $\{\alpha(k_i)\}$. The matrix (F.3) is known as the monodromy matrix and is the quantum equivalent of the scattering coefficients of the classical inverse scattering problem.

The eigenvalue equations, (6.17) and (6.100), are essentially special cases of the equation,

$$\mathbf{T}_j\Phi(\mathbf{I}) = e^{ik_j L}\Phi(\mathbf{I}), \tag{F.4}$$

for $j = 1, 2, \ldots N$, where $\mathbf{T}_j = \mathbf{A}(\alpha_j) + \mathbf{D}(\alpha_j) = \mathrm{Tr}(\mathbf{T}_c(\alpha_j))$, where the trace is taken with respect to the spin space C. The solution depends on the two particle factorizability condition or Yang–Baxter relation,

$$\mathbf{R}_{i,j}(\alpha_i - \alpha_j)\mathbf{R}_{i,m}(\alpha_i - \alpha_m)\mathbf{R}_{j,m}(\alpha_j - \alpha_m) =$$
$$\mathbf{R}_{j,m}(\alpha_j - \alpha_m)\mathbf{R}_{i,m}(\alpha_i - \alpha_m)\mathbf{R}_{i,j}(\alpha_i - \alpha_j),$$
$$(\mathrm{F.5})$$

corresponding to (6.98). By substituting (F.1) into (F.5) it can be shown that this relation is satisfied if

$$\frac{a_{i,j}}{b_{i,j}} = \frac{a_{i,m}}{b_{i,m}} + \frac{a_{m,j}}{b_{m,j}}, \qquad (\mathrm{F.6})$$

where $a_{i,j} = a(\alpha_i - \alpha_j)$ and $b_{i,j} = 1 - a_{i,j}$. If this equation is satisfied then it can be easily demonstrated that the operators $\mathbf{T}_j$ commute with one another so that equations (F.4) for $j = 1, 2, \ldots N$ can be simultaneously diagonalized. It is satisfied for $a(u) = u/(u+i/2)$, $b(u) = 1 - a(u)$, as $a(u)/b(u) = 2iu$.

The Yang–Baxter relation can be put in the form,

$$\mathbf{R}_{c,d}(u - v)\mathbf{T}_c(u)\mathbf{T}_d(v) = \mathbf{T}_d(v)\mathbf{T}_c(u)\mathbf{R}_{c,d}(u - v), \qquad (\mathrm{F.7})$$

by multiplying (F.5) with the appropriate $\mathbf{R}$ operators (F.1) and introducing a further auxiliary spin space D. If (F.7) is expressed as a matrix equation using the product space $C \otimes D$ as a basis, with representations for $\mathbf{T}_c(u)$ and $\mathbf{T}_d(u)$ of the form (F.3), and the corresponding matrix elements equated, commutation relations are generated for the operators $\mathbf{A}(u)$, $\mathbf{B}(u)$, $\mathbf{C}(u)$ and $\mathbf{D}(u)$ (see the reviews quoted and the references therein for further details). These relations are

$$\mathbf{A}(u)\mathbf{B}(u_1) = \frac{1}{a(u_1 - u)}\mathbf{B}(u_1)\mathbf{A}(u) - \frac{b(u_1 - u)}{a(u_1 - u)}\mathbf{B}(u)\mathbf{A}(u_1), \quad (\mathrm{F.8})$$

$$\mathbf{D}(u)\mathbf{B}(u_1) = \frac{1}{a(u - u_1)}\mathbf{B}(u_1)\mathbf{D}(u) - \frac{b(u - u_1)}{a(u - u_1)}\mathbf{B}(u)\mathbf{D}(u_1), \quad (\mathrm{F.9})$$

and

$$[\mathbf{B}(u), \mathbf{B}(u_1)]_- = 0 \quad [\mathbf{A}(u) + \mathbf{D}(u), \mathbf{A}(u_1) + \mathbf{D}(u_1)]_- = 0. \quad (\mathrm{F.10})$$

These can be used to generate eigenstates of $\mathbf{A}(u) + \mathbf{D}(u)$, and hence of $\mathbf{T}_j$, in a way analogous to the algebraic methods for generating the eigenstates of the harmonic oscillator. The state $|0\rangle$, with all the spins aligned in the positive z direction, the reference state, is trivially an eigenstate of $\mathbf{A}(u)$ and $\mathbf{D}(u)$ as

$$\mathbf{A}(u)|0\rangle = |0\rangle, \quad \mathbf{D}(u)|0\rangle = \prod_{i=1}^{N} a(u - \alpha_i)|0\rangle. \qquad (\mathrm{F.11})$$

Further eigenstates are generated by operating with the **B** operators on the reference state $|0\rangle$. The state $\mathbf{B}(u)|0\rangle$ is an eigenstate with one spin reversed $(M = 1)$ for a particular choice of the argument u, $u = u_1$ (this is like the creation operator for a spin wave of wavevector u_1). To demonstrate this we operate with $\mathbf{A}(u) + \mathbf{D}(u)$ on the state $\mathbf{B}(u_1)|0\rangle$ and use the commutation relations (F.8) and (F.9). We find

$$(\mathbf{A}(u)+\mathbf{D}(u))\mathbf{B}(u_1)|0\rangle = \mathbf{B}(u_1) \left(\frac{1}{a(u_1 - u)}\mathbf{A}(u) + \frac{1}{a(u - u_1)}\mathbf{D}(u) \right) |0\rangle$$
$$+ \frac{b(u - u_1)}{a(u - u_1)}\mathbf{B}(u)(\mathbf{A}(u_1) - \mathbf{D}(u_1))|0\rangle.$$

(F.12)

The second term on the right hand side can be made to vanish for an appropriate choice of u_1. Using (F.11), this second term is found to vanish if

$$\prod_{i=1}^{N} a(u_1 - \alpha_i) = 1,$$ (F.13)

which becomes an equation for u_1. Using (F.11) again but in the first term on the right hand side of (F.12) we find that $\mathbf{B}(u_1)|0\rangle$ is an eigenstate of $\mathbf{A}(u) + \mathbf{D}(u)$ with an eigenvalue $E(u)$ given by

$$E(u) = \frac{1}{a(u_1 - u)} + \frac{\prod_{i=1}^{N} a(u - \alpha_i)}{a(u - u_1)}.$$ (F.14)

On substituting $u = \alpha_j$ into (F.14), we can deduce an eigenvalue of $\mathbf{T}_j$. The second term on the right hand side vanishes on making this substitution. We then obtain from (F.4) the equation,

$$e^{ik_j L} = \frac{1}{a(u_1 - \alpha_j)},$$ (F.15)

and as $\alpha_j = \alpha(k_j)$ this becomes an equation for k_j which is dependent on the other ks through equation (F.13). On taking logarithms equation (F.15) corresponds to (6.101), and equation (F.13) to (6.102), for the case $M = 1$ with $\lambda_1 = 2u_1 + i/2$.

The general case can be proved by operating with $\mathbf{A}(u) + \mathbf{D}(u)$ on the state

$$\prod_{s=1}^{M} \mathbf{B}(u_s)|0\rangle.$$ (F.16)

by manipulating the terms and using the commutation relations in a similar way, with the u_ss chosen so as to make the extra terms, such as those in (F.12), vanish. Equation (F.14) becomes generalized to

$$E(u) = \prod_{s=1}^{M} \frac{1}{a(u_s - u)} + \frac{\prod_{i=1}^{N} a(u - \alpha_i)}{\prod_{s=1}^{M} a(u - u_s)}, \qquad (F.17)$$

and on putting $u = \alpha_j$ to find the eigenvalues of $\mathbf{T}_j$, (F.15) becomes generalized to

$$e^{ik_j L} = \prod_{s=1}^{M} \frac{1}{a(u_s - \alpha_j)}. \qquad (F.18)$$

The quasi-momenta u_s are determined by the condition

$$\prod_{i=1}^{N} a(u_s - \alpha_i) = \prod_{r \neq s}^{M} \frac{a(u_s - u_r)}{a(u_r - u_s)}, \qquad (F.19)$$

for $s = 1, 2, \ldots M$, which is the generalization of (F.13).

Solutions of the form (6.101) and (6.102) follow on making the substitution, $2u_s = \lambda_s - 1/2$, $\alpha_i = \alpha(k_i)$, with $a(u) = u/(u + i/2)$, and taking logarithms. Solutions (6.19) and (6.20) follow similarly but with $\alpha_0 = -1/2\tan J$, and $\alpha_j = 0$ for $j \neq 0$.

The generalization of this approach to the equations for the N-fold degenerate case has been given by Kulish & Reshetikhin (1981).

Appendix G

The Wiener–Hopf Solution

Here we outline the solution of equation (6.29),

$$\rho(\lambda) = \rho^0(\lambda) + \int_{-\infty}^{0} R(\lambda - \lambda')\rho(\lambda')\, d\lambda', \qquad (G.1)$$

with the kernel given by (6.28). We first of all write $\rho(\lambda)$ as $\rho^+(\lambda) + \rho^-(\lambda)$, where $\rho^\pm(\lambda) = \theta(\pm\lambda)\rho(\lambda)$, substitute into (G.1) and then take a Fourier transform, so that (G.1) becomes

$$\tilde{\rho}^+(p) + (1 - \tilde{R}(p))\tilde{\rho}^-(p) = \tilde{\rho}^0(p), \qquad (G.2)$$

where the tildes denote the Fourier transformed quantities. A solution via the Wiener–Hopf technique requires one to be able to factorize the kernel in the form, $(1 - \tilde{R}(p)) = \gamma^-(p)/\gamma^+(p)$, where $\gamma^\pm(p)$ are analytic in the upper/lower half p planes, and have at the most algebraic growth as $p \to \infty$. The equation can then be written in the form,

$$\gamma^+(p)\tilde{\rho}^+(p) + \gamma^-(p)\tilde{\rho}^-(p) = \tilde{\rho}^0(p)\gamma^+(p). \qquad (G.3)$$

The next step is to express the right hand side as the sum of terms which are either analytic in the upper or lower half planes. If $\tilde{\rho}^0(p)\gamma^+(p)$ is analytic in the strip $-s < \operatorname{Im} p < s$, $(s > 0)$, this can be accomplished using Cauchy's theorem,

$$\gamma^+(p)\tilde{\rho}^0(p) = u^+(p) - u^-(p), \qquad (G.4)$$

where

$$u^\pm(p) = \frac{1}{2\pi i} \int_{-\infty}^{\infty} \frac{\tilde{\rho}^0(p)\gamma^+(p)}{p - z \pm is}\, dp. \qquad (G.5)$$

The variables can now be completely separated so that the terms on the left hand side are analytic in the lower half plane, and those on the right hand side in the upper half plane,

$$u^-(p) + \gamma^-(p)\tilde{\rho}^-(p) = u^+(p) - \gamma^+(p)\tilde{\rho}^+(p) = e(p), \qquad (G.6)$$

so $e(p)$ must be an entire function. If $\gamma^+(p) \to$ constant as $p \to \infty$ then $e(p) = 0$, and the solution is given by

$$\tilde{\rho}^\pm(p) = \pm\frac{u^\pm(p)}{\gamma^\pm(p)}. \qquad (G.7)$$

The complete solution is obtained on applying an inverse Fourier transform.

In the case we are considering here the Fourier transform of the kernel

(6.28) is given by

$$\tilde{R}(p) = \frac{e^{-|p|/2}}{2\cosh(p/2)}. \tag{G.8}$$

The factorization into $\gamma^-(p)/\gamma^+(p)$ is straightforward on using the identities,

$$\cosh(p/2) = \frac{\pi}{\Gamma(1/2\pi - ip/2\pi)\Gamma(1/2\pi + ip/2\pi)}, \tag{G.9}$$

and

$$e^{-|p|/2} = e^{ip/2\pi}e^{(\ln(-p/2\pi + is) - \ln(p/2\pi + is))}, \qquad s \to +0. \tag{G.10}$$

We then find

$$\gamma^+(p) = \frac{1}{\gamma^-(-p)} = \frac{(2\pi)^{1/2}e^{-ip/2\pi(1 + i\pi/2 - \ln(-p/2\pi + is))}}{\Gamma(1/2\pi + ip/2\pi)}. \tag{G.11}$$

Using this result for $\gamma^+(p)$, and the result for $\tilde{\rho}^0_{\text{imp}}(p)$,

$$\tilde{\rho}^0_{\text{imp}}(p) = \frac{e^{ip(B-1/c)}}{2\cosh(p/2)}, \tag{G.12}$$

which follows from (6.25) on taking the Fourier transform, we find

$$\tilde{\rho}^0_{\text{imp}}(p)\gamma^+(p) = \frac{e^{ip(B-1/c)}}{(2\pi)^{1/2}}\Gamma\left(\frac{1}{2\pi} - \frac{ip}{2\pi}\right)e^{-ip/2\pi(1 + i\pi/2 - \ln(-p/2\pi + is))}. \tag{G.13}$$

The impurity magnetization can then be deduced from either $\tilde{\rho}^+(0)$ or $\tilde{\rho}^-(0)$ using

$$M_{\text{imp}} = g\mu_{\text{B}}\left(\frac{1}{2} - \tilde{\rho}^+(0)\right) = \frac{g\mu_{\text{B}}}{2}\tilde{\rho}^-(0), \tag{G.14}$$

which follows from (6.24).

For $B < 1/c$, it is convenient to use $\tilde{\rho}^-(0)$,

$$\tilde{\rho}^-(0) =$$

$$\frac{i}{2\pi^{3/2}}\int_{-\infty - is}^{\infty - is}e^{iz(B-1/c)}\Gamma\left(\frac{1}{2\pi} - \frac{iz}{2\pi}\right)e^{-iz/2\pi(1 + i\pi/2 - \ln(-z/2\pi + is))}\frac{dz}{z}, \tag{G.15}$$

$s \to +0$, as the contour, which is just below the real axis, can be distorted to surround the poles in the lower half plane (see contour C_- in figure G.1(i)), and then evaluated using the residue theorem. The poles of $\Gamma(1/2\pi - iz/2\pi)$ are at $z = -i\pi(2n + 1)$, and the residues may be evaluated using

$$\Gamma(z)\Gamma(1 - z) = \frac{\pi}{\sin(\pi z)}. \tag{G.16}$$

The result (6.33) for M_{imp} for $g\mu_{\text{B}}H < k_{\text{B}}T_1$, follows straightforwardly on using the relation (6.32) between B and H.

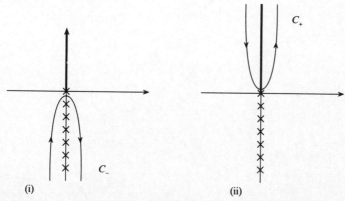

Figure G.1 The contours used in the evaluation of the integrals in the z plane: (i) the contour C_- for the evaluation of $\tilde{\rho}^-(0)$, which is distorted to surround the poles in the lower half plane, and (ii) the contour C_+ for the evaluation of $\tilde{\rho}^+(0)$ which is distorted to surround the cut in the upper half plane.

For $B > 1/c$, it is more convenient to use the solution for $\tilde{\rho}^+(0)$ and distort the contour, which in this case lies just above the real axis, around the cut in the upper half z plane (see contour C_+ in figure G.1(ii)). Using the substitution $z = 2\pi i t$, the result for M_{imp} can be expressed in the form,

$$M_{\text{imp}} = \frac{g\mu_B}{2}\left(1 - \frac{1}{\pi^{3/2}}\int_0^\infty e^{-2\pi t(B-1/c)}e^{t(\ln t - 1)}\sin(\pi t)\Gamma(t+1/2)\,\frac{dt}{t}\right).$$
(G.17)

The form (6.34) follows on making an asymptotic expansion in $1/B$ for large B, and again using the relation (6.32) between B and H.

Appendix H

Rules for Diagrams

Here we summarize the rules for drawing and evaluating the diagrams for the perturbational contributions to the resolvent $R_\alpha(z)$, defined in equation (7.8), for the degenerate Anderson model. The expansion is in powers of the hybridization parameter V. We give rules for the $U = \infty$ model and general N.

(1) Each diagram has a base line running from left to right which corresponds to the impurity state. A full line with an arrow directed from left to right and a label m indicates an occupied state f, $n_f = 1$, m. A dashed line represents the unoccupied state $n_f = 0$. The initial and final lines correspond to the state $|\alpha\rangle$.

(2) The vertices are associated with the interaction (7.10), at which a single conduction electron (full line) is created or annihilated. When the arrow is directed away from the vertex the electron is created, when directed towards the vertex it is annihilated. Contributions of the order $|V|^{2n}$ correspond to all possible diagrams with $2n$ vertices consistent with the direction of the arrows.

(3) Quantum numbers k, m, are assigned to each conduction line and have an associated factor, $1 - f(\epsilon_{k,m})$, for lines running from left to right, and a factor, $f(\epsilon_{k,m})$, for lines running in the reverse direction. The z-component of angular momentum, m, is conserved at each vertex.

(4) Each state $|\alpha'\rangle$ of the impurity, including the initial and final states, carries a factor, $(z - E_{\alpha'} - E_c)^{-1}$, where E_c is the sum of the energies of all the conduction electron lines passing from left to right through a vertical line through the state $|\alpha'\rangle$ (see figure H.1), minus the sum of the energies of all the conduction electrons states passing through the line in the opposite direction.

(5) There is a sum over all possible allowed m values in the intermediate states and a sum over all possible k values for the conduction electrons and an overall factor, $|V|^{2n}(-1)^c$, where c is the number of conduction electron crossings.

As an example of these rules we consider the fourth order diagram shown in figure H.1. This diagram gives a contribution to $R_m(z)$,

$$\sum_{k,k',m'} \frac{|V|^4}{(z - E_{1,m})^2} \frac{1 - f(\epsilon_{k,m})}{(z - E_0 - \epsilon_{k,m})^2} \frac{f(\epsilon_{k',m'})}{(z - E_{1,m'} - \epsilon_{k,m} + \epsilon_{k',m'})}.$$

$$\text{(H.1)}$$

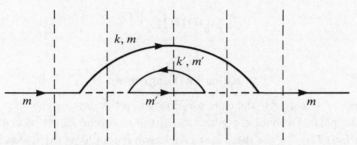

Figure H.1 A fourth order contribution to the resolvent $R_{1,m}(z)$.

To calculate the one electron Green's function, $G_m^f(i\omega_n)$, to order $|V|^{2n}$ extra vertices are introduced, which are represented by open circles, with an f electron external line carrying a label m and carrying an energy $i\omega_n$. The first vertex must be an external one at which an external f electron line enters $(X_{0,m})$ which can be subsequently removed $(X_{m,0})$ at any later stage. The rules for evaluating the diagrams are essentially as outlined above. The crossing of conduction lines only is taken into account in calculating the factor $(-1)^c$. If the total contribution using the above rules is $F(z)$, there is a final integration over z to give the contribution to $G_m^f(i\omega_n)$,

$$\frac{1}{Z_{\mathrm{f}}} \int_{\Gamma} e^{-\beta z} F(z) \, \frac{dz}{2\pi i}, \tag{H.2}$$

where Z_{f} is the partition function (8.4), and the contour Γ encircles all the singularities of $F(z)$ in a counter-clockwise sense.

A sixth order diagram for $G_m^f(i\omega_n)$ is shown in figure H.2. The total contribution to $F(z)$ from this diagram is

$$\sum_{m',m''} \sum_{k,k',k''} \frac{|V|^6}{(z-E_{1,m}+i\omega_n)^2} \frac{1-f(\epsilon_{k,m})}{(z-E_0-\epsilon_{k,m}+i\omega_n)^2}$$

$$\times \frac{f(\epsilon_{k',m'})}{(z-E_{1,m'}-\epsilon_{k,m}+\epsilon_{k',m'}+i\omega_n)} \frac{1}{(z-E_0)^2} \frac{f(\epsilon_{k',m'})}{(z-E_{1,m''}+\epsilon_{k'',m''})}. \tag{H.3}$$

This is a contribution which is included in the non-crossing approximation for $G_m^f(i\omega_n)$ in equation (8.5). A diagram which has a crossing term, and hence is not included in the sum (8.5), is shown in figure 8.5. Such terms can be included as vertex corrections to the external vertex.

The calculation of the dynamic response function $\chi_{\mathrm{imp}}(\omega)$ proceeds along similar lines. The external vertices, filled circles in this case, correspond to the operators, $X_{m,m}$ and $X_{m',m'}$. The first vertex is the external vertex corresponding to $X_{m,m}$ with an f electron line with label

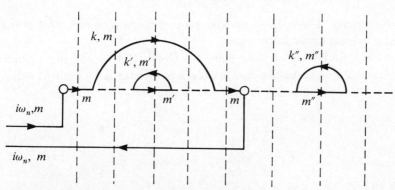

Figure H.2 A sixth order contribution to $G_m^f(i\omega_n)$.

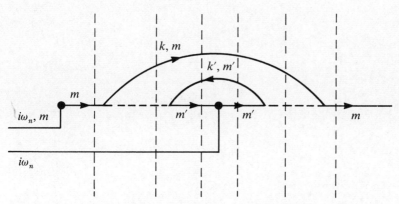

Figure H.3 A non-crossing diagram for $\chi_{\mathrm{imp}}(\omega)$ involving corrections to the external vertex. Such terms vanish, however, on summing over m and m'.

m entering and leaving. The second external vertex can be positioned on an occupied f state with z-component of angular momentum m'. The external f lines carry an energy $i\omega_n$ and the returning external line carries no m label. The prescription for evaluating the contribution to $\chi_{\mathrm{imp}}(\omega)$ is as for the Green's function apart from an extra multiplicative factor $(g\mu_B)^2 mm'$, with an additional sum over m and m'.

Diagrams contributing to $\chi_{\mathrm{imp}}(\omega)$ are shown in figures 8.8 and 8.9. The lowest order diagrams omitted from this sum are crossing diagrams of order $(1/N)^2$.

There are low order non-crossing diagrams which are not included in this sum, such as that shown in figure H.3, which have conduction electron lines across the external vertex and hence correspond to vertex corrections. These, however, give no contribution on summing over m and m' due to the factor mm'.

For a derivation of these rules see Grewe (1982), and for a general review of this perturbational approach see Keiter & Morandi (1984).

Appendix I

Perturbational Results to Order 1/N

Using the perturbational methods described in section 7.2 some exact results can be derived to order $1/N$ for the $U = \infty$ degenerate Anderson model (7.2). The exact results for χ_{imp}, γ_{imp}, and n_{f}, at $T = 0$ to order $(1/N)^0$ can be written in the form,

$$\chi_{\mathrm{imp}}^{(0)} = \frac{\mu}{(1+\mu)} \frac{(g\mu_{\mathrm{B}})^2 j(j+1)}{3k_{\mathrm{B}}T_{\mathrm{A}}}, \quad \gamma_{\mathrm{imp}}^{(0)} = \frac{\mu}{(1+\mu)} \frac{\pi^2}{3k_{\mathrm{B}}T_{\mathrm{A}}}, \tag{I.1}$$

and

$$n_{\mathrm{f}}^{(0)} = \frac{\mu}{(1+\mu)}, \tag{I.2}$$

where $\mu = N\Delta/\pi k_{\mathrm{B}}T_{\mathrm{A}}$ and T_{A} is the solution of equation (7.17).

Exact results for these quantities to order $(1/N)^1$ are

$$\chi_{\mathrm{imp}}^{(1)} = \chi_{\mathrm{imp}}^{(0)} \left\{ 1 + \frac{\mu^2 B}{N(1+\mu)} \right\}, \tag{I.3}$$

$$\gamma_{\mathrm{imp}}^{(1)} = \gamma_{\mathrm{imp}}^{(0)} \left\{ 1 + \frac{\mu^2 B}{N(1+\mu)} - \frac{1}{N}\left(1 - \frac{1}{(1+\mu)^2}\right) \right\}, \tag{I.4}$$

and

$$n_{\mathrm{f}}^{(1)} = n_{\mathrm{f}}^{(0)} \left\{ 1 + \frac{\mu}{N(1+\mu)} \left(L_2 - \mu K - \frac{\mu}{1+\mu}L_1 \right) \right\}, \tag{I.5}$$

where

$$B = \left\{ \frac{2(1+\mu)}{\mu} L_3 - L_2 - M - \left(2 - \frac{\mu}{1+\mu}L_1\right) \right\}, \tag{I.6}$$

and L_n, M and K are integrals given by

$$L_n = \int_0^{D/k_{\mathrm{B}}T_{\mathrm{A}}} \frac{dy}{y + \mu\ln(1+y)} \left(1 - \frac{1}{(1+y)^n}\right), \tag{I.7}$$

$$M = \int_0^{D/k_{\mathrm{B}}T_{\mathrm{A}}} \frac{dy}{[y + \mu\ln(1+y)]^2} \frac{y^2(2+\mu+y)}{(1+y)^3}, \tag{I.8}$$

$$K = \int_0^{D/k_{\mathrm{B}}T_{\mathrm{A}}} \frac{dy}{[y + \mu\ln(1+y)]^2} \frac{y^2}{(1+y)^2}. \tag{I.9}$$

These results are valid for $|\epsilon_{\mathrm{f}}| \ll D$, and therefore do not apply in the Coqblin–Schrieffer regime, $-\epsilon_{\mathrm{f}} \gg D$.

The Wilson or χ/γ ratio (7.47) derived from (I.3) and (I.4) is

$$R = 1 + \frac{1}{N}(1 - (1 - n_{\mathrm{f}}^{(0)})^2) + O(1/N^2), \tag{I.10}$$

and is in agreement with the Fermi liquid relation (7.49) to order $(1/N)^1$.

In the Kondo regime $-\epsilon_f \gg N\Delta/\pi$, the integrals can be evaluated analytically (Rasul & Hewson, 1984) and (I.3) becomes

$$\chi^{(1)}_{imp} = \chi^{(0)}_{imp} \left\{ 1 + \frac{1}{N} \left(C - \ln\left(\frac{N\Delta}{\pi D}\right) \right) \right\} + O(1/\mu), \qquad (I.11)$$

where C is Euler's constant. The corresponding result at high temperatures is

$$\chi^{(1)}_{imp}(T) = \frac{(g\mu_B)^2 j(j+1)}{3k_B T} \left\{ 1 - \frac{2N\Delta}{N\pi|\epsilon_f|} + \frac{2}{N} \left(\frac{N\Delta}{\pi|\epsilon_f|} \right)^2 \left(\ln\left(\frac{k_B T}{D}\right) \right. \right.$$
$$\left. \left. + \frac{1}{N}\ln\left(\frac{eD}{|\epsilon_f|}\right) - \left(1 + C - \frac{1}{2N}\right) + \ln(2\pi) \right) \right\} + O(|V|^6),$$

$$(I.12)$$

which has the $\ln T$ correction term which is of order $(1/N)^1$. The Wilson number relating the high and low energy scales can be deduced from (I.11) and (I.12) to order $(1/N)^1$ and is in agreement with (7.43).

These results are also asymptotically exact for arbitrary N in the non-magnetic regime $\epsilon_f \gg N\Delta/\pi$, $k_B T_A \to \epsilon_f$, and can be used to deduce equations (7.34) and (7.35).

For the Coqblin–Schrieffer model results are required for the regime $-\epsilon_f \gg D$. These are given by

$$\chi^{(1)}_{imp} = \frac{(g\mu_B)^2 j(j+1)}{3k_B T_L^{(0)}} \left(1 + \frac{1}{N}(2\bar{L}_3 - \bar{L}_1 - \bar{L}_2 - \bar{M} - 1) \right), \qquad (I.13)$$

and

$$\gamma^{(1)}_{imp} = \frac{\pi^2}{3k_B T_L^{(0)}} \left(1 + \frac{1}{N}(2\bar{L}_3 - \bar{L}_1 - \bar{L}_2 - \bar{M} - 2) \right), \qquad (I.14)$$

where

$$\bar{L}_n = \int_0^{D/k_B T_L^{(0)}} \frac{dy}{\ln(1+y)} \left(1 - \frac{1}{(1+y)^n} \right), \qquad (I.15)$$

and

$$\bar{M} = \int_0^{D/k_B T_L^{(0)}} \frac{dy}{[\ln(1+y)]^2} \frac{y^2}{(1+y)^3}, \qquad (I.16)$$

where $T_L^{(0)} = De^{-1/N\rho_0 J}$ and $\rho_0 J = \Delta/\pi|\epsilon_f|$.

Evaluation of the integrals, $\bar{L}_n$ and $\bar{M}$, gives

$$\chi^{(1)}_{imp} = \frac{(g\mu_B)^2 j(j+1)}{3k_B T_L^{(0)}} \left\{ 1 - \frac{1}{N}(1 - C + \ln(N\rho_0 J)) \right\} + O\left(\frac{T_L^{(0)}}{D}, \rho_0 J \right). \qquad (I.17)$$

At high temperatures the $1/N$ corrections are

$$\chi_{imp}^{(1)}(T) = \frac{(g\mu_B)^2 j(j+1)}{3k_B T} \left\{ 1 - \frac{2(N\rho_0 J)}{N} \right.$$
$$\left. + \frac{2(N\rho_0 J)^2}{N} \left(\ln\left(\frac{k_B T}{D}\right) - \left(1 + C - \frac{1}{2N}\right) + \ln(2\pi) \right) \right\} + O(\rho_0 J).$$

$$(I.18)$$

These results differ from (I.11) and (I.12) but the Wilson number deduced from them agrees with (7.43) to order $(1/N)^1$. The difference arises from the difference in the form of the Kondo temperature (7.44) in the regimes $-\epsilon_f \gg D$ and $D \gg -\epsilon_f \gg N\Delta$. The results (I.13) and (I.14) have also been obtained directly from a perturbation expansion for the Coqblin–Schrieffer model by Bickers (1987).

Appendix J

The n-Channel Kondo Model for n > 2S.

The n-channel model (9.30) with spin S for $n > 2S$ has distinctly new features arising from the fact that the strong coupling fixed point $J \to \infty$ leads to an over-compensated spin $n - S/2$ which has a residual antiferromagnetic interaction with the remaining conduction electrons (as opposed to the under compensated case $n < 2S$ which has a residual ferromagnetic coupling). As pointed out by Nozières & Blandin (1980) this leads to a non-trivial stable fixed point $J = J^*$, as both the $J = 0$ and $J = \infty$ fixed points are unstable, a situation first envisaged by Abrikosov and Migdal (1970), and critical behaviour as $T, H \to 0$. This does not imply that the low temperature behaviour corresponds to the model (9.30) with $J \sim J^*$, but that all the couplings in the effective Hamiltonian near this fixed point are functions of J^*. In the vicinity of a finite stable fixed point the scaling equation (3.48) must take the form,

$$\frac{d\tilde{J}}{d\ln\tilde{D}} = \beta^*(\tilde{J} - J^*), \tag{J.1}$$

with $\beta^* > 0$, giving $\tilde{J} - J^* \sim (\tilde{D}/T_{\mathrm{K}})^{\beta^*}$, and for $\tilde{D} \sim T$ this implies critical behaviour.

Explicit numerical renormalization group calculations for $S = \frac{1}{2}$, $n = 2$ of Cragg & Lloyd (1979) and Cragg et al (1980), confirmed the conjecture of a finite coupling fixed point. At this fixed point, unlike the $J = 0$ and $J = \infty$ fixed points, the many-body levels cannot be reproduced by filling single particle states of an effective one electron model. The correlations also appear to be long range as the fixed point does not depend on whether the chain used in the renormalization group calculation has an even or an odd number of sites. Exact Bethe ansatz solutions (Tsvelick & Wiegmann, 1984; Andrei & Destri, 1984) have confirmed the critical behaviour for $n > 2S$. The ground state is a singlet $M_{\mathrm{imp}}(0) = 0$. The limits $H \to 0$ and $T \to 0$ do not commute. For large fields such that $\mu_{\mathrm{B}}H \gg k_{\mathrm{B}}T$, the magnetization can be expressed in the form,

$$M_{\mathrm{imp}}(H) = \sum_{n=1} \frac{k^k}{k!}(-1)^k \sin\left(\frac{\pi k S}{n}\right)\left(\frac{\mu_{\mathrm{B}}H}{k_{\mathrm{B}}T_{\mathrm{H}}}\right)^{2k/n}, \tag{J.2}$$

where $T_H \propto T_K$. The leading term gives the scaling law,

$$M_{\text{imp}}(H) \propto \left(\frac{H}{T_K}\right)^{2/n} , \quad \chi_{\text{imp}}(H) \propto \left(\frac{H}{T_K}\right)^{(2-n)/n} . \quad (J.3)$$

The leading term for the specific heat in this limit is

$$C_{\text{imp}}(T,H) \propto T \left(\frac{H}{T_K}\right)^{2/n} , \quad (J.4)$$

deduced from the thermodynamic equations by Tsvelick (1985).

For $k_B T \ll \mu_B H$ the form of the free energy is

$$\frac{F_{\text{imp}}}{k_B T} = -\ln\left(\frac{\sin(\pi(S+1)/(n+2))}{\sin(\pi/(n+2))}\right) - \sum_{k=2} \alpha_k \left(\frac{H}{T}\right) \left(\frac{T}{T_K}\right)^{2k/(n+2)} , \quad (J.5)$$

where the coefficient $\alpha_k(H/T)$ can be expanded in powers of $(H/T)^2$. The leading term gives a scaling law for the specific heat and susceptibility,

$$\frac{C_{\text{imp}}(T,H)}{T} \propto \left(\frac{T}{T_K}\right)^{(n-2)/(n+2)} , \quad \chi_{\text{imp}}(T,H) \propto \left(\frac{T}{T_K}\right)^{(2-n)/(2+n)} . \quad (J.6)$$

Numerical solutions of the thermodynamic equations by Desgranges (1985) show the the critical region with exponents (J.6), independent of S, only applies for $T < 10^{-4} T_K$. For $n = 2$ the critical exponents vanish and a logarithmic dependence occurs. The exact solutions of the $n = 2$ model for $S = \frac{1}{2}$ has been extensively studied by Sacramento and Schlottman (1989), who also give further numerical results for the more general model (Sacramento & Schlottmann, 1991). Green's function and correlation functions for this model have also been studied using conformal field theory (Ludwig & Affleck, 1991).

It has not been fully established that the over-compensated n-channel model applies to any real physical system. It has been used to interpret data on V impurities in Au (Geens, Labro & Mordijck, 1987). The behaviour of $Y_{1-x}U_x Pd_3$ has been analysed in terms of the $n = 2$, $S = \frac{1}{2}$, model as discussed in section 10.6 (Seaman et al, 1991).

Appendix K

Summary of Single Impurity Results

For readers whose interest is not primarily theoretical we give a brief survey of the results of the many-body techniques described in chapters 3–8 for the behaviour of the more important single magnetic impurity models introduced in chapter 1. We give a guide to the main conclusions and the theoretical picture of single impurity behaviour which emerges.

We start with the s-d model (1.64) for an impurity spin $S = \frac{1}{2}$ coupled by an exchange interaction J to the conduction electrons of the host metal,

$$H_{sd} = \sum_{\mathbf{k},\mathbf{k'}} J(S^+ c^\dagger_{\mathbf{k},\downarrow} c_{\mathbf{k'},\uparrow} + S^- c^\dagger_{\mathbf{k},\uparrow} c_{\mathbf{k'},\downarrow} + S_z(c^\dagger_{\mathbf{k},\uparrow} c_{\mathbf{k'},\uparrow} - c^\dagger_{\mathbf{k},\downarrow} c_{\mathbf{k'},\downarrow}))$$

$$+ \sum_{\mathbf{k},\sigma} \epsilon_{\mathbf{k}} c^\dagger_{\mathbf{k},\sigma} c_{\mathbf{k},\sigma},$$

$$(K.1)$$

where $\epsilon_{\mathbf{k}}$ is the conduction electron energy for wave vector $\mathbf{k}$. For simplicity it can be assumed that the conduction electrons have a rectangular density of states of width $2D$ with the Fermi level at the centre of the band. In the perturbative scaling approach discussed in section 3.3 virtual transitions of the electrons to the conduction band edges are eliminated to give an effective model which describes the same low energy behaviour as the original model. In the perturbative regime this effective model has the same form as the original model (K.1) but modified parameters. When the band width is reduced by an amount $|\delta D|$ the coupling in the antiferromagnetic case ($J > 0$) is found to increase by an amount δJ,

$$\delta J = 2J(\rho_0 J - (\rho_0 J)^2 + O((\rho_0 J)^3)) \frac{|\delta D|}{D}, \qquad (K.2)$$

where ρ_0 is the density of conduction states at the Fermi level. This process of scaling from one effective model to another effective model can be continued as long as the perturbation theory is valid. It breaks down when the new coupling strength becomes too large. This occurs when an energy scale $\tilde{D}$ of the reduced band width becomes of the order the Kondo temperature, $k_B T_K$, where T_K is given by

$$k_B T_K \sim D|2J\rho_0|^{1/2} e^{-1/2J\rho_0}. \qquad (K.3)$$

The parameters J and D only appear in the scaling equation via the

Kondo temperature, $T_K(J, D)$ so that $k_B T_K$ is the only energy scale of the model for energies on a scale much less than the band width $2D$. As a result the thermodynamic properties for the model in a magnetic field H are universal functions of T/T_K and H/T_K. In the regime where perturbation theory is valid $T \gg T_K$ the zero field impurity susceptibility $\chi_{imp}(T)$ is given by

$$\chi_{imp}(T) = \frac{(g\mu_B)^2}{4k_B T}\left\{1 - \frac{1}{\ln(T/T_K)} - \frac{\ln(\ln(T/T_K))}{2\ln^2(T/T_K)} + O\left(\frac{1}{\ln^3(T/T_K)}\right)\right\}$$

(K.4)

behaving approximately like a Curie law with a reduced moment. The induced impurity magnetization $M_{imp}(H)$ in a magnetic field at $T = 0$ is given by

$$M_{imp} = \frac{g\mu_B}{2}\left\{1 - \frac{1}{2\ln(g\mu_B H/k_B T_H)} - \frac{\ln\ln(g\mu_B H/k_B T_H)}{4(\ln(g\mu_B H/k_B T_H))^2}\right.$$
$$\left. + O(\ln(g\mu_B H/k_B T_H))^{-3}\right\},$$

(K.5)

for large field such that $g\mu_B H \gg k_B T_H$, where T_H is proportional to T_K. The impurity resistivity $R_{imp}(T)$ for $J > 0$ and $T \gg T_K$ is well represented by the expression,

$$R_{imp}(T) = \frac{R_0}{2}\left\{1 - \frac{\ln(T/T_K)}{[(\ln(T/T_K))^2 + 3\pi^2/4]^{1/2}}\right\}.$$

(K.6)

The numerical renormalization calculations of Wilson continue this scaling beyond the perturbative regime allowing the low temperature behaviour to be calculated (details in chapter 4). Though the effective model for the low energy behaviour in the non-perturbative regime no longer corresponds to the form (K.1), in the limit as the energy scale tends to zero (appropriate only for very low temperature calculations) it corresponds to the strong coupling limit $J \to \infty$ of (K.1). In the low temperature limit the impurity spin is compensated by the conduction electrons and the impurity susceptibility is finite corresponding to a Pauli contribution. At $T \to 0$ the susceptibility is given by

$$\chi_{imp}(0) = \frac{(g\mu_B)^2 w}{4k_B T_K},$$

(K.7)

where the numerical constant w, known as the Wilson number, has the value $e^{C+1/4}/\pi^{3/2}$ and C is Euler's constant 0.5772. The definition of T_K in (K.7) and (K.8) corresponds to Wilson's definition where the constant factor, which was not specified in (K.3), is chosen such that there are no terms of order $1/\ln^2(T/T_K)$ in the high temperature expansion (K.4). A common alternative definition T_{L} is via (K.7) and is such that $1/T_{\mathrm{L}} = w/T_{\mathrm{K}}$. The specific heat coefficient γ_{imp} as $T \to 0$ in the same limit is given by

$$\gamma_{\mathrm{imp}} = \frac{k_{\mathrm{B}} \pi^2 w}{6 T_{\mathrm{K}}}. \tag{K.8}$$

The Wilson or Sommerfeld ratio for the impurity, deduced from (K.7) and (K.8), is given by

$$R = \frac{4\pi^2 k_{\mathrm{B}}^2}{3(g\mu_{\mathrm{B}})^2} \frac{\chi_{\mathrm{imp}}}{\gamma_{\mathrm{imp}}} = 2, \tag{K.9}$$

and is enhanced over that for non-interacting conduction electrons (1.61) by a factor 2.

In the low temperature limit $T \ll T_{\mathrm{K}}$ the magnetically screened impurity scatters the conduction electrons like a non-magnetic impurity with a resonance at the Fermi level with a width of order $k_{\mathrm{B}} T_{\mathrm{K}}$. The exact expression for the impurity contribution to the resistivity as $T \to 0$ is given by

$$R_{\mathrm{imp}}(T) = R_0 \left(1 - \frac{\pi^4 w^2}{16} \left(\frac{T}{T_{\mathrm{K}}} \right)^2 + O(T^4) \right). \tag{K.10}$$

The susceptibility in zero field over the full temperature range linking the perturbative high temperature and strong coupling low temperature results (K.4) and (K.7), first derived by Wilson (1975), is given in figure 4.7. The specific heat, shown over the temperature range $0 < T < 5T_{\mathrm{K}}$ in figure 6.5, has a pronounced peak at a temperature $T \sim T_{\mathrm{K}}$.

Similar results apply in the Kondo or local moment regime for the Anderson model (1.66),

$$H = \sum_{\sigma} \epsilon_d n_{d,\sigma} + U n_{d,\uparrow} n_{d,\downarrow} + \sum_{\mathbf{k},\sigma} \epsilon_{\mathbf{k}} c_{\mathbf{k},\sigma}^{\dagger} c_{\mathbf{k},\sigma}$$
$$+ \sum_{\mathbf{k},\sigma} (V_{\mathbf{k}} c_{d,\sigma}^{\dagger} c_{\mathbf{k},\sigma} + V_{\mathbf{k}}^* c_{\mathbf{k},\sigma}^{\dagger} c_{d,\sigma}),$$

$$\tag{K.11}$$

where the impurity level ϵ_d is hybridized with conduction electrons. In the regime where the Schrieffer–Wolff transformation (1.73) is valid (ϵ_d below the conduction band and $U + \epsilon_d$ above) the results for the Anderson model correspond to those of the s-d model with $J = U|V^2|/|\epsilon_d||U + \epsilon_d|$. There is also a local moment regime for the Anderson model when the d level ϵ_d (or $U + \epsilon_d$) lies within the conduction band. In this case a more general expression for the Kondo temperature T_{K} must be used.

For $U \rightarrow \infty$ this takes the form,

$$k_B T_K \sim D \left(\frac{\Delta}{\max(|\epsilon_d|, D)} \right)^{1/2} e^{\pi \epsilon_d / 2\Delta}, \qquad (K.12)$$

where $\Delta = \pi \rho_0 |V|^2$.

The thermodynamic behaviour of both the s-d and Anderson models can be calculated essentially exactly over the whole temperature range using the Bethe ansatz equations given in chapter 6.

The behaviour in the low temperature range can be understood in terms interacting quasi-particles within a Fermi liquid theory. The quasi-particles can be described by an Anderson model with a renormalized hybridization $\tilde{V}$, renormalized level $\tilde{\epsilon}_d$ and interaction $\tilde{U}$,

$$H_{\text{eff}} = \sum_{\mathbf{k},\sigma} \epsilon_{\mathbf{k},\sigma} c^\dagger_{\mathbf{k},\sigma} c_{\mathbf{k},\sigma} + \sum_{\mathbf{k},\sigma} (\tilde{V}_\mathbf{k} c^\dagger_{d,\sigma} c_{\mathbf{k},\sigma} + \tilde{V}^*_\mathbf{k} c^\dagger_{\mathbf{k},\sigma} c_{d,\sigma})$$

$$\sum_\sigma \tilde{\epsilon}_{d,\sigma} c^\dagger_{d,\sigma} c_{d,\sigma} + \tilde{U} n_{d,\uparrow} n_{d,\downarrow}.$$

$$(K.13)$$

This Hamiltonian describes quasi-particle excitations from the interacting ground state. The interaction comes into play only when these excitations are created at finite T or in a finite field H. Thermodynamic Fermi liquid theory corresponds to a mean field treatment of this interaction term. The equations for χ_{imp}, γ_{imp} and the charge susceptibility $\chi_{c,\text{imp}}$ in the mean field theory take the form,

$$\chi_{\text{imp}} = \frac{(g\mu_B)^2}{2} \tilde{\rho}_{\text{imp}}(\epsilon_F)(1 + \tilde{U} \tilde{\rho}_{\text{imp}}(\epsilon_F)), \qquad (K.14)$$

$$\chi_{c,\text{imp}} = 2\tilde{\rho}_{\text{imp}}(\epsilon_F)(1 - \tilde{U} \tilde{\rho}_{\text{imp}}(\epsilon_F)), \qquad (K.15)$$

and

$$\gamma_{\text{imp}} = \frac{2\pi^2 k_B^2}{3} \tilde{\rho}_{\text{imp}}(\epsilon_F), \qquad (K.16)$$

for $T = 0$ where $\tilde{\rho}_{\text{imp}}(\epsilon)$ is the quasi-particle density of states. If the equation for γ_{imp} is expressed in terms of T_K via (K.8) for the Kondo regime the renormalized parameters in quasi-particle Hamiltonian (K.13) can be expressed in terms of T_K via the condition that the d charge is localized $\chi_{c,\text{imp}} = 0$. For the particle–hole symmetric case $\tilde{\epsilon}_d = 0$ this gives

$$\pi \tilde{\Delta} = \tilde{U} = 4k_B T_L, \qquad (K.17)$$

with $\tilde{\rho}_{\text{imp}}(\epsilon_F) = 1/\pi\tilde{\Delta}$. When these results are substituted into (K.14) the quasi-particle susceptibility is enhanced by a factor of 2 giving the Wilson ratio (K.9).

The quasi-particle density of states $\tilde{\rho}_{\mathrm{imp}}(\epsilon)$ corresponds to a resonance of width $\tilde{\Delta}$ at the Fermi level. If particle–hole symmetry is relaxed the resonance can be asymmetric about the Fermi level. Calculation of the quasi-particle scattering diagram to order $\tilde{U}^2$ gives the exact result for the T^2 coefficient in (K.10). The resonance is a many-body resonance commonly known as the Kondo resonance. If the quasi-particle inter-action is ignored ($\tilde{U} = 0$) the model for the quasi-particles becomes a 'resonant level model' which qualitatively describes the correct low T behaviour. The Wilson ratio is unenhanced as the quasi-particles in this approximation are non-interacting so $R = 1$ and the T^2 coefficient in (K.10) is too small by a factor 3. The resonant level model was used in the interpretation of experimental results for Kondo systems before a complete theory of the low temperature behaviour was obtained.

The density of states of the impurity d electron (defined by the spectral density of the d Green's function) has a peak at the Fermi level in the Kondo regime reflecting low lying quasi-particle excitations, but the weight in this peak can be quite small for weak hybridization. The main weight in this density of states lies in the region of the 'atomic' peaks at ϵ_{d} and $\epsilon_{\mathrm{d}} + U$ which are broadened on an energy scale Δ. These can be clearly seen in figure 5.10 with the Kondo resonance at the Fermi level. As the impurity level ϵ_{d} moves to the Fermi level and one passes from the Kondo to the mixed valence regimes the Kondo resonance merges with the broadened atomic peak at ϵ_{d} as can be seen in figure 5.11. When the impurity level is moved above the Fermi level to correspond to the non-magnetic or empty orbital regime the Kondo resonance disappears (see figure 5.12). In the mixed valence regime the impurity susceptibility develops a maximum. This can be seen in the results in figure 4.13.

A similar picture emerges from many-body calculations for the N-fold degenerate Anderson model ($U \to \infty$) and Coqblin–Schrieffer models (1.84) and (1.89). These are suitable models for describing Ce and Yb impurities with $N = 2j + 1$, where j is the total angular momentum quantum number associated with the degenerate ground state multiplet of the impurity ion. In the Kondo regime with the occupation of the impurity f level $n_{\mathrm{f}} \sim 1$ there is local moment behaviour for $T \gg T_{\mathrm{K}}$ and Fermi liquid behaviour for $T \ll T_{\mathrm{K}}$. However, in the Kondo regime some new features appear as a function of N. The renormalized parameters for the Fermi liquid Hamiltonian (see (7.53)) are given by

$$\tilde{\epsilon}_{\mathrm{f}} = k_{\mathrm{B}} T_{\mathrm{L}} \frac{N^2 \sin(\pi/N) \cos(\pi/N)}{\pi(N-1)}, \tag{K.18}$$

$$\tilde{\Delta} = k_B T_L \frac{N^2 \sin^2(\pi/N)}{\pi(N-1)}, \quad \tilde{U} = \left(\frac{N}{N-1}\right)^2 k_B T_L, \quad \text{(K.19)}$$

which gives for the quasi-particle density of states at the Fermi level,

$$\tilde{\rho}_{\text{imp}}(\epsilon_F) = \frac{(N-1)}{N^2 k_B T_L}. \quad \text{(K.20)}$$

For $N > 2$ the Kondo resonance becomes asymmetric about the Fermi level. This is clearly seen in figure 7.14 where the quasi-particle density of states has been plotted for different values of N. As the resonance moves above the Fermi level with increasing N the susceptibility $\chi_{\text{imp}}(T)$ develops a peak for $N \geq 4$ (see figure 7.12) and the $T = 0$ induced magnetization $M_{\text{imp}}(H)$ has positive curvature in low field (see figure 7.10).

The Wilson ratio in the Kondo regime is given by

$$R = \frac{\pi^2 k_B^2}{j(j+1)(g\mu_B)^2} \frac{\chi_{\text{imp}}}{\gamma_{\text{imp}}} = \frac{N}{N-1}. \quad \text{(K.21)}$$

In the limit $N \to \infty$ the R tends to the non-interacting result $R = 1$. In this limit the interaction between the quasi-particles can be neglected even though $\tilde{U}$ remains finite because this term in the mean field equations is multiplied by the factor $\tilde{\rho}_{\text{imp}}(\epsilon_F)$ (K.19) which tends to zero as $N \to \infty$. The narrowing of the quasi-particle resonance with increasing N is apparent in the quasi-particle density of states shown in figure 7.14. In the limit $N \to \infty$ the resonance narrows to a delta function at $\epsilon = k_B T_L$ above the Fermi level. There are several theoretical techniques which give asymptotically exact results in the large N limit and good approximations for $N \geq 4$ (see chapters 7 and 8). The mean field theory based on the slave boson representation used in section 7.5 gives a particularly simple derivation of the Fermi liquid theory (without interactions) from a first principles calculation.

The situation becomes more complicated when orbital angular momentum and crystal field terms are taken into account. In many cases in the Kondo regime essentially the same picture emerges with a modified T_K. In other cases there may be temperature regimes with different values of T_K, and different effective degeneracies for the impurity state, with a crossover region from one regime to another. It is not a general result that the impurity moment must be compensated in the ground state though experimentally in most systems this appears to be the case. In chapter 9 some more realistic models of magnetic impurity systems are considered. The evidence that suggests the physical picture which emerges from the theory is consistent with the experimental observations is reviewed in the same chapter.

Appendix L

Renormalized Perturbation Theory

In chapter 5 where the Fermi liquid theory for magnetic impurity models was developed we considered two approaches. One was the phenomenological approach, based on the work of Landau and Nozières, where we conjectured a specific form for a quasi-particle Hamiltonian with a local interaction term. This Hamiltonian was essentially equivalent to the effective Hamiltonian near the strong coupling fixed point obtained by Wilson in his numerical renormalization group calculation (section 4.5, equation (4.49)). In the later chapters, where we used this approach for the N-fold degenerate Anderson model (section 7.4) and the n-channel Kondo model for $n = 2S$ (section 9.3), the conjectured form for the quasi-particle Hamiltonian was not backed up by any first principles renormalization group calculation. The other approach developed in chapter 5 was the microscopic Fermi liquid theory based largely on the work of Luttinger (1960, 1961) and Yamada & Yosida (1975) using a conventional perturbation expansion in powers of U. This microscopic treatment confirmed all the results based on the conjectured quasi-particle Hamiltonian. Here we develop a synthesis of the two approaches which we will refer to as 'renormalized perturbation theory' (Hewson, 1992). It is based on the general idea of renormalization used in quantum field theory. The results at low temperatures correspond essentially to our earlier calculations with the conjectured quasi-particle Hamiltonian. These are obtained from first and second order perturbation theory in powers of the renormalized interaction $\tilde{U}$. The approach, however, is not limited to low temperatures and low frequencies, as everything in the original Hamiltonian is retained. To demonstrate the approach we consider the Anderson model as defined in equation (1.66),

$$H = \sum_\sigma \epsilon_{d,\sigma} c_{d,\sigma}^\dagger c_{d,\sigma} + U n_{d,\uparrow} n_{d,\downarrow}$$
$$+ \sum_{\mathbf{k},\sigma} (V_\mathbf{k} c_{d,\sigma}^\dagger c_{\mathbf{k},\sigma} + V_\mathbf{k}^* c_{\mathbf{k},\sigma}^\dagger c_{d,\sigma}) + \sum_{\mathbf{k},\sigma} \epsilon_{\mathbf{k},\sigma} c_{\mathbf{k},\sigma}^\dagger c_{\mathbf{k},\sigma},$$

$$(\text{L.1})$$

and the corresponding retarded one particle double-time Green's function for the d-electron,

$$G_{d\sigma,d\sigma}(\omega) = \frac{1}{\omega - \epsilon_d + i\Delta(\omega) - \Sigma_\sigma(\omega)}, \qquad \text{(L.2)}$$

which is the analytic continuation of the thermal Green's function (5.44).
The function $\Sigma_\sigma(\omega)$ is the proper self-energy in the perturbation expansion in powers of the local interaction U described in section 5.2.
We will need the corresponding irreducible four point vertex function
$\Gamma_{\sigma,\sigma'}(\omega,\omega')$. Our aim is to reorganize this perturbation expansion into
a more convenient form to consider the strong correlation regime which
corresponds to a model with U large at low temperatures and in weak
magnetic fields.

Our first step is to write the self-energy in the form,

$$\Sigma_\sigma(\omega) = \Sigma_\sigma(0) + \omega\Sigma_\sigma'(0) + \Sigma_\sigma^{\text{rem}}(\omega), \qquad \text{(L.3)}$$

which is simply a definition for the remainder self-energy $\Sigma_\sigma^{\text{rem}}(\omega)$. Using
this form the Green's function given in equation (L.2) can be written in
the form,

$$G_{d\sigma,d\sigma}(\omega) = \frac{z}{\omega - \tilde\epsilon_d + i\tilde\Delta(\omega) - \tilde\Sigma_\sigma(\omega)}, \qquad \text{(L.4)}$$

where z, the wavefunction renormalization factor, is given by

$$z = \frac{1}{1 - \Sigma_\sigma'(0)}. \qquad \text{(L.5)}$$

The prime denotes a derivative with respect to ω and $\omega = 0$ corresponds
to the Fermi level. We assume that the general theorem of Luttinger,
that the imaginary part of $\Sigma(\omega)$ vanishes at $\omega = 0$ (condition (iii) following equation (5.44)), holds so that z is real. The 'renormalized'
quantities, which are denoted by a tilde, are defined by

$$\tilde\epsilon_d = z(\epsilon_d + \Sigma_\sigma(0,0)), \quad \tilde\Delta = z\Delta, \quad \tilde\Sigma_\sigma(\omega) = z\Sigma_\sigma^{\text{rem}}(\omega), \qquad \text{(L.6)}$$

as in section 5.3. Two zeros are given in the argument of Σ of (L.6) to
emphasize that this is to be evaluated at $T = 0$, and in zero magnetic
field, as well as at $\omega = 0$. A renormalized four point vertex function is
defined by $\tilde\Gamma_{\sigma,\sigma'}(\omega,\omega') = z^2\Gamma_{\sigma,\sigma'}(\omega,\omega')$, and a renormalized interaction
$\tilde U$ by the value of $\tilde\Gamma_{\sigma,\sigma'}(\omega,\omega')$ at $\omega = \omega' = 0$, $\tilde U = \tilde\Gamma_{\sigma,\sigma'}(0,0)$.

The next step is to introduce rescaled creation and annihilation operators for the d-electron via

$$c_{d,\sigma}^\dagger = \sqrt{z}\tilde c_{d,\sigma}^\dagger \quad c_{d,\sigma} = \sqrt{z}\tilde c_{d,\sigma}. \qquad \text{(L.7)}$$

We can now rewrite the Hamiltonian (L.1) in the form,

$$H = \tilde H_{\text{qp}} - \tilde H_{\text{c}}, \qquad \text{(L.8)}$$

where $\tilde H_{\text{qp}}$ will be referred to as the quasi-particle Hamiltonian which can be written as $\tilde{H}_{\text{qp}}^{(0)} + \tilde{H}_{\text{qp}}^{(I)}$. The Hamiltonian $\tilde{H}_{\text{qp}}^{(0)}$ describes non-interacting particles and is given by

$$\tilde{H}_{\text{qp}}^{(0)} = \sum_\sigma \tilde{\epsilon}_{d,\sigma} \tilde{c}_{d,\sigma}^\dagger \tilde{c}_{d,\sigma}$$

$$+ \sum_{\mathbf{k},\sigma} (\tilde{V}_\mathbf{k} \tilde{c}_{d,\sigma}^\dagger c_{\mathbf{k},\sigma} + \tilde{V}_\mathbf{k}^* c_{\mathbf{k},\sigma}^\dagger \tilde{c}_{d,\sigma}) + \sum_{\mathbf{k},\sigma} \epsilon_{\mathbf{k},\sigma} c_{\mathbf{k},\sigma}^\dagger c_{\mathbf{k},\sigma},$$

$$\text{(L.9)}$$

and $\tilde{H}_{\text{qp}}^{(I)}$ is the interaction term,

$$\tilde{H}_{\text{qp}}^{(I)} = \tilde{U} \tilde{n}_{d,\uparrow} \tilde{n}_{d,\downarrow}. \tag{L.10}$$

The second term in equation (L.8) will be known as the counter term and takes the form,

$$\tilde{H}_c = \lambda_1 \sum_\sigma \tilde{c}_{d,\sigma}^\dagger \tilde{c}_{d,\sigma} + \lambda_2 \tilde{n}_{d,\uparrow} \tilde{n}_{d,\downarrow}, \tag{L.11}$$

where λ_1 and λ_2 are given by

$$\lambda_1 = z\Sigma(0,0), \quad \lambda_2 = z^2(\Gamma_{\uparrow,\downarrow}(0,0) - U). \tag{L.12}$$

Equations (L.9)–(L.10) are simply a rewriting of the original Hamiltonian (L.1), so equation (L.8) is an identity.

We note that by construction the renormalized self-energy $\tilde{\Sigma}_\sigma(\omega)$ is such that

$$\tilde{\Sigma}_\sigma(0,0) = 0, \quad \tilde{\Sigma}'_\sigma(0,0) = 0, \tag{L.13}$$

so that $\tilde{\Sigma}_\sigma(\omega) = O(\omega^2)$ for small ω, on the assumption that it is analytic at $\omega = 0$. As $\tilde{\Gamma}_{\sigma,\sigma}(0,0) = 0$ we also have

$$\tilde{\Gamma}_{\sigma,\sigma'}(0,0) = \tilde{U}(1 - \delta_{\sigma,\sigma'}). \tag{L.14}$$

We now identify the Hamiltonian $\tilde{H}_{\text{qp}}^{(0)}$ with our quasi-particle Hamiltonian used in chapter 5 (equation (5.6)). We note that $\tilde{H}_{\text{qp}}$ is identical in form to the effective Hamiltonian near the strong coupling fixed point for the s-d model, equation (4.49), as obtained by Wilson in his numerical renormalization group calculation. The s-d model corresponds to the regime, $\epsilon_d < 0$ and $U \gg \pi\Delta$. We will show that $\tilde{H}_{\text{qp}}$ is an effective Hamiltonian about the Fermi liquid fixed point for all the parameter regimes of the Anderson model.

To develop a theory appropriate for the low temperature regime we follow the renormalization procedure used in quantum field theory so that we can make a perturbation expansion in terms of our fully dressed quasi-particles (see, for instance, Ryder, 1985). We take our renormalized parameters $\tilde{\epsilon}_d$, $\tilde{\Delta}(\omega)$ (which we take to be a constant by taking the wide band limit with a constant density of states) and $\tilde{U}$ as known and reorganize the perturbation expansion in powers of the renormalized coupling $\tilde{U}$. The full interaction Hamiltonian is $\tilde{H}_{\text{qp}}^{(I)} - \tilde{H}_{\text{c}}$. The terms λ_1, λ_2 and z are formally expressed as series in powers of $\tilde{U}$,

$$\lambda_1 = \sum_{n=0}^{\infty} \lambda_1^{(n)} \tilde{U}^n, \quad \lambda_2 = \sum_{n=0}^{\infty} \lambda_2^{(n)} \tilde{U}^n, \quad z = \sum_{n=0}^{\infty} z^{(n)} \tilde{U}^n. \quad \text{(L.15)}$$

The coefficients $\lambda_1^{(n)}$, $\lambda_2^{(n)}$ and $z^{(n)}$ are determined by the requirement that conditions (L.13) and (L.14) are satisfied to each order in the expansion. The perturbation expansion is about the free quasi-particle Hamiltonian given in equation (L.9) so that the non-interacting propagator in this expansion is

$$\tilde{G}_{\text{d}\sigma,\text{d}\sigma}^{(0)}(i\omega_n) = \frac{1}{i\omega_n - \tilde{\epsilon}_\text{d} + i\tilde{\Delta}\text{sgn}(\omega_n)}. \quad \text{(L.16)}$$

For most field theoretic models the counter terms are necessary to cancel the divergences which result where there is no high energy cut-off on the integrals. The condition that this procedure gives finite predictions in terms of renormalized parameters when the cut-off is removed from the regularized integrals is the condition that the field theory is renormalizable. In condensed matter systems there is always some form of high energy cut-off so this type of reorganization of the perturbation series is not necessary. However, for strong correlation problems, such as magnetic impurity and heavy fermion systems where the quasi-particles are strongly renormalized, it is a very suitable procedure to adopt for low temperature calculations. We propose to show that it makes direct contact with both the Landau phenomenological and the microscopic formulations of Fermi liquid theory.

The Friedel sum rule (5.47) gives the occupation of the d-level $n_{d,\sigma}$ at $T = 0$,

$$n_{\text{d},\sigma} = \frac{1}{2} - \frac{1}{\pi}\tan^{-1}\left(\frac{\epsilon_{\text{d},\sigma} + \Sigma_\sigma(0,H)}{\Delta}\right), \quad \text{(L.17)}$$

in terms of the self-energy $\Sigma(\omega)$ (in a finite magnetic field H) for the expansion in powers of U for the 'bare' Hamiltonian (L.1). It takes the same form for the renormalized expansion,

$$n_{\text{d},\sigma} = \frac{1}{2} - \frac{1}{\pi}\tan^{-1}\left(\frac{\tilde{\epsilon}_{\text{d},\sigma} + \tilde{\Sigma}_\sigma(0,H)}{\tilde{\Delta}}\right). \quad \text{(L.18)}$$

This follows directly from (L.17) by writing it in terms of $\tilde{\epsilon}_\text{d}$, $\tilde{\Delta}$ and $\tilde{\Sigma}$ from equation (L.6) as the common factor of z in the argument of (L.18) cancels.

The quasi-particle interaction plays no role as $T \to 0$ and $H \to 0$ as

$\tilde{\Sigma}(0,0) = 0$ and so $n_{d,\sigma}$ corresponds to the non-interacting quasi-particle number.

There are the two Ward identities in the unrenormalized theory which were derived in section 5.3, equations (5.77) and (5.78),

$$\frac{\partial \Sigma_\sigma(\omega)}{\partial \omega}\bigg|_{\omega=0} + \frac{\partial \Sigma_\sigma(\omega)}{\partial \mu}\bigg|_{\omega=0} = -\rho_{d,\sigma}(0)\Gamma_{\uparrow,\downarrow}(0,0), \qquad (L.19)$$

and

$$\frac{\partial \Sigma_\sigma(\omega)}{\partial h}\bigg|_{\omega=0} - \frac{\partial \Sigma_\sigma(\omega)}{\partial \omega}\bigg|_{\omega=0} = -\rho_{d,\sigma}(0)\Gamma_{\uparrow,\downarrow}(0,0), \qquad (L.20)$$

where μ is the chemical potential and $h = g\mu_B H$, and $\rho_d(0)$ is the d-density of states at the Fermi level. As the renormalized derivatives are related to the unrenormalized ones via

$$1 + \frac{\partial \tilde{\Sigma}_\sigma(\omega)}{\partial \mu}\bigg|_{\omega=0} = z\left(1 + \frac{\partial \Sigma_\sigma(\omega)}{\partial \mu}\bigg|_{\omega=0}\right), \qquad (L.21)$$

and

$$1 - \frac{\partial \tilde{\Sigma}_\sigma(\omega)}{\partial h}\bigg|_{\omega=0} = z\left(1 - \frac{\partial \Sigma_\sigma(\omega)}{\partial h}\bigg|_{\omega=0}\right), \qquad (L.22)$$

the two equations, (L.19) and (L.20), can be written in a renormalized form (all quantities with a tilde). From equation (L.13) the ω derivative of $\tilde{\Sigma}(\omega)$ vanishes at $\omega = 0$ so that the renormalized equations reduce to

$$\frac{\partial \tilde{\Sigma}_\sigma(\omega)}{\partial h}\bigg|_{\omega=0} = \frac{\partial \tilde{\Sigma}_\sigma(\omega)}{\partial \mu}\bigg|_{\omega=0} = -\tilde{\rho}_{d,\sigma}(0)\tilde{\Gamma}_{\uparrow,\downarrow}(0,0), \qquad (L.23)$$

where $\tilde{\rho}_{d,\sigma}(0) = \rho_{d,\sigma}(0)/z$ is the quasi-particle density of states at the Fermi level.

As the effects of the quasi-particle interactions go to zero as $T \to 0$ due to the cancellation with the counter term giving $\tilde{\Sigma}(0,0) = 0$ and $\tilde{\Sigma}'(0,0) = 0$, the specific heat coefficient of the impurity $\gamma_{\rm imp}$ is due to the non-interacting quasi-particles so

$$\gamma_{\rm imp} = \frac{2\pi^2 k_B^2}{3}\tilde{\rho}_d(0), \qquad (L.24)$$

as derived earlier in equations (5.54) and (5.56). In finite magnetic field the quasi-particle interaction gives an energy shift which is not cancelled by the counter term and the susceptibility is enhanced over the non-interacting quasi-particle value. The spin susceptibility is given by

$$\chi_{\rm imp} = \frac{(g\mu_B)^2}{2}\tilde{\rho}_d(0)(1 - \partial\tilde{\Sigma}/\partial h) = \frac{(g\mu_B)^2}{2}\tilde{\rho}_d(0)(1 + \tilde{U}\tilde{\rho}_d(0)). \quad (L.25)$$

and the charge susceptibility by

$$\chi_{\rm c,imp} = 2\tilde{\rho}_d(0)(1 + \partial\tilde{\Sigma}/\partial \mu) = 2\tilde{\rho}_d(0)(1 - \tilde{U}\tilde{\rho}_d(0)) \qquad (L.26)$$

using (L.17), (L.18) and (L.23).

These are equivalent to previous results (5.57), (5.60), (5.18) and (5.19) in the wide band limit, and give the Fermi liquid relation (5.81).

In the Kondo regime ($U \gg \pi\Delta$, $\epsilon_d < 0$, $\chi_{c,\mathrm{imp}} \to 0$) we can write γ_{imp} in terms of the Kondo temperature T_{L}, $\gamma_{\mathrm{imp}} = \pi^2 k_{\mathrm{B}}/6T_{\mathrm{L}}$. In this limit $n_d \to 1$ and hence from the Friedel sum rule (18) $\tilde{\epsilon}_d = 0$ and $\tilde{\rho}_d(0) = 1/\pi\tilde{\Delta}$. This gives us all the renormalized parameters in the strong correlation regime in terms of T_{L}.

$$\tilde{U} = \pi\tilde{\Delta} = 4k_{\mathrm{B}}T_{\mathrm{L}}, \tag{L.27}$$

where the value to $\tilde{U}$ was obtained by equating $\chi_{c,\mathrm{imp}}$ to zero.

For $U \ll \pi\Delta$ it is possible to calculate the renormalized parameters perturbatively in powers of the bare interaction U. For the particle-hole symmetric model to third order they are given by

$$\tilde{\Delta} = \Delta \left(1 - \frac{(12 - \pi^2)}{4} \left(\frac{U}{\pi\Delta} \right)^2 + O\left(\left(\frac{U}{\pi\Delta} \right)^4 \right) \right) \tag{L.28}$$

and

$$\tilde{U} = U \left(1 - (\pi^2 - 9) \left(\frac{U}{\pi\Delta} \right)^2 + O\left(\left(\frac{U}{\pi\Delta} \right)^4 \right) \right). \tag{L.29}$$

We can see from (L.26) that the physical requirement $\chi_{c,\mathrm{imp}} \geq 0$ implies that the renormalized perturbation expansion is always in the weak coupling regime which for the symmetric model corresponds to $\tilde{U} \leq \pi\tilde{\Delta}$.

The thermodynamic results can be obtained from the lowest order term in the renormalized perturbation for $\tilde{\Sigma}$, corresponding to the usual 'tadpole diagram', which gives

$$\tilde{\Sigma}^{(1)}(\omega, H) = \tilde{U}(n_{d,\sigma}^{(0)}(0, H) - n_{d,\sigma}^{(0)}(0, 0)). \tag{L.30}$$

There is no wavefunction renormalization to this order so $z^{(1)} = 0$ and to this order $\tilde{\Gamma}^{(1)}(\omega, \omega') = \tilde{U}$, $\lambda_2^{(1)} = 0$ and $\lambda_1^{(1)} = \tilde{U}n_{d,\sigma}^{(0)}(0,0)$. The complete cancellation by the counter term only occurs for $H = 0$ and $T = 0$. Calculation of the spin susceptibility from (L.31) gives (L.25), and similarly the charge susceptibility gives (L.26). The renormalized Ward identities in equation (L.23) show that higher order terms in $\tilde{U}$ cancel so that the first order calculations of χ_{imp} and $\chi_{c,\mathrm{imp}}$ are exact.

Exact results for the impurity Green's function (L.2) to order ω^2 follow from the calculation of the second order diagram for $\tilde{\Sigma}$ shown in figure 5.4, page 122. There is no second order counter term contribution to this diagram from λ_2 as $\lambda_2^{(1)} = 0$. There is a contribution to z which is required to eliminate the contribution linear term in ω to this order and is given by

$$z^{(2)} = \frac{(\pi^2 - 12)}{4\pi^2 \tilde{\Delta}^2}. \tag{L.31}$$

Calculation of this diagram to order ω^2 gives

$$\mathrm{Im}\,\tilde{\Sigma}_\sigma(\omega, 0) = \frac{\tilde{U}^2 \omega^2}{2\tilde{\Delta}(\pi\tilde{\Delta})^2}. \tag{L.32}$$

There is a temperature dependent contribution to $\tilde{\Sigma}_\sigma(0, T)$ to first order of the form (L.21) with T replaced by H. For the particle-hole symmetric model, and more generally in the Kondo regime $n_d \to 1$, this vanishes as $n_{d,\sigma}(0, T) = n_{d,\sigma}(0, 0) = 1$ and the leading order temperature dependence arises from the second order diagram 5.4. This gives the result

$$\mathrm{Im}\,\tilde{\Sigma}_\sigma(0, T) = \frac{(\pi\tilde{U})^2 k_B^2 T^2}{2\tilde{\Delta}(\pi\tilde{\Delta})^2}. \tag{L.33}$$

These results can be used to calculate the leading order T^2 contribution to the impurity conductivity $\sigma_{\mathrm{imp}}(T)$ and give

$$\sigma_{\mathrm{imp}}(T) = \sigma_0 \left\{ 1 + \frac{\pi^2}{3} \left(\frac{k_B T}{\tilde{\Delta}} \right)^2 (1 + 2(R-1)^2) + \mathrm{O}(T^4) \right\}, \tag{L.34}$$

where R is the Wilson ratio or 'χ/γ' ratio defined in (5.23) and is given by

$$R = 1 + \tilde{U}/\pi\tilde{\Delta}. \tag{L.35}$$

These results are identical to the exact result of Nozières (1974) in the Kondo regime, $\pi\tilde{\Delta} \to 4k_B T_L$, $R \to 2$, and also to the more general result of Yamada (1975) and Yosida & Yamada (1975) derived in section 5.4. Hence we see that all the Fermi liquid relations can be obtained within the renormalized expansion up to second order in $\tilde{U}$.

We see that the effects of the counter terms play no really significant role in the Fermi liquid regime. The same results were obtained in this regime in section 5.1 by neglecting $\tilde{H}_c$ and working solely with $\tilde{H}_{\mathrm{qp}}$. The renormalized perturbation theory with the counter terms, however, goes beyond the Fermi liquid regime. In this perturbation theory nothing has been omitted so that, in principle, calculations can be performed at high temperatures and high fields allowing the bare particles to be seen. If the primary interest is in these regimes then the more appropriate starting point is the model in terms of the bare parameters rather than the renormalized ones. Relatively low order calculations can, however, provide some estimate of the bare parameters in terms of the renormalized ones. These can be obtained by inverting

(L.6) and (L.12),

$$\epsilon_d = (\tilde{\epsilon}_d - \lambda_1)/z, \quad \Delta = \tilde{\Delta}/z, \quad U = (\tilde{U} - \lambda_2)/z^2, \tag{L.36}$$

where λ_1, λ_2 and z are implicit functions of $\tilde{\epsilon}_d$, $\tilde{\Delta}$ and $\tilde{U}$. In the very weak coupling limit $\tilde{U} \ll \pi\tilde{\Delta}$ $z \to 1$ and $\epsilon_d \to \tilde{\epsilon}_d$, $\Delta \to \tilde{\Delta}$ and $U \to \tilde{U}$. The most useful regime for the renormalized approach, however, should be in calculating corrections to Fermi liquid theory. The calculation with counter terms gives a systematic procedure for taking such corrections into account. Simple resonant level quasi-particle models of this type have been used to yield qualitatively correct behaviour for magnetic impurity models in the higher temperature regime $T \sim T_{\mathrm{L}}$ (Newns & Hewson, 1980). This procedure gives a way of extending those approaches.

From the exact Bethe ansatz results for $\chi_{\mathrm{c,imp}}$ and γ_{imp} for the symmetric Anderson model in section 6.7, equations (6.111) and (6.112), it is possible to calculate $\tilde{\Delta}$ and $\tilde{U}$ exactly. The asymptotic forms for small U, equation (L.28), and large U, equation (L.27) can be verified.

The approach can be extended to other impurity models such as the N-fold degenerate Anderson model (7.53) and the n-channel Kondo model with $n = 2S$ (9.28) and the calculations proceed along similar lines. The method justifies the form of the quasi-particle Hamiltonian used in the treatment of these models in sections 7.4 and 9.4.

To give insight into the renormalized perturbation theory we have demonstrated its application here to magnetic impurity models but it can be applied quite generally to describe Fermi liquid theory, including the dynamics due to quasi-particle scattering. More importantly it provides a framework for considering corrections to Fermi liquid theory and in certain circumstances, such as low dimensionality or in the presence of attractive interactions it could be used to investigate the breakdown of Fermi liquid theory, such as in the one-dimensional Hubbard model or superconductivity where the breakdown occurs because quasi-particle scattering becomes singular.

Addendum

Some Recent Developments

There has been much activity in recent years in the field of strongly correlated electron systems. This has been driven in part by the continued efforts to understand the behaviour of the high temperature superconductors, for which there is as yet no universally accepted theory. There are several competing 'scenarios'; of those some give a primary role to the strong antiferromagnetic fluctuations [1], spin/charge separation [2], or the presence of the van Hove singluarity [3]. It would be impossible to summarize these theories in a few lines here, or the other extensive work in this field, so I will give references to a few recent theoretical review articles [4], where the interested reader can get an overview of the subject, and find further references. Some primarily experimental reviews of the high T_c materials are given in reference [5].

There have been some developments in the theory of strongly correlated electron systems which relate rather directly to the topics covered in this book. To bring this present edition up-to-date I think it would be worthwhile to describe these briefly, as well as providing a few references. The first topic is now often referred to as 'dynamic mean field' theory, which has arisen out of work on strongly correlated lattice models, such as the Hubbard model and periodic Anderson model, in the limit of infinite dimensionality.

It was mentioned in Chapter 10, section 10.6, that in the infinite dimensionality limit the electron self-energy for lattice models with a purely on-site interaction depends only on the frequency variable ω and not on the wave-vector $\mathbf{k}$ [6]. The hopping term must be scaled appropriately with the dimension d in taking this limit. For example, for the Hubbard model (10.60) which has the on-site interaction term $Un_{i\uparrow}n_{i\downarrow}$, the hopping matrix element $t_{i,j}$ for a hypercubic lattice must be scaled as $(\sqrt{d})^{-|\mathbf{i}-\mathbf{j}|}$; where $|\mathbf{i}-\mathbf{j}|$ is the distance between sites $\mathbf{i}$ and $\mathbf{j}$ along the lattice directions. With this form for the self-energy the one electron thermal Green's function $G_\sigma(\mathbf{k},i\omega_n)$ can be written as

$$G_\sigma(\mathbf{k},i\omega_n) = \frac{1}{i\omega_n + \mu - \epsilon_\mathbf{k} - \Sigma(i\omega_n)}, \tag{1}$$

where $\epsilon_\mathbf{k}$ is the energy of the non-interacting Bloch electrons with wavevector $\mathbf{k}$. If we sum (1) over $\mathbf{k}$ we obtain the local or on-site Green's function $G_{L,\sigma}(i\omega_n)$ $(= G_{i\sigma,i\sigma}(i\omega_n))$ which can be expressed as an integal over the density of states $D_0(\epsilon)$ of the non-interacting electrons,

$$G_{L,\sigma}(i\omega_n) = \frac{1}{N_s}\sum_\mathbf{k} G_\sigma(\mathbf{k},i\omega_n) = \int_{-\infty}^{\infty}\frac{D_0(\epsilon)d\epsilon}{i\omega_n + \mu - \epsilon - \Sigma(i\omega_n)}. \tag{2}$$

The mean field equations are derived by focussing attention on one particular site which can be regarded as an impurity. With a purely on-site interaction, which we take for illustration to be of the form $Un_{i\uparrow}n_{i\downarrow}$, electrons on the 'impurity' can only be transferred to the rest of the lattice by the hopping term, $\sum_{i,j}c_{i,\sigma}^\dagger c_{j,\sigma}$, for the Hubbard model. Then by reorganizing the perturbation theory in powers of the interaction U so that explicit account is taken only of the scattering at the 'impurity' site [7], or via a functional integral approach in which the states of the electrons in the rest of the lattice are formally integrated out to generate a local effective Lagrangian [8], the local Green's function in the infinite dimensional limit can be cast into the form,

$$G_{L,\sigma}(i\omega_n) = \frac{1}{[G_{L,\sigma}^{(0)}(i\omega_n)]^{-1} - \Sigma(i\omega_n)}, \tag{3}$$

where $G_{L,\sigma}^{(0)}(i\omega_n)$ is the Green's function at the 'impurity' site when the interaction U *at the impurity site only* is set to zero. The Green's function $G_{L,\sigma}^{(0)}(i\omega_n)$ can always be expressed in the form,

$$[G_{L,\sigma}^{(0)}(i\omega_n)]^{-1} = i\omega_n - \frac{1}{\pi}\int\frac{\bar{\Delta}(\omega')d\omega'}{(i\omega_n - \omega')}, \tag{4}$$

where $\bar{\Delta}(\omega)$ can be interpreted within the framework of an Anderson model as due to an effective hybridization $\bar{V}_\mathbf{k}$ such that $\bar{\Delta}(\omega) = \pi\sum_\mathbf{k}|\bar{V}_\mathbf{k}|^2\delta(\omega - \epsilon_\mathbf{k})$.

Given the local or impurity Green's function for the Anderson model $G_{\text{imp},\sigma}^{(0)}(i\omega_n)$, and the value of U, the self-energy $\Sigma(i\omega_n)$ is completely determined, i.e. $\Sigma(i\omega_n) = F(U, G_{\text{imp},\sigma}^{(0)})$ where F is a universal functional which can be evaluated using any one of the methods for calculating the self-energy of the Anderson model described in sections 5.5, 5.6, 8.2 and 9.6. Hence given the form of $\bar{\Delta}(\omega)$, the self-energy for the effective impurity problem can be calculated using $G_{\mathrm{L},\sigma}^{(0)}$ for $G_{\mathrm{imp},\sigma}^{(0)}$. In this fomuation $\bar{\Delta}(\omega)$ is not known a priori but must be determined self-consistently by the requirement that the equations (2) and (3) for $G_{\mathrm{L},\sigma}(i\omega_n)$ should be equivalent. This then provides an algorithm, based on impurity calculations, for determining the self-energy for the lattice problem, which is exact in the $d \to \infty$ limit. The approach can be used as an approximation for the one electron self-energy for lattice models of lower dimensionality. Because of the analogy to the mean field theories used for magnetic models, such as the Ising and Heisenberg models, which are exact in similar limits, this approach has been termed a mean field theory, but in this case a dynamic one because the effective 'field' $\bar{\Delta}(\omega)$ is frequency dependent.

Dynamic mean field calculations have been performed for a number of strong correlation models using techniques to claculate the impurity self-energy such as perturbation theory (§5.5), Monte Carlo (§9.6), the NCA (§8.2), and the numerical renormalization group (§5.6). The selfconsistency constraint can induce a strong ω dependence in $\bar{\Delta}(\omega)$, rather unlike that for a typical impurity where the frequency dependence of $\Delta(\omega)$ can usually be neglected. This means that any approximation which works well for the calculation of the self-energy for an impurity has to be reassessed when the selfconsistency condition is used. A full account of this approach and a comprehensive list of references can be found in the review paper of Georges et al. [8].

As an illustration of an application of this method we look briefly at some calculations for the Hubbard model in the half-filled band case, based on a second order perturbation theory calculation of the self-energy [8]. Results are shown in figure 1 as a function of U for the spectral density of the local Green's function. It is interesting to compare these with similar calculations for the corresponding Anderson impurity model in sections 5.5 and 5.6. As U increases this spectrum develops a three peaked structure as in the impurity case. However for larger values of U the central quasi-particle peak becomes isolated with the depletion of states immediately above and below this peak. Eventually at a critical value of U there is a Mott transition to an insulating state in which the quasi-particle peak disappears completely and there is a finite band gap between the upper and lower Hubbard bands (see the lowest section in figure 1). The height of the quasi-particle peak remains unchanged up to this point as a result of the Friedel sum rule, but then

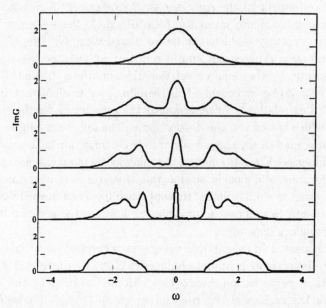

Figure 1 The spectral density for the local Green's function at $T = 0$ calculated from the iterated second order perturbation theory for the $d = \infty$ Hubbard model. Reading from top to bottom the figures correspond to $U/D = 1, 2, 2.5, 3, 4$, where $2D$ is the band width for $U = 0$ (taken from reference [8]).

breaks down at, and above, the Mott transition when the self-energy develops a zero frequency singularity. This basic picture of the Mott transition is confirmed by other dynamic mean field calculations based on Monte Carlo calculations [8].

There have also been some significant technical advances in the application of the numerical renormalization group method to strong correlation models in low dimensions. When the Wilson approach, as described in §4.4, is applied to lattice problems there are difficulties in selecting a suitable truncated set of basis states for convergence of the sequence of iterative diagonalizations. These difficulties have been overcome in the 'density matrix numerical renormalization group algorithm' (DMRGA). The approach is limited to the calculation of the ground state and a few of the low lying excitations, but these can be obtained with remarkable accuracy for one dimensional systems. The algorithm is too involved to describe briefly here but details can be found in the paper of White [9], and references to some recent work using this approach can be found in [10].

There has been extensive work on heavy fermion and other anomalous 4f and 5f systems. As well as the discovery of some new superconductors (e.g. UPd_2Al_3) there has been the discovery of a class of materials similar to SmB_6 (see §10.2), which appear to be insulating as $T \to 0$ due to a very small renormalized hybridization gap (see §10.5, equation (10.50)), and these materials have been termed 'Kondo insulators'. Examples are $Ce_3Bi_4Pt_3$ and $CeNiSn$ [11]. There has also been increasing interest in those 4f and 5f materials which have been observed to display non-Fermi liquid behaviour at low temperatures (for a review see [12]). Anderson [13] has drawn attention to the fact that there are many other materials which do not behave like conventional Fermi liquids at low temperatures, quite apart from the well known example of the high T_c materials (see §10.7). This has stimulated the study of models which have some form of non-Fermi liquid low temperature fixed point. Much of this work has focussed on the two channel Kondo model, which corresponds to the multichannel model discussed in Appendix J (see also the review in reference [14]) with $n = 2$ and $S = \frac{1}{2}$. This model gives a $\ln T$ term in the specific heat coefficient, which is often observed as the signature of non-Fermi liquid behaviour, and a resistivity contribution which behaves as $T^{1/2}$. This type of behaviour is reported to occur in $U_{0.9}Th_{0.1}Be_{13}$ [16], and may be a consequence of the quadrupolar two channel Kondo effect (Cox(1987b)). The resistivities of most non-Fermi liquid systems display a linear temperature dependence. Non-Fermi liquid behaviour can occur at the critical point when a magnetic phase transition is reduced to zero temperature either by pressure or chemical substitution, and this is believed to be the origin of the non-Fermi liquid behaviour observed in $CeCu_{5.9}Au_{0.1}$ [15]. Another possible explanation for non-Fermi liquid behaviour in disordered systems is based on the idea of a distribution $P(T_K)$ of Kondo temperatures. When an ensemble average is taken over the possible exchange couplings in a disordered material it can result in a singular form for $P(T_K)$, and consequently a singular temperature dependence in the thermodynamic behaviour, due to virtually isolated moments, and an apparent non-Fermi liquid behaviour [17].

There has been further speculation as to the origin of the superconductivity in the heavy fermion materials (for a recent review of heavy fermion superconductivity see [18]), and some exotic mechanisms have been proposed including odd frequency pairing, possibly induced by the two channel Kondo model [19].

The speculation on non-Fermi liquid theories has also stimulated a fresh look at Fermi liquid theory. Some of this work has been based on the generalization of the bosonization approach to dimensions greater than one [20]. Other work has been from a renormalization group perspective aimed at a clarification of the concept of a Fermi liquid fixed point (for a review see [21]). The renormalized perturbation theory, described in Appendix L, has also been generalized to an expansion about a Fermi liquid fixed point for a translationally invariant system [22]. This form of the renormalization group approach provides a link between the intuitive Landau phenomenological approach to Fermi liquid theory and the more formal diagramatic perturbation theory.

The recent work on heavy fermion systems is well covered in the proceedings of the SCES conferences 1993 to 1995 [23]. There is also a recent review article on the spectroscopies of Kondo systems in [24], and one on non-Fermi liquid systems in [25].

REFERENCES

[1] Monthoux, P. & Pines, D. (1994) *Phys. Rev. B* **49**, 4261.

[2] Anderson, P.W. (1995) *J. Phys. Chem. Sol.* **56**, 1593: (1994) *Physica B* **199–200**, 8.

[3] Newns, D.M., Tsuei, C.C. & Pattnaik, P.C. (1995) *Phys. Rev. B* **52**, 13611.

[4] Brenig, W. (1995) *Physics Reports* **251**, 153: Scalapino, D.J., *Physics Reports* **250**, 329: Dagotto, E. (1994) *Rev. Mod. Phys.* **66**, 763: Kampf, A.P. (1994) *Physics Reports* **249**, 219: see also Anderson, P.W. and Schrieffer, J.R. (1991) in *Physics Today*, June, 61.

[5] *Physical Properties of High Temperature Superconductors*, D.M. Ginsberg (ed.), Vols I–IV (1990–94) Singapore: World Scientific; see also *J. Phys. Chem. Sol.* **56**, 1567–1973 (Spectroscopies).

[6] Metzner, W. & Vollhardt, D. (1989). *Phys. Rev. Lett.* **62**, 324: Müller-Hartmann, E. (1989) *Z. Phys. B* **74**, 3298.

[7] Edwards, D.M. (1993) *J. Phys. Cond. Mat.* **5**, 161.

[8] Georges, A., Kotliar, G., Krauth, W. & Rozenberg, M.J. (1996) *Rev. Mod. Phys.* **68**, 13.

[9] White, S.R. (1993) *Phys. Rev. B* **48**, 10345.

[10] White, S.R., Noack, R.M. & Scalapino, D.J. (1995) *J. Low Temp. Phys.* **99**, 593.

[11] Aeppli, G. & Fisk, Z. (1992) *Comments Cond. Mat. Phys.* **16**, 155.

[12] Maple, M.B. (1995) *J. Low Temp. Phys.* **99**, 223 (there are related papers in this volume, pages 183–635).

[13] Anderson, P.W. (1995) *Physics World*, December, 37.

[14] Schlottmann, P. (1993) *Advances in Physics* **42**, 641.

[15] Aliev, F.G., Vieira, S., Villar, R., Martinez, J.L., Seaman, C.L. & Maple, M.B. (1995) *Physica B* **206 & 207**, 454.

[16] Löhneysen, H.V. (1995) *Physica B* **206 & 207**, 101.

[17] Bhatt, R.N. & Fisher, D.S. (1992) *Phys. Rev. Lett.* **68**, 3072; Dobrosavljević, V., Kirkpatrick, T.R. & Kotliar G. (1992) *Phys. Rev. Lett.* **69**, 1113.

[18] Heffner, R.H. & Norman, M.R. (1996) *Comments in Cond. Mat. Phys.* **17**, 361.

[19] Cox, D.L. & Maple, M.B. (1995) *Physics Today* **44**, June, 54; Emery, V.J. & Kivelson, S.A. (1992) *Phys. Rev. B* **46**, 10812; (1994) *Phys. Rev. Lett.* **72**, 1918: Coleman, P., Miranda, E. & Tsvelik, A.M. (1993) *Phys. Rev. Lett.* **70**, 2960.

[20] Houghton, A. & Marston, J.B. (1993) *Phys. Rev. B* **48**, 7790: Castro-Neto, A.H. & Fradkin, E.H. (1995) *Phys. Rev. B* **51**, 4084.

[21] Shankar, R. (1994) *Rev. Mod. Phys.* **66**, 129.

[22] Hewson, A.C. (1994) *Advances in Physics* **43**, 543.

[23] Proceedings of the International Conferences on Strongly Correlated Electron Systems (SCES) *Physica B* (1994) **199–200**, (1995) **206 & 207**, (1996) **223 & 224**.

[24] Malterre, D., Grioni, M., & Baer, Y. (1996) *Advances in Physics* **45**, 299.

[25] Allen, J.W., Gweon, G.-H., Claessen, R. & Matho, K. (1995) *J. Phys. Chem. Sol.* **56**, 1849.

REFERENCES

Abbas, D.C., Aton, T.J. & Slichter, C.P. (1982). *Phys. Rev. B* **25**, 1474.

Abrikosov, A.A. (1965). *Physics* **2**, 5.

Abrikosov, A.A., Gorkov, L.P. & Dzyaloshinski, I.E. (1975). *Methods of Quantum Field Theory in Statistical Physics*. New York:Dover.

Abrikosov, A.A.& Migdal, A.B. (1970). *J. Low Temp. Phys.* **3**, 519.

Adenwalla, S., Lin, S.W., Zhao, Z., Ran, Q.Z., Ketterson, J.B., Sauls, J.A., Taillefer, L., Honks, D.G., Levy, M. & Sarma, B.K. (1990). *Phys. Rev. Lett.* **65**, 2298.

Aeppli, G., Bucher, E., Goldman, A.I., Shirane, G., Broholm, C. & Kjems, J.K. (1988). *J. Mag. Mag. Mat.* **36 & 77**, 385.

Aligia, A.A., Balseiro, C.A. & Proetto, C.R. (1986). *Phys. Rev. B* **33**, 6476.

Allen, J.W., Batlogg, B. & Wachter, P. (1979). *Phys. Rev. B* **20**, 4807.

Allen, J.W. & Martin, R.W. (1982). *Phys. Rev. Lett.* **49**, 1106.

Allen, J.W., Oh, S-J, Gunnarsson, O., Schönhammer, K., Maple, M.B., Torikachvili, M.S. & Lindau I. (1986). *Adv. in Phys.* **35**, 275.

Alloul, H. (1975). *Phys. Rev. Lett.* **35**, 460.

Alloul, H. (1977). *Physica B* **86–88**, 449.

Amato, A., Jaccard, D., Flouquet, J., Lapierre, F., Tholence, J.L., Fisher, R.A., Lacy, S.E., Olsen, J.A. & Phillips, N.E. (1987). *J. Low Temp. Phys.* **68**, 371.

Andersen, O.K. (1975). *Phys. Rev. B* **12**, 3060.

Anderson, P.W. (1961). *Phys. Rev.* **124**, 41.

Anderson, P.W. (1967). *Phys. Rev.* **164**, 352.

Anderson, P.W. (1970). *J. Phys. C* **3**, 2439.

Anderson, P.W. (1987). *Science* **235**, 1196.

Anderson, P.W. & Ren, Y. (1990). In *The Los Alamos Symposium 1989, High Temperature Superconductivity Proceedings*, ed. K.S. Bedell, D. Coffey, D.E. Meltzer, D. Pines, & J.R. Schrieffer, p. 3, New York:Addison-Wesley.

Anderson, P.W. & Yuval, G. (1969). *Phys. Rev. Lett.* **23**, 89.

Anderson, P.W. & Yuval, G. (1973). In *Magnetism*, eds. G.T. Rado & H. Suhl, vol V, p. 217. New York:Academic Press.

Anderson, P.W., Yuval, G. & Hamann, D.R. (1970). *Phys. Rev. B* **1**, 4464.
Andrei, N. (1980). *Phys. Rev. Lett.* **45**, 379.
Andraka, B. & Tsvelick, A.M. (1991). *Phys. Rev. Lett.* **67**, 2886.
Andrei, N. (1982). *Phys. Lett.* **87A**, 299.
Andrei, N. & Destri, C. (1984). *Phys. Rev. Lett.* **52**, 364.
Andrei, N., Furuya, K. & Lowenstein, J.H. (1983). *Rev. Mod. Phys.* **55**, 331.
Andrei, N. & Lowenstein, J.H. (1981). *Phys. Rev. Lett.* **46**, 356.
Andres, K., Graebner, J.E. & Ott H.R. (1975). *Phys. Rev. Lett.* **55**, 1979.
Anisimov, V.I. & Gunnarsson, O. (1991). *Phys. Rev. B* **43**, 7570.
Appelbaum, J.A. & Kondo, J. (1967) *Phys. Rev. Lett.* **19**, 906.
Auerbach, A. & Levin, K. (1986a). *Phys. Rev. Lett.* **57**, 877.
Auerbach, A. & Levin, K. (1986b). *Phys. Rev. B* **34**, 3524.
Bader, S.D., Phillips, N.E., Maple, M.B. & Luengo, C.A. (1975). *Sol. Stat. Com.* **16**, 1263.
Balian, R. & de Dominicis C. (1971). *Ann. Phys. NY* **62**, 229.
Baliña, M. & Aligia, A.A. (1990). *Europhys. Lett.* **13**, 739.
Bangalore (1987). *Theoretical and Experimental Aspects of Valence Fluctuators and Heavy Fermions*, New York & London: Plenum Press.
Batlogg, B. (1991) *Physica B* **169**, 7.
Batlogg, B., Kaldis, E., Schlegel, A. & Wachter, P. (1976). *Phys. Rev. B* **14**, 5053.
Baxter, R.J. (1982). *Exactly Solved Models in Statistical Mechanics.* London: Academic Press.
Bednorz, J.G. & Müller, K.A. (1986). *Z. Phys. B* **64**, 189.
Bednorz, J.G. & Müller, K.A. (1988). *Rev. Mod. Phys.* **60**, 585.
Bethe, H. (1931). *Z. Phys.* **71**, 205.
Bickers, N.E. (1987). *Rev. Mod. Phys.* **59**, 845.
Bickers, N.E., Cox, D.L. & Wilkins, J.H. (1985). *Phys. Rev. Lett.* **54**, 230.
Bickers, N.E., Cox, D.L. & Wilkins, J.H. (1987). *Phys. Rev. B* **36**, 2036.
Blandin, A. & Friedel, J. (1959). *J. Phys. Radium* **20**, 160.
Blankenbecler, R., Scalapino, D.J. & Sugar, R.L. (1981). *Phys. Rev. D* **24**, 2278.
Bloch, C. (1965). In *Studies in Statistical Mechanics*, eds. J. de Boer & G.E. Uhlenbeck, **111**, p. 1. Amsterdam:North-Holland.
Bloomfield, P.E. & Hamann, D.R. (1967). *Phys. Rev.* **164**, 856.
Boato, G. & Vig, J. (1967). *Sol. St. Com.* **5**, 649.
Boppart, H., Treindl, A., Wachter, P. & Roth, S. (1980). *Sol. St. Com.* **35**, 483.
Boyce, J. & Slichter, C. (1974). *Phys. Rev. Lett.* **32**, 61.
Brandow, B.H. (1986). *Phys. Rev. B* **33**, 215.
Brandow, B.H. (1988). *Phys. Rev. B* **37**, 250.
Brandt, U., Keiter, H. & Liu, F.S. (1985). *Z. Phys. B* **58**, 267.
Bredl, C.D. (1987). *J. Mag. Mag. Mat.* **63** & **64**, 355.
Bredl, C.D., Horn, S., Steglich, F., Lüthi, B. & Martin, R.M. (1984). *Phys. Rev. Lett.* **52**, 1982.
Brenig, W. & Zittartz, J. (1973). In *Magnetism*, eds. G.T. Rado & H. Suhl, vol V, p. 185. NewYork:Academic Press.
Brito, J.J.S. & Frota, H.O. (1990). *Phys. Rev. B* **42**, 6378.

Broholm, C., Kjems J.K., Aeppli, G., Fisk, Z., Smith, J.L., Shapiro, S.M., Shirane, G. & Ott, H.R. (1987). *Phys. Rev. Lett* **58**, 917

Campagna, M. & Wertheim, G.K. (1974). *AIP Conf. Proc. Magnetism and Magnetic Materials* **24**, eds. C.D. Graham, Jr., G.H. Lander & J.J. Rhyne. p. 22. New York:Institute of Physics.

Caplin, A.D. & Rizzuto, C. (1968). *Phys. Rev. Lett.* **21**, 746.

Chapman, S., Hunt, M., Meeson, D., Reinders, P.H.P. & Springford, M. (1990). *J. Phys. Cond. Mat.* **2**, 8123.

Clogston, A.M., & Anderson, P.W. (1961). *Bull. Am. Phys. Soc.* **6**, 124.

Clogston, A.M., Matthias, B.T., Peter, M., Williams, H.J., Corenzwit, E. & Sherwood, R.C. & (1962). *Phys. Rev.* **125**, 541.

Coleman, P. (1984). *Phys. Rev. B* **29**, 3035.

Coleman, P. (1985). In *Theory of Heavy Fermions and Valence Fluctuations*, eds. Kasuya & Saso, p. 163. Berlin:Springer Verlag.

Coleman, P. (1987). In *Theoretical and Experimental Aspects of Valence Fluctuators and Heavy Fermions*, p. 581. New York & London: Plenum Press.

Coleman, P., Anderson, P.W. & Ramakrishnan, T.V. (1985). *Phys. Rev. Lett.* **55**, 414.

Coleman, P. & Andrei, N. (1986). *J. Phys. C* **19**, 3211.

Cooper, J.R., Vučić, Z. & Babić, E. (1974). *J. Phys. F* **4**, 1489.

Coqblin, B. (1977). *The Electronic Structure of Rare-Earth Metals and Alloys: the Magnetic Heavy Rare-Earths*. London & New York:Academic Press.

Coqblin, B. & Schrieffer, J.R. (1969). *Phys. Rev.* **185**, 847.

Costi, T.A. & Hewson, A.C. (1990). *Physica B* **163**, 179 & unpublished.

Costi, T.A. & Hewson, A.C. (1991). Unpublished.

Costi, T.A. & Hewson, A.C. (1992a). *J. Mag. Mag. Mat.* **108**, 129.

Costi, T.A. & Hewson, A.C. (1992b). *Phil. Mag. B* **65**, 1165.

Cox, B.N., Coulthard, M.A.. & Lloyd, P. (1974). *J. Phys. F* **4**, 807.

Cox, D. L. (1987a). *Phys. Rev. B* **35**, 4561.

Cox, D. L. (1987b). *Phys. Rev. Lett.* **59**, 1240

Cox, D. L. (1988). *Physica C* **153**, 1642.

Cox, D.L. (1991). Preprint.

Cox, D. L. & Grewe, N. (1988). *Z. Phys. B* **71**, 321.

Cox, P.A., Lang, J.K. & Baer, Y. (1981). *J. Phys. F* **11**, 113.

Cragg, D.M. & Lloyd, P. (1979). *J. Phys. C* **12**, 3301.

Cragg, D.M., Lloyd, P. & Nozières, P. (1980). *J. Phys. C* **13**, 803.

Daybell, M.D. (1973). In *Magnetism*, eds. G.T. Rado & H. Suhl, vol V, p. 121. New York:Academic Press.

Daybell, M.D. & Steyert, W.A. (1968). *Rev. Mod. Phys.* **40**, 380.

Dederichs, P.H., Blügel, S., Zeller, R. & Akai, H. (1984). *Phys. Rev. Lett.* **53**, 2512.

Desgranges, H-U. (1985). *J. Phys. C* **18**, 5481.

Desgranges, H-U. & Rasul, J.W. (1985). *Phys. Rev. B* **32**, 6100.

Desgranges, H-U. & Rasul, J.W. (1987). *Phys. Rev. B* **36**, 328.

Desgranges, H-U. & Schotte, K.D. (1982). *Phys. Lett. A* **91**, 240.

Doniach, S. (1987). *Phys. Rev. B* **35**, 1814.

Doniach, S. & Šunjić, M. (1970). *J. Phys. C* **3**, 285.

Drew, H.D. & Doezema, R.E. (1972). *Phys. Rev. Lett.* **28**, 1581.

Edwards, D.M. (1988). In *Narrow Band Phenomenoma*, p. 23, NATO ASI Series. New York:Plenum.

Edwards, D.M. (1991). *Physica B* **169**, 271.

Evans, S.M.M. (1991). *J. Phys. Cond. Mat.* **3**, 8441.

Evans, S.M.M. (1992). *J. Mag. Mag. Mat.* **108**, 129.

Evans, S.M.M. & Coqblin, B. (1991). *Phys. Rev. B* **43**, 12790.

Evans, S.M.M., Chung, T. & Gehring, G.A. (1989). *J. Phys. Cond. Mat.* **1**, 3095

Evans, S.M.M. & Gehring, G.A. (1989). *J. Phys. Cond. Mat.* **1**, 10487.

Fano, U. (1961). *Phys. Rev.* **124**, 1866.

Felsch, W. & Winzer, K. (1973). *Sol. Stat. Com.* **13**, 569.

Felsch, W., Winzer, K., & von Minnigerode, G. (1975). *Z. Phys. B* **21**, 151.

Fert, A. & Levy, P.M. (1987). *Phys. Rev. B* **36**, 1907.

Fetter, A.L. & Walecka, J.D. (1971). *Quantum Theory of Many Particle Systems.* New York:McGraw Hill.

Filyov, V.M. Tsvelick, A.M. & Wiegmann, P.B. (1981). *Phys. Lett. A* **81**, 115.

Fisk, Z., Ott, H.R., Rice, T.M. & Smith, J.L. (1986). *Nature* **320**, 124.

Flouquet, J. (1970) *Phys. Rev. Lett.* **25**, 288.

Flouquet, J. (1978) In *Progress in Low Temperature Physics,* **VII**, ed. D.F. Brewer, p. 650. Amsterdam:North Holland.

Frankel, R.B., Blum, N.A. Schwartz, B.B. & Kim, D.K. (1967). *Phys. Rev. Lett.* **18**, 1051.

Frankowski, I. & Wachter, P. (1982). *Sol. Stat. Com.* **41**, 577.

Friedel, J. (1952). *Phil. Mag.* **43**, 153.

Friedel, J. (1956). *Can. J. Phys.* **34**, 1190.

Friedel, J. (1958). *Nuovo Cimento (Suppl)* **7**, 287.

Friedel, J. (1989). *J. Phys. Cond. Mat.* **1**, 7757.

Frossati, G., Mignot, J.M., Thoulouze, D. & Tournier, R. (1976). *Phys. Rev. Lett* **36**, 203.

Frota, H.O. & Oliveira L.N. (1986). *Phys. Rev. B* **33**, 7871.

Fulde, P. (1988). In *Narrow Band Phenomenoma*, p. 27, NATO ASI Series. New York:Plenum.

Fulde, P., Keller, J. & Zwicknagl, G. (1988). *Solid State Physics* **41**, eds. F. Seitz, D. Turnbull & H. Ehrenreich, p. 2, New York:Academic Press.

Geens, R., Labro, M. & Mordijck, A. (1987). *Phys. Rev. Lett.* **59**, 2345.

Grewe, N. (1982). *Z. Phys. B* **52**, 193.

Grewe, N. & Steglich, F. (1991). In *Handbook on the Physics and Chemistry of Rare Earths*, eds. K.A. Gschneider Jr. & L. Eyring, **14**, p. 343. Amsterdam:North-Holland.

Gruhl, H. & Winzer, K. (1986). *Sol. St. Com.* **57**, 67.

Grüner, G. (1974). *Adv. Phys.* **23**, 941.

Grüner, G. & Minier, M. (1977). *Adv. Phys.* **26**, 231.

Grüner, G. & Zawadowski A. (1974). *Rep. Prog. Phys.* **37**, 1497.

Grüner, G. & Zawadowski A. (1978). In *Progress in Low Temperature Physics*, ed. D.F. Brewer, **V11B**, p. 593. Amsterdam: North-Holland.

Gubernatis, J.E., Hirsch, J.E. & Scalapino, D.J. (1987). *Phys. Rev. B* **35**, 8478.

Gunnarsson, O., Andersen, O.K., Jepson, O. & Zaanan, J. (1988). In *Proceedings of the Tenth Taniguchi Symposium on Core Level Spectroscopies*, eds. J. Kanamori, & A. Kotani, p. 82. Berlin:Springer-Verlag.

Gunnarsson, O. & Schönhammer, K. (1983). *Phys. Rev. Lett.* **50**, 604; *Phys. Rev. B* **28**, 4315.

Gunnarsson, O. & Schönhammer, K. (1985). *Phys. Rev. B* **31**, 4815.

Gunnarsson, O. & Schönhammer, K. (1987). In *Handbook on the Physics and Chemistry of Rare Earths*, eds. K.A. Gschneider Jr. & L. Eyring, **10**. Amsterdam: North-Holland.

Gutzwiller, M.C. (1965). *Phys. Rev.* **137**, A1726.

de Haas, W.J., de Boer J.H. & van den Berg, G.J. (1934). *Physica* **1**, 1115.

Haen, P., Flouquet, J., Lapierre, F., Lejay, P. & Remenyi, F. (1987). *J. Low Temp. Phys.* **67**, 391.

Haldane, F.D.M. (1978). *Phys. Rev. Lett.* **40**, 416.

Haldane, F.D.M. (1981). *J. Phys. C* **14**, 2585.

Hamann, D.R. (1967). *Phys. Rev.* **158**, 570.

Hanzawa, K., Yamada, K. & Yosida K. (1985). *J. Mag. Mag. Mat.* **47**, 357.

Hedgcock, F.T. & Li, P.I. (1970). *Phys. Rev. B* **2**, 1342.

Hedgcock, F.T. & Rizzuto, C. (1967). *Phys. Rev.* **163**, 517.

Heeger, A.J. (1969). In *Solid State Physics* **23**, eds. F. Seitz, D. Turnbull & H. Ehrenreich, p. 283. New York:Academic Press.

Herbst, J.F., Watson, R.E. & Wilkins, J.W. (1978). *Phys. Rev. B* **17**, 3089.

Herring, C. (1966). *Magnetism*, eds. G.T. Rado & H. Suhl, vol IV. New York:Academic Press.

Hershfield, S., Davies, J.H. & Wilkins, J.W. (1991). *Phys. Rev. Lett.* **67**, 3720.

Hewson, A.C. (1982). *J. Phys. C* **15**, L611.

Hewson, A.C. (1992). Preprint.

Hewson, A.C. & Rasul, J.W. (1983). *J. Phys. C* **16**, 6799.

Hewson, A.C., Rasul, J.W. & Newns, D.M. (1983). *Sol. Stat. Com.* **47**, 59.

Hirsch, J.E. (1983). *Phys. Rev. B* **28**, 4059.

Hirsch, J.E. & Fye, R.M. (1986). *Phys. Rev. Lett.* **56**, 2521.

Hirst, L.L. (1978). *Adv. Phys.* **27**, 231.

Hohenberg, P.C. & Kohn, W. (1964). *Phys. Rev. B* **136**, 864.

Holland-Moritz, E. (1983). *J. Mag. Mag. Mat.* **38**, 253.

Holland-Moritz, E., Loewenhaupt, M., Schmatz, W. & Wohlleben, D.K. (1976). *Phys. Rev. Lett.* **38**, 983.

Holland-Moritz, E., Wohlleben, D.K. & Loewenhaupt, M. (1982). *Phys. Rev. B* **25**, 7482.

Horvatić, B. & Zlatić, V. (1985). *J. Physique* **46**, 1459.

Horvatić, B., Šokčević, D. & Zlatić, V. (1987). *Phys. Rev. B* **36**, 365.

Houghton, A., Read, N. & Won, H. (1987). *Phys. Rev. B* **35**, 5213.

Houghton, A., Read, N. & Won, H. (1988). *Phys. Rev. B* **37**, 3782.

Hubbard, J. (1963). *Proc. Roy. Soc. A* **276**, 238.

Hubbard, J. (1964). *Proc. Roy. Soc. A* **277**, 237 & **281**, 401.

Hubbard, J. (1965). *Proc. Roy. Soc. A* **285**, 542.

Hurd, C.M. (1967). *Phys. Rev. Lett.* **18**, 1127.

Hybertson, M.S., Schlüter, M. & Christensen, N.E. (1989). *Phys. Rev. B* **39**, 9028.

Ishii, H. (1976). *Prog. Theor. Phys.* **55**, 1373.

Ishii, H. (1978). *J. Low. Temp. Phys.* **32**, 457.

Ishii, H. & Yosida, K. (1967). *Prog. Theor. Phys.* **38**, 61.

Jarrell, M., Gubernatis, J. & Silver, R.N. (1991). *Phys. Rev. B* **44**, 5347.

Jarrell, M., Gubernatis, J., Silver, R.N. & Sivia, D.S. (1991). *Phys. Rev. B* **43**, 1206.

Jayaraman, A., Dernier, P.D. & Longinotti, L.D. (1975). *High Temp. High Pressure* **7**, 1.

Jefferson, J.H. (1977). *J. Phys. C* **10**, 3589.

Jin, B. & Kuroda, Y. (1988). *J. Phys. Soc. Jap.* **57**, 1687.

Jin, B., Matsuura., T. & Kuroda, Y. (1991). *J. Phys. Soc. Jap.* **60**, 580.

Johanson, W.R., Crabtree, G.W., Edelstein, A.S. & McMasters, O.D. (1981). *Phys. Rev. Lett.* **46**, 504.

Joss, W., van Ruitenbeck, J.M., Crabtree, G.W., Tholence, J.L., van Deursen, A.J.P. & Fisk, Z. (1987). *Phys. Rev. Lett.* **59**, 1609.

Joyce, J.J., Arko, A.J., Canfield, P.C., Fisk, Z., Bartlett, R.J., Smith, J.L., Thompson, J.D., Lawrence, J. & Tang, J. (1992). *Preprint.*

Jullien, R., Fields, J. & Doniach, S. (1977). *Phys. Rev. Lett.* **38**, 1500.

Kadanoff, L.P. (1966). *Ann. Phys. N.Y.* **2**, 263.

Kadanoff, L.P. (1967). *Rev. Mod. Phys.* **39**, 395.

Kadowaki, K. & Woods, S.B. (1986). *Sol. St. Com.* **58**, 507.

Kampf, A. & Schrieffer, J.R. (1990). *Phys. Rev. B* **42**, 4064.

Kang, J-S, Allen, J.W., Maple, M.B., Torikachivili, Pate, B.B., Ellis, W.P. & Lindau I. (1987). *Phys. Rev. Lett.* **59**, 493.

Kawakami, N. & Okiji, A. (1981). *Phys. Lett. A* **86**, 483.

Kawakami, N. & Okiji, A. (1986). *J. Mag. Mag. Mat.* **54–57**, 327.

Kawakami, N. & Okiji, A. (1990). *Prog. Theor. Phys. Suppl.* **101**, 429.

Kawakami, N. & Yang, S.K. (1991). *J. Phys. Cond. Mat.* **3**, 5983.

Kawakami, N., Usuki, T. & Okiji, A. (1987). *J. Phys. Soc. Jap.* **56**, 1539.

Keiter, H. & Kimball, J.C. (1971). *Int. J. Magn.* **1**, 233.

Keiter, H. & Morandi, G. (1984). *Phys. Rep.* **109**, 227.

Kitchens, T.A., Steyert, W.A. & Taylor, R.D. (1965). *Phys. Rev. A* **238**, 467.

Kondo, J. (1964). *Prog. Theor. Phys.* **32**, 37.

Kojima, H., Kuromoto, Y. & Tachiki, M. (1984). *Z. Phys. B* **54**, 293.

Köln (1984) *Proceedings of the 4th International Conference on Valence Fluctuations*, eds. E. Müller-Hartmann, B. Roden & D. Wohlleben. Amsterdam:North-Holland.

Kuromoto, Y. (1983). *Z. Phys. B* **53**, 37.

Kondo, J. (1969). In *Solid State Physics* **23**, eds. F. Seitz, D. Turnbull & H. Ehrenreich, p. 183. New York: Academic Press.

Korringa, J. (1950). *Physica* **19**, 601.

Kotliar, G. & Ruckenstein, A.E. (1982). *Phys. Rev. Lett.* **57**, 1362.

Krishna-murthy, H.R. (1978). Ph.D. Thesis, Cornell University.

Krishna-murthy, H.R., Wilkins, J. W. & Wilson K.G. (1980). *Phys. Rev. B* **21**, 1003 & 1044.

Kubo, R. (1957). *J. Phys. Soc. Japan* **12**, 570.

Kulish, P.P. & Reshetikhin N.Yu. (1981). *Sov. Phys. JETP* **53**, 108.

Kuromoto, Y. & Kojima, H. (1984). *Z. Phys. B* **57**, 95.

Kuromoto, Y. & Müller-Hartmann, E. (1985). *J. Mag. Mag. Mat.* **57**, 122.

Landau, L.D. (1957). *Sov. Phys. JETP* **3**, 920; **5**, 101.

Landau, L.D. (1958). *Sov. Phys. JETP* **8**, 70.

Lang, J.K., Baer, Y. & Cox, P.A. (1981) *J. Phys. F* **11**, 121.

Langer, J.S. & Ambegaokar, V. (1961). *Phys. Rev.* **121**, 1090.

Langreth, D.C. (1966). *Phys. Rev.* **150**, 516.

Lavagna, M., Millis, A.J. & Lee, P.A. (1987). *Phys. Rev. Lett.* **58**, 266.

Lawrence, J. (1979). *Phys. Rev. B* **20**, 3770.

Lawrence, J., Arko, A.J., Joyce, J.J., Canfield, P.C., Fisk, Z., Thompson, J.D. & Bartlett, R.J. (1991) *J. Mag. Mag. Mat.* **108**, 215.

Lee, P.A., Rice, T.M., Serene, J.W., Sham, L.J. & Wilkins, J.H. (1986). *Comments in Solid State Physics* **12**, 99.

Leggett, A.J. (1975). *Rev. Mod. Phys.* **47**, 331.

Levin, K., Kim J.H., Lu, J.P. & Si, Q. (1991). *Physica C* **175**, 449.

Lieb, E.H. & Wu, F.Y. (1968). *Phys. Rev. Lett.* **20**, 1445.

Lin, C.L., Wallash, A., Crow, J.E., Mihalisin, T. & Schlottmann, P. (1987). *Phys. Rev. Lett.* **58**, 1232.

von der Linden, W. & Edwards, D.M. (1991). *J. Phys. Cond. Mat.* **3**, 4917.

Loewenhaupt, M. & Holland-Moritz, E. (1978). *J. Mag. Mag. Mat.* **9**, 50.

Loewenhaupt, M. & Schmatz, W.E. (1977). In *Neutron Inelastic Scattering 1977*, Int. Atomic Energy Agency, Vienna, p. 227.

Lonzarich, G.G. (1988). *J. Mag. Mag. Mat.* **74–77**, 1.

Loram, J.W. Whall, T.E. & Ford, P.J. (1970). *Phys. Rev. B* **2**, 857.

Los Alamos (1989). *Proceedings of the Los Alamos Symposium -1989 on High Temperature Superconductivity*, eds. K.S. Bedell, D. Coffey, D.E. Meltzer, D. Pines & J.R. Schrieffer. New York:Addison Wesley.

Ludwig, A.W.W. & Affleck, I. (1991). *Phys. Rev. Lett.* **67**, 3160.

Luttinger, J.M. (1960). *Phys. Rev.* **119**, 1153.

Luttinger, J.M. (1961). *Phys. Rev.* **121**, 942.

Luttinger, J.M. (1963). *J. Math. Phys.* **4**, 1154.

MacDonald D.K.C., Pearson, W.B. & Templeton, I.M. (1962). *Proc. Roy. Soc.* **266**, 161.

Mahan, G.D. (1974). In *Solid State Physics*, eds. F. Seitz, D. Turnbull & H. Ehrenreich, vol 29, p. 75. New York:Academic Press.

Mahan, G.D. (1990). *Many Particle Physics* (2nd edition). New York:Plenum.

van der Marel, D. & Sawatzky, G.A. (1988). *Phys. Rev. B* **37**, 10674.

Mattens, W.C.M. (1980). Ph.D. Thesis, University of Amsterdam.

Mattis, D.C. (1967). *Phys. Rev. Lett.* **19**, 1478.

Mattis, D.C. & Lieb, E.H. (1965). *J. Math. Phys.* **6**, 304.

McMahan, A.K., Martin, R.M. & Satpathy, S. (1988). *Phys. Rev. B* **38**, 6650.

Metzner, W., & Vollhardt, D. (1989). *Phys. Rev. Lett* **62**, 324.

Mihály, N. & Zawadowski, A. (1978). *J. de Physique Lett.* **39**, L483.

Millis, A.J. & Lee, P.A. (1987). *Phys. Rev. B* **35**, 3394.

Miyake, K. & Kuramoto, Y. (1990). *J. Mag. Mag. Mat.* **90–91**, 438.

Monnier, R., Degiorgi, L. & Koelling, D.D. (1986). *Phys. Rev. Lett.* **56**, 2744.

Mook, H.A., Nicklow, R.M., Penney, T., Holtzberg, F. & Schafer, M.W. (1978). *Phys. Rev. B* **18**, 2925.

Murani, A.P. (1987). In *Theoretical and Experimental Aspects of Valence Fluctuators and Heavy Fermions*, p. 287. New York & London: Plenum Press.

Murani, A.P., Mattens, W.C.M., de Boer, F.R. & Lander, G.H. (1985). *Phys. Rev. B* **31**, 52.

Murani, A.P. & Tholence, J.L. (1977). *Sol. St. Com.* **22**, 25.

Myers, H.P., Walldén, L. & Karlsson, A. (1968). *Phil. Mag.* **18**, 725.

Nagaoka, Y. (1965). *Phys. Rev. A* **138**, 1112.

Nagaoka, Y. (1966). *Phys. Rev.* **147**, 392.

Nagaoka, Y. (1967). *Prog. Theor. Phys.* **37**, 13.

Narath, A. (1969). *Sol. Stat. Com.* **10**, 521.

Narath, A. (1972). *Crit. Rev. in Solid State Science* **3**, 1.

Narath, A. (1973). In *Magnetism*, eds. G.T. Rado & H. Suhl, p. 149. New York:Academic Press.

Narath, A. & Gossard, A.C. (1969). *Phys. Rev.* **183**, 391.

Narath, A. & Weaver, H. T. (1969). *Phys. Rev. Lett.* **23**, 233.

Newns, D.M. & Hewson, A.C. (1980). *J. Phys. F* **10**, 2429.

Newns, D.M. & Read, N. (1987). *Adv. in Phys.* **36**, 799.

Norman, M.R. (1987). *Phys. Rev. Lett.* **59**, 232.

Nozières, P. (1964). *The Theory of Interacting Fermi Systems*. New York:Benjamin.

Nozières, P. (1974). *J. Low Temp. Phys.* **17**, 31.

Nozières, P. (1975). *Low Temperature Physics Conference Proceedings LT14* **5**, eds. Krusius & Vuorio. p. 339. Amsterdam, North Holland/Elsevier,

Nozières, P. (1985). *Ann. de Phys.* **10**, 19.

Nozières, P. & Blandin, A. (1980). *J. Physique* **41**, 193.

Nozières, P. & De Dominicis, C.T. (1969). *Phys. Rev.* **178**, 1097.

Ogievetskii, E., Tsvelick, A.M. & Wiegmann, P.B. (1983). *J. Phys. C* **16**, L797.

Okada, I. & Yosida, K. (1973). *Prog. Theor. Phys.* **49**, 1483.

Okiji, A. & Kawakami, N. (1982). *Sol. Stat. Com.* **43**, 365: *J. Phys. Soc. Japan* **51**, 3192.

Okiji, A. & Kawakami, N. (1983). *Phys. Rev. Lett.* **50**, 1157.

Okiji, K. & Kawakami N. (1986). *J. Mag. Mag. Mat.* **54–57**, 327.

Oliveira, L.N. & Wilkins J. (1981). *Phys. Rev. Lett.* **47**, 1553.

Ott, H.R. (1987). *Progress in Low Temp. Physics* **XI**, 215.

Ott, H.R., Rudiger, H., Fisk, Z. & Smith J.L. (1983). *Phys. Rev. Lett.* **52**, 1551.

Ott, H.R., Rudiger, H., Fisk, Z. & Smith J.L. (1984a). In *Moment Formation in Solids*, ed. W.J.L. Buyers, p. 305. New York:Plenum.

Ott, H.R., Rudiger, H., Fisk, Z. & Smith J.L. (1984b). *Physica B* **127**, 359.

Palstra, T.T.M., Menovsky, A.A., van den Berg, J., Dirkmaat, P.J., Kes, P.H., Nieuwenhuys, G.J. & Mydosh, J.A. (1985). *Phys. Rev. Lett.* **55**, 2727.

Patthey, F., Imer, J.M., Schneider, W.D., Beck, H., Baer, Y. & Delley, B. (1990). *Phys. Rev. B* **42**, 8864.

Pethick, C.J., Pines, D., Quader, K.F., Bedell, K.S. & Brown G.E. (1986). *Phys. Rev. Lett.* **57**, 1955.

Pickett, W.E. (1989). *Rev. Mod. Phys.* **61**, 433.

Potts, J.E. & Welsh, L.B. (1972). *Phys. Rev. B* **5**, 3421.

Rajan, V.T. (1983). *Phys. Rev. Lett.* **51**, 308.

Rajan, V.T., Lowenstein, J.H., & Andrei. N. (1982). *Phys. Rev. Lett.* **49**, 497.

Ramakrishnan, T.V. & Sur, K. (1982). *Phys. Rev. B* **26**, 1798.

Rasul, J. (1989). *Phys. Rev. B* **39**, 663.

Rasul, J. & Desgranges, H-U. (1986). *J. Phys. C* **19**, L671

Rasul, J.W. & Harrington, A.P. (1987). *J. Phys. C* **20**, 4783.

Rasul, J.W. & Hewson, A.C. (1984). *J. Phys. C* **17**, 2555 & 3332.

Razafimandimby, H., Fulde, P. & Keller, J. (1984) *Z. Phys. B* **54**, 111.

Read, N. (1985). *J. Phys. C* **18**, 2651.

Read, N. (1986). Ph.D. Thesis, University of London.

Read, N., Dharamvir, K., Rasul, J.W. & Newns, D.M. (1986). *J. Phys. C* **19**, 1597.

Read, N. & Hewson, A.C. (1984). Unpublished.

Read, N. & Newns, D.M. (1983a). *J. Phys. C* **16**, 3273.

Read, N. & Newns, D.M. (1983b). *J. Phys. C* **16**, L1055.

Rizzuto, C. (1974). *Rep. Prog. Phys.* **37**, 147.

Ruderman, M.A. & Kittel, C. (1954). *Phys. Rev.* **96**, 99.

Ryder, L.H. (1985). Quantum Field Theory. Cambridge: Cambridge University Press.

Sacramento, P.D. & Schlottmann, P. (1989). *Phys. Lett. A* **142**, 245.

Sacramento, P.D. & Schlottmann, P. (1990a). *Sol. Stat. Com.* **73**, 747.

Sacramento, P.D. & Schlottmann, P. (1990b). *Phys. Rev. B* **42**, 743.

Sacramento, P.D. & Schlottmann, P. (1991). *J. Phys. Cond. Mat.* **3**, 9687.

Sakai, O., Shimizu, Y. & Kasuya T. (1989). *J. Phys. Soc. Japan* **58**, 162.

Samwer, K. & Winzer, K. (1976). *Z. Phys. B* **25**, 269.

Sanchez-Castro, C. & Bedell, K.S. (1991). *Phys. Rev. B* **43**, 12874.

Santa Fe (1989) *Proceedings of the International Conference on the Physics of Highly Correlated Electron Systems*, eds. J.O. Willis, J.D. Thompson, R.P. Guertin & J.E. Cross. Amsterdam: North-Holland.

Sarachik, M., Corenzwit, E. & Longinotti, L.D. (1964). *Phys. Rev. A* **135**, 1041.

Saso, T. (1989). *J. Phys. Soc. Jap.* **58**, 4468.

Scheuer, H., Loewenhaupt, M. & Schmatz, M. (1977). *Physica B* **86–88**, 842.

Schlottmann, P. (1983). *Z. Phys. B* **51**, 49.

Schlottmann, P. (1984a). *J. Phys. C* **17**, L267.

Schlottmann, P. (1984b). *Phys. Rev. B* **30**, 1454.

Schlottmann, P. (1984c). *Phys. Rev. B* **29**, 630 & 4468.

Schlottmann, P. (1985). *Z. Phys. B* **59**, 391.

Schlottmann, P. (1987a). *J. Mag. Mag. Mat.* **63–64**, 205.

Schlottmann, P. (1987b). *Phys. Rev. B* **35**, 5279.

Schlottmann, P. (1989). *Physics Reports* **181**, 1.

Schotte, K.D. & Schotte U. (1971). *Phys. Rev. B* **4**, 2228.

Schotte, K.D. & Schotte U. (1975). *Phys. Lett. A* **55**, 38.

Schrieffer, J.R (1967). *J. Appl. Phys.* **38**, 1143.

Schrieffer, J.R. & Wolff P.A. (1966). *Phys. Rev.* **149**, 491.

Schweitzer, H., & Czycholl, G. (1990). *Sol. Stat. Com.* **74**, 735.

Schweitzer, H., & Czycholl, G. (1992). *J. Mag. Mag. Mat.* **108**, 150.

Seaman, C.L., Maple, M.B., Lee, B.W., Gamaty, S., Torikachvili, M.S., Kang, J-S, Liu, L.Z., Allen, J.W. & Cox, D.L. (1991). *Phys. Rev. Lett.* **67**, 2882.

Shiba, H. (1975). *Prog. Theor. Phys.* **54**, 967.

Shiba, H. & Fazekas, P. (1990). *Prog. Theor. Phys. Suppl.* **101**, 403.

Sklyanin, E.K. & Faddeev, L.D. (1978). *Sov. Phys. Dokl.* **23(12)**, 902.

Šokčević, D., Zlatić, V. & Horvatić, B. (1989). *Phys. Rev. B* **39**, 603.

Star, W.M., Basters, F.B., Nap, G.M., de Vroede, E. & van Baarle, C. (1972). *Physica* **58**, 585.

Star, W.M. & Nieuwenhuys G.J. (1969) *Phys. Lett. A* **30**, 22.

Steglich, F., Aarts, J., Bredl, C.D., Liecke, W., Meschede, D., Franz, W. & Schäfer, J. (1979). *Phys. Rev. Lett.* **43**, 1892.

Steiner, P., Hüfner, S.& Zdrojewska, W.V. (1974). *Phys. Rev. B* **10**, 4704.

Steiner, P., Zdrojewska, W.V., Gumprecht, D. & Hüfner, S. (1973). *Phys. Rev. Lett.* **31**, 355.

Stewart, G.R. (1984). *Rev. Mod. Phys.* **56**, 755.

Stewart, G.R., Fisk, Z., Willis, J.O. & Smith, J.L. (1984). *Phys. Rev. Lett.* **52**, 679.

Sticht, J., d'Ambrumenil, N. & Kübler, J. (1986). *Z. Phys. B* **65**, 149.

Strange, P., & Newns, D.M. (1986). *J. Phys. F* **16**, 335.

Suhl, H. (1965). *Phys. Rev.* **138**, A515.

Suhl, H. & Wong, D. (1967). *Physics* **3**, 17.

Sutherland, B. (1968). *Phys. Rev. Lett.* **20**, 98.

Taillefer, L., Flouquet, J. & Lonzarich, G.G. (1991). *Physica B* **169**, 257.

Taillefer, L., Newbury, R, Lonzarich, G.G., Fisk, Z. & Smith J.L. (1987). *J. Mag. Mag. Mat.* **63–64**, 372.

Takahashi, M. (1970–4). *Prog. Theor. Phys.* **44**, 899, 348: **46**, 401, 1388: **47**, 1388: **50**, 1519: **51**, 1348.

Takano, F. & Ogawa, T. (1966). *Prog. Theor. Phys.* **35**, 343.

Takhatajan, L.A. (1985). In *Exactly Solved Problems in Condensed Matter and Relativistic Field Theory*, eds. B.S. Shastry, S.S. Jha & V. Singh, p. 175. Berlin:Springer Verlag.

Thacker, H.B. (1981). *Rev. Mod. Phys.* **53**, 253.

Thalmeier, P. & Lüthi, B. (1991). In *Handbook on the Physics and Chemistry of Rare Earths*, eds. K.A. Gschneider Jr. & L. Eyring, **14**, p. 225. Amsterdam: North-Holland.

Tholence, J.L. & Tournier, R. (1970). *Phys. Rev. Lett.* **25**, 867.

Ting, C.S. (1970). *J. Phys. Chem. Solids* **31**, 777.

Triplett, B.B. & Phillips, N.E. (1971). *Phys. Rev. Lett.* **27**, 1001.

Trugman, S.A. (1990). *Phys. Rev. Lett.* **65**, 500.

Tsvelick, A.M. (1985). *J. Phys. C* **18**, 159.

Tsvelick, A.M. & Wiegmann, P.B. (1983). *Adv. Phys.* **32**, 453.

Tsvelick, A.M. & Wiegmann, P.B. (1984). *Z. Phys. B* **54**, 201.

Varma, C.M. (1976). *Rev. Mod. Phys.* **48**, 219.

Varma, C.M., Littlewood, P.B., Schmidt-Rink, S., Abrahams, E. & Ruckenstein, A.E. (1989). *Phys. Rev. Lett.* **63**, 1996.

Vollhardt, D. (1984). *Rev. Mod. Phys.* **56**, 99.

Walker, M.B. (1970). *Phys. Rev. B* **1**, 3690.

Wasserman, A., Springford, M. & Hewson A.C. (1989). *J. Phys. Cond. Mat.* **1**, 2669

Wiegmann, P.B. (1980). *Sov. Phys. JETP Lett.* **31**, 392.

Wiegmann, P.B. (1981). *Phys. Lett. A* **31**, 163.

Wiegmann, P.B. & Tsvelick, A.M. (1983). *J. Phys. C* **12**, 2281, 2321.

Wilson, A.H. (1953). *The Theory of Metals.* Cambridge:Cambridge University Press.

Wilson, K.G. (1974). *Nobel Symposia* **24**, p. 68. New York:Academic Press.

Wilson, K.G. (1975). *Rev. Mod. Phys.* **47**, 773.

Wilson, K.G. & Kogut, J. (1974). *Phys. Rep. C* **12**, 75.

Winzer, K. (1975). *Sol. St. Com.* **16**, 521.

Witten, E. (1978). *Nucl. Phys. B* **145**, 110.

Wohlleben, D.K. & Coles, B.R. (1973). In *Magnetism*, eds. G.T. Rado & H. Suhl, vol V, p. 3. New York:Academic Press.

Wu, M.K., Ashburn, J.R., Torng, C.J., Hor, P.H., Meng, R.L., Gao, L., Huang, Z.J., Wang, Y.Q. & Chu, C.W. (1987). *Phys. Rev. Lett.* **58**, 908.

Wuilloud, E., Baer, Y., Ott, H.R., Fisk, Z. & Smith J.L. (1984). *Phys. Rev. B* **29**, 5228.

Yafet, Y., Varma, C.M. & Jones, B. (1985). *Phys. Rev. B* **32**, 360.

Yamada, K. (1975). *Prog. Theor. Phys.* **53**, 970; *Prog. Theor. Phys.* **54**, 316.

Yamada, K. (1976). *Prog. Theor. Phys.* **55**, 1345.

Yamada, K. & Yosida, K. (1985). In *Theory of Heavy Fermions and Valence Fluctuations*, eds. T. Kasuya & T. Saso, Springer Series in Science, **62**, p. 183. Berlin:Springer Verlag.

Yamada, K. & Yosida, K. (1986). *Prog. Theor. Phys.* **76**, 621.

Yang, C.N. (1967). *Prog. Theor. Phys.* **55**, 67.

Yang, C.N. & Yang, C.P. (1969). *J. Math. Phys.* **10**, 1115.

Yoshimori, A. (1976). *J. Phys. C* **15**, 5241.

Yoshimori, A. & Zawadowski A. (1982). *J. Phys. C* **15**, 5241.

Yosida, K. (1966). *Phys. Rev.* **147**, 223.

Yosida, K. & Yamada, K. (1970). *Prog. Theor. Phys. Suppl.* **46**, 244.

Yosida, K. & Yamada, K. (1975). *Prog. Theor. Phys.* **53**, 1286.

Yosida, K. & Yoshimori, A. (1973). In *Magnetism* , eds. G.T. Rado & H. Suhl, p. 253. New York:Academic Press.

Yuval, G. & Anderson, P.W. (1970). *Phys. Rev. B* **1**, 1522.

Zener, C. (1951). *Phys. Rev.* **81**, 440.

Zhang, F.C. & Lee, T.K. (1983). *Phys. Rev. B* **28**, 33.

Zhang, F.C. & Rice, T.M. (1988). *Phys. Rev. B* **37**, 3759.

Zittartz, J. & Müller-Hartmann, E. (1968). *Z. Physik* **212**, 280.

Zlatić, V. & Rivier, N. (1974). *J. Phys. F* **4**, 732.

Zubarev, D.N. (1960). *Sov. Phys. Usp.* **3**, 320.

Index